ADAPTIVE RADAR DETECTION AND ESTIMATION

Edited by

Simon Haykin
Communications Research Laboratory
McMaster University
Hamilton, Ontario, Canada

and

Allan Steinhardt
Department of Electrical Engineering
Cornell University
Ithaca, New York

A WILEY-INTERSCIENCE PUBLICATION

JOHN WILEY & SONS, INC.

New York • Chichester • Brisbane • Toronto • Singapore

Library of Congress Cataloging in Publication Data:

Adaptive radar detection and estimation / edited by Simon Haykin and Allan Steinhardt.
p. cm. -- (Wiley series in remote sensing)
"A Wiley-Interscience publication."
ISBN 0-471-54468-X
1. Radar. 2. Adaptive signal processing. 3. Adaptive antennas.
I. Haykin, Simon S., 1931– . II. Steinhardt, Allan Otto.
III. Series.
TK6580.A32 1992
621.3848--dc20 91-42278
CIP

Printed and bound in the United States of America by Braun-Brumfield, Inc.

10 9 8 7 6 5 4 3 2 1

CONTRIBUTORS

ANASTASIOS DROSPOPOULOS, Defence Research Establishment Ottawa, Ottawa, Ontario, Canada K1A 0Z4

SIMON HAYKIN, Communications Research Laboratory, McMaster University, Hamilton, Ontario, Canada L8S 4K1

JAMES P. REILLY, Communications Research Laboratory, McMaster University, Hamilton, Ontario, Canada L8S 4K1

ALLAN STEINHARDT, School of Electrical Engineering, Cornell University, Ithaca, NY 14853, USA

BARRY VAN VEEN, Department of Electrical and Computer Engineering, University of Wisconsin, Madison, WI 53706-1691, USA

K.M. WONG, Department of Electrical Engineering, McMaster University, Hamilton, Ontario, Canada L8S 4K1

Q.T. ZHANG, Communications Research Laboratory, McMaster University, Hamilton, Ontario, Canada L8S 4K1

MICHAEL D. ZOLTOWSKI, School of Electrical Engineering, Purdue University, West Lafayette, IN 47907, USA

PREFACE

This edited book is devoted to a detailed mathematical treatment of some important issues in *adaptive radar detection and estimation*. The need for the use of adaptive processing in a radar environment arises because of the inherently nonstationary nature of the environment.

The book is organized in 7 chapters. Chapter 1 provides an overview of the subject matter of the book. Chapter 2 is devoted to the development of model-based techniques for the detection of radar targets in the presence of clutter as experienced in a radar surveillance environment. Chapter 3 presents a detailed theoretical treatment of the basic issues involved in adaptive radar detection using an array of sensors. Chapter 4 discusses minimum variance beamforming techniques with emphasis on partial adaptivity and design procedures. Chapter 5 considers maximum likelihood bearing estimation in beamspace for an adaptive phased array radar. Chapter 6 presents an algorithm for angle-of-arrival estimation (direction finding) based on maximum a posteriori probability considerations. Finally, Chapter 7 describes the method of multiple windows for spectrum estimation and applies it to direction finding in a low-angle tracking radar environment.

Much of the material presented here has not appeared in book form before. Moreover, it fills an important gap in the literature on radar signal processing. It is our hope that the book will be found useful by newcomers to the field as well as radar engineers and scientists who are seeking an up-to-date treatment of adaptive radar detection and estimation techniques.

SIMON HAYKIN
ALLAN STEINHARDT

Hamilton, Ontario, Canada
Ithaca, New York
March, 1992

CONTENTS

CHAPTER 5
BEAMSPACE ML BEARING ESTIMATION FOR ADAPTIVE PHASED ARRAY RADAR **237**

Michael D. Zoltowski

CHAPTER 6

THE MAP ALGORITHM FOR DIRECTION OF ARRIVAL ESTIMATION IN UNKNOWN COLORED NOISE 333

James P. Reilly and K. M. Wong

CHAPTER 7

ADAPTIVE RADAR PARAMETER ESTIMATION WITH THOMSON'S MULTIPLE-WINDOW METHOD 381

A. Drosopoulos and Simon Haykin

ADAPTIVE RADAR DETECTION AND ESTIMATION

1

INTRODUCTION

SIMON HAYKIN
Communications Research Laboratory, McMaster University, Hamilton, Ontario, Canada

ALLAN STEINHARDT
Department of Electrical Engineering Cornell University Ithaca, New York, USA

The signal processing operations of detection and estimation are basic to every radar system application. *Detection* refers to the decision made by a radar receiver regarding the presence or absence of a radar target, given a set of observations that constitute the received signal. If it is established that a target is present, we may then proceed with *estimation*, which refers to the estimation of parameters that characterize the target. The radar parameters of interest include the *range* (distance) of a target from the radar site, *angle-or-arrival* (direction) of the target, and *size* (shape) of the target.

The radar may be active or passive. In an *active radar*, a transmitted signal is used purposely to probe the environment, and a two-dimensional image (in range and azimuth) or a three-dimensional image (in range, azimuth, and elevation) of the surrounding environment (including the targets that lie in it) is thereby formed. A *passive radar*, on the other hand, *listens* to the surrounding environment and information about the targets of interest is extracted by processing the signal picked up by the radar.

The radar may be classified in another way; it may be a surveillance or tracking radar. In a *surveillance radar*, the antenna scans the surrounding environment on a continuous basis and therefore dwells on any one target for a limited period of time only; the purpose here is to develop an overall picture of the environment. In a *tracking radar*, on the other hand, attention

Adaptive Radar Detection and Estimation, Edited by Simon Haykin and Allan Steinhardt.
ISBN 0-471-54468-X

is fixed on a target of interest and its trajectory is tracked by the radar on a continuous basis so long as the target is in view of the radar.

In this book, we study *adaptive* techniques for performing the operations of radar detection and estimation. By adaptive we mean that the radar is capable of adjusting its operations automatically in response to statistical variations in environmental conditions. The purpose of this introductory chapter is to provide an overview of the underlying theory and framework of the subject of the book.

1-1 DETECTION

The target detection problem is *statistical* in nature, made so by the random fluctuations in the additive *noise* that (in one form or another) corrupts the received signal. The noise may arise because of thermal agitation at the front end of the receiver. This form of noise is commonly referred to as *thermal noise*, which is well modeled as a *white Gaussian process*. Another form of noise that is often found in a received radar signal is *clutter* that refers to unwanted radar echoes from objects (e.g., trees, buildings, and weather storms) lying along the path of the transmitted radar signal. Clutter is modeled as a *colored*, and possibly non-Gaussian process. Yet another form of noise to be considered is that of *multipath* which arises in a low-angle tracking radar environment. Multipath itself may be specular or diffuse. *Specular multipath* is deterministic, whereas *diffuse multipath* is stochastic; both of these forms of multipath are spatial in character. The forms of noise referred to thus far are all natural in origin. In contrast, the noise may be artificial, examples of which include *interference* or *jamming* that is of a directional nature.

Earlier we remarked that detection is a decision-making operation. It is therefore not surprising that much of the well-established *statistical decision theory* built on Bayes formalism applies to the mathematical analysis of radar detection [1]. Here particular mention should be made of the Neymann–Pearson criterion that is particularly appropriate for radar detection. According to the *Neymann–Pearson criterion*, the probability of target detection is optimized subject to the constraint that the probability of false alarm is held below a preset level [1].

We may identify the following detection hierarchy:

1. *Single channel* (*temporal case*). In this case, the detection problems of interest include (a) *target detection in white noise* and (b) *target detection in colored noise*. When the target signal is known, the solution to the first problem is the well-known *matched filter* [1]. The solution to the second problem is obtained by *prewhitening* the colored noise component of the received signal. When, however, the target signal is unknown, we have a more difficult problem to deal with. Single channel detection is the focus of Chapter 2.

2. *Multi-channel* (*spatial case*). In this case, the radar receiver is equipped with a *sampled-aperture* (*array*) *antenna*, and we have a single snapshot of data to work with, which is collected by the radar at a particular instant of time. Here again, depending on the characteristics of the additive noise component, we may identify the two special cases: (a) target detection in white noise and (b) target detection in colored noise [2]. The coloring here refers to correlation in the spatial domain. The single snapshot multichannel problem is studied in Chapter 7.
3. *Multichannel, multiple snapshot* (*spatio-temporal case*). In this third and final class of radar detection problems, addressed in Chapters 3 and 4, the received signal consists of a multiplicity of snapshots of data collected at different instants of time. Here also, we may identify the two special cases of target detection in white noise and target detection in colored noise [2]. In the latter case the noise is now correlated in either space, time, or both.

1-2 ESTIMATION

The estimation problem is closely linked with that of target detection. In particular, we may identify two kinds of estimation, namely, parameter estimation and waveform estimation, depending on the application of interest. In *parameter estimation*, the issue is one of estimating one or more of the parameters that characterize the target. For example, the problem may be that of estimating the range, Doppler, or direction of the target. The range estimation problem is also referred to in the literature as *time-delay estimation*. The direction (angle-of-arrival) estimation is commonly referred to as the *direction finding* problem. The Doppler estimate provides information on the relative velocity of the target. In *waveform estimation*, on the other hand, the requirement is to estimate the waveform of the target signal; the motivation here may be that of *target recognition or imaging*.

In any event, we may identify the following estimation hierarchy:

1. *Single channel* (*temporal case*) (Chapters 2, 3, 6). As with the detection problem, in this case we have (a) waveform and parameter estimation in white noise and (b) waveform and parameter estimation in colored noise.
2. *Multichannel* (*spatial case*) (Chapter 7). In this case, the received signal consists of a single snapshot of data, which means that the radar receiver has an array antenna. Under this class of parameter estimation problems, we have (a) angle-of-arrival estimation in white noise, (b) angle-of-arrival estimation in colored noise (e.g., diffuse multipath in a low-angle tracking radar environment), (c) angle-of-arrival estimation in directional interference that may be man-made or natural.

3. *Multichannel, multiple-snapshot* (*spatio-temporal*) (Chapters 3–6, 7). In this third and final class of estimation problems, the estimator involves processing in both *space* (based on snapshots of data at different points in space) and *time* (with the snapshots of data collected at different instants of time). The use of multiple snapshots might arise due to wideband targets or because of a desire to obtain more accurate estimates than is feasible with spatial-only (single snapshot) processing. Specific problems of interest here include (a) angle-of-arrival estimation in white noise, (b) angle-of-arrival estimation in colored noise, (c) angle-of-arrival estimation in narrowband directional interference, (d) angle-of-arrival estimation in wideband directional interference, (e) waveform and parameter estimation in spatially correlated noise.

1-3 ADAPTIVITY

Typically, a radar environment is subject to statistical variations that may arise due to natural or artificial factors. Natural factors include changes in weather conditions and, in the case of ocean surveillance, variations in the ocean surface. Artificial factors include *jammers* of fairly simple and sophisticated kinds, that are designed to fool the radar receiver or disrupt its operation.

In any event, we may identify the following factors that call for the use of adaptivity:

1. Covariance matrix of the noise is unknown. This arises in temporal, spatial, or spatio-temporal cases.
2. Direction of (narrowband) interference is unknown.
3. The interference is wideband, and both its direction and waveform are unknown.

There is no unique method to deal with these unknown situations. Rather, we have a number of different approaches that may be used to deal with them, as outlined here:

1. *Stochastic approach* (Chapters 2–7). This class of adaptive techniques includes (a) *ensemble averaging* considerations as in Wiener filters, maximum entropy method (MEM), and multiple signal classification (MUSIC) algorithm, and (b) *probability density* considerations as in likelihood ratio (Bayesian) detectors and maximum a posteriori probability (MAP) estimators.
2. *Nonstochastic approach* (Chapter 3). This second class of adaptive techniques includes (a) *singular value decomposition* (SVD), (b) *recursive least squares* (RLS) estimation in its various forms, and (c) *chaotic theoretic methods*.

Another issue that arises in the study of adaptive radar techniques is that of classification of adaptivity. Here, we may identify the following classes of adaptive techniques in the context of a radar system using array antennas:

1. *Full versus partial adaptivity*. Specifically, a *fully adaptive radar antenna* (Chapters 3, 6, 7) means that the array consists of a set of uniformly spaced antenna elements, and that all of the elements are in actual use. In a *partially adaptive radar antenna* (Chapters 4, 5), on the other hand, the array is *thinned* out by purposely leaving out some of the antenna elements, or preprocessing the array data to reduce dimensionality. This is done to save cost, particularly when the size of the array is very large.
2. *Unconstrained versus constrained adaptivity*. In an *unconstrained adaptive* radar system, the array antenna performs its function with no constraints imposed on the adaptive filtering algorithm responsible for the operation of the system. Examples of unconstrained adaptive techniques include the ubiquitous *least mean square* (LMS) *algorithm*, and the *recursive least squares* (RLS) *algorithm*. On the other hand, in *constrained* forms of adaptive radar techniques, specific constraints are imposed on the operation of the adaptive filtering algorithms. Examples of constrained adaptive techniques include the *minimum variance distortionless response* (MVDR) *beamformer* (Chapter 4).
3. *Adaptivity in data versus beam space*. The adaptivity may be performed in data space by having the adaptive filtering algorithm operate on the incoming data directly. On the other hand, it may be performed in *beamspace* (Chapters 4, 5) by first transforming the incoming date through the use of a *fixed* beamformer, and then applying the adaptive filtering algorithm to the transformed data at the beamformer output.

1-4 ORGANIZATION AND SCOPE

Nonadaptive systems detect, and/or estimate, desired signals when the statistics of the signal and resident interference are known a priori. In contrast, an adaptive system must infer the stochastic structure of the desired signal and interference prior to excising the former from the latter. An example of a nonadaptive detector is a matched filter, which provides an optimal detection and false alarm rate tradeoff for Gaussian noise [1, 3]. An example of a nonadaptive estimator is a Wiener filter, which requires second order statistics of all attendant quantities, and is optimal in the least mean square sense for Gaussian noise. (While this text focuses on adaptivity, nonadaptive systems, both single and multichannel, are reviewed variously in Chapters 2, 3).

This book contains a collection of contributed chapters pertaining to adaptive detection and estimation, that is, statistical inference, procedures, specifically as they arise in the radar context. A number of texts, such as Van Trees' classic [1] and Scharf's more modern treatment [3], develop detection and estimation, including adaptivity, from an applied engineering perspective. Likewise various texts, beginning with Skolnik's classic [4], and more recently [5], [6], and [7], provide overviews of signal processing procedures as applied to radar systems. Our text is unique in that we consider advanced adaptive inference algorithms specifically as they apply in the radar context. We thus offer a merger between a typical radar text (which usually only briefly touches upon signal processing, and briefer still on adaptivity) and a typical detection and estimation test (which by design cannot develop aspects of adaptivity peculiar to the radar environment). Whenever possible, we present the results of the described adaptive signal processing procedures as applied to actual radar data.

Additive noise, correlated among sensors, or in time, is ubiquitous in radar systems. If it so happens that we can observe this noise in isolation, for an extended period of time, then we can estimate it and proceed nonadaptively as if it were known. If, however, we cannot measure the noise apart from the desired signal, then we are confronted with the more challenging task of joint noise-target estimation. This presence of unknown noise, observed only during target signal presence, is a common thread recurring throughout the text.

The book can be roughly broken into two segments. First, Chapters 2–4, address the topic of *waveform and parameter estimation* of a signal with known bearing. This class of problems arises in active radar, where a search beam by design illuminates potential targets in a specified sector. Chapter 2 covers single sensor detection (where the issue of bearing is actually mute). Chapter 3 provides a survey of adaptivity for antenna arrays. This chapter is transitional, in that all subsequent chapters address radar systems comprised of multiple antennas. Chapter 4 discusses minimum variance beamforming, a widely used and exceedingly flexible means of endowing multisensor systems with adaptivity.

In contrast, in the second half of the book, Chapters 5–7 address various aspects of the *bearing estimation* problem. Chapter 5 addresses the problem of tracking, particularly in the presence of multipath as it occurs in low grazing angle situations. Here we assume partial prior knowledge of bearing based on our initial surveillance detection and exploit this prior structure to obtain computational reductions. In the following two chapters we consider bearing estimation without employing prior knowledge on bearing location. As such, these methods are applicable to passive surveillance as well as the active application discussed above.

Here is a more detailed description of the four major inference problems addressed in this text.

1. *Single sensor adaptive detection of a harmonic signal in unknown*

noise. In this classic (active) airborne radar problem, one seeks to detect a target return from the broadband clutter return. The harmonic (sinusoidal) signal arises from the reflection of the transmitted radar waveform off the target. In addition to the uncertainty regarding the noise correlation, the target return generally has unknown (Doppler) frequency and amplitude parameters which need to be estimated should a detection arise. Target range is often unknown as well. Range estimation is typically done by simply handling separate range gates as separate detection problems. Range is then estimated by simply tabulating where detections occurred. (Of course refined estimates can be subsequently derived.) In some cases one can treat the noise as approximately white, at least in some small spectral band surrounding the target Doppler. This then leads to the time honored "cell averaging constant false alarm rate" radar detection [1, 4, 5, 6]. After reviewing this scheme we consider in this book two alternatives which are more robust (with respect to both detection probability and constancy of false alarm) to deviations from white noise. The first method, discussed in Chapter 2, relies on parametric models for the noise, followed by whitening and thresholding. The second method, discussed in Chapter 7, takes a nonparametric point of view. Whitening and thresholding is again employed, but the spectral estimate used to infer the statistics of the noise here is the multiple window method recently devised by Thomson. Both chapters provide target detection examples using data gathered from actual radar testbeds.

2. *Multisensor adaptive detection of a harmonic signal in unknown noise*. In this problem we assume that the noise is uncorrelated in time (for sufficiently long records this can be often approximately satisfied by processing in the Fourier domain), but that the interference is colored spatially. Such is the case in arrays of antennas when the interference is comprised of one or several propagating signals each of which arrives, at one time or another, at all the sensors. Such signals naturally arise from clutter, jamming, etc.* We also assume that the direction of the harmonic target return is known. This will be the case (approximately) in any active system, where a narrow beam is transmitted to test for targets in a specified, small, angular sector. This is the first of several multisensor detection and estimation problems, and Chapter 3 provides an overview of those aspects of detection and estimation (both parameter and waveform) unique to an array environment. As in the scalar harmonic detection problem, we present two methods of solution. The first, considered in Chapter 3, is based on (non-Bayesian) likelihood theory as applied to multivariate data. The second is based on minimum variance beamforming. Here one seeks to collapse array data into a scalar time series so as to minimize beamforming. Here one seeks to

* Because the number of sensors is generally rather limited, one often cannot successfully use a spatial Fourier transform to decorrelate the spatial structure of the noise as one can for the temporal structure.

collapse array data into a scalar time series so as to minimize output variance, subject to appropriate target-signal preserving constraints. Beamforming is popular in surveillance radars, where it is used for detection, and range, Doppler, and waveform estimation. Partially adaptive, narrowband, and wideband beamforming are all considered in Chapter 4.

3. *Bearing estimation of point sources in unknown spatially coherent noise*. One of the distinguishing features of adaptive signal processing research in the last decade or so has been the flurry of activity surrounding eigenanalysis based bearing estimation procedures. These procedures, discussed in Chapters 3 and 7, reply on the dominant invariant subspace decomposition of array snapshots to determine bearing. One deficiency of eigenanalysis is that it assumes white, or at least known, background noise. We detail several recent variants/alternatives to eigenanalysis which adapt to the background noise color. First, Chapter 5 is concerned with angle of arrival estimation in beamspace. This is a problem which can arise in tracking radars. Once one locates a target, one often seeks to track its range and bearing. The bearing is roughly known, to within one beamwidth. If multipath is present, tracking is complicated by the occurrence of a (possibly nearby) ghost target, correlated with the true target, and exhibiting its own distinct bearing. Chapter 5 addresses this problem, generalizing the sum and difference beam procedure [5] to a maximum likelihood formulation in beamspace. The beams selected are clustered around the target of interest, thereby reducing complexity (and possibly estimation error) over a fully adaptive approach. In Chapter 6 the bearing estimation problem is revisited from a Bayesian, maximum, a priori perspective. The approach here is to treat the bearings as unknown, but to attach a prior density to the unknown colored noise; a discussion of computational aspects and systolic implementations is included. Chapter 7 develops a spatial spectral estimation approach to bearing estimation. The specific spectral estimate employed is the multiwindow approach devised by Thomson. This estimate allows one not only to estimate bearings but to detect point sources with an approximately constant false alarm rate. Thomson's scheme allows for bearing estimation with either a single or multiple snapshots. When operating in the presence of diffused multipath, one is confronted with bearing estimation of a point source with a nearby distributed source. This translates to harmonic detection in colored noise, with the harmonic now in space rather than time. Thomson's method is unique in that it can be employed to recover the structure of the diffused multipath return as well as determining the direct path.

4. *Multisensor adaptive waveform estimation of signals in spatially coherent noise*. The task here is to reconstruct a signal arriving from a specified bearing. This is a nonparametric estimation problem. While in analog communications, waveform recovering, or estimation, is an end in itself, in radar such a task is usually a precursor to target classification. Waveform estimation is considered in Chapters 3 and 4. Minimum variance, maximum

likelihood, Bayesian, and Wiener formulations are all considered. Note that we do not mention single sensor adaptive waveform recovery. Although it is feasible to detect a target with unknown parameters in unknown noise with a single sensor, it is not feasible to recover the entire waveform. Indeed, the ability of an array of sensors to adaptively recover a waveform from unknown noise is a primary motivation for their use.

REFERENCES

[1] H. L. Van Trees (1968) *Detection, Estimation, and Modulation Theory*, Part I, Wiley, New York.

[2] Z. Zhu and S. Haykin (1991) "Radar detection using array processing", in *Radar Array Processing*, S. Haykin, J. Litva, and T. Shepherd (Eds), Springer-Verlag, Berlin.

[3] L. L. Scharf (1991) *Statistical Signal Processing*, Addison-Wesley, New York.

[4] M. Skolnik (1987) *Introduction to Radar Systems*, 2nd edition, McGraw-Hill, New York.

[5] J. Eaves and E. Reedy (1987) *Principles of Modern Radar*, Van Nostrand Reinhold, New York.

[6] E. Brookner (1982) *Radar Technology*, Artech House, Dedham, MA.

[7] E. Nathanson (1969) *Radar Design Principles*, McGraw-Hill, New York.

2

MODEL-BASED DETECTION

Q.T. ZHANG and SIMON HAYKIN
Communications Research Laboratory,
McMaster University,
Hamilton, Ontario, Canada

2-1 INTRODUCTION

Detection of random signals in additive noise is often encountered in diverse areas of engineering. The randomness of received signals is either due to the nature of the signals or caused by the complexity of propagation media. In radar systems, a fluctuating target produces a random reflected signal whose squared amplitude is normally assumed to obey a chi-square distribution [1]. Such radar echoes are usually pulse-to-pulse correlated for a slowly or moderately fluctuating target [2]. The target echo is usually corrupted by clutter that refers to radar returns from unwanted objects. The unwanted objects, such as mountains, watertowers, rain, snow, and sea surfaces (depending on the application of interest) may have cross sections much larger than that of a wanted target such as an aircraft or ship. This results in a small target signal immersed in a strong clutter background, thereby presenting a difficult issue for detection. An important problem in radar signal processing is therefore to combat clutter echoes, so that target detection can be enhanced.

The difficulty is further aggravated by the following facts:

1. Only a very limited number of samples are available for a target decision, typically 10–20 samples depending on the beamwidth of the antenna beam pattern, the scan rate, and the pulse repetition frequency of the radar.

Adaptive Radar Detection and Estimation, Edited by Simon Haykin and Allan Steinhardt.
ISBN 0-471-54468-X © 1992 John Wiley & Sons, Inc.

2. Radar clutter is usually temporarily and spatially nonstationary, which makes it difficult to experimentally measure the clutter covariance matrix in advance.
3. The covariance statistics of a target depend on both of its physical and geometric properties, and thus cannot be determined *a priori*.

The last two facts indicate that both signal and clutter covariance matrices must be treated as being unknown. Therefore, the target detection problem is essentially one of testing a composite hypothesis which involves many unknown parameters. This, together with the first fact, strongly suggests that highly parameterized statistical models be used for the received data. An appropriate model with the minimum number of parameters will lead to a high estimation efficiency. Thus the model for the data should be chosen such that it contains as small a number of parameters as possible. On the other hand, for ease of implementation, the model should be such that its parameters can be estimated easily. A model satisfying these two requirements is represented by the *autoregressive-moving-average* (ARMA) *process* which we will use in the following development.

Once the model is determined for the data description, we can then formulate the detection problem as testing a hypothesis about some model parameters, for which a likelihood ratio test can be constructed. This technique is known as *model-based detection*. In this chapter, three detection schemes based on the autoregressive (AR) model (a special case of the ARMA model) are presented. One of them is for a signal in additive white noise; it is based on the AR power spectrum analysis. The two remaining schemes are for the detection of a signal in colored noise. One scheme exploits information in the clutter data obtained from the range cells surrounding the data under test, while the other scheme does not.

The remainder of this chapter is organized as follows: some mathematical preliminaries are briefly reviewed in Sections 2-2 and 2-3. Specifically, Section 2-2 introduces the theory of complex stationary processes and finite parameter models. One of the most important finite parametric models is the autoregressive (AR) process, the characterization of which is addressed in Section 2-3. Sample distributions of complex AR parameters, which are essential to the theoretical performance analysis of different AR-based signal detection schemes, are derived in Section 2-4. The problem of signal in white noise is investigated in Section 2-5, followed by the detection of signal in colored noise in Sections 2-6 and 2-7. Section 2-8 completes the chapter with concluding remarks.

2-2 STATIONARY PROCESSES AND FINITE PARAMETRIC MODELS

Wide-sense stationary processes constitute a category of stochastic processes of practical importance. The processes of this kind have an important structure. Such a structure is revealed by the well-known *Wold Decomposition Theorem*, which asserts that a wide-sense stationary process can be generally parameterized by a moving average model of infinite order. Under the assumption of a rational power spectrum, the process can be simply described by a finite parameter ARMA model. Although a practical signal process may have an irrational power spectrum, an ARMA or AR process can provide an accurate approximation to it. This is because a rational power spectrum can be made arbitrarily close to an irrational one, as long as appropriate model order and model parameters are chosen [16]. Thus, from a practical point of view, we may concentrate on ARMA or AR processes.

2-2-1 Historical Notes

The history of time series analysis has been well-documented in the literature. Here we only give a brief overview of this history, based on the material of Koopmans [13].

At the beginning of the twentieth century, time series studies were based on an implicit model consisting of an almost periodic trend component in white noise. This model was used to search for hidden periodicities, primarily in geophysical data such as the Wolfer sunspot series. However, more careful investigations revealed that some important properties could not be readily accounted for by such a model. Observations of this nature caused early criticism of the scheme of hidden periodicities and prompted the search for models which better described the observed data.

Yule [4] considered the consecutive differences of a purely random series and noted a "tendency towards regularity" in the resulting data similar to the Wolfer sunspot data. Slutsky [5] studied sums and differences of purely random series and formulated the observed regular behavior as one of the probabilistic limit theorems for stochastic processes. He referred to the series resulting from the summing and differencing operations as processes of moving summation. The more recent terminology, apparently dating from the definite study of discrete models by Wold [6], is "moving average processes". This class of processes remains one of the most important collections of finite parameter models.

In a study of the Wolfer sunspot data, Yule [4] used techniques of regression analysis of a fixed number of previous values. The scheme implicitly defined by this procedure was termed *linear autoregression* by Wold. Finite autoregressive processes, as they are commonly called today, constitute another important class of finite parameter models.

Autoregressive process appeared in economics during the 1930's in the work of Frisch [7] and Tinbergen [8]. Because of the great interest in forecasting future values of economic variables, the unique predictive feature of autoregressive processes played a significant role in the applications of these models.

In his 1938 monograph [6], Wold established a key decomposition theorem of the general wide-sense stationary process which led to the formulation and solution of the linear prediction problem by Kolmogorov during the period 1939–1941 [9]. Independently, during roughly the same period, Wiener solved the linear prediction problem in an important special case and later extended his solution to a larger class of problems including the filtering problem [10].

Kolmogorov also presented, for the first time, a geometric interpretation of wide-sense stationary processes, which has unified the subject and has proved most useful in the theoretical development of time series analysis [11]. The extension of the univariate time series theory to the multivariate counterpart is amply described in Hannan [12].

2-2-2 The Wold Decomposition Theorem

One of the most important theorems that reveals the structure of a wide-sense stationary process is that due to Wold, which can be stated as follows [13, pp. 323–324]:

Theorem (Wold) Let $X(t), t = 0, \pm 1, \ldots$ be a zero-mean wide-sense stationary stochastic process. Then $X(t)$ can be expressed as the sum of two zero-mean wide-sense stationary processes,

$$X(t) = U(t) + V(t) \tag{2-1}$$

such that

(i) The process $U(t)$ is uncorrelated with the process $V(t)$.

(ii) $U(t)$ has a one-sided moving average representation

$$U(t) = \sum_{k=1}^{\infty} a_k \xi(t-k) \tag{2-2}$$

with

$$a_0 = 1 \quad \text{and} \quad \sum_{k=0}^{\infty} a_k^2 < \infty$$

and the subspace $\mathcal{U}_t^{\xi}$, generated by the unique white noise process satisfies

$$\mathcal{U}_t^{\xi} = \mathcal{U}_t^{u} \tag{2-3}$$

for all t, where $\mathcal{U}_t^{u}$ is the subspace generated by $U(t)$.

(iii) The $V(t)$ process is completely linearly predictable from its past data.

The component $U(t)$ is known as a regular, or nondeterministic process and correspondingly $V(t)$ is known as a nonregular, or deterministic process. The Wold decomposition theorem shows that the regular process can be expressed as a linear sum of an innovation process, or simply as a moving average process.

The Wold decomposition is usually performed in the time domain. A similar decomposition may be performed in the frequency domain, which is referred to as the *Lebesque decomposition* [13, 14]. Let $F_x(d\lambda)$ denote the power spectrum of the process $X(t)$. Then there are an absolutely continuous component $F_u(d\lambda)$ and a discrete component $F_v(d\lambda)$, such that

$$F_x(d\lambda) = F_u(d\lambda) + F_v(d\lambda) \tag{2-4}$$

where $F_u(d\lambda)$ corresponds to the regular process $U(t)$, and $F_v(d\lambda)$ corresponds to the nonregular process $V(t)$.

There is a simple criterion to determine whether a stationary process is regular or not. Let $f(\lambda)$ denote the power spectral density of the process $X(t)$. Then the process is deterministic if

$$\int_{-1/2}^{1/2} \log f(\lambda)\, d\lambda = -\infty \tag{2-5}$$

and it is regular if

$$\int_{-1/2}^{1/2} \log f(\lambda)\, d\lambda < -\infty \tag{2-6}$$

This result is due to Doob [14]. The analogous result for a continuous time process was given by Paley and Wiener [12, p. 151].

In most practical situations, the deterministic component is simply a stochastic, almost period function. An example often encountered in communications is a sinusoid with random phase. It is easily checked that the condition described in (2-5) is satisfied for such a signal, since its power spectral density is zero except at the sinusoidal frequency. The deterministic component can be estimated with increasing reliability, as more and more data are accumulated. Thus, we may remove it from the received process and concentrate on the regular process.

With this in mind, an important conclusion drawn from the Wold decomposition theorem is that any wide-sense stationary process can be expressed as a moving average process.

2-2-3 AR Modeling

The moving average model described above usually involves the use of an infinite number of parameters. Our objection, however, is to obtain a compactly parameterized model for signal description. We are therefore more interested in those stationary processes which posses rational power spectra since they admit an ARMA expression of finite parameters. However, in the following, we are mainly concerned with detection schemes in which the received data can be modeled by AR processes. There are several reasons for the use of AR modeling for the data description:

1. As shown by Arato, among all stationary Gaussian processes with a rational spectral density function, only the AR processes admit a sufficient statistic that does not grow in dimension as the number of observations increases. This fact suggests that AR approximations of general stationary correlated Gaussian processes may be useful in obtaining simple near-optimal detectors [15].
2. Stochastic processes encountered in practice typically have a nonrational power spectrum. For example, the time series representing clutter obtained from a surveillance radar usually has a power spectrum of Gaussian shape, which is caused by the modulation by the antenna beampattern. For a wide-sense stationary process of nonrational power spectrum $S(f)$, Fuller [16] shows that for any $\delta > 0$, there is an AR power spectrum $S_{\mathrm{AR}}(f)$ such that at every frequency the error made by replacing $S(f)$ by $S_{\mathrm{AR}}(f)$ is at most δ.
3. The parameters of an AR model can be estimated using various computationally efficient algorithms. These algorithms can be implemented recursively, rendering themselves particularly suitable for adaptive signal processing in a real-time fashion.
4. In many situations of practical interest, received data processes can be modeled by an AR process of relatively low order, say M. In radar echo modeling, M ranges from 2 to 5 [17, 18]. In active sonar environment, $M = 8$ is suggested in [19].

2-3 CHARACTERIZATION OF AN AR PROCESS

Given that the use of AR modeling is justified, there are still a number of parameter sets that can be used to characterize an AR process. Since all these parameter sets describe the same statistical properties of the process, there are connections among them. A thorough understanding of such connections is helpful to the model-based detection described herein.

Let $y(k)$ be a stationary complex AR process described by the difference equation

$$y(k) = \sum_{m=1}^{p} a_m y(k-m) + e(k) \tag{2-7}$$

where $\{e(k)\}$ is a complex white noise process with mean zero and variance σ_e^2. Let $S_y(f)$ denote the power spectrum of $\{y(k)\}$, and let $r_0, r_1, \ldots, r_p$

denote the correlations of $\{y(k)\}$ with lags $0, 1, \ldots, p$, respectively. Then the $(p+1)$-by-$(p+1)$ covariance matrix $\mathbf{R}_{p+1}$ of $\{y(k)\}$ can be written as

$$\mathbf{R}_{p+1} = \begin{bmatrix} r_0 & r_1^* & r_2^* & \cdots & r_p^* \\ r_1 & r_0 & \ddots & & \vdots \\ r_2 & r_1 & \ddots & & r_1^* \\ \vdots & \ddots & & & \\ r_p & \cdots & r_2 & r_1 & r_0 \end{bmatrix}, \tag{2-8}$$

where the asterisk denotes complex conjugation.

Let the *partial correlation coefficients* (PARCOR) of the process be denoted by $\rho_1, \rho_2, \ldots, \rho_p$. Define

$$\rho = \begin{pmatrix} \rho_1 \\ \rho_2 \\ \vdots \\ \rho_p \end{pmatrix}, \qquad \mathbf{a} = \begin{pmatrix} a_1 \\ a_2 \\ \vdots \\ a_p \end{pmatrix} \tag{2-9}$$

Then we have four sets of parameters that can be used for the description of the process $\{y(k)\}$. They are

Set I: AR coefficients, $\{\mathbf{a}, \sigma_e^2\}$.
Set II: Power spectrum, $S_y(f)$.
Set III: Covariance matrix, $\mathbf{R}_{p+1}$.
Set IV: PARCORs, $\{\rho, \sigma_e^2\}$.

These parameter sets are equivalent to each other in the sense that any one of them can be derived from the others. These relationships are summarized as follows.

Sets I and II. The power spectrum $S_y(f)$ can be obtained from the parameter set I by [20]

$$S_y(f) = \frac{\sigma_e^2}{\left|1 - \sum_{k=1}^{p} a_k e^{-jk\omega}\right|^2} \tag{2-10}$$

This formula is easily obtained by viewing the process $\{y(k)\}$ as the output of an all-pole filter, defined by

$$H(z) = \frac{1}{1 - \sum_{k=1}^{p} a_k z^{-k}}$$

which is driven by a white noise sequence $\{e(k)\}$. The power spectrum given above also has an interpretation as a *maximum entropy spectrum* [21, 22].

Conversely, we may evaluate the parameters $\mathbf{a}$ and σ_e^2 from the given power spectrum $S_y(f)$ through the *Doob decomposition* [14, p. 363]. In particular, we factor $S_y(z)$ such that

$$S_y(z) = e^{c_0} g(z) g(z^{-1}) \tag{2-11}$$

where $g(z)$ is defined by

$$g(z) = \exp\left\{ \sum_{k=1}^{\infty} c_k z^k \right\}$$

with c's given by

$$\log S_y(z) = \sum_{k=-\infty}^{\infty} c_k z^k$$

$$c_k = \frac{1}{2\pi i} \int_{|z|=1} z^{-k-1} \log\{S_y(z)\}\, dz$$

Next, we expand $g^{-1}(z)$ into a Laurent series to get the AR coefficient, the a's, as shown by

$$g^{-1}(z) = \sum_k a_k z^{-k} \tag{2-12}$$

The variance of the model error can be determined by the well-known *Kolmogorov–Szego formula* [14, 23], given by

$$\sigma_e^2 = \exp\left\{ \int_{-0.5}^{0.5} \ln S_y(f)\, df \right\} \tag{2-13}$$

Sets I and III. Let $\mathbf{R}_{p+1}$ be partitioned, so that

$$\mathbf{R}_{p+1} = \begin{bmatrix} r_0 & \mathbf{p}^H \\ \mathbf{p} & \mathbf{R}_p \end{bmatrix} \tag{2-14}$$

where the superscript H denotes the conjugate transpose, and the vector $\mathbf{p}$ is defined as

$$\mathbf{p} = \begin{bmatrix} r_1 \\ r_2 \\ \vdots \\ r_p \end{bmatrix} \tag{2-15}$$

Then given the covariance matrix $\mathbf{R}_{p+1}$, the parameter set $\{\mathbf{a}, \sigma_e^2\}$ can be calculated as

$$\mathbf{a} = \mathbf{R}_p^{-1}\mathbf{p}$$
$$\sigma_e^2 = \frac{\det(\mathbf{R}_{p+1})}{\det(\mathbf{R}_p)} \tag{2-16}$$

The first line of (2-16) is the well-known *normal equation* written in matrix form [20], and the second line can be found in [12, p. 149].

Likewise, the covariance matrix $\mathbf{R}_{p+1}$ can be evaluated explicitly from $\{\mathbf{a}, \sigma_e^2\}$ through the *Gohberg–Semencul formula* [24, 25]

$$\mathbf{R}_{p+1}^{-1} = (\mathbf{A}_1\mathbf{A}_1^H - \mathbf{A}_2\mathbf{A}_2^H)/\sigma_e^2 \tag{2-17}$$

where $\mathbf{A}_1$ and $\mathbf{A}_2$ are defined by

$$\mathbf{A}_1 = \begin{bmatrix} 1 & & & \mathbf{0} \\ a_1 & \ddots & & \\ \vdots & \ddots & \ddots & \\ a_p & \cdots & a_1 & 1 \end{bmatrix}, \quad \mathbf{A}_2 = \begin{bmatrix} 0 & & & & \mathbf{0} \\ a_p & \ddots & & & \\ \vdots & \ddots & \ddots & & \\ a_1 & a_2 & \cdots & a_p & 0 \end{bmatrix}$$

Sets I and IV. The functional relation between the parameter sets $\{\mathbf{a}, \sigma_e^2\}$ and $\{\rho, \sigma_e^2\}$ is defined by the well known *Levinson–Durbin algorithm*. Let $a_m(k)$, $k = 0, \ldots, m$ with $a_m(0) = 1$ denote the coefficients of an order m linear prediction filter (LPF). These coefficients are related to our AR coefficients by $a_p(k) = a_k$, $k = 1, \ldots, p$. With these notations, the Levinson–Durbin algorithm can be written as [20, p.144]

For $m = 1, p$

For $k = 1, m - 1$

$$a_m(k) = a_{m-1}(k) + \rho_m a_{m-1}^*(m-k) \tag{2-18}$$

End

End

This algorithm can be used to evaluate $a_1, \ldots, a_p$ for the given set of partial correlation coefficients. In case we need to calculate $\rho_1, \ldots, \rho_p$ from $a_1, \ldots, a_p$, we may use the *inverse Levinson–Durbin algorithm*, given by [20, p.146]

1. $\rho_p = a_p$
2. For $m = p, 1$

   For $k = 0, m-1$

$$a_{m-1}(k) = \frac{a_m(k) - a_m(m)a_m^*(m-k)}{1 - |a_m(m)|^2} \tag{2-19}$$

   End

$$\rho_{m-1} = \alpha_{m-1}(m-1)$$

   End

Sets II and III. The power spectrum and covariance constitute a Fourier transform pair, usually known as the *Einstein–Wiener–Khintchine theorem.*

Sets II and IV. There are no explicit expressions between the power spectrum and the partial correlation coefficients.

Sets III and IV. Let P_m denote the output power of an order-m linear prediction-error filter, and let $R(\tau)$ denote the autocovariance of the AR process for lag τ. The relation between the autocovariance and the partial correlation coefficients are described by [20, p.147]

$$\begin{aligned} R(m) &= -\rho_m^* P_{m-1} - \sum_{k=1}^{m-1} a_{m-1}^*(k) R(m-k) \\ \rho_m &= -\frac{1}{P_{m-1}} \sum_{k=0}^{m} a_{m-1}(k) R(k-m) \end{aligned} \tag{2-20}$$

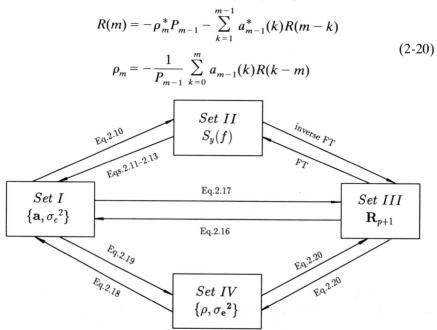

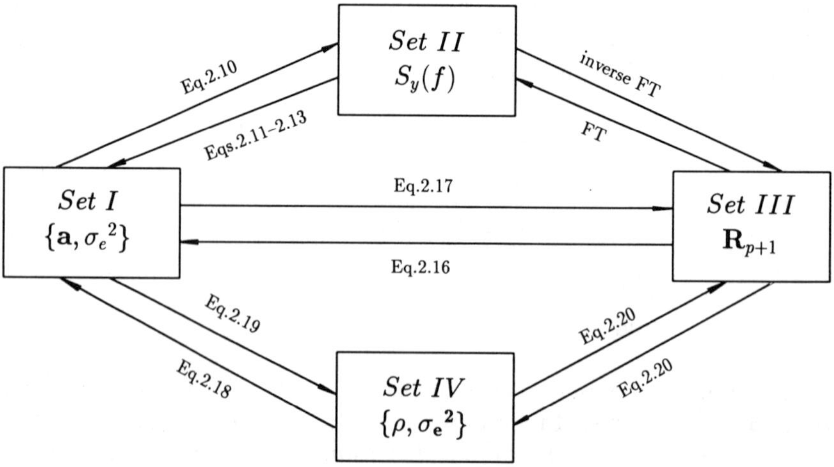

Figure 2-1 Diagram illustrating the relationships among various parameter sets that can be used to characterize an AR process.

The various relationships described above are summarized in diagrammatic form in Fig. 2-1.

2-4 SAMPLE DISTRIBUTIONS OF COMPLEX AR PARAMETERS

The sample distributions of the parameters of a real AR process have been well established in the literature. The joint *probability density function* (PDF) of the sample estimates of the AR coefficients can be found in [27] and [28]. The PDF of a single PARCOR is given in [12], while that for multiple PARCORs is given in [29]. However, relatively little work has been done for complex AR processes, although it is essential for analyzing many signal processors arising in radar, sonar, and communications where complex baseband signals need to be considered. The extension of the results for a real AR process to their complex counterparts should be straightforward, at least in principle. However, the derivations involved are usually very lengthy. This situation can be illustrated by [30], where the sample distribution of a single PARCOR of a complex AR process was derived. The distributions for the general complex case were recently established in [31] and [32]. Here we follow the approach by Zhang [31, 32].

2-4-1 Notations

As a notational convention, throughout the chapter, we use $CN_m(\boldsymbol{\mu}, \mathbf{R})$ to denote an m-dimensional complex Gaussian distribution with mean $\boldsymbol{\mu}$ and covariance $\mathbf{R}$. For the real case, we use the symbol $N_m(.,.)$. Superscripts T, H and $*$ stand for transposition, conjugate transposition, and complex conjugation, respectively. A mathematical symbol with a hat atop of it signifies the *maximum likelihood estimate* (MLE) of the corresponding parameter. Symbol $\sim$ implies distributed as. In the expression

$$\mathbf{a} = \alpha + j\beta \tag{2-21}$$

α and β represent the real and imaginary parts of $\mathbf{a}$. Sometimes, we express α and β more explicitly as

$$\alpha = \mathrm{Re}(\mathbf{a}) , \qquad \beta = \mathrm{Im}(\mathbf{a}) \tag{2-22}$$

Suppose that a complex received process described by $y_1, \ldots, y_n$ can be modeled as an AR(p) process with $p \ll n$, such that

$$y_k = \sum_{i=1}^{p} a_i y_{k-i} + e_k \tag{2-23}$$

where the error process $\{e_k\}$ is complex Gaussian with

$$e_k \sim CN(0, \sigma_p^2), \qquad k = 1, \ldots, n \tag{2-24}$$

Again we use $\rho_1, \ldots, \rho_p$ to denote the PARCORs of the received process. We further define the following notations:

$$\begin{aligned}
a &= [a_1 \quad \cdots \quad a_p]^T = \alpha + j\beta \\
\rho &= [\rho_1 \quad \cdots \quad \rho_p]^T = \mathbf{u} + j\mathbf{v} \\
\theta_a &= \begin{pmatrix} \alpha \\ \beta \end{pmatrix} \\
\theta_p &= \begin{pmatrix} \mathbf{u} \\ \mathbf{v} \end{pmatrix} \\
\mathbf{y}_p &= [y_1 \quad \cdots \quad y_p]^T \\
\mathbf{R}_p &= E[\mathbf{y}_p \mathbf{y}_p^H] \\
\Omega &= \begin{bmatrix} \mathrm{Re}(\mathbf{R}_p) & -\mathrm{Im}(\mathbf{R}_p) \\ \mathrm{Im}(\mathbf{R}_p) & \mathrm{Re}(\mathbf{R}_p) \end{bmatrix} \\
\mathbf{y} &= (y_{p+1}, \ldots, y_n)^T = \mathbf{y}_R + j\mathbf{y}_I \\
\mathbf{Y} &= \begin{bmatrix} y_p & y_{p-1} & \cdots & y_1 \\ y_{p+1} & y_p & \cdots & y_2 \\ y_{n-1} & y_{n-2} & \cdots & y_{n-p} \end{bmatrix} = \mathbf{Y}_R + j\mathbf{Y}_I
\end{aligned} \tag{2-25}$$

2-4-2 The PDF of $\hat{\mathbf{a}}$

We first establish the result for the maximum likelihood estimate of **a**.

Theorem 2-1 As the sample size $n \to \infty$, we have

$$\hat{\mathbf{a}} \sim CN_p(\mathbf{a}, \sigma_p^2[n\mathbf{R}_p]^{-1}) \tag{2-26}$$

Proof. Since $\hat{\mathbf{a}}$ is the MLE of **a**, then according to the asymptotic properties of maximum likelihood estimators, it is asymptotically Gaussian distributed with mean

$$E[\hat{\mathbf{a}}] = \mathbf{a}$$

and the covariance defined by the *Fisher information matrix*. Thus, only the Fisher information matrix remains to be determined. To this end, we write the log likelihood of **a** and σ_p^2 as

$$
\begin{aligned}
L(\mathbf{a}, \sigma_p^2) &= \log p(y_1, \ldots, y_n \mid \mathbf{a}, \sigma_p^2) \\
&= \log\{p(\mathbf{y}_p)p(\mathbf{y} \mid \mathbf{y}_p, \mathbf{a}, \sigma_p^2)\} \\
&= \log p(\mathbf{y}_p) + \log p(\mathbf{y} \mid \mathbf{y}_p, \mathbf{a}, \sigma_p^2)
\end{aligned}
$$

For large n, the first term can be ignored since it is independent of n, leading to

$$
\begin{aligned}
L(\mathbf{a}, \sigma_p^2) &= -(n-p)\log(\pi\sigma_p^2) - \sigma_p^{-2}(\mathbf{y} - \mathbf{Ya})^H(\mathbf{y} - \mathbf{Ya}) \\
&= -(n-p)\log(\pi\sigma_p^2) \\
&\quad -\sigma_p^{-2}\{(\mathbf{y}_R^T - \alpha^T\mathbf{Y}_R^T + \beta^T\mathbf{Y}_I^T)(\mathbf{y}_R - \mathbf{Y}_R\alpha + \mathbf{Y}_I\beta) \\
&\quad + (\mathbf{y}_I^T - \beta^T\mathbf{Y}_R^T - \alpha^T\mathbf{Y}_I^I)(\mathbf{y}_I - \mathbf{Y}_R\beta - \mathbf{Y}_I\alpha)\} \qquad (2\text{-}27)
\end{aligned}
$$

Through computations, we obtain

$$
\begin{aligned}
\frac{1}{n}\frac{\partial^2 L}{\partial\alpha\,\partial\alpha^T} &= \frac{1}{n}\frac{\partial^2 L}{\partial\beta\,\partial\beta^T} \\
&= -\frac{2}{\sigma_p^2}\frac{\mathbf{Y}_R^T\mathbf{Y}_R + \mathbf{Y}_I^T\mathbf{Y}_I}{n} \\
&\to -\frac{2}{\sigma_p^2}\operatorname{Re}(\mathbf{R}_p) \qquad (2\text{-}28)
\end{aligned}
$$

$$
\begin{aligned}
\frac{1}{n}\frac{\partial^2 L}{\partial\beta\,\partial\alpha^T} &= -\frac{1}{n}\frac{\partial^2 L}{\partial\alpha\,\partial\beta^T} \\
&= -\frac{2}{\sigma_p^2}\frac{\mathbf{Y}_R^T\mathbf{Y}_I - \mathbf{Y}_I^T\mathbf{Y}_R}{n} \\
&\to -\frac{2}{\sigma_p^2}\operatorname{Im}(\mathbf{R}_p) \qquad (2\text{-}29)
\end{aligned}
$$

$$
\frac{1}{n}\frac{\partial^2 L}{\partial\alpha\,\partial\sigma_p^2} = \frac{2}{n\sigma_p^4}\{(\mathbf{Y}_R^T\mathbf{Y}_R + \mathbf{Y}_I^T\mathbf{Y}_I)\alpha + (\mathbf{Y}_I^T\mathbf{Y}_R - \mathbf{Y}_R^T\mathbf{Y}_I)\beta - (\mathbf{Y}_R^T\mathbf{y}_R - \mathbf{Y}_I^T\mathbf{y}_I)\} \qquad (2\text{-}30)
$$

$$
\frac{1}{n}\frac{\partial^2 L}{\partial\beta\,\partial\sigma_p^2} = \frac{2}{n\sigma_p^4}\{\mathbf{Y}_R^T\mathbf{Y}_I - \mathbf{Y}_I^T\mathbf{Y}_R)\alpha + (\mathbf{Y}_R^T\mathbf{Y}_R + \mathbf{Y}_I^T\mathbf{Y}_I)\beta - (\mathbf{Y}_R^T\mathbf{y}_I - \mathbf{Y}_I^T\mathbf{y}_R)\} \qquad (2\text{-}31)
$$

As $n \to \infty$, both (2-30) and (2-31) tend to zero, since

$$\frac{1}{n}\left(\frac{\partial^2 L}{\partial \alpha\, \partial \sigma_p^2} + j\frac{\partial^2 L}{\partial \beta\, \partial \sigma_p^2}\right) = \frac{2}{\sigma_p^4}\left(\frac{1}{n}\mathbf{Y}^H\mathbf{Y}\right)(\mathbf{a} - \hat{\mathbf{a}})$$
$$\rightarrow 0 \tag{2-32}$$

Here we have used the fact that $\hat{\mathbf{a}}$ approaches $\mathbf{a}$ in probability. This indicates that $\hat{\mathbf{a}}$ is asymptotically uncorrelated with σ_p^2. Thus we may consider the covariance of $\hat{\mathbf{a}}$ and the variance of σ_p^2 separately.

From (2-28) and (2-29), the Fisher information matrix of both α and β is equal to

$$J(\alpha, \beta) = E\begin{bmatrix} \dfrac{\partial^2 L}{\partial \alpha\, \partial \alpha^T} & \dfrac{\partial^2 L}{\partial \alpha\, \partial \beta^T} \\ \dfrac{\partial^2 L}{\partial \beta\, \partial \alpha^T} & \dfrac{\partial^2 L}{\partial \beta\, \partial \beta^T} \end{bmatrix}$$
$$= -2n\sigma_p^{-2}\Omega \tag{2-33}$$

which implies that the covariance matrix of $\hat{\theta}_a$ is $\sigma_p^2(2n\Omega)^{-1}$. The matrix Ω has been defined in (2-25).

It follows therefore that

$$\hat{\theta}_a \sim N_{2p}(\theta_a, \sigma_p^2[2n\Omega]^{-1}) \tag{2-34}$$

Note that the symmetrical matrix Ω as defined in (2-25) is isomorphic to the Hermitian matrix $\mathbf{R}_p$ [33]. Thus, according to Miller [34], the above expression can be written in a complex form:

$$\hat{\mathbf{a}} \sim CN_p(\boldsymbol{a}, \sigma_p^2[n\mathbf{R}_p]^{-1}) \tag{2-35}$$

which completes the proof.

2-4-3 The PEF of $\hat{\rho}$

The sample PARCOR $\hat{\theta}_p$ is related to the MLE $\hat{\theta}_a$ of θ_a through a transform defined by the Levinson–Durbin algorithm. According to the invariance principle of maximum likelihood estimators, $\hat{\theta}_p$ is asymptotically Gaussian with mean θ_p and covariance matrix dependent upon both the covariance of θ_a and the Jacobian matrix associated with the transformation [35].

Let us elaborate on this issue in detail. In the Levinson–Durbin algorithm given in (2-18), defining

$$a_m(i) = \alpha_m(i) + j\beta_m(i)$$

$$\rho_m = u_m + jv_m$$

and then separating the real and imaginary parts of (2-18), we have

$$\begin{aligned} \alpha_m(i) &= \alpha_{m-1}(i) + u_m \alpha_{m-1}(m-i) + v_m \beta_{m-1}(m-i) \\ \beta_m(i) &= \beta_{m-1}(i) + v_m \alpha_{m-1}(m-i) - u_m \beta_{m-1}(m-i) \end{aligned} \tag{2-36}$$

where $i = 1, \ldots, m$; $m = 1, \ldots, p$. This transformation defines a $2p$-by-$2p$ matrix $\mathbf{L}$, such that

$$\mathbf{L} = \begin{bmatrix} \dfrac{\partial \theta_a(1)}{\partial \theta_\rho(1)} & \cdots & \dfrac{\partial \theta_a(1)}{\partial \theta_\rho(2p)} \\ \vdots & & \\ \dfrac{\partial \theta_a(2p)}{\partial \theta_\rho(1)} & \cdots & \dfrac{\partial \theta_a(2p)}{\partial \theta_\rho(2p)} \end{bmatrix} \tag{2-37}$$

With $\theta_a(k)$ and $\theta_\rho(k)$ denoting the kth elements of θ_a and θ_ρ, respectively, the covariance matrices of θ_a and θ_ρ are then related by

$$\text{cov}(\hat{\theta}_a) = \mathbf{L}\, \text{cov}(\hat{\theta}_\rho) \mathbf{L}^T$$

or

$$\begin{aligned} \text{cov}(\hat{\theta}_\rho) &= \mathbf{L}^{-1}\, \text{cov}(\hat{\theta}_a)(\mathbf{L}^T)^{-1} \\ &= \frac{\sigma_p^2}{2n} (\mathbf{L}^T \Omega \mathbf{L})^{-1} \end{aligned} \tag{2-38}$$

Summarizing the above analysis, we have:

Theorem 2-2 As the sample size $n \to \infty$, we have

$$\hat{\theta}_\rho \sim N_{2p}\left(\theta_\rho, \frac{\sigma_p^2}{2n} [\mathbf{L}^T \Omega \mathbf{L}]^{-1}\right) \tag{2-39}$$

2-4-4 An Algorithm for L

We need an algorithm to calculate $\mathbf{L}$ from the given AR coefficients. We first define an $2m$-by-$2m$ matrix

$$\mathbf{L}_m = \begin{bmatrix} \mathbf{A}_m & \mathbf{B}_m \\ \mathbf{C}_m & \mathbf{D}_m \end{bmatrix} \tag{2-40}$$

where the elements of the m-by-m submatrices $\mathbf{A}_m$, $\mathbf{B}_m$, $\mathbf{C}_m$, and $\mathbf{D}_m$ are such that for $i, j = 0, 1, \ldots, m$ we have

$$
\begin{aligned}
(\mathbf{A}_m)_{ij} &= \frac{\partial \alpha_m(i)}{\partial u_j} \\
(\mathbf{B}_m)_{ij} &= \frac{\partial \alpha_m(i)}{\partial v_j} \\
(\mathbf{C}_m)_{ij} &= \frac{\partial \beta_m(i)}{\partial u_j} \\
(\mathbf{D}_m)_{ij} &= \frac{\partial \beta_m(i)}{\partial v_j}
\end{aligned}
\tag{2-41}
$$

Note that $\mathbf{L}_m$ has a similar structure to $\mathbf{L}$. The only difference is in the dimension. However, as the algorithm evolves from $m = 1$ to p, the dimension of $\mathbf{L}_m$ will increase from 2-by-2 to $2p$-by-$2p$. The matrix $\mathbf{L}_p$ is the $\mathbf{L}$ matrix we are looking for.

To obtain a recursive algorithm to evaluate the above derivatives, we differentiate both sides of (2-36). Through direct calculation and simplifying the recursions can be written in a compact matrix form. This is now summarized in Algorithm 2-1 as described below.

The recursions can be written in more compact matrix form.

ALGORITHM 2-1

1. For $m = 1$, $A_1 = 1$, $B_1 = 0$, $C_1 = 0$, $D_1 = 1$, and hence $\mathbf{L}_1 = \text{diag}(1, 1)$.
2. For $m \geq 2$, define $(m\text{-}1)$-by-1 vectors

$$
\boldsymbol{\alpha}_{m-1} = [\alpha_{m-1}(m-1) \quad \cdots \quad \alpha_{m-1}(1)]^T
$$

$$
\boldsymbol{\beta}_{m-1} = [\beta_{m-1}(m-1) \quad \cdots \quad \beta_{m-1}(1)]^T
$$

and define $(m-1)$-by-$(m\text{-}1)$ matrices

$$
\mathscr{A}_{m-1} = [(\mathbf{A}_{m-1})_{ij} + u_m(\mathbf{A}_{m-1})_{m-i,j} + v_m(\mathbf{C}_{m-1})_{m-i,j}]
$$

$$
\mathscr{B}_{m-1} = [(\mathbf{B}_{m-1})_{ij} + u_m(\mathbf{B}_{m-1})_{m-i,j} + v_m(\mathbf{D}_{m-1})_{m-i,j}]
$$

$$
\mathscr{C}_{m-1} = [(\mathbf{C}_{m-1})_{ij} + v_m(\mathbf{A}_{m-1})_{m-i,j} - u_m(\mathbf{C}_{m-1})_{m-i,j}]
$$

$$
\mathscr{D}_{m-1} = [(\mathbf{D}_{m-1})_{ij} + v_m(\mathbf{B}_{m-1})_{m-i,j} - u_m(\mathbf{D}_{m-1})_{m-i,j}]
$$

$$
i, j = 1, 2, \ldots, m-1
$$

Here we have used symbol $(\mathbf{A}_m)_{i,j}$ to denote the (i, j)th element of the matrix $\mathbf{A}_m$. Using these notations, the matrices $\mathbf{A}_m$, $\mathbf{B}_m$, $\mathbf{C}_m$, and $\mathbf{D}_m$ can be updated as follows: For $m = 1, \ldots, p$,

$$\mathbf{A}_m = \begin{bmatrix} \mathscr{A}_{m-1} & \boldsymbol{\alpha}_{m-1} \\ \mathbf{0} & 1 \end{bmatrix}$$

$$\mathbf{B}_m = \begin{bmatrix} \mathscr{B}_{m-1} & \boldsymbol{\beta}_{m-1} \\ \mathbf{0} & 0 \end{bmatrix}$$

$$\mathbf{C}_m = \begin{bmatrix} \mathscr{C}_{m-1} & -\boldsymbol{\beta}_{m-1} \\ \mathbf{0} & 0 \end{bmatrix}$$

$$\mathbf{D}_m = \begin{bmatrix} \mathscr{D}_{m-1} & \boldsymbol{\alpha}_{m-1} \\ \mathbf{0} & 1 \end{bmatrix}$$

3. Compute $\mathbf{L}$ using

$$\mathbf{L} = \begin{bmatrix} \mathbf{A}_p & B_p \\ \mathbf{C}_p & D_p \end{bmatrix}$$

2-4-5 Examples

To illustrate the use of this algorithm, we give two examples.

Example 2-1 $\quad p = 2$

$$\mathbf{L}_2 = \begin{bmatrix} 1+u_2 & \alpha_1(1) & v_2 & \beta_1(1) \\ 0 & 1 & 0 & 0 \\ v_2 & -\beta_1(1) & 1-u_2 & \alpha_1(1) \\ 0 & 0 & 0 & 1 \end{bmatrix}$$

Example 2-2 $\quad p = 3$

$$\mathbf{L}_3 = \begin{bmatrix} A_3 & B_3 \\ C_3 & D_3 \end{bmatrix}$$

where

$$\mathbf{A}_3 = \begin{bmatrix} 1+u_2 & \alpha_1(1)+u_3 & \alpha_2(2) \\ u_3(1+u_2)+v_3v_2 & 1+u_3\alpha_1(1)-v_3\beta_1(1) & \alpha_2(1) \\ 0 & 0 & 1 \end{bmatrix}$$

$$\mathbf{B}_3 = \begin{bmatrix} v_2 & \beta_1(1)+v_3 & \beta_2(2) \\ u_3v_2+v_3(1-u_2) & u_3\beta_1(1)+v_3\alpha_1(1) & \beta_2(1) \\ 0 & 0 & 0 \end{bmatrix}$$

$$\mathbf{C}_3 = \begin{bmatrix} v_2 & -\beta_1(1)+v_3 & -\beta_2(2) \\ v_3(1+u_2)-u_3v_2 & v_3\alpha_1(1)+u_3\beta_1(1) & -\beta_2(1) \\ 0 & 0 & 0 \end{bmatrix}$$

$$\mathbf{D}_3 = \begin{bmatrix} 1-u_2 & \alpha_1(1)-u_3 & \alpha_2(2) \\ v_3v_2-u_3(1-u_2) & 1+v_3\beta_1(1)-u_3\alpha_1(1) & \alpha_2(1) \\ 0 & 0 & 1 \end{bmatrix}$$

2-4-6 Special Cases

When the received signal is a white noise sequence, we have for $m = 1, 2, \ldots, p$,

$$\alpha_m(i) = 0\,, \qquad i = 1, \ldots, m$$

$$\rho_m = 0$$

and

$$\sigma_p^2 = \sigma_{p-1}^2 = \cdots = \sigma_0^2 \tag{2-42}$$

$$\mathbf{R}_p = \sigma_0^2 \mathbf{I}_p$$

with $\mathbf{I}_p$ denoting the identity matrix of order p.

Application of the conditions given in (2-42) to Algorithm 2-1 yields

$$\mathbf{L} = \mathbf{I}_{2p} \tag{2-43}$$

Inserting this result in (2-39), we obtain

$$\begin{aligned} \mathrm{cov}(\hat{\theta}_\rho) &= \frac{\sigma_p^2}{2n} (\sigma_0^2 \mathbf{I}_{2p})^{-1} \\ &= (2n)^{-1} \mathbf{I}_{2p} \end{aligned} \tag{2-44}$$

Combining this result with Theorem 2-2, we obtain:

COROLLARY 2-2-1 If $y_1, \ldots, y_n$ is a complex white Gaussian sequence of mean zero, then for large n,

$$\hat{\theta}_\rho \sim N_{2p}\left(\mathbf{0}, \frac{1}{2n} \mathbf{I}_{2p}\right)$$

or

$$\hat{\rho} \sim CN_p\left(\mathbf{0}, \frac{1}{n} \mathbf{I}_p\right) \tag{2-45}$$

As we see, in the case of a white noise sequence, the PDF of $\hat{\rho}$ is complex Gaussian distributed. A natural question arises: Is this observation valid in general? The answer is no. Let us explain. According to Miller [34], a real-valued Gaussian distribution can be expressed as a complex Gaussian one if and only if there exists an isomorphism between its covariance matrix and a complex one. A real covariance matrix Ω is said to be isomorphic to a complex one $\mathbf{R}$ if Ω is such that

$$\Omega = \begin{pmatrix} \mathrm{Re}(\mathbf{R}) & -\mathrm{Im}(\mathbf{R}) \\ \mathrm{Im}(\mathbf{R}) & \mathrm{Re}(\mathbf{R}) \end{pmatrix}$$

Goodman [33] shows that this isomorphism remains invariant under operations of matrix multiplication and inversion. Now

$$\operatorname{cov}(\hat{\theta}_p) = \frac{\sigma_p^2}{2n}[\mathbf{L}^T\Omega\mathbf{L}]^{-1}$$

and Ω is isomorphic to $\mathbf{R}$. Thus in order for the PDF of $\hat{\theta}_p$ to have a complex Gaussian expression, $\mathbf{L}$ must have a structure like Ω. Unfortunately, this is not the case in general, as we have observed in the previous section.

Sometimes, we may need the PDF of a single PARCOR ρ_p, the pth PARCOR of an AR(p) process. In this particular case, it has a simple expression given by

$$\hat{\rho}_p \sim CN\left(\rho_p, \frac{1}{n}[1 - |\rho_p|^2]\right) \tag{2-46}$$

For a proof, the reader is referred to [32]. However, it should be pointed out that the PDFs of other sample PARCORs do not have such a simple expression.

2-5 DETECTION OF SIGNALS IN WHITE NOISE: SPECTRUM-BASED APPROACH

The use of power spectrum as a means of signal analysis has a long history and can be traced back to the early studies of light in physics. The modern study of spectral theory was motivated by the work of Sir Arthur Schuster in geophysics [36, 37]. Its solid theoretical foundation was laid by Wiener [38], Khintchine [39], Kolmogorov [11] and Cramér [40].

A significant advantage of spectral analysis over other methods of signal processing is its intuition and physical interpretability.

2-5-1 Formulation

Let $s(k)$ and $v(k)$ denote the signal and white noise components of the received data $x(k)$. Given n observation samples $x(k)$, $k = 1, \ldots, n$, signal detection can be formulated as testing between the following hypotheses:

$$\begin{aligned} H_0&: \quad x(k) = v(k) \\ H_1&: \quad x(k) = s(k) + v(k) \end{aligned} \tag{2-47}$$

To use high resolution spectral analysis techniques to solve this problem, the first step is to model the received data with an AR process of appropriate order, say M. That is,

$$x(k) = \sum_{m=1}^{M} a_m x(k-m) + e(k) \tag{2-48}$$

Here $\mathbf{a} = [a_1 \quad \cdots \quad a_M]^T$ are the model coefficients and $\{e(k)\}$ is a zero-mean complex white Gaussian process with variance σ_e^2. The model order M depends on the particular application of interest and is assumed to be known in advance. The power spectrum of the received process is then given by [20]

$$S_x(\omega) = \frac{\sigma_e^2}{|1 - a_1 e^{-j\omega} - \cdots - a_M e^{-jM\omega}|^2}$$

The power spectrum contains information necessary for signal detection. Under hypothesis H_0, the process $y(k)$ is a white noise sequence, thus $\mathbf{a} = \mathbf{0}$ and the power spectrum is constant over $\omega \in [-\pi, \pi]$. Specifically,

$$H_0: \quad S_x(\omega) = \sigma_e^2, \qquad -\pi \leqslant \omega \leqslant \pi$$

Under hypothesis H_1, $\mathbf{a} \neq \mathbf{0}$ and the power spectrum is peaky with some peaks higher than σ_e^2. Thus a reasonable decision rule is

$$S_x(\omega) \begin{cases} = \sigma_e^2, & H_0 \text{ is true} \\ > \sigma_e^2, & H_1 \text{ is true} \end{cases} \tag{2-49}$$

In practice, however, the model parameters $\mathbf{a}$ and σ_e^2 are not known and have to be estimated from the data. Thus signal detection must be based on a sample spectrum estimator:

$$\hat{S}_x(\omega) \underset{H_0}{\overset{H_1}{\gtrless}} V_1 \tag{2-50}$$

where V_1 is a preset threshold and the spectrum estimator

$$\hat{S}_x(\omega) = \frac{\hat{\sigma}_e^2}{|1 - \hat{a}_1 e^{-j\omega} - \cdots - \hat{a}_M e^{-jM\omega}|^2} \tag{2-51}$$

with $\hat{\mathbf{a}} = [\hat{a}_1 \quad \cdots \quad \hat{a}_M]^T$ and $\hat{\sigma}_e^2$ denoting the MLEs of $\mathbf{a}$ and σ_e^2. This test has been studied in [41]. The main drawback of such an approach is that its false alarm properties depend on the noise variance, leading to a processor of nonconstant false alarm rate. To alleviate this problem, Kay [42] suggests the use of a modified power spectrum for signal detection, which is defined as

$$\hat{S}(\omega) = \frac{1}{|1 - \hat{a}_1 e^{-j\omega} - \cdots - \hat{a}_M e^{-jM\omega}|^2} \tag{2-52}$$

which differs from (2-51) in that it ignores $\hat{\sigma}_e^2$.

The decision rule is now given by

$$\hat{S}(\omega) \underset{H_0}{\overset{H_1}{\gtrless}} V_2 \tag{2-53}$$

where again V_2 is a preset threshold. Expressions (2-50) and (2-53) are the two most popular forms of AR spectral analysis for signal detection.

2-5-2 The PDF of the Test Statistic

In spite of their practical importance and extensive applications, the detection performance of the above spectrum-based signal detectors are not well known. Most performance analysis has been based on computer simulations [43, 44]. The first attempt to obtain an analytical performance evaluation of the modified spectrum-based detector was made in [42]. Due to the theoretical difficulties, reference [42] considers only a simple case in which the model is assumed to be AR(1). Even in this case, the resulting performance expression is very complicated. The statistical variability of the AR spectrum estimator has been studied by Kromer [45], Akaike [46], Berle [47], and Baggeroer [48], in an attempt to obtain confidence bounds on the AR spectral density. The asymptotic distribution of spectral estimator obtained by Kromer and Berke involves assuming not only that the sample size tends to infinity but also that the AR model order approaches infinity [22, pp. 194–195]. Akaike obtains a similar result. Baggeroer obtains his results by assuming that the sample covariance matrix (multiplied by the sample size) is *Wishart distributed*. This assumption requires that the vector samples be independent. Obviously, the above results provide only a rough approximation to many practical situations. In the following we follow the approach in [49], which will lead to relatively simple but accurate results for the signal detection problem.

The test given in (2-53) is equivalent to the test defined by

$$l(\omega) \underset{H_1}{\overset{H_0}{\gtrless}} \Lambda_0 \tag{2-54}$$

where

$$\begin{aligned} l(\omega) &= |1 - \hat{a}_1 e^{-j\omega} - \cdots - \hat{a}_M e^{-jM\omega}|^2 \\ \Lambda_0 &= \text{preset threshold} \end{aligned} \tag{2-55}$$

To evaluate the detection performance of this test, we need to determine the PDF of $l(\omega)$. To this end, define

$$\mathbf{w} = \begin{pmatrix} e^{j\omega} \\ \vdots \\ e^{jM\omega} \end{pmatrix}, \qquad \hat{\mathbf{a}} = \begin{pmatrix} \hat{a}_1 \\ \vdots \\ \hat{a}_M \end{pmatrix} \tag{2-56}$$

and rewrite $l(\omega)$ as

$$l(\omega) = (1 - \mathbf{w}^H\hat{\mathbf{a}})^H(1 - \mathbf{w}^H\hat{\mathbf{a}}) \tag{2-57}$$

From Section 2-4, we know that

$$\hat{\mathbf{a}} \sim CN_M(\mathbf{a}, \sigma_e^2[n\mathbf{R}_M]^{-1})$$

with $\mathbf{R}_M$ denoting the M-by-M covariance matrix of the received data. Here we note that, as a linear combination of $\hat{\mathbf{a}}$, the variable $y \overset{\Delta}{=} 1 - \mathbf{w}^H\hat{\mathbf{a}}$ is also complex Gaussian. Its mean and variance can be easily calculated as

$$\begin{aligned} \mu &\overset{\Delta}{=} E[y] = 1 - \mathbf{w}^H\mathbf{a} \\ \sigma_y^2 &\overset{\Delta}{=} \mathrm{Var}(y) = \mathbf{w}^H \,\mathrm{Var}(\hat{\mathbf{a}})\mathbf{w} \\ &= \mathbf{w}^H\sigma_e^2[n\mathbf{R}_M]^{-1}\mathbf{w} \end{aligned} \tag{2-58}$$

It follows that

$$y = (1 - \mathbf{w}^H\hat{\mathbf{a}}) \sim CN(\mu, \sigma_y^2) \tag{2-59}$$

Combining the results in (2-57) and (2-59), we have

$$l = \frac{\sigma_y^2}{2} z \tag{2-60}$$

where z is defined as

$$z = 2|y|^2/\sigma_y^2 \tag{2-61}$$

According to Giri [52], the variable z is noncentral chi-square distributed with 2 degrees of freedom and noncentrality parameter

$$\lambda = 2|\mu|^2/\sigma_y^2 \tag{2-62}$$

or simply denoted by $z \sim \chi^2(2, \lambda)$. Explicitly we write

$$p(z) = \frac{1}{2} I_0(\sqrt{\lambda z}) \exp\left[-\frac{1}{2} (\lambda + z) \right], \qquad y > 0 \qquad (2\text{-}63)$$

where $I_0(.)$ denotes the modified Bessel function of order 0.

It follows from (2-60) and (2-63) that the PDF of l is given by

$$p(l) = \frac{1}{\sigma_y^2} I_0\left(2\sqrt{\frac{|\mu|^2}{\sigma_y^2} \frac{l}{\sigma_y^2}} \right) \exp\left[-\left(\frac{|\mu|^2}{\sigma_y^2} + \frac{l}{\sigma_y^2} \right) \right], \qquad l > 0 \quad (2\text{-}64)$$

This is recognized to be the type I Bessel function distribution [50]. We may use either (2-63) or (2-64) to determine the detection performance of the spectrum-based test.

2-5-3 The False Alarm Rate

Under hypothesis H_0, the received signal consists of white noise alone. It turns out that

$$\mathbf{a} = \mathbf{0}$$

$$\mathbf{R}_M = \sigma_e^2 \mathbf{I}$$

Inserting these results into (2.58), we have

$$\mu = 1 , \qquad \sigma_y^2 = M/n \qquad (2\text{-}65)$$

Combining (2-60)–(2-62) and (2-65), we obtain the PDF of l under hypothesis H_0:

$$l \sim \frac{M}{2n} \chi^2\left(2, \frac{2n}{M}\right) \qquad (2\text{-}66)$$

For a constant false alarm rate (CFAR), $P_{\text{fa}} = \alpha$, let χ_α^2 denote the lower $100\alpha\%$ point of the PDF of $\chi^2(2, 2n/M)$. Then the threshold for l is given by

$$l_\alpha = \frac{M}{2n} \chi_\alpha^2 \qquad (2\text{-}67)$$

2-5-4 The Detection Probability

The detection probability can be calculated from (2-54), (2-60), and (2-63), giving

$$
\begin{aligned}
P_d &= \{l < l_\alpha\} \\
&= \Pr\left\{ z < \frac{2}{\sigma_y^2} l_\alpha \right\} \\
&= \Pr\left\{ z < \frac{M}{n\sigma_y^2} \chi_\alpha^2 \right\} \\
&= \int_0^{(M/n\sigma_y^2)\chi_\alpha^2} \frac{1}{2} I_0(\sqrt{\lambda z}) \exp\left[-\frac{1}{2}(\lambda + z) \right] dz
\end{aligned} \tag{2-68}
$$

where the parameter λ has been defined in (2-62), and σ_y^2 defined in (2-58).

To examine the analytical results obtained above, we apply the spectrum-based detection procedure to the case of a sinusoid in white noise. For this situation, the received signal model is

$$
x(k) = ae^{j(\omega k + \varphi)} + n(k)\,, \qquad k = 1, \ldots, N
$$

where

$n(k)$ = complex white Gaussian sequence with zero mean and unknown variance

a = unknown amplitude

$\omega = 0.24\pi$

φ = random phase uniformly distributed over $(-\pi, \pi)$

$N = 25$

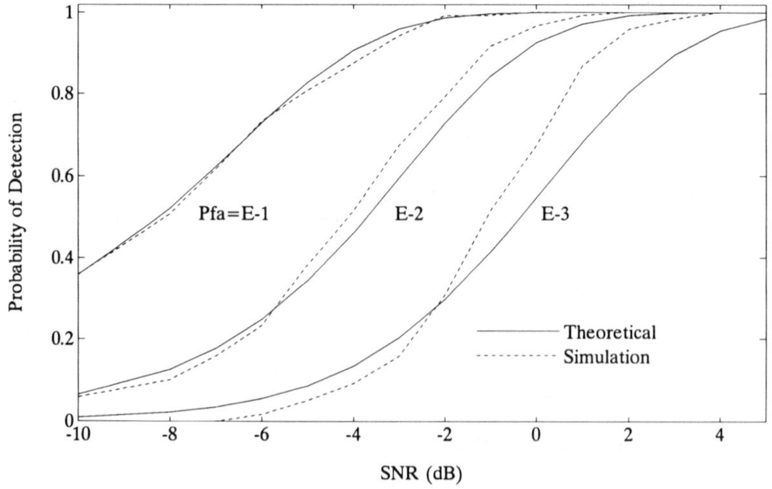

Figure 2-2 Theoretical performance of the AR spectrum-based detector compared with simulation results.

Since the power spectrum of $x(k)$ is unimodal, an AR model of order 3 will provide a good description of it. Thus we choose $M = 3$. We may use (2-67) to determine the theoretical threshold values and then use (2-68) to evaluate the detection probability.

Figure 2-2 compares the theoretical performance with that obtained through computer simulations. The simulation results were obtained by 150 independent trials for each signal-to-noise ratio (SNR) point. The results show good agreement between theory and simulated performance.

2-6 SIGNAL DETECTION IN COLORED NOISE: METHOD I

The type of interference encountered in most practical situations is correlated, and has unknown statistics. Under these circumstances, there are different ways to formulate the detection problem, and the detection performance will strongly depend on how much information is extracted from the environment of interest. In this and the subsequent sections, we describe two such detection schemes. The first scheme pertains to the processing of radar signals obtained along a range ring without exploiting useful information from the neighboring range cells. The second scheme, on the other hand, treats radar echoes as a set of two-dimensional data and exploits the information from neighboring range cells to enhance signal detection. As a result, the performance of Scheme II will be superior. Nevertheless, Scheme I has its own virtues. In particular, it provides a Gram–Schmidt-type of orthogonalization procedure to formulate the detection problem, supplying not only a deeper insight into the problem at hand, but also providing a computationally efficient way to implement the optimum receiver for the detection of a random signal in additive colored noise.

The test statistics for both schemes are constructed, assuming the use of AR modeling of the received data; it is therefore computationally efficient.

2-6-1 An Innovation Expression of the Optimum Receiver

When the covariance matrices of both signal and noise are completely specified, the optimum receiver for Gaussian signal in Gaussian noise is well defined and can be derived from the likelihood ratio principle, ending up with a receiver called the *estimator-correlator* [72]. This name comes from the fact that the operations involved in the receiver are first to estimate the signal component and then correlate it with the received data. One of the major drawbacks of such a receiver is its computational inefficiency. The implementation becomes even more difficult when some of the above statistics are unknown.

On the other hand, modern filter theory is so developed today that linear adaptive prediction and filtering operations are just simple routines. Consequently, it is desirable to express the optimal test, such that it can be easily implemented with an adaptive algorithm on a computer. One such version is the *innovations-based detection algorithm* (IBDA), described next.

Let $\mathbf{x} = [x_N \quad \cdots \quad x_1]^T$ be a sequence of complex-valued baseband data. Two hypotheses are postulated concerning the presence of a random signal vector $\mathbf{s} = [s_N \quad \cdots \quad s_1]^T$ in the received data $\mathbf{x}$.

$$\begin{aligned} H_0&: \quad \mathbf{x} = \mathbf{c} \\ H_1&: \quad \mathbf{x} = \mathbf{s} + \mathbf{c} \end{aligned} \tag{2-69}$$

where $\mathbf{c}$ represents the interference vector characterized by

$$\mathbf{c} \sim CN_N(\mathbf{0}, \mathbf{R}_c)$$

and $\mathbf{s}$ is Gaussian, independent of $\mathbf{c}$, such that

$$\mathbf{s} \sim CN_N(\mathbf{0}, \mathbf{R}_s)$$

Thus the PDF of $\mathbf{x}$ under the two hypotheses can be expressed respectively, as

$$P_0(\mathbf{x}|H_0) = \frac{1}{\pi^N \det(\mathbf{R}_c)} \exp(-\mathbf{x}^H \mathbf{R}_c^{-1} \mathbf{x})$$

and

$$P_1(\mathbf{x}|H_1) = \frac{1}{\pi^N \det(\mathbf{R}_s + \mathbf{R}_c)} \exp(-\mathbf{x}^H [\mathbf{R}_s + \mathbf{R}_c]^{-1} \mathbf{x}) \tag{2-70}$$

If we take the ratio of these two PDFs, perform log-operation, and then simplify, we obtain an estimator-correlator described by

$$\begin{aligned} l_{\mathrm{EC}} &= \mathbf{x}^H (\mathbf{R}_c^{-1} - [\mathbf{R}_s + \mathbf{R}_c]^{-1}) \mathbf{x} \\ &= \hat{\mathbf{s}}^H \mathbf{R}_c^{-1} \mathbf{x} \end{aligned} \tag{2-71}$$

where $\hat{\mathbf{s}}$ is a smoothed minimum mean-square estimate of $\mathbf{s}$, given by

$$\hat{\mathbf{s}} = \mathbf{R}_s (\mathbf{R}_s + \mathbf{R}_c)^{-1} \mathbf{x} \tag{2-72}$$

If, however, we convert the correlated received data to an innovation process before constructing a likelihood ratio test, we obtain an alternative expression for the optimum receiver. This corresponds to a *Cholesky decomposition* of the covariance matrix of the received data. Consider first the case of hypothesis H_1. For the positive-definite Hermitian matrix $(\mathbf{R}_s + \mathbf{R}_c)^{-1}$, there exists a unit upper triangular matrix $\mathbf{P}_1$ such that

$$(\mathbf{R}_s + \mathbf{R}_c)^{-1} = \mathbf{P}_1^H \mathbf{D}_1^{-1} \mathbf{P}_1 \tag{2-73}$$

where $\mathbf{D}_1$ is a diagonal matrix defined by

$$\mathbf{D}_1 = \text{diag}(d_1^2(N), \ldots, d_1^2(1))$$

The innovation vector $\mathbf{e}_1 = [e_1(N) \quad \cdots \quad e_1(1)]^T$ of $\mathbf{x}$, under hypothesis H_1, is

$$\mathbf{e}_1 = \mathbf{P}_1 \mathbf{x} \tag{2-74}$$

Note that the Jacobian of this transformation is unity. Thus from (2-70), we have

$$p_1(\mathbf{e}_1 | H_1) = \left\{ \prod_{k=1}^{N} [\pi d_1^2(k)]^{-1} \right\} \exp\left(-\sum_{k=1}^{N} \frac{|e_1(k)|^2}{d_1^2(k)} \right) \tag{2-75}$$

Likewise, we may obtain an innovation expression of the likelihood function under hypothesis H_0, as shown by

$$p_0(\mathbf{e}_0 | H_0) = \left\{ \prod_{k=1}^{N} [\pi d_0^2(k)]^{-1} \right\} \exp\left(-\sum_{k=1}^{N} \frac{|e_0(k)|^2}{d_0^2(k)} \right) \tag{2-76}$$

where $\mathbf{e}_0$ is the innovation vector of received data under H_0, and $d_0^2(k)$ is defined in a way similar to $d_1^2(k)$ through the Cholesky decomposition of $\mathbf{R}_c^{-1}$. From (2-75) and (2-76), the log-likelihood ratio is obtained as

$$L = \sum_{k=1}^{N} \left[\ln\left(\frac{d_0^2(k)}{d_1^2(k)} \right) + \frac{|e_0(k)|^2}{d_0^2(k)} - \frac{|e_1(k)|^2}{d_1^2(k)} \right] \tag{2-77}$$

The Cholesky decomposition described above is essentially the Gram–Schmidt orthogonalization procedure. This procedure can be easily implemented using a linear prediction-error filter (PEF). In particular, by writing the upper triangular matrix $\mathbf{P}_1$ explicitly in terms of the coefficients of linear prediction filters of varying orders, we have

$$\mathbf{P}_1 = \begin{pmatrix} 1 & -a_{N-1}(1) & -a_{N-1}(2) & \cdots & -a_{N-1}(N-1) \\ & 1 & -a_{N-2}(1) & \cdots & -a_{N-2}(N-2) \\ & & \ddots & & \vdots \\ & \mathbf{0} & & \ddots & \\ & & & & 1 \end{pmatrix} \tag{2-78}$$

With this interpretation, $e_1(1), e_1(2), \ldots, e_1(N)$ are respectively the outputs of the PEFs of orders $0, 1, \ldots, N-1$, and $d_1^2(1), \ldots, d_1^2(N)$ represent their corresponding variances. Therefore, $\mathbf{P}_1$ represents a *linear time-varying filter*.

2-6-2 The IBDA Algorithm for AR Signal Processes

The test statistic given in (2-77) provides a general innovation expression of the optimum receiver; hence the name innovations-based detection algorithm (IBDA). It is applicable to a stationary process of an arbitrary covariance matrix. However, the IBDA becomes simple if the received data under the two hypotheses can be described by AR processes. In such a situation, we may write

$$\begin{aligned} H_0\colon\ x(k) &= \sum_{i=1}^{M_0} b_{M_0}(i)x(k-i) + e_0(k) \\ H_1\colon\ x(k) &= \sum_{i=1}^{M} a_M(i)x(k-i) + e_1(k) \end{aligned} \tag{2-79}$$

where

$$\begin{aligned} e_0(k) &\sim \text{i.i.d. } CN(0, \sigma_0^2) \\ e_1(k) &\sim \text{i.i.d. } CN(0, \sigma_1^2) \end{aligned} \tag{2-80}$$

Without loss of generality, we assume that $M > M_0$. Moreover, under the AR assumption, $\mathbf{P}_1$ and $\mathbf{P}_0$ become *banded matrices*. In particular, we have

$$\mathbf{P}_1 = \begin{pmatrix} 1 & -\mathbf{a}_M^T & & \mathbf{0} \\ & \ddots & \ddots & \\ & & 1 & -\mathbf{a}_M^T \\ \mathbf{0} & & & \mathbf{Q}_1 \end{pmatrix}, \qquad \mathbf{P}_0 = \begin{pmatrix} 1 & -\mathbf{b}_{M_0}^T & & \mathbf{0} \\ & \ddots & \vdots & \\ & & 1 & -\mathbf{b}_{M_0}^T \\ \mathbf{0} & & & \mathbf{Q}_0 \end{pmatrix} \tag{2-81}$$

with

$$\mathbf{Q}_1 = \begin{pmatrix} 1 & -a_M(1) & \cdots & -a_M(M) \\ & \ddots & & \vdots \\ & \mathbf{0} & 1 & -a_1(1) \\ & & & 1 \end{pmatrix}, \quad \mathbf{Q}_0 = \begin{pmatrix} 1 & -b_{M_0}(1) & \cdots & -b_{M_0}(M_0) \\ & \ddots & & \vdots \\ & \mathbf{0} & 1 & -b_{M_0}(1) \\ & & & 1 \end{pmatrix}$$

$$\mathbf{a}_M = [a_M(1) \quad \cdots \quad a_M(M)]^T, \qquad \mathbf{b}_{M_0} = [b_{M_0}(1) \quad \cdots \quad b_{M_0}(M_0)]^T \tag{2-82}$$

Obviously, the innovation process remains stationary until the order $N - M + 1$ PEF filter, resulting in

$$d_1^2(k) = \sigma_1^2, \qquad k = 1, \ldots, N - M \tag{2-83}$$

Similarly, we can show that

$$d_0^2(k) = \sigma_0^2, \qquad k = 1, \ldots, N - M_0 \tag{2-84}$$

Thus, if we define

$$\gamma_1^2(k) = \begin{cases} 1, & 1 \leqslant k \leqslant N - M \\ d_1^2(k)/\sigma_1^2, & N - M < k \leqslant N \end{cases} \tag{2-85}$$

$$\gamma_0^2(k) = \begin{cases} 1, & 1 \leqslant k \leqslant N - M_0 \\ d_0^2(k)/\sigma_0^2, & N - M_0 < k \leqslant N \end{cases} \tag{2-86}$$

then (2-77) can be expressed as

$$L = -\sum_{k=1}^{N} \left\{ \ln\left(\frac{\sigma_1^2}{\sigma_0^2}\right) + \ln\left(\frac{\gamma_1^2(k)}{\gamma_0^2(k)}\right) + \frac{|e_1(k)|^2}{\sigma_1^2 \gamma_1^2(k)} - \frac{|e_0(k)|^2}{\sigma_0^2 \gamma_0^2(k)} \right\}$$

In general, we have $\max\{M_0, M\} \ll N$; thus L is dominated by the terms with index k less than (N-$\max\{M_0, M\}$) and the second term inside the braces can be ignored, yielding

$$L = -\sum_{k=1}^{N} \left\{ \ln\left(\frac{\sigma_1^2}{\sigma_0^2}\right) + \frac{|e_1(k)|^2}{\sigma_1^2} - \frac{|e_0(k)|^2}{\sigma_0^2} \right\} \tag{2-87}$$

2-6-3 Three Cases

In expression (2-87), the innovation process $\{e_i(k), i = 0, 1\}$ is evaluated using the true AR coefficients under the two hypotheses, H_0 and H_1. In many situations, however, certain approximations to the innovation processes must be used. Let $\{\hat{e}_i(k)\}$ denote the innovation process obtained using estimated AR coefficients under hypothesis H_i, $i = 0, 1$. To distinguish the true innovation process and its estimate, we denote

$$\begin{aligned} \hat{\sigma}_i^2 &= \frac{1}{N} \sum_{k=1}^{N} |e_i(k)|^2/\gamma_i^2(k) \\ \hat{\hat{\sigma}}_i^2 &= \frac{1}{N} \sum_{k=1}^{N} |\hat{e}_i(k)|^2/\gamma_i^2(k) \end{aligned} \tag{2-88}$$

The innovation expression of the optimum receiver provides a method for its approximate implementation. Three cases of practical interest follow:

Case 1. The AR parameters under H_0 are completely specified, with coefficient vector $\mathbf{b}_{M_0} = \boldsymbol{\beta}_0$ and error variance σ_0^2. In this case, replacing $\{e_1(k)\}$ with $\{\hat{e}_1(k)\}$, and setting $\sigma_1^2 = \hat{\hat{\sigma}}_1^2$, we obtain

$$L_1 = N\left\{\ln\left(\frac{\sigma_0^2}{\hat{\hat{\sigma}}_1^2}\right) + \frac{\hat{\sigma}_0^2}{\sigma_0^2} - 1\right\} \tag{2-89}$$

Case 2. Only σ_0^2 is specified. In this case we use $\hat{e}_0(k)$ and $\hat{e}_1(k)$ to replace $e_0(k)$ and $e_1(k)$, and set $\sigma_1^2 = \hat{\hat{\sigma}}_1^2$, yielding the result

$$L_2 = N\left\{\ln\left(\frac{\sigma_0^2}{\hat{\hat{\sigma}}_1^2}\right) + \left(\frac{\hat{\hat{\sigma}}_0^2}{\sigma_0^2}\right) - 1\right\} \tag{2-90}$$

Case 3. All AR parameters are unknown, except for the AR orders. In this third and final case, we use $\{\hat{e}_i(k)\}$ in place of $\{e_i(k)\}$, and set $\sigma_i^2 = \hat{\hat{\sigma}}_i^2$, $i = 0, 1$, obtaining

$$L_3 = N\left\{\ln\left(\frac{\hat{\hat{\sigma}}_0^2}{\hat{\hat{\sigma}}_1^2}\right)\right\} \tag{2-91}$$

The three cases described above represent different extents of prior information that we have about the underlying process. Clearly, the more prior information we have, the better will the detection performance be.

2-6-4 Relation to the Generalized Likelihood Ratio Test

It is noteworthy that the test statistics L_1, L_2, and L_3 can also be obtained by using the *generalized likelihood ratio test* (GLRT). For the detection problem of interest, the signal model under hypothesis H_0 can be viewed as a restricted version of that for H_1. Thus, all three cases described above can be formulated as testing certain parameters in an AR model. To simplify notation, we use AR$(M, \mathbf{b}, \sigma^2)$ to denote an AR model for hypothesis H_1, with coefficient vector $\mathbf{b} = [a_M(1) \quad \cdots \quad a_M(M)]^T$ and error variance σ^2. Let the vector $\mathbf{b}$ be partitioned, such that

$$\mathbf{b} = \begin{bmatrix} \mathbf{b}_0 \\ \mathbf{b}_1 \end{bmatrix} \tag{2-92}$$

where $\mathbf{b}_0$ is an M_0 vector. Thus the three cases can be rewritten as

Case 1.

$$\begin{aligned} H_0&: \quad \text{AR}(M, \mathbf{b}, \sigma^2), \qquad \mathbf{b} = \begin{pmatrix} \boldsymbol{\beta}_0 \\ \mathbf{0} \end{pmatrix}, \qquad \sigma^2 = \sigma_0^2 \\ H_1&: \quad \text{AR}(M, \mathbf{b}, \sigma^2), \qquad \mathbf{b} \neq \begin{pmatrix} \boldsymbol{\beta}_0 \\ \mathbf{0} \end{pmatrix}, \qquad \sigma^2 \neq \sigma_0^2 \end{aligned} \tag{2-93}$$

Case 2.

$$\begin{aligned} H_0&: \ \mathrm{AR}(M, \mathbf{b}, \sigma^2), \quad \mathbf{b}_1 = \mathbf{0}, \quad \sigma^2 = \sigma_0^2 \\ H_1&: \ \mathrm{AR}(M, \mathbf{b}, \sigma^2), \quad \mathbf{b}_1 \neq \mathbf{0}, \quad \sigma^2 \neq \sigma_0^2 \end{aligned} \tag{2-94}$$

Case 3.

$$\begin{aligned} H_0&: \ \mathrm{AR}(M, \mathbf{b}, \sigma^2), \quad \mathbf{b}_1 = \mathbf{0} \\ H_1&: \ \mathrm{AR}(M, \mathbf{b}, \sigma^2), \quad \mathbf{b}_1 \neq \mathbf{0} \end{aligned} \tag{2-95}$$

With these restrictions, it is easy to show that the *log-likelihood ratio tests* for cases 1, 2, and 3 have exactly the same forms as L_1, L_2, and L_3 given above. More specifically, we have:

Case 1.

$$\ln\left\{\frac{\max_{\mathbf{b},\sigma^2} p(\mathbf{x}|\mathbf{b}, \sigma^2)}{p(\mathbf{x}|M_0, \beta_0, \sigma_0^2)}\right\} = N\left\{\ln\left(\frac{\sigma_0^2}{\hat{\hat{\sigma}}_1^2}\right) + \frac{\hat{\sigma}_0^2}{\sigma_0^2} - 1\right\} \tag{2-96}$$

Case 2.

$$\ln\left\{\frac{\max_{\mathbf{b},\sigma^2} p(\mathbf{x}|\mathbf{b}, \sigma^2)}{\max_{\mathbf{b}_0} p(\mathbf{x}|M_0, \mathbf{b}_0, \sigma_0^2)}\right\} = N\left\{\ln\left(\frac{\sigma_0^2}{\hat{\hat{\sigma}}_1^2}\right) + \frac{\hat{\hat{\sigma}}_0^2}{\sigma_0^2} - 1\right\} \tag{2-97}$$

Case 3.

$$\ln\left\{\frac{\max_{\mathbf{b},\sigma^2} p(\mathbf{x}|\mathbf{b}, \sigma^2)}{\max_{\mathbf{b}_0,\sigma^2} p(\mathbf{x}|M_0, \mathbf{b}_0, \sigma^2)}\right\} = N \ln\left(\frac{\hat{\hat{\sigma}}_0^2}{\hat{\hat{\sigma}}_1^2}\right) \tag{2-98}$$

For a reason that will become apparent later, we may modify L_1, L_2, and L_3 as

$$l_{\mathrm{IBDA1}} = 2N\left\{\ln\left(\frac{\sigma_0^2}{\hat{\hat{\sigma}}_1^2}\right) + \frac{\hat{\sigma}_0^2}{\sigma_0^2} - 1\right\} \tag{2-99}$$

$$l_{\mathrm{IBDA2}} = 2N\left\{\ln\left(\frac{\sigma_0^2}{\hat{\hat{\sigma}}_1^2}\right) + \frac{\hat{\hat{\sigma}}_0^2}{\sigma_0^2} - 1\right\} \tag{2-100}$$

$$l_{\mathrm{IBDA3}} = \left(\frac{\hat{\sigma}_0^2 - \hat{\sigma}_1^2}{\hat{\sigma}_1^2} \right) \frac{N - M}{M - M_0} \tag{2-101}$$

The decision rule is simply

$$l_{\mathrm{IBDA}i} \underset{H_0}{\overset{H_1}{\gtrless}} TH_i\,, \qquad i = 1, 2, 3 \tag{2-102}$$

Obviously, a detection strategy based on $l_{\mathrm{IBDA}i}$ is equivalent to that based on L_i, since the former is a monotonic function of the latter. However, the theoretical analysis based on $l_{\mathrm{IBDA}i}$, $i = 1, 2, 3$ is easier to handle.

2-6-5 An Asymptotic Expression of the Likelihood Function

To evaluate the statistical performance of the likelihood ratio tests derived in the previous section, it is necessary to determine the PDF of the likelihood ratios in (2-99)–(2-101) under both H_0 and H_1. It is however, difficult to directly evaluate the PDFs of the statistics $l_{\mathrm{IBDA}i}$, $i = 1, 2, 3$. This difficulty can be avoided if we derive an alternative expression for l. Specifically, the original statistics, the l are obtained from the likelihood functions of the received data under hypotheses H_0 and H_1. For large samples, these likelihood functions can be expressed as a symptotic functions of the sample estimates of the unknown parameters, from which an alternative expression can be derived.

This approach is originally due to Wilks [60] and generalized by Wald [61]. In both references, only real-valued data are used. The extension to complex autoregressive data is treated in [51, 56].

Since the likelihood function for hypothesis H_1 is the same for the three cases and since it contains hypothesis H_0 as a special case, it therefore suffices to consider its asymptotic expression. Denote the parameter vector by

$$\theta = \begin{pmatrix} \mathbf{b} \\ \sigma^2 \end{pmatrix} \tag{2-103}$$

and its MLE by

$$\hat{\theta} = \begin{pmatrix} \hat{\mathbf{b}} \\ \hat{\sigma}^2 \end{pmatrix} \tag{2-104}$$

Let Ω denote the M-by-M covariance matrix of the received data normalized by σ^2. From Section 2-2, we recognize that Ω depends only on $\mathbf{b}$. To emphasize this, we write $\Omega(\mathbf{b})$. Then for large samples, the likelihood function can be expressed in terms of θ.

Theorem 2-3 As $N \to \infty$, the likelihood function is given by

$$p(\mathbf{x}|\mathbf{b}, \sigma^2) = c \exp\{-(\hat{\mathbf{b}} - \mathbf{b})^H \mathbf{I}(\bar{\mathbf{b}})(\hat{\mathbf{b}} - \mathbf{b}) - (\hat{\sigma}^2 - \sigma^2)^2 \mathbf{I}(\bar{\sigma}^2)\} \tag{2-105}$$

where c is a constant independent of θ, $\bar{\mathbf{b}}$ is a point on the line segment linking $\mathbf{b}$ and its MLE $\hat{\mathbf{b}}$, and $\bar{\sigma}^2$ is located between σ^2 and $\hat{\sigma}^2$. The matrices $\mathbf{I}(\bar{\mathbf{b}})$ and $\mathbf{I}(\bar{\sigma}^2)$ are the sample estimates of the Fisher information matrices evaluated at $\bar{\mathbf{b}}$ and $\bar{\sigma}^2$, which are given by

$$\begin{aligned} \mathbf{I}(\mathbf{b}) &= N\mathbf{R}_M/\sigma^2 = N\Omega \\ \mathbf{I}(\sigma^2) &= \frac{1}{2} N\sigma^{-4} \end{aligned} \tag{2-106}$$

The procedure to establish this result is similar to that by which we derive a more general result in the next section; it is therefore omitted here.

2-6-6 Statistical Performance: Case 1

We are now in a position to establish the alternative test statistics by using the asymptotic expression given in (2-105).

Case 1. It is easily seen that under hypothesis H_1,

$$\max_{\mathbf{b}, \sigma^2} \{p(\mathbf{x} \mid \mathbf{b}, \sigma^2)\} = c \tag{2-107}$$

Under H_0, all parameters are specified; hence the likelihood is simply equal to

$$p(\mathbf{x} \mid \beta_0, \sigma_0^2) = c\{\exp(-D)\} \exp\{-(\hat{\sigma}^2 - \sigma_0^2)^2 \mathbf{I}(\bar{\sigma}^2)\} \tag{2-108}$$

where D is a scaler defined as

$$D = \left[\hat{\mathbf{b}} - \begin{pmatrix} \beta_0 \\ \mathbf{0} \end{pmatrix}\right]^H \mathbf{I}(\bar{\mathbf{b}}) \left[\hat{\mathbf{b}} - \begin{pmatrix} \beta_0 \\ \mathbf{0} \end{pmatrix}\right] \tag{2-109}$$

From the above, the log-likelihood ratio multiplied by two is obtained as

$$l_1 = 2[D + (\hat{\sigma}^2 - \sigma_0^2)^2 \mathbf{I}(\bar{\sigma}^2)] \tag{2-110}$$

Note that $N \to \infty$, $\bar{\mathbf{b}} \to \mathbf{b}$ and $\bar{\sigma}^2 \to \sigma^2$, and $\mathbf{I}(\bar{\mathbf{b}})$ and $\mathbf{I}(\bar{\sigma}^2)$ approach the true values of the Fisher information matrices, given by

$$\mathbf{I}(\mathbf{b}) = N\Omega$$

$$\mathbf{I}(\sigma^2) = \frac{1}{2} N\sigma^{-4}$$

Combining these results with (2-109) and (2-110), an alternative expression of l_{IBDA1} is obtained:

$$l_{\text{IBDA1}} = 2\left[\hat{\mathbf{b}} - \begin{pmatrix} \beta_0 \\ \mathbf{0} \end{pmatrix}\right]^H [N\Omega]\left[\hat{\mathbf{b}} - \begin{pmatrix} \beta_0 \\ \mathbf{0} \end{pmatrix}\right] + \frac{(\hat{\sigma}^2 - \sigma_0^2)^2}{\sigma^4/N} \tag{2-111}$$

We recall from Section 2-4-2 that

$$\text{cov}(\hat{\mathbf{b}}) = (N\Omega)^{-1}$$

$$\text{cov}(\hat{\sigma}^2) = \sigma^4/N$$

and $\hat{\mathbf{b}}$ and $\hat{\sigma}^2$ are independent for large N. Thus both terms on the right side of (2-111) are chi-square distributed, and so is the l_{IBDA1}. More specifically, we may write

$$l_{\text{IBDA1}} \sim \chi^2(2M+1, \lambda_1) \tag{2-112}$$

where λ_1 is the noncentrality parameter given by

$$\lambda_1 = 2N\left[\mathbf{b} - \begin{pmatrix} \beta_0 \\ \mathbf{0} \end{pmatrix}\right]^H \Omega\left[\mathbf{b} - \begin{pmatrix} \beta_0 \\ \mathbf{0} \end{pmatrix}\right] + \frac{N(\sigma^2 - \sigma_0^2)^2}{\sigma^4} \tag{2-113}$$

The first term on the right side of (2-113) is the noncentrality parameter of the first term on the right side of (2-111). This noncentrality parameter is obtained by directly using the results of Giri [52]. The second term on the right side of (2-113) is the noncentrality parameter of the second term on the right side of (2-111).

Let $f_{\chi^2}(\chi^2, \nu, \lambda)$ denote the PDF of a chi-square variable $\chi^2(\nu, \lambda)$. Note that under hypothesis H_0,

$$\mathbf{b} = \begin{pmatrix} \beta \\ \mathbf{0} \end{pmatrix}$$

and $\sigma^2 = \sigma_0^2$, and thus $\lambda_1 = 0$. This implies that under hypothesis H_0, l_{IBDA1} becomes a central chi-square distributed variable. Now we can determine the probabilities of false alarm and detection for a given threshold TH_1 as follows:

$$\begin{aligned} P_{\text{fa}} &= \int_{TH_1}^{\infty} f_{\chi^2}(\chi^2; 2M+1, 0)\, d\chi^2 \\ P_d &= \int_{TH_1}^{\infty} f_{\chi^2}(\chi^2; 2M+1, \lambda_1)\, d\chi^2 \end{aligned} \tag{2-114}$$

2-6-7 Statistical Performance: Case 2

Proceeding as before, we have under H_1,

$$\max_{\mathbf{b},\sigma^2}\{p(\mathbf{x}\,|\,\mathbf{b},\sigma^2)\} = c \tag{2-115}$$

The evaluation of the maximum likelihood under H_0 is similar but needs a few more steps. In particular, by substituting the conditions specified in (2-94) into (2-105), we have

$$\begin{aligned}\max_{H_0}\{p(\mathbf{x}\,|\,\mathbf{b},\sigma^2)\} &= c\max_{\mathbf{b}_0}\{\exp(-D)\}\exp\{-(\hat{\sigma}^2-\sigma_0^2)^2 I(\bar{\sigma}^2)\}\\ &= c\exp(-\min_{\mathbf{b}_0} D)\exp\{-(\hat{\sigma}^2-\sigma_0^2)^2 I(\bar{\sigma}^2)\}\end{aligned} \tag{2-116}$$

where D is a scalar defined by

$$D = \left[\begin{pmatrix}\hat{\mathbf{b}}_0\\ \hat{\mathbf{b}}_1\end{pmatrix} - \begin{pmatrix}\mathbf{b}_0\\ \hat{\mathbf{0}}\end{pmatrix}\right]^H \mathbf{I}(\bar{\mathbf{b}})\left[\begin{pmatrix}\hat{\mathbf{b}}_0\\ \hat{\mathbf{b}}\end{pmatrix} - \begin{pmatrix}\mathbf{b}_0\\ \hat{\mathbf{0}}\end{pmatrix}\right] \tag{2-117}$$

To optimize D, we partition the M-by-M matrix such that

$$\Omega = \begin{bmatrix}\Omega_{00} & \Omega_{01}\\ \Omega_{10} & \Omega_{11}\end{bmatrix} \tag{2-118}$$

where Ω_{00} is M_0-by-M_0. Substituting $\mathbf{I}(\mathbf{b}) = N\Omega$ given in (2-106) into D, taking the derivative with respect to $\boldsymbol{b}_0$, and simplifying, we obtain

$$\min_{\mathbf{b}_0} D = \hat{\mathbf{b}}_1^H \mathbf{V}^{-1}(\bar{\mathbf{b}})\hat{\mathbf{b}}_1 \tag{2-119}$$

where $\mathbf{V}(\mathbf{b})$ is defined as

$$\mathbf{V}(\mathbf{b}) = [N(\Omega_{11} - \Omega_{01}^H\Omega_{00}^{-1}\Omega_{01})]^{-1} \tag{2-120}$$

We recognize that $\mathbf{V}(\mathbf{b})$ is the covariance matrix of the MLE $\mathbf{b}_1$. Combining (2-120), (2-119), (2-116), and (2-115), a new log-likelihood ratio statistic results, namely,

$$l_{\text{IBDA2}} = 2\hat{\mathbf{b}}_1^H \mathbf{V}^{-1}(\bar{\mathbf{b}})\hat{\mathbf{b}}_1 + (\hat{\sigma}^2 - \sigma_0^2)^2 \mathbf{I}(\bar{\sigma}^2)$$

Proceeding as before, and passing N to infinity, we have $\bar{\mathbf{b}} \to \mathbf{b}$ and $\bar{\sigma}^2 \to \sigma^2$, and hence

$$l_{\text{IBDA2}} = 2\hat{\mathbf{b}}_1^H \mathbf{V}^{-1}(\mathbf{b})\hat{\mathbf{b}}_1 + \frac{N(\hat{\sigma}^2 - \sigma_0^2)^2}{\sigma^4} \tag{2-121}$$

From general statistical theory and using the result by Giri [52] again, it can be shown that

$$l_{\text{IBDA2}} \sim \chi^2(2M - 2M_0 + 1, \lambda_2) \tag{2-122}$$

where

$$\lambda_2 = 2N\mathbf{b}_1^H(\Omega_{11} - \Omega_{01}^H \Omega_{00}^{-1} \Omega_{01})\mathbf{b}_1 + \frac{N(\sigma^2 - \sigma_0^2)^2}{\sigma^4} \tag{2-123}$$

Thus for a given threshold TH_2, the probabilities of false alarm and detection can be determined by using the respective formulas

$$\begin{aligned} P_{\text{fa}} &= \int_{TH_2}^{\infty} f_{\chi^2}(\chi^2; 2M - 2M_0 + 1, 0)\, d\chi^2 \\ P_d &= \int_{TH_2}^{\infty} f_{\chi^2}(\chi^2; 2M - 2M_0 + 1, \lambda_2)\, d\chi^2 \end{aligned} \tag{2-124}$$

2-6-8 Statistical Performance: Case 3

The performance analysis for l_{IBDA3} is relatively easy. Case 3 is essentially the well-known test in the theory of variance analysis. We therefore leave out the details of derivation and only state the main results here. In particular, the test statistic l_{IBDA3} is distributed as noncentral F with noncentrality parameter λ_3 specified by

$$\lambda_3 = 2N\mathbf{b}_1^H(\Omega_{11} - \Omega_{01}^H \Omega_{00}^{-1} \Omega_{01})\mathbf{b}_1 \tag{2-125}$$

and degrees of freedom ν_1 and ν_2 equal to

$$\nu_1 = 2(M - M_0)\,, \qquad \nu_2 = 2(N - M) \tag{2-126}$$

If we denote this PDF by $f_F(F; \nu_1, \nu_2, \lambda_3)$, then the probabilities of false alarm and detection are

$$\begin{aligned} P_{\text{fa}} &= \int_{TH_3}^{\infty} f_F(F; \nu_1, \nu_2, 0)\, dF \\ P_d &= \int_{TH_3}^{\infty} f_F(F; \nu_1, \nu_2, \lambda_3)\, dF \end{aligned} \tag{2-127}$$

Here TH_3 is a preset threshold, and the fact that $\lambda_3 = 0$ under H_0 is easily checked.

2-6-9 Experimental Results

2-6-9-1 Tests with Simulation Data

We now present simulation results for the probability of detection. The three detection scenarios considered herein correspond to different prior information regarding the interference. The first scheme requires complete knowledge of the noise; in practice, this is seldom the case. We will therefore concentrate on Schemes 2 and 3. We briefly outline the steps involved in the calculation of theoretical performance in both l_{IBDA2} and l_{IBDA3} tests.

l_{IBDA2} Test

1. Based on the model parameters **a** and σ^2, calculate the parameters for the noncentral chi-squared distribution using (2-122) and (2-123).
2. Determine the appropriate threshold for a given P_{fa} using (2-124).
3. Compute the P_d using (2-124).

l_{IBDA3} Test

1. Based on the model parameter **a** and σ^2, calculate the parameters for the noncentral F distribution using (2-125) and (2-126).
2. Determine the appropriate threshold for a given P_{fa} using (2-127).
3. Compute the P_d using (2-127).

For this simulation, we have chosen to model the received data under hypothesis H_0 by an AR(1) process such that

$$x(k) = (0.402 + j0.553)x(k-1) + e(k) \tag{2-128}$$

where $\{e(k)\}$ is an i.i.d. white complex Gaussian process with zero mean and unit variance. The received data under hypothesis H_1 are modeled by an AR(4) process, such that

$$x(k) = \sum_{i=1}^{4} a_i x(k-i) + e(k) \tag{2-129}$$

where $\{e(k)\}$ is an i.i.d white complex Gaussian process with mean zero and variance σ^2, and the AR coefficients are as follows:

$$\begin{aligned} a_1 &= 0.625 + j0.587 \\ a_2 &= -0.034 - j0.536 \\ a_3 &= -0.182 + j0.221 \\ a_4 &= 0.111 - j0.014 \end{aligned} \tag{2-130}$$

These coefficients have been selected such that the power spectra of the AR(4) process under H_1 and the AR(1) process under H_0 overlap to a large extent. Then we may clearly see the variation of P_d with small and moderate sample sizes. (Otherwise, P_d would be very close to unity even for a sample size $n = 20$.) This is done to simulate reasonably realistic situations. The

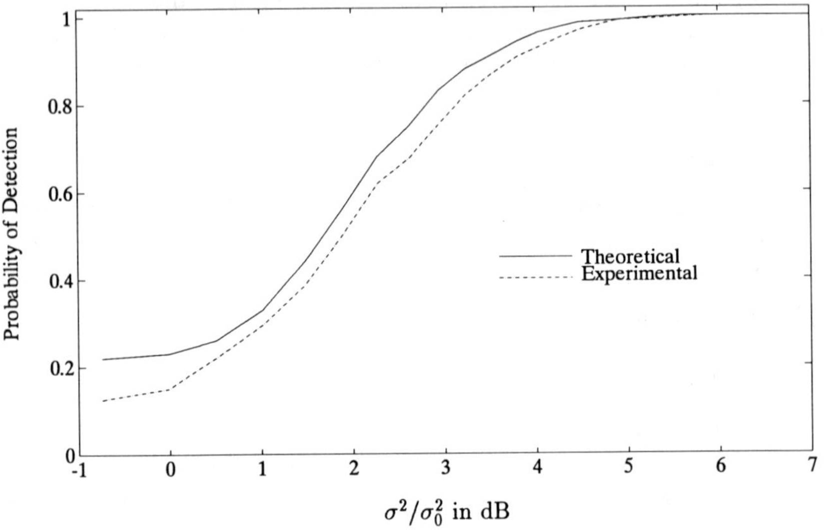

(a)

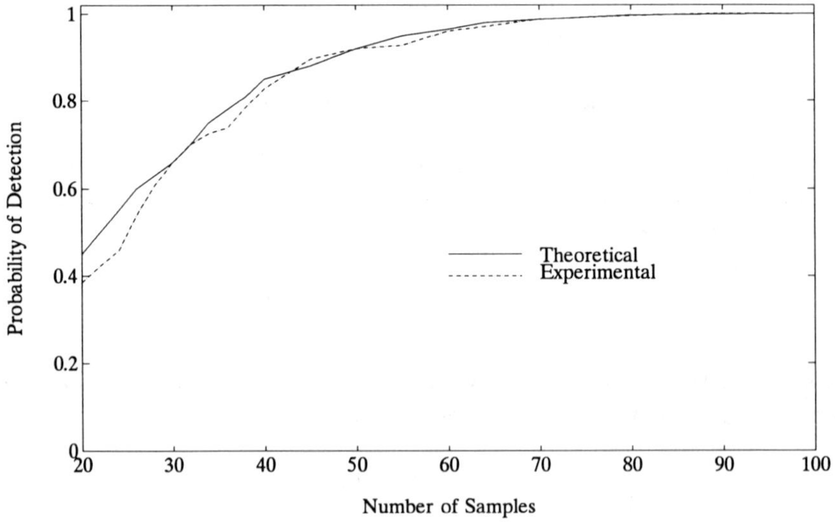

(b)

Figure 2-3 Comparison of the theoretical and experimental performance of the l_{IBDA2} under the conditions that $P_{\text{fa}} = 10^{-2}$; AR(1) for H_0 and AR(4) for H_1. (a) P_d vs. σ^2/σ_0^2 for sample size = 20; (b) P_d vs. sample size for $\sigma^2/\sigma_0^2 = 1.48$ dB.

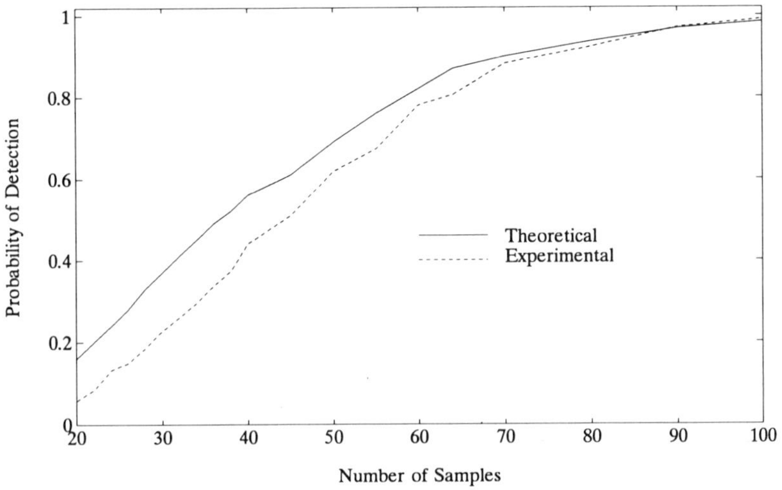

Figure 2-4 Comparison of theoretical and experimental performance of the l_{IBDA3} test. P_d is drawn as a function of the sample size for the following conditions: AR(1) for H_0 and AR(4) for H_1; $P_{\mathrm{fa}} = 10^{-2}$ and $\sigma^2/\sigma_0^2 = 1.48$ dB.

statistical properties of $\{x(k)\}$ are uniquely determined by the parameter set $\{\mathbf{a}, \sigma^2\}$ or equivalently, by its covariance matrix of corresponding order. Once $(\mathbf{a}, \sigma^2)$ are given, the covariance matrix can be determined, which is needed for the calculation of the noncentrality parameters.

The experimental value of P_d is obtained by counting the number of correct decisions (H_1) in 400 independent computer trials. A comparison between the experimental results for the test l_{IBDA2} with their theoretical counterparts are shown in Fig. 2-3. In this and the subsequent figures, for simplicity, we only draw the curves for $P_{\mathrm{fa}} = 10^{-2}$. In Fig. 2.3(a), the detection probability P_d is shown as a function of the ratio σ^2/σ_0^2, for a fixed sample size equal to 20 in this case. P_d as a function of the sample size for some fixed σ^2/σ_0^2, say equal to 1.48 dB in this case, is depicted in Fig. 2.3(b). Both figures show excellent agreement between theoretical and simulation results.

Like l_{IBDA2}, the P_d of l_{IBDA3} varies with the sample size. This relation is drawn in Fig. 2-4 for the case where $\sigma^2/\sigma_0^2 = 1.48$ dB. From these curves we see that the experimental results agree quite well with their theoretical counterpart albeit not as well as with the l_{IBDA2} test.

2-6-9-2 Application to Real Radar Data

The three IBDA detection schemes require different amounts of prior information about radar clutter. In particular l_{IBDA1} requires a complete statistical knowledge of the clutter and this is impossible in practice. l_{IBDA3}, on the other hand, requires only knowledge of the AR model orders to model the data under the two hypotheses at the cost of a poor detection performance. Therefore, we will only consider the application of l_{IBDA2} to actual radar data.

The complex baseband radar data used for our experiments are obtained using the Westinghouse TRACS radar system, in conjunction with an Ampex HBR-3000 digital recorder. The Westinghouse TRACS is a fully coherent radar system. The radar returns are uniformly sampled, and both in-phase and quadrature channels are digitized using 10 bits. The major operating characteristics of the TRACS are summarized in Table 2-1.

Since both signal-to-clutter ratio and target Doppler shift are required to change in the experimental study, an artificial target signal is more flexible than a true one. The target signal used for our simulations is generated by a model of the form

$$s(t) = \exp\left\{ -\frac{1}{2\sigma_\theta^2} (\Omega t)^2 \right\} \exp[j2\pi f_d t] , \qquad t = 0, \pm 1, \pm 2, \ldots , \pm m$$

where

$\sigma_\theta = \Delta\theta\pi/360$
$\Delta\theta$ = two sided antenna azimuth beamwidth
Ω = antenna rotation rate normalized by the PRF
f_d = target Doppler shift normalized by the PRF
t = time index

The sample size we use for signal processing is $N = 20$, which is approximately equal to the duration of a target return. The signal-to-clutter ratio is defined as the ratio of the signal power evaluated using 20 samples (centered at the signal envelope peak) to the clutter power estimated from 20 clutter samples under test before a target signal is added. To use l_{IBDA2}, we need knowledge of the model-error power for the clutter alone and this was done by estimating the error power from 15 previous scans and saving them as a clutter map for reference. The detection performance for ground clutter and weather clutter are shown in Figs. 2-5 and 2-6, respectively, where the dash-dot line and solid-line are for $P_{\text{fa}} = 10^{-4}$ and 10^{-5}.

TABLE 2-1 The Operating Characteristics of the Westinghouse TRACS Radar System

Transmitter frequency	1.23 GHz
Pulse repetition frequency (PRF)	657.2 Hz
Pulse width	0.5 μs
Antenna revolution rate	12.5 rev/min
Antenna beamwidth	1.35°
Sampling rate	1.35 MHz
Sample word size	10 bits

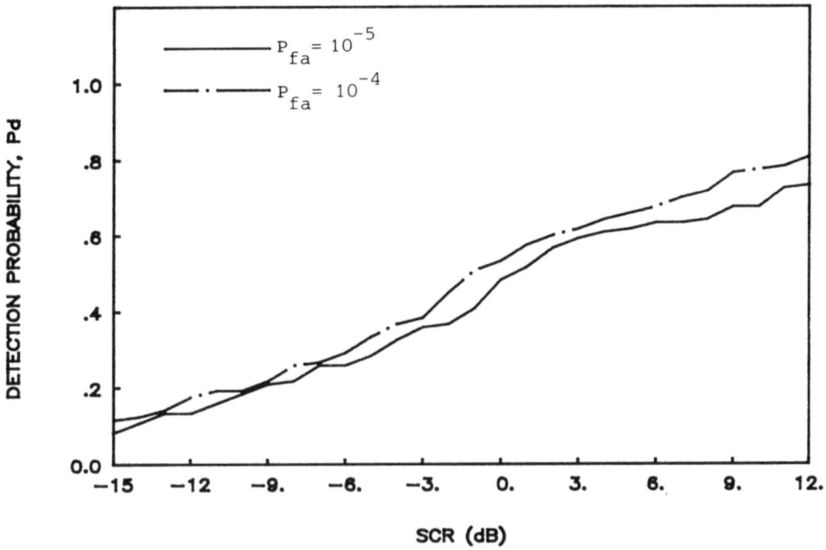

Figure 2-5 Detection performance of the l_{IBDA2} in a ground clutter dominated environment.

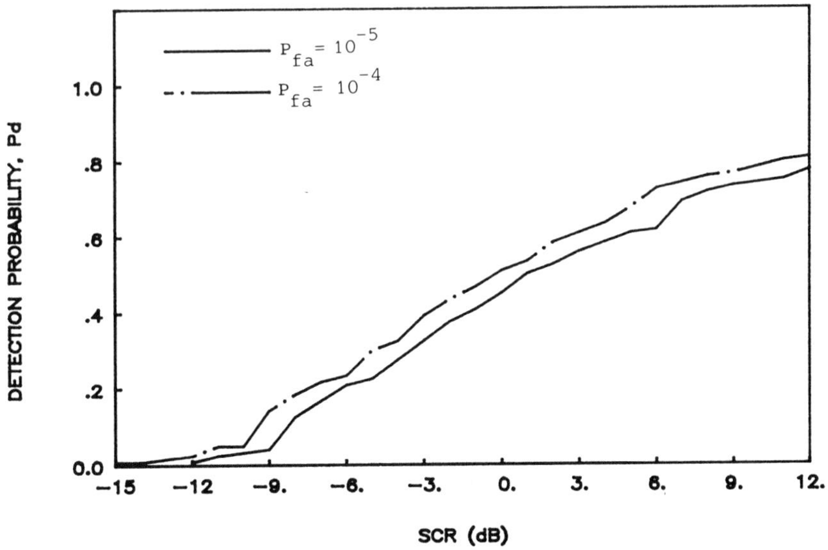

Figure 2-6 Detection performance of the l_{IBDA2} in a weather clutter dominated environment.

2-7 SIGNAL DETECTION IN COLORED NOISE: METHOD II

2-7-1 Introduction

When a radar antenna scans around, the echoes from the scatterers form a two-dimensional random field, one dimension representing range and the other representing azimuth. The detection scheme described in the preceding section, like many other conventional methods, treats the data as if they were one-dimensional by confining the processing along the azimuthal direction. Obviously, these processing schemes do not exploit the information from the adjacent cells which surround the cells under test.

In this section we present a new AR-based detection scheme which can avoid the above drawback. The new scheme performs two-dimensional processing; it is motivated by the following observations:

a. The rule for describing a target is different from the rule for describing clutter echoes. A target usually has different geometry and radial velocity from radar scatterers which produce clutter echoes.
b. When a target is present in the clutter field, the local statistics should change.

The strategy used here is to compare the statistical similarity of the data under test and the data obtained from the neighborhood. The second set of data will be referred to as the reference data for convenience. In this case, the AR modeling technique is still applicable. In fact, we may model the data under test and the reference data as two distinct AR processes, and the detection problem becomes one of testing the equality between the two sets of AR parameters. This can be easily done by using a maximum likelihood ratio test.

The new test exploits the full information of the second order statistics in both the test and reference data. This is in contrast to the two commonly used techniques: normalization and spectrum analysis. A normalization technique employs the amplitude information in the data. It compares the power of the data under test with the power of the background noise estimated from the neighboring cells. Signal processor based on spectral analysis, on the other hand, uses mainly the Doppler information. The new technique may be therefore considered as a natural extension of the normalization technique. It exploits, through the use of an AR model, not only the power level but also the Doppler shift implicitly in data.

2-7-2 Formulation of the Test Statistic

Suppose that $x = F(\xi, \eta)$ is a complex-valued two-dimensional random field where variables ξ and η represent azimuth and range, respectively. We make the following assumptions:

a. The random field is sampled, so that the data are correlated in the azimuth but uncorrelated in the range direction. Although this assumption is made for theoretical convenience, it provides a good approximation to many practical situations. In a radar environment, the correlation of the received data in range is introduced mainly by the pulse width of the transmit signal. Therefore, the data samples would be considered uncorrelated if the sampling interval is chosen to be greater than the pulse width. This is often the case for many applications of interest. The correlation of the data along an azimuthal ring, on the other hand, is caused by the modulation of the antenna beam pattern, and also by the physical properties of the scatterers. As such, when no targets are present, this correlation provides a way of investigating the physical properties of the scatterers, which in turn, supply a background to discriminate a target from clutter. Likewise, when a target is present, its properties will be reflected, implicitly or explicitly, by a corresponding correlation structure. Appropriate exploitation and use of these correlation structures will not only reduce the influence of clutter to a minimum extent, but also integrate the target echoes coherently. As a result, target detection is enhanced.

b. The random field is locally stationary. This assumption is often satisfied approximately in practice and (as indicated in the introduction to this section) has been accepted by many signal processing schemes.

c. The correlation of data in the azimuth direction is described by an autoregressive process of relatively lower order.

In light of these assumptions, we can formulate our procedure of hypothesis testing, using the local statistics of the random field described by the AR coefficients and the variance of the model error process.

Consider the set of K data vectors

$$\{\mathbf{x}_k\} = \{F(\xi, \eta_k),\ \xi = \xi_1, \ldots, \xi_N\}\,, \qquad k = 1, \ldots, K \tag{2-131}$$

in which $\{\mathbf{x}_K\}$ is the data vector to be processed, and $\{x_k;\, k \neq K\}$ are the $(K-1)$ reference sequences obtained from the ranges adjacent to η_K. Suppose that the reference vector set $\{\mathbf{x}_k;\, k \neq K\}$ contains clutter only. Then the processing of $\mathbf{x}_K$ for target detection means that we have to decide whether $\mathbf{x}_K$ is composed of clutter only or is it composed of a target echo plus clutter. Since the random field is locally stationary, it follows that without a target component, $\mathbf{x}_K$ should have the same statistics as the reference sequences $\{\mathbf{x}_k;\, k \neq K\}$. On the other hand, when a target is present, the statistics of $\mathbf{x}_K$ should deviate from those of its neighbors. Thus our decision becomes one of establishing whether the statistics of $\mathbf{x}_K$ are the same as or different from those of $\{\mathbf{x}_k;\, k \neq K\}$. Accordingly, to formulate the hypothesis testing, we characterize each of the two hypotheses in terms of the parameters of a probabilistic model. This model should be chosen so as to provide a good fit to the data for a good detection performance. However, it should be compact in the sense that it has as small a number of parameters as possible; we say this, because we know from estimation theory that for a given number of data samples, the smaller the number of unknown parameters the higher will the estimation accuracy be. AR modeling satisfies this requirement since, in general, the number of coefficients and the variance of the model error are the total number of parameters needed and is much less than the sample size.

Suppose that we can model the elements of the vector $\mathbf{x}_k$ containing clutter only by an AR process defined by

$$x_k(n) = \sum_{m=1}^{M} a_0(m)x_k(n-m) + e_k(n)\,, \qquad n = 1, \ldots, N,\ k = 1, \ldots, K-1 \tag{2-132}$$

where the error process $\{e_k(n)\}$ is an independent, identically distributed Gaussian process with zero mean and variance σ_0^2. Since the radar clutter data are assumed to be locally stationary, then (2-132) applies to vectors at any range if they contain clutter only. Denote the coefficient vector by

$$\mathbf{a}_0 = [a_0(1) \quad \cdots \quad a_0(M)]^T \tag{2-133}$$

Symbolically, we can write the parameterization of $\mathbf{x}_k$, which contains clutter only, as

$$\mathbf{x}_k \sim \mathrm{AR}(M, \mathbf{a}_0, \sigma_0^2)\,, \qquad k = 1, \ldots, K-1 \tag{2-134}$$

Suppose next that the data vector $\mathbf{x}_k$ to be processed contains a target echo plus clutter; then we can write

$$\mathbf{x}_K \sim \mathrm{AR}(M, \mathbf{a}, \sigma^2) \tag{2-135}$$

Usually, this model is different from that for $\mathbf{x}_k$ which contains clutter only. We may use the same order M for both models by simply choosing M to be the larger of the two orders. Then, the two hypotheses may be written as follows:

$$\begin{aligned} H_0&: \begin{pmatrix} \mathbf{a} \\ \sigma_2 \end{pmatrix} = \begin{pmatrix} \mathbf{a}_0 \\ \sigma_0^2 \end{pmatrix} \\ H_1&: \begin{pmatrix} \mathbf{a} \\ \sigma^2 \end{pmatrix} \neq \begin{pmatrix} \mathbf{a}_0 \\ \sigma_0^2 \end{pmatrix} \end{aligned} \tag{2-136}$$

Now let,

$$\mathbf{x}_k = [\mathbf{v}_k^T \,\vdots\, \mathbf{u}_k^T] = [x_k(1), \ldots, x_k(M) \,\vdots\, x_k(M+1), \ldots, x_k(N)]^T \tag{2-137}$$

and

$$\mathbf{X}_k = \begin{bmatrix} x_k(M) & x_k(M-1) & \ldots & x_k(1) \\ x_k(M+1) & x_k(M) & \ldots & x_k(2) \\ \vdots & & & \\ x_k(N) & x_k(N-1) & \ldots & x_k(N-M+1) \end{bmatrix} \tag{2-138}$$

Also let

$$E[\mathbf{v}_k \mathbf{v}_k^H] = \begin{cases} \mathbf{R}_0 = \sigma_0^2 \Omega_0 \,, & k \neq K \\ \mathbf{R} = \sigma^2 \Omega \,, & k = K \end{cases} \tag{2-139}$$

Under hypothesis H_0, $\mathbf{R} = \mathbf{R}_0$, and the joint PDF of $\mathbf{x}_1, \ldots, \mathbf{x}_K$ is given by

$$\begin{aligned} p(\mathbf{x}_1, \ldots, \mathbf{x}_K \mid H_0) &= \prod_{k=1}^{K} p(\mathbf{v}_k \mid H_0) \cdot p(\mathbf{u}_k \mid H_0, \mathbf{v}_k) \\ &= \left\{ \prod_{k=1}^{K} \frac{1}{(\pi\sigma_0^2)^M \det(\Omega_0)} \exp\left(-\frac{1}{\sigma_0^2} \mathbf{v}_k^H \Omega_0^{-1} \mathbf{v}_k \right) \right\} \times \\ &\quad \left\{ \prod_{k=1}^{K} \frac{1}{(\pi\sigma_0^2)^{N-M}} \exp\left[-\frac{1}{\sigma_0^2} (\mathbf{u}_k - \mathbf{X}_k \mathbf{a}_0)^H (\mathbf{u}_k - \mathbf{X}_k \mathbf{a}_0) \right] \right\} \\ &= (\pi\sigma_0^2)^{-KN} \exp\left[-\sigma_0^{-2} \sum_{k=1}^{K} (\mathbf{u}_k - \mathbf{X}_k \mathbf{a}_0)^H (\mathbf{u}_k - \mathbf{X}_k \mathbf{a}_0) \right] \times \\ &\quad (\det \Omega_0)^{-K} \exp\left(-\sigma_0^{-2} \sum_{k=1}^{K} \mathbf{v}_k^H \Omega_0^{-1} \mathbf{v}_k \right) \end{aligned} \tag{2-140}$$

from which the log-likelihood function is written as

$$\begin{aligned} &\log\{ p(\mathbf{x}_1, \ldots, \mathbf{x}_K \mid H_0) \} \\ &\quad = \left\{ -NK \log(\pi\sigma_0^2) - \sigma_0^{-2} \sum_{k=1}^{K} (\mathbf{u}_k - \mathbf{X}_k \mathbf{a}_0)^H (\mathbf{u}_k - \mathbf{X}_k \mathbf{a}_0) \right\} \\ &\qquad + \left\{ -K \log(\det \Omega_0) - \sigma_0^{-2} \sum_{k=1}^{K} \mathbf{v}_k^H \Omega_0^{-1} \mathbf{v}_k \right\} \\ &\quad \simeq -KN \log(\pi\sigma_0^2) - \sigma_0^{-2} \sum_{k=1}^{K} (\mathbf{u}_k - \mathbf{X}_k \mathbf{a}_0)^H (\mathbf{u}_k - \mathbf{X}_k \mathbf{a}_0) \end{aligned} \tag{2-141}$$

The last step in (2-141) is an approximation made possible for $N \gg M$.

Maximizing this expression with respect to $\mathbf{a}_0$ and σ_0^2, we obtain the maximum likelihood estimates under H_0, such that

$$\tilde{\mathbf{a}}_0 = \left(\sum_{k=1}^{K} \mathbf{X}_k^H \mathbf{X}_k\right)^{-1} \left(\sum_{k=1}^{K} \mathbf{X}_k^H \mathbf{u}_k\right)$$

and (2-142)

$$\tilde{\sigma}_0^2 = \frac{1}{KN} \sum_{k=1}^{M} (\mathbf{u}_k - \mathbf{X}_k \tilde{\mathbf{a}}_0)^H (\mathbf{u}_k - \mathbf{X}_k \tilde{\mathbf{a}}_0)$$

Inserting this result into (2-141) yields the maximum log-likelihood function under H_0 as follows:

$$\max\{\log[p(\mathbf{x}_1, \ldots, \mathbf{x}_K \mid H_0)]\} = -KN[\log(\pi\tilde{\sigma}_0^2) + 1] \qquad (2\text{-}143)$$

Next, consider the maximum likelihood function of the data under H_1. We write

$$p(\mathbf{x}_1, \ldots, \mathbf{x}_{K-1}, \mathbf{x}_K) = p(\mathbf{x}_1, \ldots, \mathbf{x}_{K-1} \mid H_1) p(\mathbf{x}_K \mid H_1) \qquad (2\text{-}144)$$

where the two PDFs, when subject to a derivation similar to (2-141), are given by

$$p(\mathbf{x}_1, \ldots, \mathbf{x}_{K-1} \mid H_1) = (\pi\sigma_0^2)^{-N(K-1)} \exp\left[-\sigma_0^{-2} \sum_{k=1}^{K-1} (\mathbf{u}_k - \mathbf{X}_k \mathbf{a}_0)^H \times (\mathbf{u}_k - \mathbf{X}_k \mathbf{a}_0)\right] \times (\det \Omega_0)^{-(K-1)} \exp\left(-\sigma_0^{-2} \sum_{k=1}^{K-1} \mathbf{v}_k^H \Omega_0^{-1} \mathbf{v}_k\right)$$

and

$$p(\mathbf{x}_K \mid H_1) = (\pi\sigma^2)^{-N} \exp[-\sigma^{-2}(\mathbf{u}_K - \mathbf{X}_K \mathbf{a})^H (\mathbf{u}_K - \mathbf{X}_K \mathbf{a})] \times (\det \Omega)^{-1} \exp\left(-\sigma^{-2} \mathbf{v}_K^H \Omega^{-1} \mathbf{v}_K\right)$$

Taking the logarithm of (2-144) and dropping terms independent of N, we have

$$\log p(\mathbf{x}_1, \ldots, \mathbf{x}_{K-1}, \mathbf{x}_K \mid H_1) \simeq \left\{-N(K-1)\log(\pi\sigma_0^2) - \sigma_0^{-2} \sum_{k=1}^{K-1} (\mathbf{u}_k - \mathbf{X}_k \mathbf{a}_0)^H (\mathbf{u}_k - \mathbf{X}_k \mathbf{a}_0)\right\} + \{-N\log(\pi\sigma^2) - \sigma^{-2}(\mathbf{u}_K - \mathbf{X}_K \mathbf{a})^H (\mathbf{u}_K - \mathbf{X}_K \mathbf{a})\} \qquad (2\text{-}145)$$

Maximizing (2-145) with respect to the unknown parameters $\mathbf{a}_0$, $\mathbf{a}$, σ_0^2, and σ^2 by equating the respective derivatives to zero, gives the following MLE of the parameters:

$$\hat{\mathbf{a}}_0 = \left(\sum_{k=1}^{K-1} \mathbf{X}_k^H \mathbf{X}_k\right)^{-1} \left(\sum_{k=1}^{K-1} \mathbf{X}_k^H \mathbf{u}_k\right)$$

$$\hat{\mathbf{a}} = \left(\mathbf{X}_K^H \mathbf{X}_K\right)^{-1} (\mathbf{X}_K^H \mathbf{u}_K)$$

$$\hat{\sigma}_0^2 = \frac{1}{N(K-1)} \sum_{k=1}^{K-1} (\mathbf{u}_k - \mathbf{X}_k \hat{\mathbf{a}}_0)^{\mathrm{H}} (\mathbf{u}_k - \mathbf{X}_k \hat{\mathbf{a}}_0) \tag{2-146}$$

$$\hat{\sigma}^2 = \frac{1}{N} (\mathbf{u}_K - \mathbf{X}_K \hat{\mathbf{a}})^H (\mathbf{u}_K - \mathbf{X}_K \hat{\mathbf{a}})$$

Substituting (2-146) into (2-145), the maximum likelihood function under H_1 is given by

$$\max\{\log p(\mathbf{x}_1, \ldots, \mathbf{x}_K \mid H_1)\} = -N(K-1)[\log(\pi\hat{\sigma}_0^2) + 1] - N[\log(\pi\hat{\sigma}^2) + 1] \tag{2-147}$$

We are now in a position to obtain the likelihood ratio, described by

$$\begin{aligned} \Lambda &= \log[\max p(\mathbf{x}_1, \ldots, \mathbf{x}_K \mid H_1) / \max p(\mathbf{x}_1, \ldots, \mathbf{x}_K \mid H_0)] \\ &= \max\{\log p(\mathbf{x}_1, \ldots, \mathbf{x}_K \mid H_1)\} - \max\{\log p(\mathbf{x}_1, \ldots, \mathbf{x}_K \mid H_0)\} \end{aligned} \tag{2-148}$$

Interchanging the order of the operators, max(.) and log(.) in (2-148) is permissible since log(.) is a monotonic increasing function. Using (2-143) and (2-147) in (2-148), the likelihood ratio becomes

$$\Lambda = N \log\left[\frac{(\tilde{\sigma}_0^2)^K}{(\hat{\sigma}_0^2)^{K-1} \hat{\sigma}^2}\right] \tag{2-149}$$

For reasons that will become clear later, we use the likelihood ratio

$$l = 2\Lambda \tag{2-150}$$

The test statistics l is relatively simple to compute, and our likelihood ratio test may be formulated as

$$l \underset{H_0}{\overset{H_1}{\gtrless}} \Lambda_T \tag{2-151}$$

where Λ_T is a pre-set threshold dependent on a specified value of the false alarm rate. To compute the value of Λ_T, however, we need to know the probability distribution of l which is not easy to evaluate when l is given by its present form. This problem can be overcome by deriving an alternative expression for l from which the probability distribution can be readily obtained.

2-7-3 An Alternative Expression for the Likelihood Ratio

To evaluate the statistical performance of the likelihood ratio test derived in Subsection 2-7-2, it is necessary to determine the PDF of the likelihood ratio in (2-150) under H_0 and H_1. It is, however, difficult to directly evaluate the PDF of the statistic l. This difficulty can be avoided if we derive an alternative expression for l. Specifically, the original statistic l is obtained from the likelihood function under hypotheses H_0 and H_1 for large samples. These likelihood functions can be expressed as asymptotic functions of the sample estimates of the unknown parameters, from which an alternative expression can be derived. This can be done by using the techniques we used in the preceding section. Let

$$L = \log p(\mathbf{x}_1, \ldots, \mathbf{x}_K \mid H_1) \tag{2-152}$$

$$L_1 = \log p(\mathbf{x}_1, \ldots, \mathbf{x}_{K-1} \mid H_1) \tag{2-153}$$

$$L_2 = \log p(\mathbf{x}_K \mid H_1) \tag{2-154}$$

Then, from (2-144), we have

$$L = L_1 + L_2 \tag{2-155}$$

where the appropriate values of L_1 and L_2 are given by the terms in the first and second sets of braces in (2-145), respectively. Furthermore, let the real and imaginary parts of $\mathbf{x}_k$, $\mathbf{X}_k$, $\mathbf{a}_0$, and $\mathbf{a}$ be written as

$$\begin{aligned} \mathbf{u}_k &= \mathbf{u}_{k1} + j\mathbf{u}_{k2} \\ \mathbf{X}_k &= \mathbf{X}_{k1} + j\mathbf{X}_{k2}, \qquad k = 1, \ldots, K \\ \mathbf{a}_0 &= \alpha_0 + j\beta_0 \\ \mathbf{a} &= \alpha + j\beta \end{aligned}$$

Also, let

$$\begin{aligned} \mathbf{b} &= \begin{pmatrix} \alpha \\ \beta \end{pmatrix}, \qquad \mathbf{b}_0 = \begin{pmatrix} \alpha_0 \\ \beta_0 \end{pmatrix} \\ \boldsymbol{\theta} &= \begin{pmatrix} \mathbf{b} \\ \sigma^2 \end{pmatrix}, \qquad \boldsymbol{\theta}_0 = \begin{pmatrix} \mathbf{b}_0 \\ \sigma_0^2 \end{pmatrix} \end{aligned} \tag{2-156}$$

Then L_1 and L_2 may be rewritten as

$$
\begin{aligned}
L_1(\theta_0) = & -N(K-1)\log(\pi\sigma_0^2) \\
& -\frac{1}{\sigma_0^2}\sum_{k=1}^{K-1}[(\mathbf{u}_{k1}-\mathbf{X}_{k1}\alpha_0+\mathbf{X}_{k1}\beta_0)+j(\mathbf{u}_{k2}-\mathbf{X}_{k1}\beta_0+\mathbf{X}_{k2}\alpha_0)]^H \\
& \times[(\mathbf{u}_{k1}-\mathbf{X}_{k1}\alpha_0+\mathbf{X}_{k2}\beta_0)+j(\mathbf{u}_{k2}-\mathbf{X}_{k1}\beta_0-\mathbf{X}_{k2}\alpha_0)]\} \\
L_2(\theta) = & -N\log(\pi\sigma^2) \\
& -\frac{1}{\sigma^2}[(\mathbf{u}_{K1}-\mathbf{X}_{K1}\alpha+\mathbf{X}_{K2}\beta)+j(\mathbf{u}_{K2}-\mathbf{X}_{K1}\beta-\mathbf{X}_{K2}\alpha)]^H \\
& \times[(\mathbf{u}_{k1}-\mathbf{X}_{K1}\alpha+\mathbf{X}_{K2}\beta)+j(\mathbf{u}_{K2}-\mathbf{X}_{K1}\beta-\mathbf{X}_{K2}\alpha)]\}
\end{aligned}
\tag{2-158}
$$

Consider the Taylor expansion of $L_1(\theta)$ at $\theta_0=\hat{\theta}_0$, as shown by

$$
L_1(\theta_0)=L_1(\hat{\theta}_0)+\left[\frac{\partial L_1}{\partial\theta_0}\right]_{\hat{\theta}_0}^T(\theta_0-\hat{\theta}_0)+\frac{N(K-1)}{2}(\theta_0-\hat{\theta}_0)^T\mathbf{I}(\bar{\theta}_0)(\theta_0-\hat{\theta}_0)
\tag{2-159}
$$

where $\bar{\theta}_0$ is a point on the line segment connecting θ_0 and $\hat{\theta}_0$, and $\mathbf{I}$ is the sample information matrix defined by

$$
\mathbf{I}(\bar{\theta}_0)=\frac{1}{N(K-1)}\left.\frac{\partial^2 L_1(\theta_0)}{\partial\theta_0\,\partial\theta_0^T}\right|_{\bar{\theta}_0}
$$

Using (2-156), we have

$$
\mathbf{I}(\bar{\theta}_0)=\frac{1}{N(K-1)}\begin{bmatrix}\mathbf{I}_{11} & \mathbf{I}_{12}\\ \mathbf{I}_{21} & \mathbf{I}_{22}\end{bmatrix}
\tag{2-160}
$$

where

$$
\mathbf{I}_{11}=\frac{1}{N(K-1)}\begin{bmatrix}\dfrac{\partial^2 L_1}{\partial\alpha_0\,\partial\alpha_0^T} & \dfrac{\partial^2 L_1}{\partial\alpha_0\,\partial\beta_0^T}\\[2ex] \dfrac{\partial^2 L_1}{\partial\beta_0\,\partial\alpha_0^T} & \dfrac{\partial^2 L_1}{\partial\beta_0\,\partial\beta_0^T}\end{bmatrix}
$$

and

$$
\begin{aligned}
\mathbf{I}_{21}=\mathbf{I}_{21}^T &= \frac{1}{N(K-1)}\left[\frac{\partial^2 L_1}{\partial\sigma_0^2\,\partial\alpha_0^T}\quad\frac{\partial^2 L_1}{\partial\sigma_0^2\,\partial\beta_0^T}\right]\\
\mathbf{I}_{22} &= \frac{1}{N(K-1)}\frac{\partial^2 L_1}{\partial(\sigma_0^2)^2}
\end{aligned}
\tag{2-161}
$$

Theorem 2-4 The submatrices $\mathbf{I}_{11}, \mathbf{I}_{12}, \mathbf{I}_{21}$, and $\mathbf{I}_{22}$ have the following properties:

(i)

$$\mathbf{I}_{11} = -\frac{2}{\sigma_0^2}\begin{bmatrix} \text{Re}(\hat{\mathbf{R}}_0) & \text{Im}(\hat{\mathbf{R}}_0) \\ \text{Im}(\hat{\mathbf{R}}_0) & \text{Re}(\hat{\mathbf{R}}_0) \end{bmatrix}$$

where $\hat{\mathbf{R}}_0$ is the sample estimate of $\mathbf{R}_0$, given by

$$\hat{\mathbf{R}}_0 = \frac{1}{(N-M+1)(K-1)} \sum_{k=1}^{K-1} \mathbf{X}_k^H \mathbf{X}_k$$

$$\simeq \frac{1}{N(K-1)} \sum_{k=1}^{K-1} \mathbf{X}_k^H \mathbf{X}_k\,, \qquad \text{for } N \gg M$$

(ii)

$$\mathbf{I}_{22} = \sigma_0^{-4} - \frac{2\sigma_0^{-6}}{N(K-1)} \sum_{k=1}^{K-1} (\mathbf{u}_k - \mathbf{X}_k \mathbf{a}_0)^{\text{H}} (\mathbf{u}_k - \mathbf{X}_k \mathbf{a}_0)$$

(iii)

$$\lim_{N\to\infty} \mathbf{I}_{21} = \lim_{N\to\infty} \mathbf{I}_{12}^T = \mathbf{0}$$

That is, the maximum likelihood estimates of the model coefficients are asymptotically uncorrelated with that of the variance of model error.

The proof of this theorem is similar to that of Theorem 2-1 and is therefore omitted here.

With the properties described in Theorem 2-4, together with the fact that the second term in (2-159) is zero since $\hat{\theta}_0$ is the MLE of θ_0, we find that L_1 in (2-159) can be simplified as

$$L_1(\theta_0) = L_1(\hat{\theta}_0) + \frac{N(K-1)}{2}\begin{bmatrix} \alpha_0 - \hat{\alpha}_0 \\ \beta_0 - \hat{\beta}_0 \end{bmatrix} I_{11}(\hat{\theta}_0)\begin{bmatrix} \alpha_0 - \hat{\alpha}_0 \\ \beta_0 - \hat{\beta}_0 \end{bmatrix} + \frac{N(K-1)}{2} I_{22}(\bar{\theta}_0)(\sigma_0^2 - \hat{\sigma}_0^2)^2 \qquad (2\text{-}162)$$

Similarly, we can write

$$L_2(\theta) = L_2(\hat{\theta}) + \frac{N}{2}\begin{bmatrix} \alpha - \hat{\alpha} \\ \beta - \hat{\beta} \end{bmatrix}^T J_{11}(\bar{\theta})\begin{bmatrix} \alpha - \hat{\alpha} \\ \beta - \hat{\beta} \end{bmatrix} + \frac{N}{2} J_{22}(\bar{\theta})(\sigma^2 - \hat{\sigma}^2)^2 \qquad (2\text{-}163)$$

where

$$J_{11} = -\frac{2}{\sigma^2}\begin{bmatrix} \mathrm{Re}(\hat{\mathbf{R}}) & -\mathrm{Im}(\hat{\mathbf{R}}) \\ \mathrm{Im}(\hat{\mathbf{R}}) & \mathrm{Re}(\hat{\mathbf{R}}) \end{bmatrix}$$

$$J_{22} = \sigma^{-4} - \frac{2}{N\sigma^6}(\mathbf{u}_K - \mathbf{X}_K\mathbf{a})^H(\mathbf{u}_K - \mathbf{X}_K\mathbf{a}) \tag{2-164}$$

$$\hat{\mathbf{R}} = \frac{1}{(N-M+1)}\mathbf{X}_K^H\mathbf{X}_K \simeq \frac{1}{N}\mathbf{X}_K^H\mathbf{X}_K, \qquad \text{for } N \gg M$$

and $\bar{\theta}$ is a point on the line segment linking θ and its MLE $\hat{\theta}$. Substituting (2-162) and (2-163) into (2-155), the log-likelihood under H_1, can be expressed as

$$\begin{aligned} L &= L_1(\hat{\theta}_0) + L_2(\hat{\theta}) \\ &+ \left\{\frac{N(K-1)}{2}\begin{bmatrix}\alpha_0 - \hat{\alpha}_0 \\ \beta_0 - \hat{\beta}_0\end{bmatrix}^T I_{11}(\bar{\theta}_0)\begin{bmatrix}\alpha_0 - \hat{\alpha}_0 \\ \beta_0 - \hat{\beta}_0\end{bmatrix} + \frac{N}{2}\begin{bmatrix}\alpha - \hat{\alpha} \\ \beta - \hat{\beta}\end{bmatrix}^T J_{11}(\bar{\theta})\begin{bmatrix}\alpha - \hat{\alpha} \\ \beta - \hat{\beta}\end{bmatrix}\right. \\ &+ \left.\frac{N(K-1)}{2} I_{22}(\bar{\theta}_0)(\sigma_0^2 - \hat{\sigma}_0^2)^2 + \frac{N}{2} J_{22}(\bar{\theta})(\sigma^2 - \hat{\sigma}^2)^2\right\} \end{aligned} \tag{2-165}$$

Note that all the terms inside the braces in (2-165) are nonpositive definite. Therefore, under hypothesis H_1, the maximum log-likelihood occurs when everyone of the terms inside the braces is zero; that is,

$$\max L(H_1) = \max_{\theta_0} L_1 + \max_{\theta} L_2 = L_1(\hat{\theta}_0) + L_2(\hat{\theta}) \tag{2-166}$$

Equation (2-165) may also be used for the likelihood function under hypothesis H_0, since this case is only a restricted version of that under H_1. Thus putting $\theta = \theta_0$ in (2-165), we have

$$\begin{aligned} \max L(H_0) &= \max_{\theta=\theta_0}\{L\} \\ &= L_1(\hat{\theta}_0) + L_2(\hat{\theta}_0) + \max_{\sigma_2^0}\left\{\frac{N(K-1)}{2} I_{22}(\bar{\theta}_0)(\sigma^2 - \hat{\sigma}_0^2)^2\right. \\ &\quad \left. + \frac{N}{2} J_{22}(\bar{\theta})(\sigma_0^2 - \hat{\sigma}^2)^2\right\} \\ &\quad + \max_{\alpha_0, \beta_0}\left\{\frac{N(K-1)}{2}\begin{bmatrix}\alpha_0 - \hat{\alpha}_0 \\ \beta_0 - \hat{\beta}_0\end{bmatrix}^T I_{11}(\bar{\theta}_0)\begin{bmatrix}\alpha_0 - \hat{\alpha}_0 \\ \beta_0 - \hat{\beta}_0\end{bmatrix}\right. \\ &\quad \left. + \frac{N}{2}\begin{bmatrix}\alpha_0 - \hat{\alpha} \\ \beta_0 - \hat{\beta}\end{bmatrix}^T J_{11}(\bar{\theta})\begin{bmatrix}\alpha_0 - \hat{\alpha} \\ \beta_0 - \hat{\beta}\end{bmatrix}\right\} \end{aligned} \tag{2-167}$$

It is a straightforward matter to show that the values of α_0, β_0, and σ_0^2 which maximize (2-167) are given by

$$\hat{b}_0 = [(K-1)I_{11}(\hat{\theta}_0) + J_{11}(\bar{\theta}]^{-1}[(K-1)I_{11}(\bar{\theta}_0)\hat{\mathbf{b}}_0 + J_{11}(\theta)\hat{\mathbf{b}}]$$
$$\hat{\hat{\sigma}}_0^2 = [(K-1)I_{22}(\bar{\theta}_0) + J_{22}(\bar{\theta})]^{-1}[(K-1)\hat{\sigma}_0^2 I_{22}(\theta_0) + \hat{\sigma}^2 J_{22}(\bar{\theta})] \tag{2-168}$$

where

$$\hat{\mathbf{b}}_0 = \begin{bmatrix} \hat{\alpha}_0 \\ \hat{\beta}_0 \end{bmatrix}, \qquad \hat{\mathbf{b}} = \begin{bmatrix} \hat{\alpha} \\ \hat{\beta} \end{bmatrix} \tag{2-169}$$

In view of these results, (2-167) may be rewritten as

$$\max L(H_0) = -\frac{1}{2}(\hat{\mathbf{b}} - \hat{\mathbf{b}}_0)^T \mathbf{K}_b (\hat{\mathbf{b}} - \hat{\mathbf{b}}_0) - \frac{1}{2} K_\sigma (\hat{\sigma}^2 - \hat{\sigma}_0^2)^2 + L_1(\bar{\theta}_0) + L_2(\bar{\theta}) \tag{2-170}$$

where $\mathbf{K}_b$ and $\mathbf{K}_\sigma$, after simplification, are given by

$$\mathbf{K}_b = -\left[\frac{1}{N}\mathbf{J}_{11}^{-1}(\bar{\theta}) + \frac{1}{N(K-1)}\mathbf{I}_{11}^{-1}(\bar{\theta}_0)\right]^{-1} \tag{2-171}$$

and

$$\mathbf{K}_\sigma = -N(K-1)[(K-1)I_{22}(\bar{\theta}_0) + J_{22}(\bar{\theta})]^{-1} I_{22}(\bar{\theta}_0) J_{22}(\bar{\theta}) \tag{2-172}$$

Combining (2-166) and (2-170) yields an alternative expression of the log-likelihood ratio statistic of (2-150) such that

$$\begin{aligned} l &= 2\{\max L(H_1) - \max L(H_0)\} \\ &= (\hat{\mathbf{b}} - \hat{\mathbf{b}}_0)^T \mathbf{K}_b (\hat{\mathbf{b}} - \hat{\mathbf{b}}_0) + \mathbf{K}_\sigma (\hat{\sigma} - \hat{\sigma}_0^2)^2 \end{aligned} \tag{2-173}$$

Here, the factor 2 has been introduced for convenience of presentation.

It now becomes clear that the log-likelihood ratio statistic is a quadratic function of the differences $(\hat{\mathbf{b}} - \hat{\mathbf{b}}_0)$ and $(\hat{\sigma}^2 - \hat{\sigma}_0^2)$. This implies that the test computes the model parameters based on the reference data and on the data under test, and then compares the distance (using a quadratic function) between these two sets of parameters. A target is said to be present if the distance measure is larger than a preset threshold; otherwise, we say that there is no target present. The threshold is set according to the prescribed probability of false alarm.

We further note that since

$$\lim_{N\to\infty} \hat{\mathbf{R}} = \lim_{N\to\infty} \frac{1}{N} \mathbf{X}_K^H \mathbf{X}_K = \mathbf{R}$$

$$\lim_{N\to\infty} \hat{\mathbf{R}}_0 = \lim_{N\to\infty} \frac{1}{N(K-1)} \sum_{k=1}^{K-1} \mathbf{X}_k^H \mathbf{X}_k = \mathbf{R}_0$$

$$\lim_{N\to\infty} \frac{1}{N(K-1)} \sum_{k=1}^{K-1} (\mathbf{u}_k - \mathbf{X}_k\mathbf{a}_0)^H(\mathbf{u}_k - \mathbf{X}_k\mathbf{a}_0) = \sigma_0^2$$

$$\lim_{N\to\infty} \frac{1}{N} (\mathbf{u}_K - \mathbf{X}_K\mathbf{a})^H(\mathbf{u}_K - \mathbf{X}_K\mathbf{a}) = \sigma^2 \tag{2-174}$$

then it can easily be seen that the asymptotic behavior of $\mathbf{K}_b$ and $\mathbf{K}_\sigma$ is such that

$$\lim_{N\to\infty} \mathbf{K}_b^{-1} \stackrel{\Delta}{=} \mathbf{V}_b = \frac{1}{2N}\begin{bmatrix} \mathrm{Re}(\Omega) & -\mathrm{Im}(\Omega) \\ \mathrm{Im}(\Omega) & \mathrm{Re}(\Omega) \end{bmatrix}^{-1} + \frac{1}{2N(K-1)}\begin{bmatrix} \mathrm{Re}(\Omega_0) & -\mathrm{Im}(\Omega_0) \\ \mathrm{Im}(\Omega_0) & \mathrm{Re}(\Omega_0) \end{bmatrix}^{-1}$$

and

$$\lim_{N\to\infty} \mathbf{K}_\sigma^{-1} \stackrel{\Delta}{=} V_\sigma = \frac{\sigma^4}{N} + \frac{\sigma_0^4}{N(K-1)} \tag{2-175}$$

Thus for large samples, the alternative expression for the likelihood ratio test statistic is given by

$$l = (\hat{\mathbf{b}} - \hat{\mathbf{b}}_0)^T \mathbf{V}_b^{-1}(\hat{\mathbf{b}} - \hat{\mathbf{b}}_0) + (\hat{\sigma}^2 - \hat{\sigma}_0^2)^2/V_\sigma \tag{2-176}$$

2-7-4 Performance Analysis

Recall that by definition, the real-valued vectors $\hat{\mathbf{b}}$ and $\hat{\mathbf{b}}_0$ are related to the AR model coefficient vectors by

$$\hat{\mathbf{b}} = \begin{pmatrix} \mathrm{Re}(\hat{\mathbf{a}}) \\ \mathrm{Im}(\hat{\mathbf{a}}) \end{pmatrix}, \qquad \hat{\mathbf{b}}_0 = \begin{pmatrix} \mathrm{Re}(\hat{\mathbf{a}}_0) \\ \mathrm{Im}(\hat{\mathbf{a}}_0) \end{pmatrix}$$

As shown in Section 2-4, the MLEs $\hat{\mathbf{a}}$ and $\hat{\mathbf{a}}_0$ are complex Gaussian distributed, that is

$$\hat{\mathbf{a}} \sim CN_M(\mathbf{a}, [N\Omega]^{-1})$$

$$\hat{\mathbf{a}}_0 \sim CN_M(\mathbf{a}_0, [N(K-1)\Omega_0]^{-1})$$

By the property of isomorphism of the complex Gaussian distribution, $\mathbf{b}$ and $\mathbf{b}_0$ are real Gaussian distributed [34]. Specifically, we have

$$\hat{\mathbf{b}} \sim N_{2M}\left(\mathbf{b}, \frac{1}{2}[N\boldsymbol{\Gamma}]^{-1}\right)$$

$$\hat{\mathbf{b}}_0 \sim N_{2M}\left(\mathbf{b}_0, \frac{1}{2}[N(K-1)\boldsymbol{\Gamma}]^{-1}\right) \tag{2-177}$$

where the $2M$-by-$2M$ matrices $\mathbf{\Gamma}$ and $\mathbf{\Gamma}_0$ are defined by

$$\mathbf{\Gamma} = \begin{bmatrix} \mathrm{Re}(\Omega) & -\mathrm{Im}(\Omega) \\ \mathrm{Im}(\Omega) & \mathrm{Re}(\Omega) \end{bmatrix}$$
$$\mathbf{\Gamma}_0 = \begin{bmatrix} \mathrm{Re}(\Omega_0) & -\mathrm{Im}(\Omega_0) \\ \mathrm{Im}(\Omega_0) & \mathrm{Re}(\Omega_0) \end{bmatrix} \tag{2-178}$$

From these distributions and the fact that $\hat{\mathbf{b}}$ and $\hat{\mathbf{b}}_0$ are mutually independent, since they are estimated from independent populations, we can easily calculate the first two moments of $(\hat{\mathbf{b}} - \hat{\mathbf{b}}_0)$, as shown by

$$\begin{aligned} E[\hat{\mathbf{b}} - \hat{\mathbf{b}}_0] &= (\mathbf{b} - \mathbf{b}_0) \\ \mathrm{Var}(\hat{\mathbf{b}} - \hat{\mathbf{b}}_0) &= \mathrm{Var}(\hat{\mathbf{b}}) + \mathrm{Var}(\hat{\mathbf{b}}_0) \\ &= \mathbf{V}_b \end{aligned} \tag{2-179}$$

Thus, according to a well-known result in statistics,

$$(\hat{\mathbf{b}} - \hat{\mathbf{b}}_0)^T \mathbf{V}_b^{-1} (\hat{\mathbf{b}} - \hat{\mathbf{b}}_0) \sim \chi^2(2M, \lambda_b) \tag{2-180}$$

with the noncentrality parameter λ_b given by

$$\lambda_b = (\mathbf{b} - \mathbf{b}_0)^T \mathbf{V}_b^{-1} (\mathbf{b} - \mathbf{b}_0) \tag{2-181}$$

In a similar manner, we can argue that

$$(\hat{\sigma}^2 - \hat{\sigma}_0^2)^2 / V_\sigma \sim \chi^2(1, \lambda_\sigma) \tag{2-182}$$

with the noncentrality parameter λ_σ given by

$$\lambda_\sigma = (\sigma^2 - \sigma_0^2) / V_\sigma \tag{2-183}$$

Since $(\hat{\mathbf{b}} - \hat{\mathbf{b}}_0)$ and $(\hat{\sigma}^2 - \hat{\sigma}_0^2)$ are independent, using the reproductive property of chi-square distribution, the test statistic l is also chi-square distributed. In particular, we have:

Theorem 2-4

$$l \sim \chi^2(2M + 1, \lambda) \tag{2-184}$$

where

$$\begin{aligned} \lambda &= \lambda_b + \lambda_\sigma \\ &= (\mathbf{b} - \mathbf{b}_0)^T \mathbf{V}_b^{-1} (\mathbf{b} - \mathbf{b}_0) + (\sigma^2 - \sigma_0^2)^2 / V_\sigma \end{aligned} \tag{2-185}$$

We are now in a position to evaluate the performance of the scheme by calculating the probabilities of detection and false alarm. Under hypothesis H_1, for a given threshold Λ_T, the probability of detection is given by

$$P_d = \int_{\Lambda_T}^{\infty} f_{\chi^2}(z; 2M+1, \lambda)\, dz \tag{2-186}$$

Under hypothesis H_0, with no target present in $\mathbf{X}_K$, we have

$$\begin{aligned} E[\hat{\mathbf{b}}] &= E[\hat{\mathbf{b}}_0] = \mathbf{b}_0 \\ E[\hat{\sigma}^2] &= E[\hat{\sigma}_0^2] = \sigma_0^2 \end{aligned} \tag{2-187}$$

which, together with (2-185), implies that $\lambda = 0$. Thus the false alarm rate is given by

$$P_{\text{fa}} = \int_{\Lambda_T}^{\infty} f_{\chi^2}(z; 2M+1, 0)\, dz \tag{2-188}$$

Obviously, for a given model order M, the detection probability P_d is an increasing function of the noncentrality parameter λ, which can be viewed as the distance between the two sets of second order statistics. From (2-186), the following observations on P_d can be made:

1. When a target is present, the distance between the parameter sets $(\mathbf{b}, \sigma^2)$ and $(\mathbf{b}_0, \sigma_0^2)$ is measured by λ whose first component corresponds mainly to the Doppler-shift of the signal and the second component corresponds to the power level.
2. When all the model parameters are given, the noncentrality parameter λ is a monotonically increasing function of the same size N. This implies that even for a weak target signal, detection probability can approach unity if a large number of samples are available.

2-7-5 Simulation Results

We next examine the theoretical performance of the new detection scheme. We have chosen two 4th order AR ($M = 4$) processes to represent the noise only and the target plus noise processes separately. These two processes have covariance matrices given, respectively by

$$\mathbf{R}_0 = [r_0(i, k)]$$

and

$$\mathbf{R} = [r_0(i, k) + \rho r_s(i, k)] = [r(i, k)], \qquad 1 \leq i, k \leq M$$

where the (i, k)th elements of the matrices are given by

$$r_s(i, k) = (0.5)^{|k-i|}(e^{-j0.3\pi})^{k-i} + (0.64)^{|k-i|}(e^{0.1\pi})^{k-i}$$

and

$$r_0(i, k) = (0.5)^{|k-i|}(e^{j0.6\pi})^{k-i}$$

and the parameter ρ is used to control the signal-to-noise ratio (SNR). We note that $\mathbf{R}$ and $\mathbf{R}_0$ are Toeplitz matrices.

For linear production theory [20], we can obtain the coefficient vectors such that

$$\mathbf{a}_0 = \mathbf{R}_0^{-1}\mathbf{r}_0 \tag{2-189}$$

and

$$\mathbf{a} = \mathbf{R}^{-1}\mathbf{r} \tag{2-190}$$

where

$$\mathbf{r}_0 = [r_0(2, 1) \quad r_0(3, 1) \quad \cdots \quad r_0(M+1, 1)]^T$$

and

$$\mathbf{r} = [r(2, 1) \quad r(3, 1) \quad \cdots \quad r(M+1, 1)]^T$$

The variance of the errors can be respectively calculated as

$$\sigma_0^2 = \det \mathbf{R}_0^A / \det \mathbf{R}_0 \tag{2-191}$$

and

$$\sigma^2 = \det \mathbf{R}^A / \det \mathbf{R} \tag{2-192}$$

where $\mathbf{R}_0^A$ and $\mathbf{R}^A$ are the augmented matrix of $\mathbf{R}_0$ and $\mathbf{R}$ defined respectively as

$$\mathbf{R}_0^A = \left[\begin{array}{c|c} r_0(1, 1) & \mathbf{r}_0^\dagger \\ \hline \mathbf{r}_0 & \mathbf{R}_0 \end{array}\right] \quad \text{and} \quad \mathbf{R}_A = \left[\begin{array}{c|c} \mathbf{r}(1, 1) & \mathbf{r}^\dagger \\ \hline \mathbf{r} & \mathbf{R} \end{array}\right]$$

Having computed all the parameters from (2-189) to (2-192), we are in the position to calculate the noncentrality parameter λ in (2-185) and then the probability of detection can be evaluated from (2-186) for a threshold predetermined by the allowable probability of false alarm in (2-188).

Figures 2-7(a), (b), and (c) show the theoretical probability of error at various SNR for the new method when the allowable false alarm rate is set at 10^{-2}, 10^{-3}, and 10^{-4}. For the sake of comparison, the same set of data is processed by two conventional detection schemes [63]. These schemes are respectively the normalization method in which the decision is made on the basis of a single test cell, and the combined normalization and integration method in which the decision is made by extracting the information from multiple test cells. The performance analyses of these conventional schemes for uncorrelated noise samples have been carried out [63], and have been extended to correlated noise [72]. It can be observed from Figs. 2-7(a), (b), and (c) that the new proposed method is far superior to the conventional methods theoretically. In our example, for the said P_d, the new method presents a SNR improvement of at least 8 dB compared to the combined normalization and integration, and a SNR improvement of at least 12 dB compared to the normalization method. Many other examples have been worked out and similar improvements have been observed.

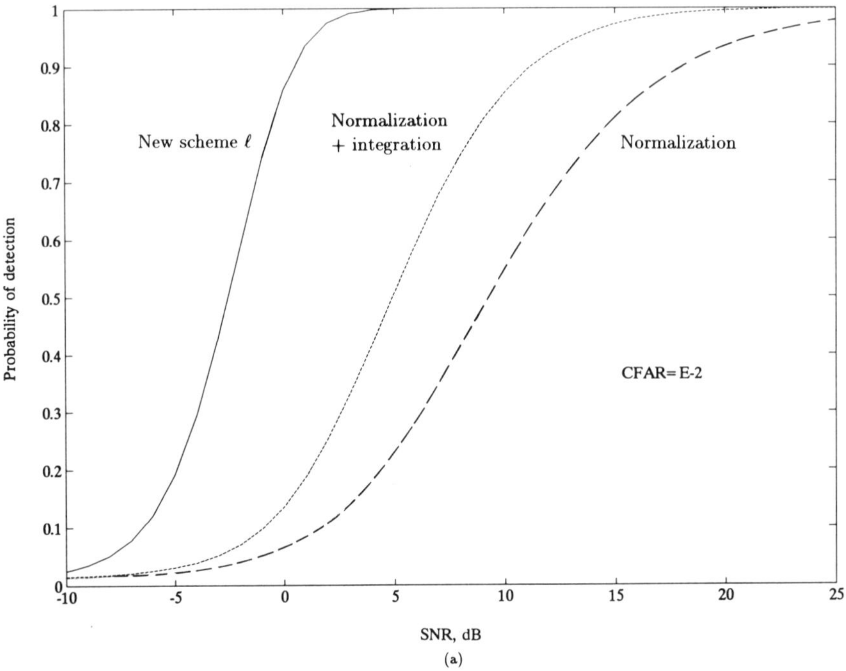

Figure 2-7 Performance comparison of the various methods. (a) $P_{fa} = 10^{-2}$; (b) $P_{fa} = 10^{-3}$; (c) = $P_{fa} = 10^{-4}$.

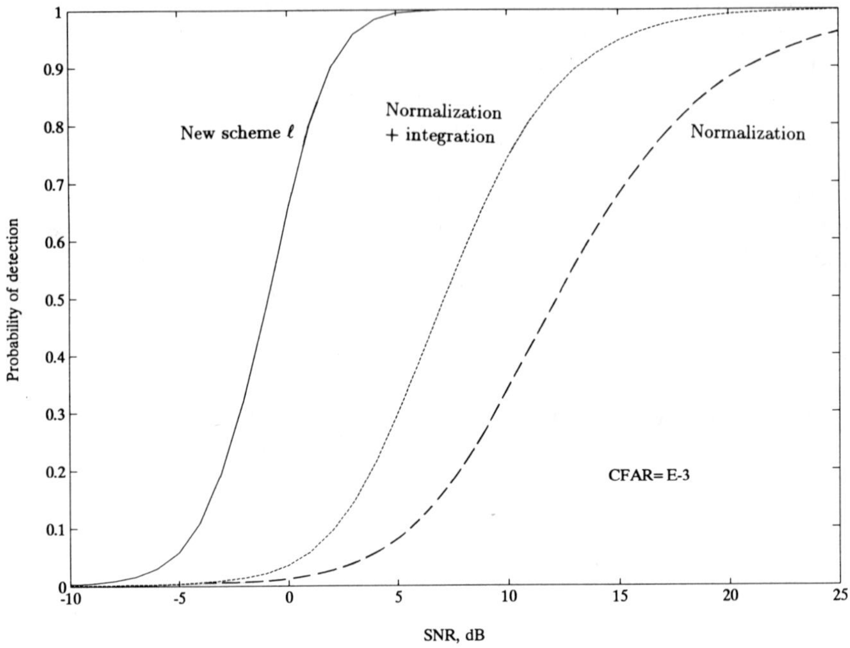

(b)

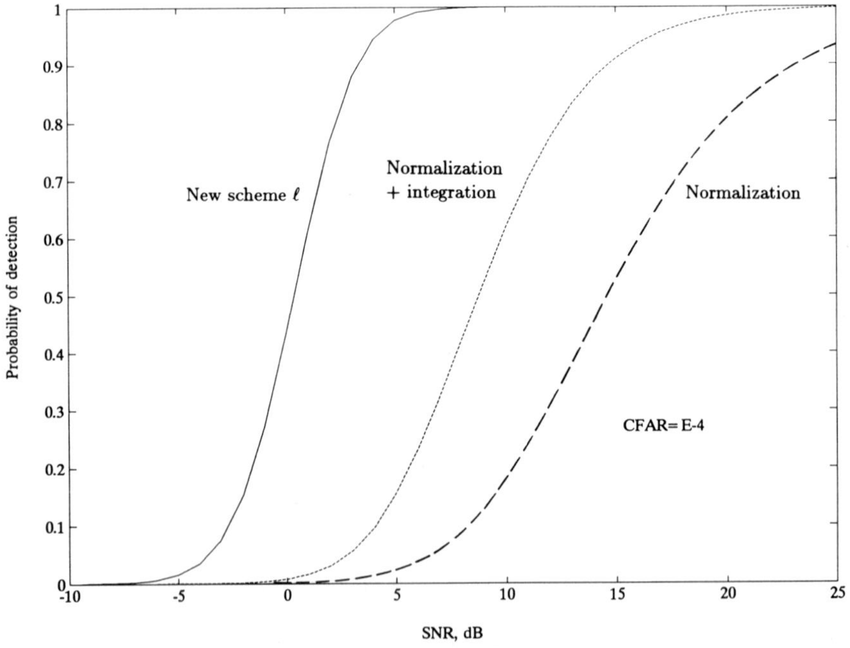

(c)

Figure 2-7 (*Continued*)

2-8 APPLICATION TO RADAR

We next examine how the technique described in Section 2-7 can be applied to practical situations of radar detection. We first show how a simple correction factor can be applied to the test statistic under the condition of limited data samples. We then examine the performance of the corrected test statistic using computer simulations so that the theoretical analysis can be verified. Finally, we apply the new method to real radar data of targets embedded in sea clutter.

2-8-1 A Modified Test Statistic for Limited Data

As shown in the previous sections, the test statistic l is asymptotically χ^2 distributed with $2M+1$ degrees of freedom. This implies that we need a large number of samples for the distribution to be true. In an actual radar target detection, however, we only have a theoretical limited data set. Under such circumstances, the χ^2-distribution is a biased approximate of the theoretical test statistic. Thus, if the new detection method is to be applied to actual radar data, a correction factor has to be introduced. In this section, we examine the bias involved in the estimation of the test statistic l using χ^2-distribution under H_0. From the evaluation of the bias, a correction factor is proposed so that the threshold X_T can be adjusted accordingly, thereby allowing the acceptable false alarm rate more accurately satisfied.

From (2-150), the test statistic is given by

$$l = 2N \log\{(\tilde{\sigma}_0^2)^K / [(\hat{\sigma}_0^2)^{K-1}(\hat{\sigma}^2)]\} \tag{2-193}$$

where $\tilde{\sigma}_0^2$ and $\hat{\sigma}_0^2$ are the estimated variances of the error process of the AR model for the clutter vectors under H_0 and H_1 respectively, and $\hat{\sigma}^2$ is the estimated variance of the error process of the AR model for the data vector containing signal and clutter under H_1. Expressions for these quantities are found in (2-142) and (2-146). Note that under H_0, $\tilde{\sigma}_0^2$, $\hat{\sigma}_0^2$, and $\hat{\sigma}^2$ are all maximum likelihood estimates of σ_0^2.

We next evaluate the bias that results if we assume l to be χ^2-distributed with $2M+1$ degrees of freedom under H_0, operating with a finite number of samples. To do this, we first introduce the following lemma.

Lemma 2-1 [66] If $x \sim \chi^2(k)$, then

$$E[\log x] = \log k - \frac{1}{k} - \frac{1}{3k^2} + \frac{2}{15k^4} + \mathrm{O}(k^{-6}) \tag{2-194}$$

The proof of this lemma is shown in [66].

Now we define the variables s_1, s_2, and s_3 such that

$$
\begin{aligned}
s_1 &= 2NK\tilde{\sigma}^2/\sigma_0^2 \\
s_2 &= 2N(K-1)\hat{\sigma}_0^2/\sigma_0^2 \\
s_3 &= 2N\hat{\sigma}^2/\sigma_0^2
\end{aligned}
\tag{2-195}
$$

We see that since s_1 involves NK complex random error variables and is subject to the constraints of M complex AR coefficients, then [12, 16]

$$
s_1 \sim \chi^2(2NK - 2M) \tag{2-196}
$$

The factor 2 in s_1 and in the degrees of freedom of its distribution is due to the fact that the model error is a complex process [52]. Similarly,

$$
s_2 \sim \chi^2(2N(K-1) - 2M) \tag{2-197}
$$

and

$$
s_3 \sim \chi^2(2N - 2M) \tag{2-198}
$$

Using (2-195) in (2-193), we can rewrite the expression of l in terms of s_1, s_2, and s_3 so that

$$
l = 2N \log\left\{\left[\frac{s_1^K}{s_2^{K-1} s_3}\right]\left[\frac{(K-1)^{K-1}}{K^K}\right]\right\} \tag{2-199}
$$

To show that $x^2(2M+1)$ is a biased approximation to the distribution of l under H_0, we evaluate the mean of l, that is,

$$
\begin{aligned}
E[l] = {} & 2N\{KE[\log s_1] - (K-1)E[\log s_2] - E[\log s_3]\} \\
& + 2N\log\{(M-1)^{M-1}/K^K\}
\end{aligned}
\tag{2-200}
$$

Now, applying Lemma 2-1 to the terms inside the first set of braces in (2-200) and simplifying, we obtain

$$
\begin{aligned}
E(l) = 2N\Big\{ & K\log\left(1 - \frac{M}{NK}\right) - (K-1)\log\left(1 - \frac{M}{N(K-1)}\right) - \log\left(1 - \frac{M}{N}\right) \\
& + \frac{1}{2}\left[\frac{1}{N-M} + \frac{K-1}{N(K-1)-M} - \frac{K}{NK-M}\right] \\
& + \frac{1}{12}\left[\frac{1}{(N-M)^2} + \frac{K-1}{(N(K-1)-M)^2} - \frac{K}{(NK-M)^2}\right]\Big\} + O(N^{-4})
\end{aligned}
\tag{2-201}
$$

Taking the series expansion of the log terms and the terms involving N^{-1} in (2-201), we find that for any constant α and $0 \leqslant \alpha \leqslant N$,

$$\log\left(1 - \frac{\alpha}{N}\right) = -\sum_{k=1}^{\infty} \frac{1}{k}\left(\frac{\alpha}{N}\right)^k \tag{2-202}$$

and

$$\frac{1}{N-\alpha} = \frac{1}{N}\sum_{k=0}^{\infty}\left(\frac{\alpha}{N}\right)^k \tag{2-203}$$

After simplification, we find that the expected value of l is

$$E[l] = (2M+1) + \beta = E[\chi^2(2M+1)] + \beta \tag{2-204}$$

where we have identified $(2M+1)$ to be the mean of a random variable having a χ^2-distribution of $(2M+1)$ degrees of freedom, and the constant β is given by

$$\begin{aligned}\beta = 2N \sum_{k=2}^{\infty} \frac{1}{k}\left\{-K\left(\frac{M}{NK}\right)^k + (K-1)\left(\frac{1}{N(K-1)}\right)^k + \left(\frac{M}{N}\right)^k\right\} \\ + \sum_{k=1}^{\infty}\left(\frac{M}{N}\right)^k\left\{1 + \frac{1}{(K-1)^k} - \frac{1}{K^k}\right\} \\ + \frac{N}{6}\left\{\frac{1}{(N-M)^2} + \frac{K-1}{(N(K-1)-K)^2} - \frac{K}{(NK-M)^2}\right\} + \mathrm{O}(N^{-4})\end{aligned}$$

β is the amount of bias if the test statistic l is approximated by a variable of χ^2-distribution with $2M+1$ degrees of freedom. This bias is of $\mathrm{O}(N^{-1})$ and tends to zero as $N \to \infty$. For a limited number of samples, this bias can be significant.

To correct this bias we multiply the test statistic l by a constant $1/c$ such that

$$\begin{aligned}c = \frac{2N}{2M+1}\left\{K\log\left(1 - \frac{M}{MK}\right) - (K-1)\log\left(1 - \frac{M}{N(K-1)}\right) - \log\left(1 - \frac{M}{N}\right)\right. \\ + \frac{1}{2}\left[\frac{1}{N-M} + \frac{K-1}{N(K-1)-M} - \frac{K}{NK-M}\right] \\ \left. + \frac{1}{12}\left[\frac{1}{(N-M)^2} + \frac{K-1}{(N(K-1)-M)^2} - \frac{K}{(NK-M)^2}\right]\right\}\end{aligned} \tag{2-205}$$

Comparing (2-205) and (2-201), we see that

$$c = \frac{1}{2M+1}\{E[l] + O(N^{-4})\}$$

so that

$$E[l/c] = (2M+1) - O(N^{-4}) \tag{2-206}$$

Equation (2-206) tells us that if the test statistic l obtained under H_0 and under a limited number of samples is divided by the constant c, then the resulting test statistic is a variable much better approximated by a χ^2-distribution with $2M+1$ degrees of freedom. The bias for this approximation is of $O(N^{-4})$, which is negligible for most practical cases. With this correction factor in place, we can establish a more accurate detection threshold X_T from the allowable probability of false alarm P_{fa}, that is

$$P_{fa} = \int_{\bar{\Lambda}_T/c}^{\infty} f_{\chi^2}(x; 2M+1, 0)\, dx \tag{2-207}$$

from which $\bar{\Lambda}_T/c$ and hence $\bar{\Lambda}_T$ can be determined. The likelihood ratio test given by (2-151) is then modified as follows:

$$l \underset{H_0}{\overset{H_1}{\gtrless}} \bar{\Lambda}_T \tag{2-208}$$

Comparing (2-188) with (2-207), it is easily seen that the corrected threshold Λ_T is related to the uncorrected one Λ_T by $\bar{\Lambda}_T = c\Lambda_T$. In this case, the probability of detection is

$$P_d = \int_{\Lambda_T}^{\infty} f_{x^2}(x; 2M+1, \lambda)\, dx \tag{2-209}$$

2-8-2 Computer Simulation Tests

In this subsection, we verify the theory using computer generated data. We first examine the theoretical prediction for Λ_T and $\bar{\Lambda}_T$ which are, respectively, the uncorrected and corrected threshold levels for the test statistic l. In our verification, we employ a second order AR model ($M = 2$) the parameters of which are

$$\begin{aligned} a &= a_0 = [1 \quad -0.72 - j0.5 \quad 0.30 + j0.25] \\ \sigma^2 &= \sigma_0^2 = 2 \\ K &= 4 \end{aligned} \tag{2-210}$$

A total of 5×10^4 independent trials were performed and the values of l

evaluated. The values of the allowable probability of false alarm P_{fa} were chosen to be 10^{-1}, 10^{-2}, 10^{-3}, and 10^{-4}. The experimental values of thresholds for the statistic l were evaluated for the sample sizes $N = 25$ and $N = 40$. The results are shown respectively in Tables 2-2(a) and 2-2(b). The theoretical threshold values predicted by Λ_T and $\bar{\Lambda}_T$ which are evaluated by using a $\chi^2(5)$ distribution are also shown in the tables for comparison. It can be seen that in all cases, the corrected threshold $\bar{\Lambda}_T$ is much closer to the simulation value. This is especially true for a small number of samples. As N increases from 25 to 40, the discrepancies between Λ_T and $\bar{\Lambda}_T$ decreases confirming that the bias of the approximation has less effect for large N.

Next, we verify the theoretical evaluation of the probability of detection P_d given by (2-209) using computer simulations. The theoretical value of P_d is calculated as follows:

1. Based on the order of the AR model, an appropriate threshold for a given P_{fa} is selected using (2-207).
2. Based on the model parameters $(\mathbf{b}, \sigma^2)$ and $(\mathbf{b}_0, \sigma_0^2)$, the noncentrality parameter λ is calculated using (2-185) in the previous section.

TABLE 2-2 Comparison of Corrected and Uncorrected Thresholds for Different P_{fa}

(a) $N = 25$

P_{fa}	Threshold Values for l Computed from Simulated Data	Theoretical Threshold Predicted Using: Λ_T (Eq. 2-188)	Theoretical Threshold Predicted Using: $\bar{\Lambda}_T$ (Eq. 2-207)
10^{-1}	10.15811	9.2364	9.7634
10^{-2}	16.4199	15.086	15.947
10^{-3}	21.8874	20.515	21.686
10^{-4}	26.3631	25.745	27.214

(b) $N = 40$

P_{fa}	Threshold Values for l Computed from Simulated Data	Theoretical Threshold Predicted Using: Λ_T (Eq. 2-188)	Theoretical Threshold Predicted Using: $\bar{\Lambda}_T$ (Eq. 2-207)
10^{-1}	9.4407	9.2364	9.3617
10^{-2}	15.3965	15.086	15.9291
10^{-3}	20.8951	20.515	20.794
10^{-4}	26.6344	25.745	26.095

3. The probability of detection, P_d is then evaluated using (2-209).

For computer simulations, we use a second order AR model described in (2-210) to generate the noise process. Again we choose three reference sequences ($K = 4$). The data process under H_1 is modeled by another second order AR process with model coefficients

$$\mathbf{a} = [1 \quad -0.64 - j0.55 \quad 0.32 + j0.27]$$

We leave the error power σ^2 to be variable. The coefficient vector **a** has been selected so that the power spectrum of the received data under H_1 is only slightly different from that under H_0. Figure 2-8 shows the two spectra under the two hypotheses. The similarity of the spectra is purposely chosen so that a reasonably difficult task is in hand and that the improvement of P_d can be more discernible for a change in sample size.

Figure 2-9(a) shows the theoretical and the simulation values of P_d as a function of σ^2/σ_0^2 for $P_{\text{fa}} = 10^{-1}$ and 10^{-2} respectively and $N = 100$. The simulation value is obtained from 200 trials. We observe that there is a reasonably good argument between the theoretical and simulation results.

Figure 2-9(b) shows the theoretical and simulation values of P_d as a function of N with $\sigma^2/\sigma_0^2 = 1.2$. Again, the comparison is performed for $P_{\text{fa}} = 10^{-1}$ and 10^{-2} and the simulation value obtained from 200 trials. As expected, P_d increases with N, and reasonable agreement between the theoretical and simulation results is observed.

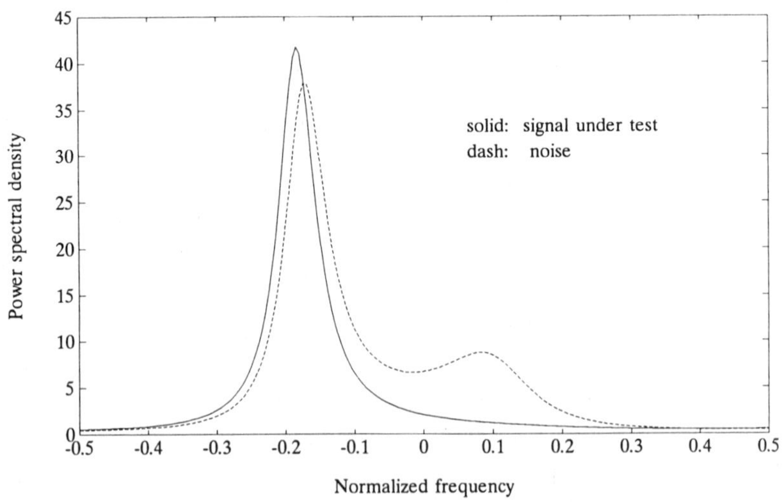

Figure 2-8 Power spectra of the simulated data under the two hypotheses.

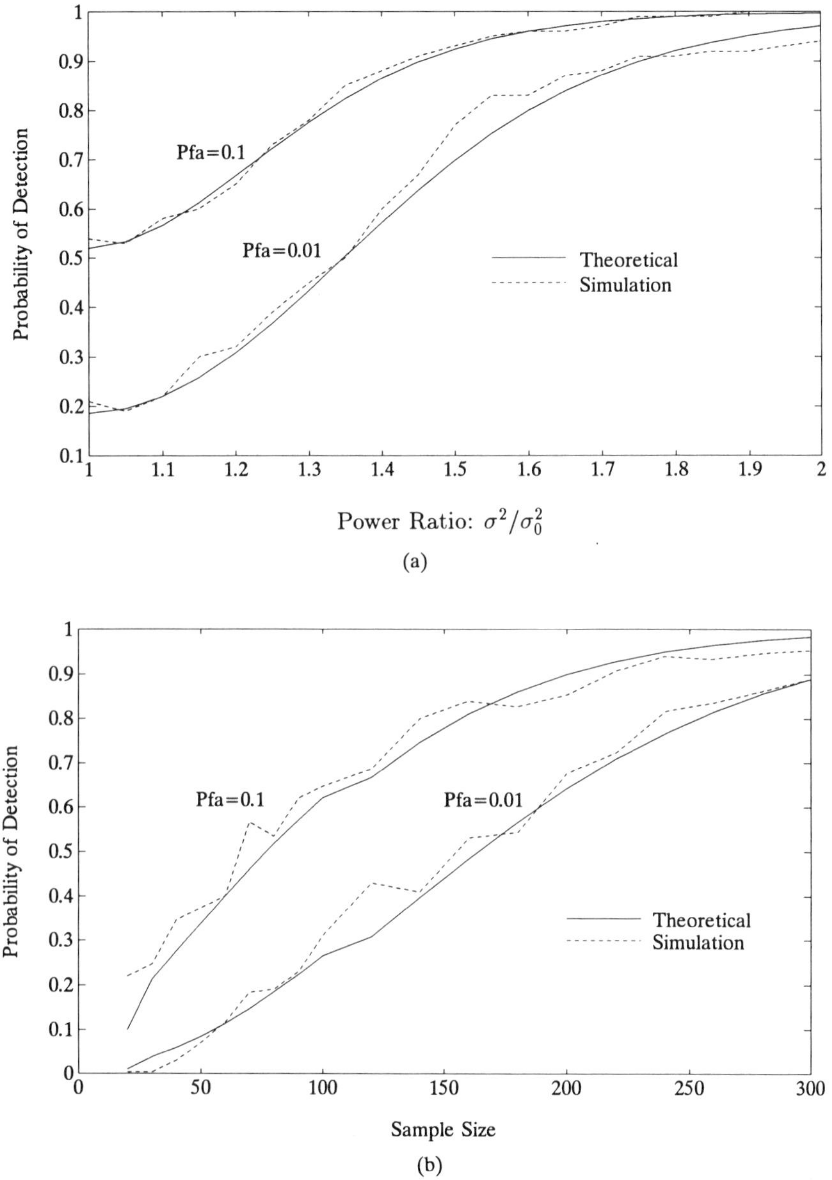

Figure 2-9 Comparison of theoretical and simulated performance of the l test. (a) P_d vs. σ^2/σ_0^2; (b) P_d vs. sample size.

2-8-3 Application to Actual Radar Data

In this subsection, we apply our method of detection on actual radar data. The radar data used for our experiments were collected using an instrument-quality X-band radar known as the IPIX radar [67]. The IPIX radar system was designed and constructed at the Communications Research Laboratory, McMaster University, and is used for research in the application of radar to navigation in ice-infested waters. In heavy seas, iceberg fragments called *bergy bits* (about the size of a small house) and *growlers* (about the size of a concert grand piano) are very difficult to detect using a conventional marine radar, and yet they pose a significant hazard to vessels. The reliable detection of such iceberg fragments in open water is vital to safe navigation. The IPIX radar was mounted at a site located at Cape Bonavista, Newfoundland, looking over the North Atlantic Ocean. Digital control allows experimental parameters to be easily and accurately varied. Built-in calibration equipment permits quantitative measurements to be made with high confidence. The data acquisition system delivers digital radar data at real-time rate into the address space of a computer where it may be validated, processed and stored on removable media for off-line analysis. The antenna of the radar was mounted 22 m above the sea level to simulate the actual relative antenna position of a shipborne radar. Details of the radar features are reported in [67]. The operating specifications are summarized as follows:

Transmitter frequency	9.39 GHz
Pulse width	200 ns
Sample rate	30 MHz
Pulse repetition frequency (PRF)	2000 Hz
Antenna beamwidth (θ)	2°
Scan rate	30 resolution/min (or = π rad/s)
Polarization	Horizontal

From this table, we may calculate some useful parameters:

$$\text{Number of hits on a point target/scan} = \theta\text{PRF/scan rate} = 22$$

Wavelength of the transmitted signal is

$$\lambda_0 = \frac{3 \times 10^8}{9.39 \times 10^9} = 0.03195\text{M}$$

If we denote the radial velocity of a target by V_r, then the normalized Doppler frequency is given by

$$f_d = \frac{2V_r}{\lambda_0 \times \text{PRF}} = \frac{2V_r}{63.9}$$

The data to be processed are those of sea clutter containing a growler. The significant wave height is 1.31 M. The clutter data were returned from a swath of sea surface, spanning 95–109° in azimuth and 4.71–5.71 km in range. This corresponds to 101 samples in range and 150 samples in azimuth, thereby producing 101 × 150 samples for each scan. The amplitude contours of 4 scans of sea clutter data are shown in Fig. 2-10(a)–(d) in which the locations having dense contour lines indicate increasing clutter powers.

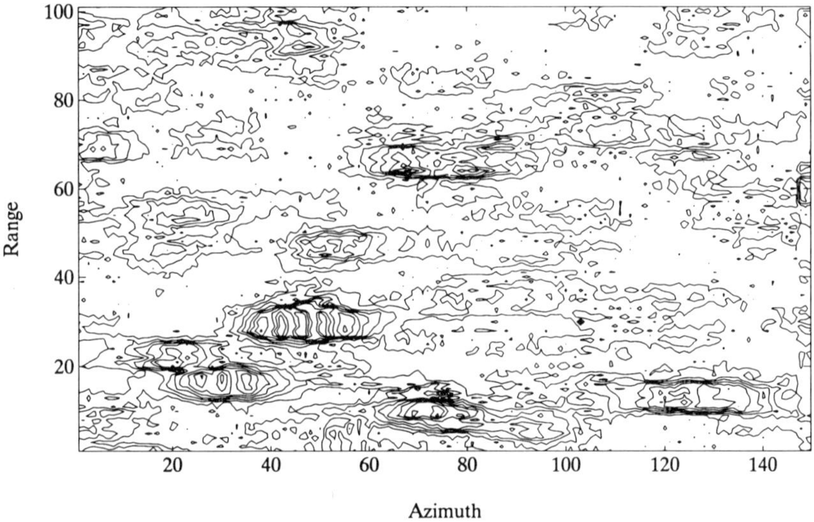

(a)

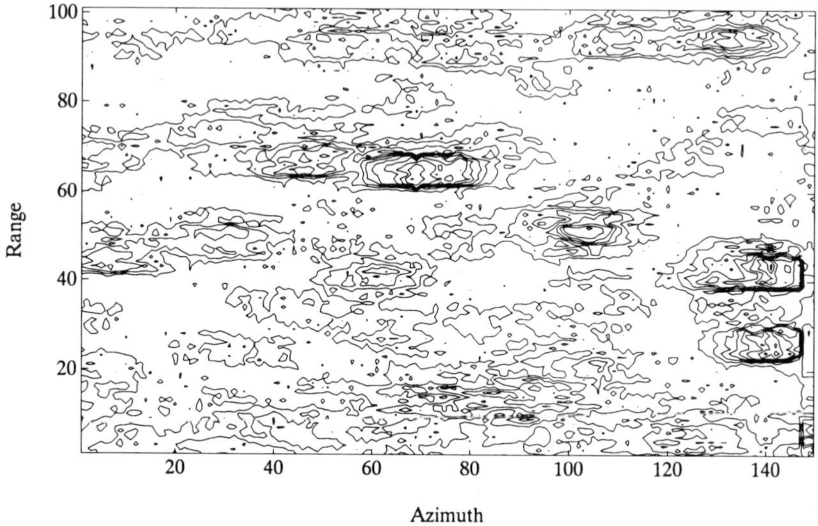

(b)

Figure 2-10 Amplitude contours for sea clutter. (a) Scan 1; (b) scan 2; (c) scan 3; (d) scan 4.

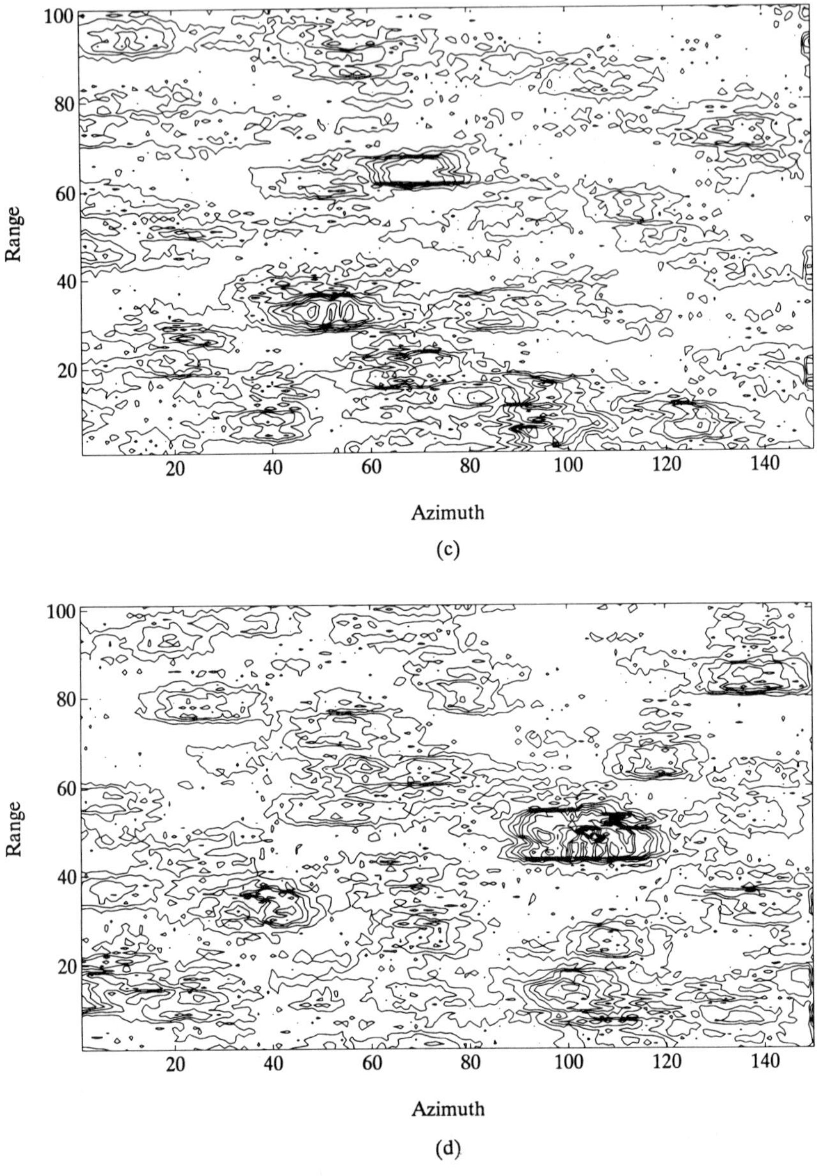

Figure 2-10 *(Continued)*

Figure 2-11 shows the spatial nonstationarity of the sea clutter. Here, p_c denotes the average local power of the sea clutter and p_g are calculated by averaging 25 local returned samples. We observe from Fig. 2-11 that the sea clutter ranges in local power from 128.27 units to 10,255.00 units. The power of the growler, on the other hand, is 4747.2 units, being lower than the power of many sea waves.

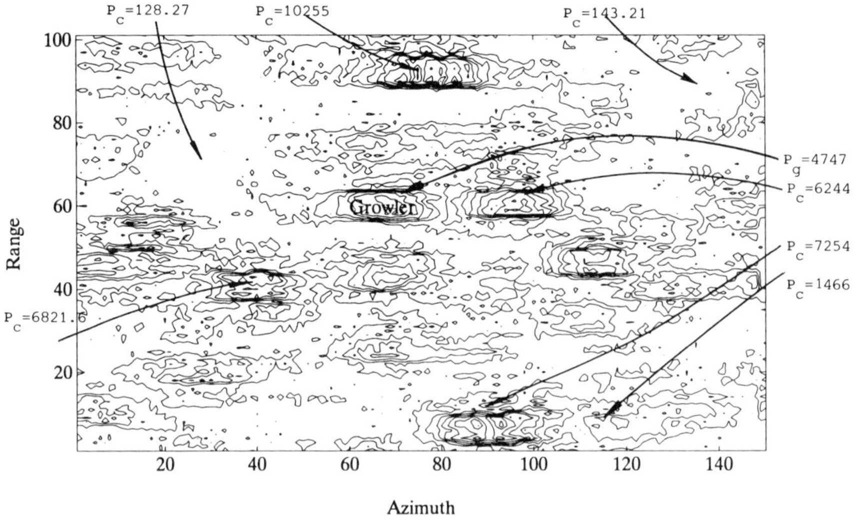

Figure 2-11 Comparison of the power at various parts in a typical scan of sea clutter.

The Doppler peak of sea clutter can be visualized by plotting the power spectrum of the data. The latter can be estimated through the use of the maximum entropy method [21, 68]. The AR coefficients and the variance of model error required for the spectrum estimation were obtained from 25 data samples along a range ring. The spectrum so obtained reflects, to some extent, the second order information of sea clutter.

Figure 2-12(a) shows the variation of Doppler shift with range in the same scan. The spectra are drawn for the data obtained on the same azimuth but different range. It is observed that Doppler frequency changes little with range. This is because the Doppler peak stands for the radial velocity of the sea waves and this velocity is subject to little change in the range direction.

Figure 2-12(b) shows the change of Doppler shift with azimuth, also in the same scan. The spectra are drawn for the data obtained on the same range, but different azimuth. In this case, Doppler frequency undergoes a small change showing that the radial velocities of sea waves in different azimuths are slightly different.

Figure 2-12(c) shows the variation of Doppler frequency with scan. The data used to estimate the spectra are obtained in the same location but different scans. Since the velocities of sea waves observed at different scans experience considerable changes, so would be the Doppler frequency, as shown in Fig. 2.12(c).

Figure 2.12(d) shows the power spectra of the data containing the growler at fixed azimuth but at different ranges. We observe that the variation of the Doppler frequency for various ranges is quite significant, showing that the growler is probably having rotational motion and also that the current velocity around the growler is varying quite considerably.

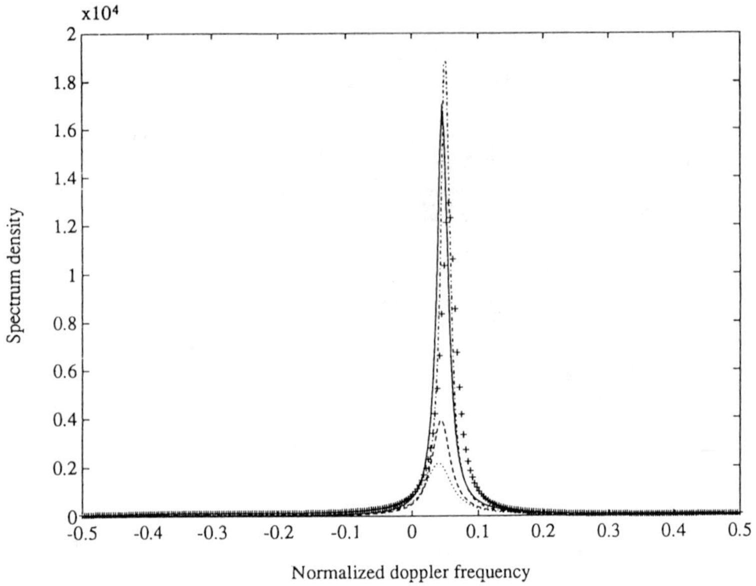

(a)

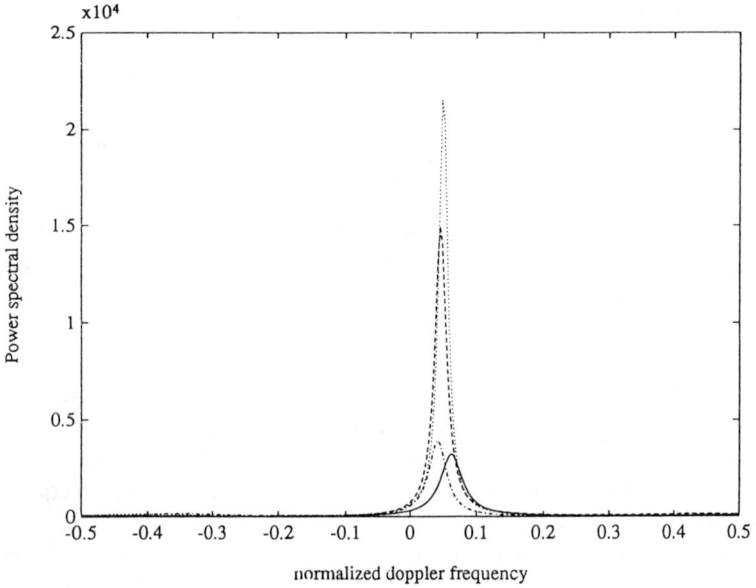

(b)

Figure 2-12 Variation of Doppler frequency. (a) Change with range; (b) change with azimuth; (c) change with scan; (d) growler Doppler.

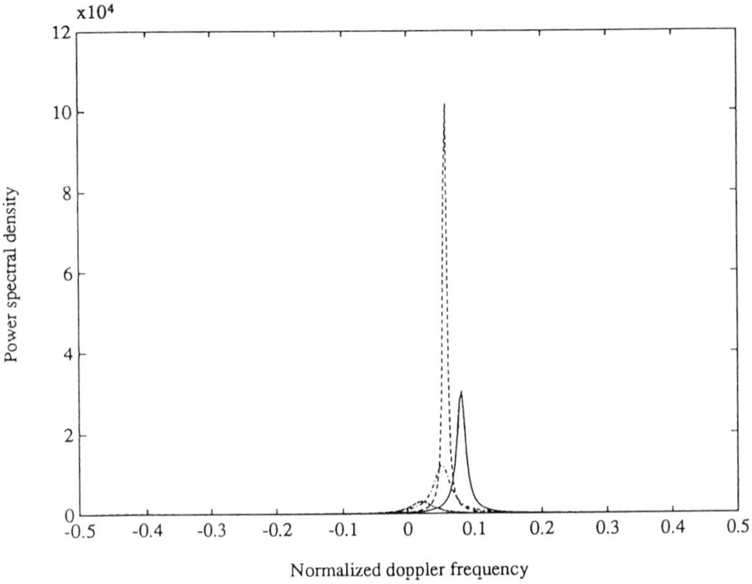

(c)

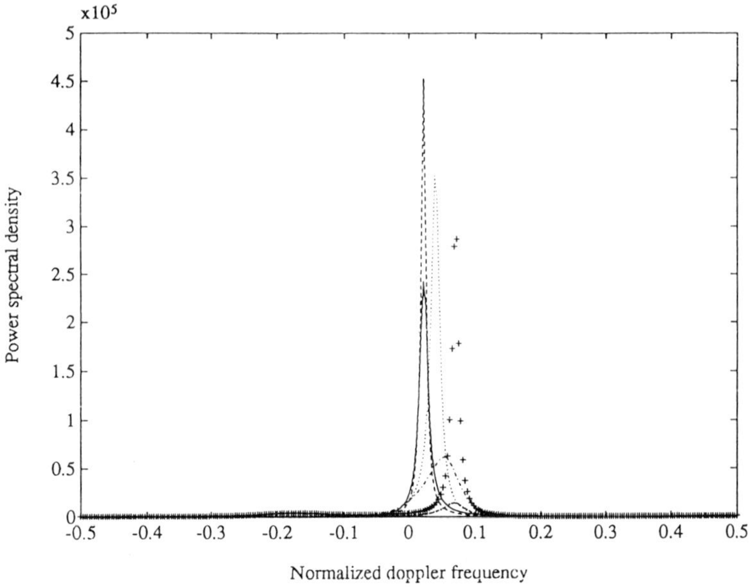

(d)

Figure 2-12 (*Continued*)

To facilitate an examination of the variation of radar detection performance with the signal-to-clutter ratio (SCR), we put in an artificial target which is embedded in sea clutter. The target signal is given by

$$s(t) = A \exp(-0.005t^2 + j2\pi f_c t)$$

where A is a constant controlling the amplitude (and therefore the SCR) of the signal and $f_c = 0.2$.

The Gaussian envelope is chosen to model the modulation effect of the antenna beam pattern. The coefficient of 0.005 is chosen, so that the number of significant nonzero samples is approximately equal to the number of hits on a point target. We apply the new method to the detection of this artificial target embedded in actual sea clutter data. In order to provide a comparison of detection performance, we apply two other detection methods in our experiments, one based on power information only and the other based on Doppler information. The test statistics for these two methods are respectively given by

$$l_p = 2 \sum_{i=1}^{25} |y_i|^2 / p_N$$

$$l_D = \omega_{DT} - \hat{\omega}_{DC}$$

where $\{y_i\}$ are the samples under test, p_N is the clutter power estimated from neighboring ranges, ω_{DT} and $\hat{\omega}_{DC}$ are respectively the Doppler frequencies of the test data and clutter estimated from neighboring ranges. The decision rules for these methods are similar to that for the new method, that is,

$$l_p \underset{H_0}{\overset{H_1}{\gtrless}} X_{Tp}$$

$$l_D \underset{H_0}{\overset{H_1}{\gtrless}} X_{TD}$$

where X_{Tp} and X_{TD} are respectively the threshold levels for the two methods. While X_{Tp} can be determined theoretically by assuming a Gaussian distribution of the sea clutter which has been shown [69] to be an inaccurate assumption, the determination of X_{TD} is usually done by subjective judgment. In our experiments, the two thresholds are determined by examining 100 scans of data from areas which are known to contain no target and thereby setting the threshold levels for which the respective test statistics satisfy the prescribed false alarm rate.

Comparisons of the detection performance of the three methods using l, l_p, and l_D for various false alarm rates are shown in Figs. 2.13(a), (b) and

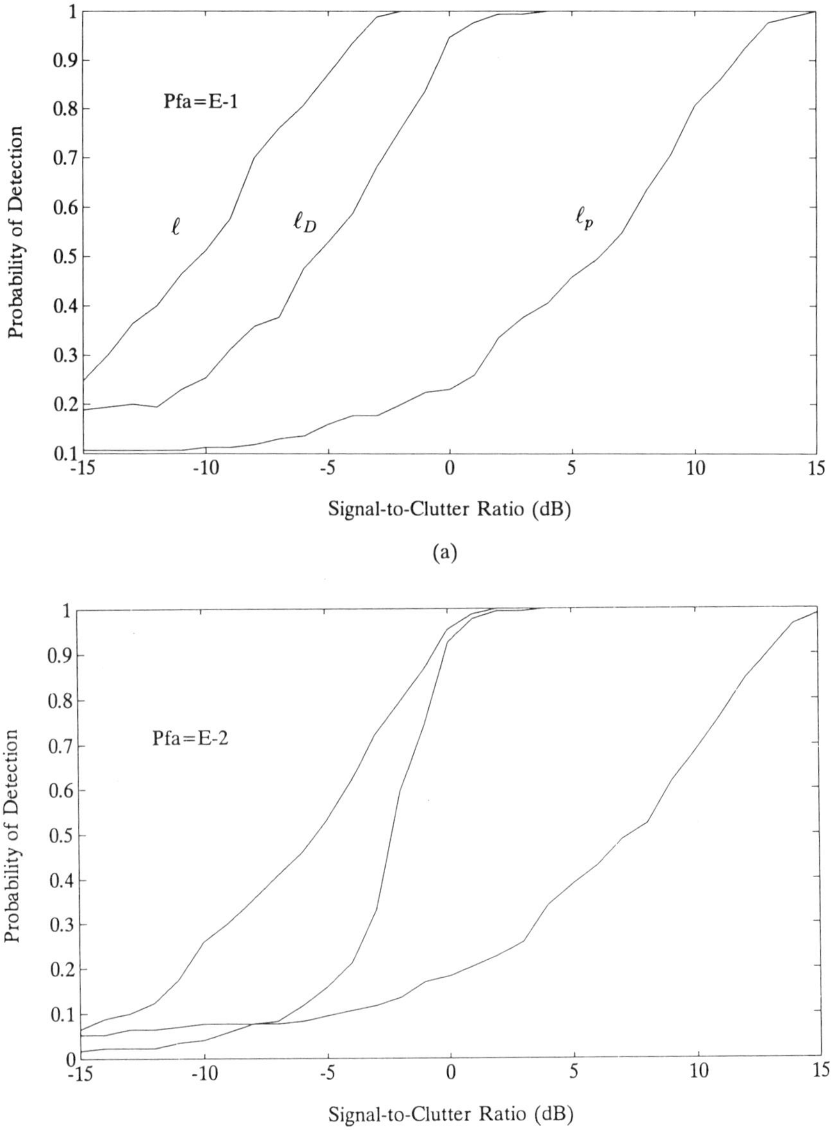

(a)

(b)

Figure 2-13 Performance comparison of the various methods. (a) $P_{fa} = 10^{-1}$; (b) $P_{fa} = 10^{-2}$; (c) $P_{fa} = 10^{-3}$.

(c). Each point of these curves is calculated from 170 trials. In all three cases of different false alarm rates, the method employing Doppler information performs much better than the method using power information. Thus we may conclude that the spatial nonstationarity of sea clutter renders the use of target amplitude information for detection more ineffective than the use

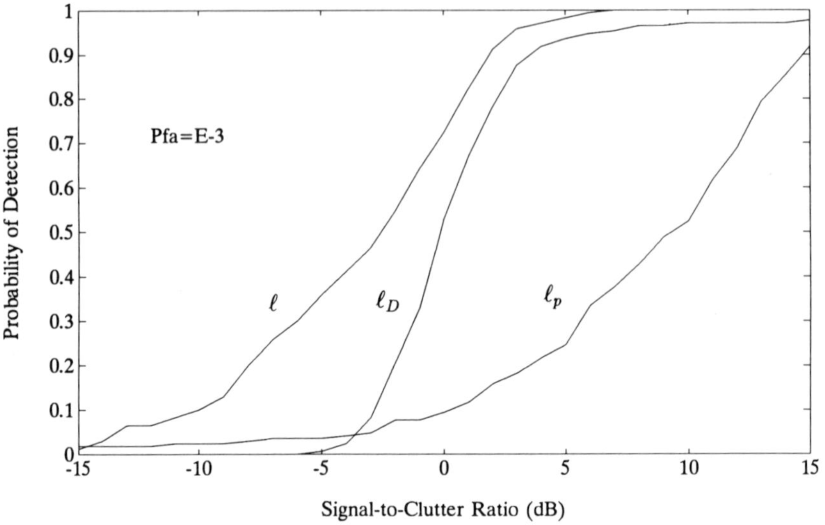

(c)

Figure 2-13 (*Continued*)

of phase information. The new method with test statistic l combines both amplitude and phase information. Hence, as evidenced in Figs. 2.13(a), (b) and (c), it exhibits a detection performance that is superior to either of the other two methods.

2-9 SUMMARY

Wold's Decomposition Theorem reveals one of the most important structures of wide-sense stationary processes, providing a theoretical foundation for modern filtering and linear prediction. From a practical point of view, stationary processes of rational power spectra are particularly important, since they can be expressed as ARMA processes of finite parameters. However, among all stationary Gaussian processes with a rational power spectrum, only the AR process admits a sufficient statistic that does not grow in dimension as the number of observations increases. Furthermore, the AR process with a rational spectrum can be used to approximate a stationary process of arbitrary irrational power spectrum. Therefore, in this chapter, we have concentrated on the construction of signal detection schemes using AR modeling techniques.

Target detection is a process of extracting information about a target of interest from the given data. Under an AR assumption, the information in the data can be characterized by different sets of parameters. We have discussed in detail the connections between these different sets of parameters.

Target detection can be formulated as a hypothesis testing problem in terms of a set of AR parameters. A theoretical performance analysis of these detection schemes requires knowledge of the sample distributions of the AR parameters sets. Some of these distributions have been established in Section 2-3.

We have studied three AR-based detection schemes, one for signal-in-white noise and two for signal-in-colored noise. In all these schemes, the target signal and noise are both assumed to have unknown statistics. The scheme for white noise is actually a detector using an AR spectral analyzer, which has been used for a long time. What is new in this chapter is its theoretical analysis. Computer simulations have been used to confirm validity of the theory described herein.

The first scheme for detecting a target in colored noise is a detection scheme closely related to the Gram-Schmidt orthogonalization procedure. The idea is to express the likelihood ratio in terms of the innovation process. The detection performance of this scheme depends on how much statistical knowledge we have about the noise. In actual radar environments, only little prior information is available, rendering this scheme to have a relatively low detection capability. However, this scheme has its own merit. It provides insight into the optimum estimator-correlator receiver and it is easy to implement.

The second scheme for target detection in colored noise is a much more powerful detection technique. This scheme exploits not only the full information of second order statistics in the data under test but also that from the reference data. The idea is to compare the statistical similarity between the data under test and the data from adjacent range rings. This is easily done by using a maximum likelihood ratio test. The test is simple, but the performance analysis is quite involved. It has been shown that the test statistic is asymptotically distributed as central chi-square under hypothesis H_0, and distributed as noncentral chi-square under H_1. These results provide a basis for determining the threshold for a given false alarm rate, and calculating the detection probability. In order for the proposed scheme to be useful for cases in which only a limited number of data samples are available, a modified test statistic has been established. This scheme has been tested with both simulated and actual radar data, confirming the correctness of our theory and demonstrating a superior performance compared to conventional techniques.

REFERENCES

[1] P. Swerling (1960) "Probability of detection of fluctuating targets", *IRE Trans. Inform. Theory*, **6**, 269–308.

[2] T. S. Edrington (1965) "The amplitude statistics of aircraft radar echoes", *IEEE Trans. Military Electron.*, **9** 10–16.

[3] M. A. Weiner (1988) "Detection probability for partially correlated chi-square targets", *IEEE Trans. Aerosp. Electron. Systems*, **24**, 411–416.

[4] G. U. Yule (1921) "On the time-correlation problem, with especial reference to variate difference correlation method", *J. R. Statist. Soc.*, **84**, 497–526.

[5] E. Slutsky (1937) "The summation of random causes as the source of cyclic processes", Problems of Economic Conditions, Vol. 3 (English translation), *Econometrica*, **5**, 105–147.

[6] H. Wold (1954) *A Study in the Analysis of Stationary Time Series*, 2nd edition, Almqvist & Wiksell, Stockholm.

[7] R. Frisch, (1933) "Propagation problems and impulse problems in dynamic economics", in *Economic Essays in Honor of Gustav Cassal*, Allen & Unwin, London, pp. 171–205.

[8] J. Tinbergen (1937) *An Economic Approach to Business Cycle Problems*, Hermann, Paris.

[9] A. N. Kolmogorov (1941) "Interpolation and extrapolation von stationaren zufalligen folgen", *Bull Acad. Sci. USSR Math. Ser.*, **5**, 3–14.

[10] N. Wiener (1970) *Extrapolation, Interpolation, and Smoothing of Stationary Time Series*, MIT Press, Cambridge, MA.

[11] A. N. Kolmogorov (1941) "Stationary sequences in Hilbert space", *Bull. Math. Univ. Moscow*, **2**, No. 6.

[12] E. J. Hannan (1970) *Multiple Time Series*, Wiley, New York.

[13] L. H. Koopmans (1974) *The Spectral Analysis of Time Series*, Academic Press, New York.

[14] J. L. Doob (1953) *Stochastic Processes*, Wiley, New York.

[15] B. W. Dickinson (1981) "Properties and applications of Gaussian autoregressive process in detection theory", *IEEE Trans. Inform. Theory*, **27**, 343–347.

[16] W. A. Fuller (1976) *Introduction to Statistical Time Series*, Wiley, New York.

[17] S. Haykin, B. W. Currie, and S. B. Kesler (1982) "Maximum-entropy spectral analysis of radar clutter", *Proc. IEEE*, **70**, (1986) 953–962.

[18] P. H. Thomas and S. Haykin (1986) "Stochastic modeling of radar clutter", *IEE Proc.*, Part F, **133**, No. 5.

[19] D. Alexandrou (1987) "Signal recovery in a reverberation-limited environment", *IEEE J. Oceanic Eng.*, **12**, 553–559.

[20] S. Haykin (1991) Adaptive Filter Theory, 2nd edition, Prentice Hall, Englewood Cliffs, NJ.

[21] S. Haykin (Ed.) (1983) *Nonlinear Methods of Spectral Analysis*, 2nd edition, Wiley, New York.

[22] S. M. Kay (1988) *Modern Spectral Estimation: Theory and Application*, Prentice Hall, Englewood Cliffs, NJ.

[23] U. L. F. Grenader and M. Rosenblatt (1957) *Statistical Analysis of Stationary Time Series*, Wiley, New York.

[24] J. P. LeCadre (1989) "Parametric methods for spatial signal processing in the presence of unknown colored noise fields", *IEEE Trans. Acoust., Speech, Signal Process.*, **37**, 965–979.

[25] T. Kailath, A. Vieira, and M. Morf (1978) "Inverse of Toeplitz operators, innovations and orthogonal polynomials", *SIAM Rev.*, **20**, 106–119.

[26] M. Kendall and A. Stuart (1979) *The Advanced Theory of Statistics, Vol. 2: Inference and Relationship*, 4th edition, Charles Griffin, London.

[27] G. E. P. Box and G. M. Jenkins (1976) *Time Series Analysis: Forecasting and Control*, revised edition, Holden-Day, San Francisco, CA.

[28] P. J. Brockwell and R. A. David (1987) *Time Series: Theory and Methods*, Springer-Verlag, New York.

[29] S. M. Kay and J. Makhoul (1983) "On the statistics of the estimated reflection coefficients of an autoregressive process", *IEEE Trans. Acoust., Speech, Signal Process.*, **31**, 1447–1455.

[30] A. S. Arslanian and T. S. Durrani (1987) "Detection performance of complex lattice processors", *IEEE Trans. Acoust., Speech, Signal Process.*, **35**, 1539–1145.

[31] Q. T. Zhang (1992) "On Performance analysis of complex lattice processors for signal detection", *IEEE Trans. Acoust., Speech, Signal Process.*, to appear in the November issue.

[32] Q. T. Zhang and H. S. Miao (1991) "Sample distributions of the partial correlation coefficients of a complex autoregressive process", in *Proc. IEEE ICASSP-91*, Toronto.

[33] N. R. Goodman (1963) "Statistical analysis based on a certain multivariate complex Gaussian distribution (an introduction)", *Ann. Math. Statist.*, **34**, 152–177.

[34] K. S. Miller (1974) *Complex Stochastic Processes: An Introduction to Theory and Applications*, Addison-Wesley, Reading, MA.

[35] A. Zacks (1971) *The Theory of Statistical Inference*, Wiley, New York.

[36] A. Schuster (1898) "On the investigation of hidden periodicities with application to supposed 26-day period of meterological phenomena", *Terr. Magn.*, **3**, 13–41.

[37] A. Schuster (1907) "On the periodicities of sunspots", *Philos. Trans. R. Soc. London, Ser. A*, **206**, 69–100.

[38] N. Wiener (1964) "Generalized harmonic analysis", *Acta. Math.*, **55**, 117–258; reprinted in *Selected Papers of Norbert Wiener*, MTI Press, Cambridge, MA.

[39] A. Khintchine (1934) "Korrelationstheorie der stationare stochastischen processe", *Math. Ann.*, **109**, 604–615.

[40] H. Cramer (1942) "On harmonic analysis in certain functional spaces", *Ark. Mat. Astron. Fys.*, **283**, No. 12.

[41] G. W. Lank et al. (1973) "A semicoherent detection and Doppler estimation statistic", *IEEE Trans. Aerosp. Electron. Systems*, **9**, 151–165.

[42] S. M. Kay (1982) "Robust detection by autoregressive spectrum analysis", *IEEE Trans. Acoust., Speech, Signal Process.*, **30**, 256–269.

[43] S. Haykin and H. C. Chan (1980) "Computer simulation study of a radar Doppler processor using the maximum entropy method", *IEE Proc.*, **27**, 464–470.

[44] E. K. L. Hung and R. W. Herring (1981) "Simulation experiments to compare the signal detection properties of DFT and MEM spectra", *IEEE Trans. Acoust., Speech, Signal Process.*, **29**, 1084–1089.

[45] R. E. Kromer (1970) "Asymptotic properties of the autoregressive spectral estimator", Ph.D. dissertation, Stanford University.

[46] A. Akaike (1969) "Power spectrum estimation through autoregressive model fitting", *Ann. Inst. Statist. Math., Tokyo*, 407–419.

[47] K. N. Berk (1974) "Consistent autoregressive spectral estimates", *Ann. Statist.*, **2**, 489–502.

[48] A. B. Baggeroer (1976) "Confidence intervals for regression (MEM) spectral estimates", *IEEE Trans. Inform. Theory*, **22**, 534–545.

[49] Q. T. Zhang (1991) "Performance analysis of a signal detector using the maximum entropy method", *Signal Process.*, submitted.

[50] M. D. Springer (1973) *Algebra of Random Variables*, Wiley, New York.

[51] Q. T. Zhang (1989) "An entropy-based receiver for the detection of random signals and its application to radar", *Signal Process.*, **18**, 387–397.

[52] N. Giri (1965) "On the complex analysis of T^2-and R^2-tests", *Ann. Math. Statist.*, **36**, 665–670.

[53] E. S. Pearson and H. O. Hartley (Eds.) (1976) *Biometrika Tables for Statisticians*, Vol. II, Cambridge University Press, Cambridge, UK.

[54] B. D. Ripley (1989) *Stochastic Simulation*, Wiley, New York.

[55] B. Porat and B. Friedlander (1984) "Parametric techniques for adaptive detection of Gaussian signals", *IEEE Trans. Acoust., Speech, Signal Process.*, **32**, 780–790.

[56] Q. T. Zhang, S. Haykin, and P. Yip (1989) "Performance limits of the innovations-based detection algorithm", *IEEE Trans. Inform. Theory*, **35**, 1213–1222.

[57] P. A. S. Metford and S. Haykin (1985) "Experimental analysis of an innovations-based detection algorithm for radar surveillance", *IEE Proc.*, Part F, **132**, 18–26.

[58] Q. T. Zhang, K. M. Wong, and S. Haykin (1991) "A new method of signal detection of the complete information of second order statistics, Part I: asymptotic theory", *IEE Proc.*, Part F, submitted.

[59] Q. T. Zhang, K. M. Wong, and S. Haykin (1991) "A new method of signal detection of the complete information of second order statistics, Part II: application to radar", *IEE Proc.*, Part F, submitted.

[60] S. S. Wilks (1938) "The large-sample distribution of the likelihood ratio for testing composite hypotheses", *Ann. Math. Statist.*, **9**, 60.

[61] A. Wald (1943) "Tests of statistical hypotheses concerning several parameters when the number of observations is large", *Trans. Am. Math. Soc.*, **54**, 426.

[62] C. Muehe, L. Cartledge, et al. (1974) "New Techniques applied to air traffic control radars", *Proc. IEEE*, **62**, 716–723.

[63] D. C. Schleher (Ed.) (1980) *Automatic Detection and Radar Data Processing*, Artech House, Dedham, MA.

[64] Q. T. Zhang and K. M. Wong (1990) "Performance of radar detection schemes", Research Report CRL-227, McMaster University.

[65] M. B. Priestly (1981) *Spectral Analysis and Time Series*, Vol. I, Academic Press, New York.

[66] B. T. Porteous (1985) "Improved likelihood ratio statistics for covariance selection models", *Biometrika*, **72**, 97–101.

[67] S. Haykin, C. Krasnor, T. Nohara, B. Currie, and D. Hamburger (1991) "A coherent dual-polarized radar for studying the ocean environment", *IEEE Trans. Geosci. Remote Sensing*, **29**, 188–191.

[68] J. P. Burg (1975) "Maximum entropy spectral analysis", Ph.D. Dissertation, Stanford University, Stanford, CA.

[69] T. Nohara, S. Haykin, B. Currie, and C. Krasnor (1989) "Towards the improved detection of small ice targets in K-distributed sea clutter", in *Proc. Int. Symp. on Noise and Clutter Rejection in Radar and Imaging Sensors*, Japan.

[70] S. M. Kay (1983) "Asymptotically optimal detection in unknown colored noise via autoregressive modeling", *IEEE Trans. Acoust., Speech, Signal Process.*, **31**, 927–940.

[71] D. C. Schleher (Ed.) (1978) *MTI Radar*, Artech House, Dedham, MA.

[72] W. L. Root (1970) "Introduction to the theory of the detection of signals in noise", *Proc. IEEE*, **58**, 610–623.

3

ADAPTIVE MULTISENSOR DETECTION AND ESTIMATION

A. STEINHARDT
School of Electrical Engineering,
Cornell University,
Ithaca, New York, USA

3-1 INTRODUCTION

Many of the recent advances in radar signal processing surround the coordinated use of multiple sensors. Multisensor radars possess spatial adaptivity. This salient virtue enables such systems to outperform monolithic radars in situations where a substantial component of the interference is directional and partially unknown a priori. Such is the case in clutter suppression, jamming, etcetera. In this chapter we lay the theoretical foundation for subsequent chapters which describe various applications and procedures for multisensor adaptivity. In single sensor, that is, scalar time series processing, Wiener filtering, or stochastic mean squared estimation, provides a framework for constructing filters which optimally suppress interference. In Section 3-2 we develop the multivariate theory of Wiener filtering and discuss its application to array signal processing. A key feature of our discussion is that we proceed in the frequency domain, avoiding convolutional equations entirely.

Wiener theory assumes that the second order statistics of the noise and desired signal are known. This is often not the case, and one must infer the structure of the signal and noise data adaptively. In such cases one can

Adaptive Radar Detection and Estimation, Edited by Simon Haykin and Allan Steinhardt.
ISBN 0-471-54468-X

either (i) estimate the second order statistics and substitute these estimates into the Wiener equations, or (ii) reformulate the problem, using a framework that explicitly allows for unknown, or partially unknown, statistics. Our discussion will favor the latter more modern and fundamental approach which relies heavily on the theory of multivariate statistical inference. In some cases we shall see that the two approaches lead to equivalent data processing procedures. The application of multivariate analysis to sensor array processing is a relatively recent phenomena, and represents an exciting (and challenging!) development for the radar engineering community.

To begin this data adaptive development, in Section 3-3 we address the problem of adaptive target recovery from a nonstatistical viewpoint. We show that under idealized assumptions, data adaptive approaches can yield results indistinguishable from the clairvoyant case, that is, where prior statistical knowledge is available. This optimistic result serves to illustrate the potential power of adaptive methods. It has the pedagogic virtue that it provides preliminary insights into the behavior of adaptive estimators in the absence of the unavoidably cumbersome mathematics accompanying stochastic estimation.

Next in Section 3-4 we introduce the notion of adaptive target-parameter estimation. We begin with a survey of competing theoretic formulations for adaptive multivariate estimation, and then focus on maximum likelihood theory, based on a multivariate Gaussian model for the interference.

We then study the PDF of various estimates of interference covariance, weight vectors, and target-waveforms (Section 3-5). This analysis hinges on properties of the distribution of the partitioned inverse of a random sample covariance matrix, which we derive from first principles.

Finally, in Section 3-6 we address the issue of adaptive detection, including a discussion of threshold selection and pd/pfa calculation. We conclude with a discussion of extensions and a survey of the current literature.

Our treatment of adaptive detection/estimation (in Sections 3-3 through 3-6) makes use of some modern developments in the area, and differs from conventional approaches in that we treat the interference covariance as unknown. It is therefore essentially a nuisance parameter, which must be estimated in order to address the target, which is all we actually care about. The formulation of a detection problem at an early stage (prior to the formation of a post-beamformer scalar waveform) can lead to significant improvements in detection.

This chapter presents the theoretical foundation for much of the material in subsequent chapters which all involve applications concerning estimation and/or detection. Therefore, to facilitate the use of this chapter as a reference we have enclosed key equations in boxes, and we present lengthy derivations as self-contained theorems.

3-2 STOCHASTIC MEAN SQUARE ESTIMATION

3-2-1 Derivation

In this section we explore the optimal recovery of a target signal of known bearing observed in multichannel interference. We model the interference and target as random and we use mean squared error (MSE) as our optimality criterion. This approach is known as the multichannel stochastic MSE, or multichannel Wiener filtering, problem.[1] The MSE criterion is a good one if the power in the desired interference and error waveforms is a physically useful measure of the size of the attendant waveform. This is often the case, and it is an implicit assumption in most discussions surrounding SNR.[2]

Let $\mathbf{x}(t)$ denote the vector valued time series, that is snapshots, obtained from an array of sensors, and denote the Fourier transform of $\mathbf{x}(t)$ by $\mathbf{x}(\omega)$,[3] $\mathbf{x}(\omega) = \mathcal{F}(\mathbf{x}(t))$, where $\mathcal{F}$ denotes the Fourier integral. One can express $\mathbf{x}(\omega)$ as

$$\boxed{\mathbf{x}(\omega) = \mathbf{d}(\omega)s(\omega) + \mathbf{i}(\omega) + \mathbf{u}(\omega)} \tag{3-1}$$

where $\mathbf{d}(\omega)$ is the direction vector[4] of the desired (target) signal $s(t)$, $\mathbf{i}$ is the spatially correlated interference (such as jamming, clutter, nonintentional man-made interference, etc.) and $\mathbf{u}$ is spatially uncorrelated noise, such as thermal receiver noise. We use lower case bold characters to denote vectors and upper case bold characters to denote matrices. The superscript H will denote conjugate (Hermitian) transposition. We seek the best multichannel (i.e., vector valued) linear filter $\mathbf{h}(\omega) = [h_1(\omega) \quad \cdots \quad h_N(\omega)]^T$ which will find $\hat{s}(t)$, the least mean squared error estimate of $s(t)$ given the observations $\mathbf{x}(t)$. We will consider here only the noncausal Wiener filter. If $\mathbf{h}(\omega)$ is restricted to be causal (which leads to the *Wiener Hopf* equations), or further restricted to be a finite impulse response filter, the performance will deteriorate from the ideal noncausal filter by an extent that can usually be determined only via simulation. We assume the target and interference are all stationary random processes with zero mean.

[1] In honor of Norbert Wiener's pioneering work on the estimation of unobserved random processes, from observed random processes (see [39, 31]).

[2] Many quantities in physics involve energy or power. This provides a strong justification for variance as an objective functional.

[3] This notation is somewhat abusive but not we trust ambiguous; the argument, t or ω, will indicate which domain we are in.

[4] We omit the dependence of the direction vector on the fixed target bearing θ_0 for notational expedience.

Let us denote the power spectra, and cross power spectral matrices, of $s(t)$, $\mathbf{i}(t)$ and $\mathbf{u}(t)$ as $S_{ss}(\omega)$, $\mathbf{S}_{ii}(\omega)$, and $S_{uu}(\omega)\mathbf{I}$ respectively.[5] Note that the latter two quantities are matrices, with the thermal noise matrix being diagonal since this interference source is by assumption spatially uncorrelated. Furthermore we assume that the spectral density of the thermal noise is equal on all channels. We assume that the direction vector of the target signal is known.

Expressed in the time domain we now seek the multichannel filter $[h_1(t) \quad \cdots \quad h_N(t)]^T = \mathcal{F}^{-1}(\mathbf{h}(\omega))$ which minimizes

$$E[|\hat{s}(t) - s(t)|^2], \; \hat{s}(t) \equiv \sum_{k=1}^{N} h_k^*(t) * x_k(t) \tag{3-2}$$

where $*$ denotes (scalar) convolution and * denotes complex conjugation. We consider the case of complex valued snapshots, and therefore the need for complex valued impulse response functions $h_i(t)$, because of the use of in phase and quadrature data in most radar processing schemes [26]. We employ a conjugate in the sum in (3-2) to allow us to interpret this sum as a complex valued inner product, simplifying subsequent analysis. Note that the entries of the solution vector $\mathbf{h}(\omega)$, or its time domain counterpart, are conjugated prior to multiplication against, or convolution with, the array data. (The conjugation can be viewed as a de-phasing. For example suppose we set $\mathbf{h}(\omega)$ equal to the target steering vector. Then the conjugation will reverse the phasing the target signal experiences while sweeping across the array.)

The celebrated projection theorem [39], [31], which is the basis for solving any stochastic linear least square estimation problem, states that error and data are orthogonal:

$$\boxed{E[\text{error}(t) \times \text{data}(t)^H] = 0} \tag{3-3}$$

where $\times$ denotes pointwise multiplication, and the error is the difference between the desired quantity and its estimate using the available data. The error is a scalar function of time, the data may be either a scalar or a vector time function, depending on the problem at hand. The projection theorem tells us that if the data is uncorrelated with the error, we have performed the best (linear) processing possible to reduce the error. This is intuitively appealing for if the error and data are correlated, the error could be reduced

[5] The cross spectrum of the vector valued complex stationary processes $\mathbf{a}(t)$ and $\mathbf{b}(t)$ is defined [26] via $\mathbf{S}_{ab}(\omega) = \mathcal{F}(E[\mathbf{a}(t+\tau)\mathbf{b}(t)^H])$. When $\mathbf{a} = \mathbf{b}$ there still arises in the multidimensional case the appearance of cross correlations, thus the term cross spectrum to describe $\mathbf{S}_{ab}(\omega)$ regardless of whether a single or a pair of processes is involved.

by studying the sign of the correlation and either amplifying or attenuating the estimate accordingly. We have expressed the projection theorem in the time domain. The error and the data can just as well be functions of frequency, space, etcetera, and the theorem applies equally well to continuous and discrete variables.

One can show (see [31, p. 418] or Appendix E) that for any two jointly stationary random processes $x(t)$ and $y(t)$ with transforms $x(\omega)$, $y(\omega)$,

$$E[x(\omega)y^*(\omega')] = 2\pi S_{xy}(\omega)\delta(\omega - \omega') \tag{3-4}$$

To appreciate (3-4) consider the special case that $x(t) = y(t)$. Then we discover that the Fourier transform of a stationary random process is itself a (nonstationary) white noise process. Furthermore the variable intensity is given by the power spectral density.[6] We now pose the problem of estimating $s(\omega_0)$, the Fourier transform of $s(t)$, at a fixed frequency ω_0. Our data are $\mathbf{x}(\omega')$, $\omega' \in [-\infty, \infty]$, the Fourier transform of $\mathbf{x}(t)$ at all frequencies. We do not need to employ frequencies $\omega' \neq \omega_0$ when attempting to estimate $s(\omega_0)$ from $\mathbf{x}(\omega')$; the estimate $\hat{s}(\omega_0) = \mathbf{h}(\omega_0)^H\mathbf{x}(\omega_0)$ suffices. This follows from the decorrelation of $x(\omega)$ and $x(\omega')$, $\omega \neq \omega'$. The Fourier domain projection theorem gives the relation

$$E[(s(\omega_0) - \mathbf{h}(\omega_0)^H\mathbf{x}(\omega_0))\mathbf{x}(\omega')^H] = 0\,, \qquad \forall \omega' \in [-\infty, \infty] \tag{3-5}$$

It follows from (3-4) that

$$[\mathbf{S}_{sx}(\omega_0) - \mathbf{h}(\omega_0)^H\mathbf{S}_{xx}(\omega_0)]\delta(\omega_0 - \omega') = 0 \tag{3-6}$$

To match the right hand side, the left hand side must be identically zero, which can only happen if the term in brackets vanishes. Allowing ω_0 to vary over the entire target spectrum, this result implies that

$$\boxed{\mathbf{h}(\omega) = \mathbf{S}_{xx}^{-1}(\omega)\mathbf{S}_{xs}(\omega)} \tag{3-7}$$

There are at least two other distinct derivations of $\mathbf{h}(\omega)$ both of which offer their own insights. We sketch them below.

One derivation is based on a generalization of the time domain, convolution based approach to Wiener filtering favored in [31]. One can readily verify that the essence of the derivation in 13–37 of Papoulis survives when (as here) the attendant processes take on vector values.

The second alternate derivation originates by noting from (3-4) that

[6] The Fourier integral converges in the mean sense, not everywhere or even almost everywhere [31].

$$\mathbf{S}_{xy}(\omega) = \frac{1}{2\pi} \int_{-\varepsilon}^{\varepsilon} E[\mathbf{x}(\omega)\mathbf{y}^*(\omega + \xi)] \, d\xi \,, \qquad \forall \varepsilon > 0 \tag{3-8}$$

This states that the cross spectrum equals the average of the cross correlation function of the Fourier transform of $x(t)$ and $y(t)$ in the neighborhood of ω. This is the dual of the well known time domain result that the intensity of a continuous time white noise process is the average of its correlation function around the origin.

The projection theorem informs us that the cross spectrum $S_{\mathrm{error},x}(\omega)$ vanishes (almost) everywhere. One can then use (3-8) to quickly re-derive the multichannel Wiener filter in (3-5). (A variant of this approach, based on so called orthogonal increments is described in [39].)

It is instructive to study the effects of the various components $(\mathbf{i}, \mathbf{u}, \mathbf{s})$ of $\mathbf{x}(\omega)$ on the weight vector $\mathbf{h}(\omega)$. To this end we can express the power spectrum of $\mathbf{x}$ in terms of the target's spectrum and that of the two types of interference, yielding

$$\boxed{\mathbf{h}(\omega) = [S_{ss}(\omega)\mathbf{d}(\omega)\mathbf{d}(\omega)^H + \mathbf{S}_{ii}(\omega) + S_{uu}(\omega)\mathbf{I}]^{-1} S_{ss}(\omega)\mathbf{d}(\omega)} \tag{3-9}$$

Note that the inverse matrix in (3-9) is of the form of the sum of two nonnegative definite matrices and a (positively) scaled identity. The rank of a sum of nonnegative definite matrices is bounded below by the rank of each constituent matrix [12]. We conclude that, provided S_{uu} does not vanish, the matrix in point, mainly $\mathbf{S}_{xx}$, does indeed possess an inverse.

3-2-2 The Optimal MSE

One can show from (3-3) that for a general MSE estimation problem,

$$\boxed{\text{Optimal MSE} = E[\text{error}(t) \times \text{desired signal}(t)^*]} \tag{3-10}$$

It will be convenient for us to evaluate the error in the Fourier domain. To this end we introduce the notion of the error spectral density $S_{ee}(\omega)$, which from (3-8) is given by

$$S_{ee}(\omega) = \frac{1}{2\pi} \int_{-\varepsilon}^{\varepsilon} E[e(\omega)e^*(\omega + \xi)] \, d\xi \,, \qquad e(\omega) = s(\omega) - \hat{s}(\omega) \tag{3-11}$$

Since the error is orthogonal to the data it follows that

$$E[e(\omega)\hat{s}^*(\omega + \xi)] = 0 \tag{3-12}$$

After some algebra, we then find that

$$S_{ee}(\omega) = S_{ss}(\omega)[1 - S_{ss}(\omega)\mathbf{d}^h(\omega)\mathbf{S}_{xx}^{-1}(\omega)\mathbf{d}(\omega)] \tag{3-13}$$

and the net error then becomes

$$\text{Optimal MSE} \equiv E[|\text{error}|^2] = \frac{1}{2\pi}\int_{-\infty}^{\infty} S_{ee}(\omega)\, d\omega \tag{3-14}$$

By means of the Woodbury inversion lemma (see Eq. (3-97)] and (3-9), the optimal multichannel filter can be expressed as

$$\mathbf{h}(\omega) = \mathbf{h}_0(\omega)g(\omega)\,, \qquad \mathbf{h}_0(\omega) \equiv [\mathbf{S}_{ii}(\omega) + S_{uu}(\omega)\mathbf{I}]^{-1}\mathbf{d}(\omega) \tag{3-15}$$

$$g(\omega) \equiv \left[\frac{1}{S_{ss}(\omega)} + \mathbf{d}(\omega)^H\mathbf{h}_0(\omega)\right]^{-1} \tag{3-16}$$

This is a remarkable result. It states that the target spectrum enters only in the scalar gain term $g(\omega)$. In other words multichannel Wiener filtering can be viewed as a two step process: (i) perform multichannel filtering, dependent only on the interference cross spectra and the target direction vector, and then (ii) perform scalar filtering on the output of the multichannel sum. This has enormous practical relevance. It states that we can get by without a knowledge of the target spectra when forming the multichannel filters. The scalar output sequence we get by arbitrarily setting the target spectra to an arbitrary value, say unity, across the target bandwidth, and solving (3-15), need only be passed through a filter with transfer function $g(\omega)$ to obtain the optimal Wiener solution.

This fortuitous partitioning of the multivariate Wiener solution into a two step process ((i) perform multichannel processing independent of the target spectrum and then (ii) perform scalar Wiener processing on the subsequent scalar output) was first observed by E.J. Kelly in the early 1960s [21]. (This is also one of the earliest papers on adaptive beamforming.)

The optimal error found by substituting (3-15), (3-16) into (3-14) can be rewritten (after more algebra) simply as

$$\text{Optimal MSE} = \frac{1}{2\pi}\int_{-\infty}^{\infty} g(\omega)\, d\omega \tag{3-17}$$

with g as in (3-16). In other words $g(\omega)$ exactly equals the error density $S_{ee}(\omega)$!

Now let us consider the minimum variance distortionless response (MVDR) beamformer. The MVDR processor, of interest in its own right, and discussed in detail in Chapter 4, also sheds light on the behavior of the multivariate Wiener filter. The MVDR processor is characterized by the multichannel weight vector, $\mathbf{w}(\omega)$, which satisfies

$$\min_{\mathbf{d}(\omega)^H\mathbf{w}(\omega)=1} (\mathbf{w}(\omega)^H[\mathbf{S}_{ii}(\omega) + S_{uu}(\omega)\mathbf{I}]\mathbf{w}(\omega)) \tag{3-18}$$

The quantity to be minimized is the output interference power spectral density. The constraint insures that the target signal $s(t)$ is undistorted by the weight vector, that is, the estimate $\hat{s}(t)$ is unbiased, $E[\hat{s}(t)] = s(t)$. (The expectation here being over the interference, conditioned on $s(t)$.) The solution of (3-18) can be found by Lagrange multipliers, or, more elegantly, by using whitening transformations (see Appendix D):

$$\boxed{\mathbf{w}(\omega)_{\mathrm{MVDR}} = \frac{[\mathbf{S}_{ii}(\omega) + S_{uu}(\omega)\mathbf{I}]^{-1}\mathbf{d}(\omega)}{\mathbf{d}(\omega)^H[\mathbf{S}_{ii}(\omega) + S_{uu}(\omega)\mathbf{I}]^{-1}\mathbf{d}(\omega)}} \tag{3-19}$$

By simple algebra we obtain from (3-19) an output interference power spectral density of

$$E[|\mathbf{w}^H(\omega)(\mathbf{i}(\omega) + \mathbf{u}(\omega))|^2] = \frac{1}{\mathbf{d}(\omega)^H\mathbf{h}_0(\omega)} \tag{3-20}$$

Note that if we pass to the limit $S_{ss}(\omega)\to\infty$ in the multivariate Wiener filter, we obtain the MVDR beamformer. This suggests that the MVDR beamformer is suboptimal (in the mean squared error sense), by a frequency dependent gain term. In fact for finite target power density, the MVDR beamformer always lets in more net power than it should, since a positive quantity is missing in the denominator of the scale factor applied to the unnormalized filter $\mathbf{h}_0(\omega)$. Wiener theory tells us it is better to distort the signal by attenuating it (in a frequency dependent fashion) to let in less noise at the expense of sacrificing some target energy. In contrast the MVDR beamformer cannot lessen the target energy, even if this helps attenuate noise, since to do so violates its commitment to providing a nondistorted (though perhaps highly noisy) signal. Interestingly, the MVDR beamformer is however optimal in another sense, mainly that it provide the maximum likelihood estimate of the unknown desired signal. This intriguing connection is developed in Section 3-4-6. Extensions of the MVDR beamformer, involving multiple constraints, are discussed further in Chapter 4.

A scalar Wiener filter has an MSE of simply twice the harmonic mean of the signal power density and the noise power density [39]. When $\mathbf{h}_0(\omega)$ becomes a scalar we clearly get back (in the usual case of a direction vector with unit magnitude entries) this mean. Now consider the present multichannel, array case. What is meant by noise power now? How do we obtain a scalar measure to describe the vector valued noise?

To answer these questions note that the second term in the denominator of $g(\omega)$ is the reciprocal of the output noise power for an MVDR beamformer steered towards the target. We conclude that $g(\omega)$ is twice the harmonic mean of the signal spectra and this post adaptive noise power spectral density. Thus the output interference power from an MVDR beamformer is the generalization of noise power we need to interpret the MSE for a vector valued Wiener filter as a harmonic mean between noise and signal. Note that in the scalar case of $N = 1$ the MVDR constraint $\mathbf{d}^H\mathbf{w} = 1$ leads to no degrees of freedom and hence no filtering, and the MVDR output interference reverts back to the input noise power.

To summarize, we can interpret the Wiener filter error density $S_{ee}(\omega)$ as a harmonic mean, this time the noise being interpreted as the post adaptive noise, after distortionless beamforming. Since spatial filtering can only decrease the noise power of the interference impinging on an array, we see directly that the multichannel Wiener filter error is bounded above by the error in the scalar Wiener solution. The ratio of the post adaptive noise in (3-20) to the initial noise spectral density at an isolated sensor, $(S_{\mathbf{n}_i,\mathbf{n}_i}(\omega))$, will determine the degree of superiority of the multisensor scheme over its scalar counterpart.

The MVDR and Wiener schemes provide, respectively, classic examples of unbiased (but not minimum variance) and minimum variance (but biased) target signal estimators. The incompatibility of variance minimization and bias elimination is a well known feature of many stochastic estimation problems [28].[7]

We restricted ourselves above to the case where the target estimator is a linear function of the data. We mention in passing that one can show that for Gaussian processes, a linear estimator is optimal; one need not consider the larger class of nonlinear estimators when attempting to minimize the MSE [31, 39].

In many radar detection problems the target spectrum's shape is unknown, but is known to be narrowband, with a center frequency given by Doppler and a spread given by acceleration. It is then common to perform spectral analysis on the post adaptive scalar series (with $\mathbf{h}$ simply set to $\mathbf{h}_0$ as described earlier) to ascertain S_{ss}. One can then do detection, normalizing by the residual noise spectral density to get a scenario independent false alarm rate (CFAR) detector [39]. Although popular, this approach is ad hoc because initial calculations are performed as if one is merely interested in mean square estimation, and one introduces detection concepts only after the vector valued data has been collapsed into a scalar waveform $y(t)$. A more direct approach, whereby a detection problem is formulated at the outset, is the topic of Section 3-6. We find that when applicable these decision theoretic approaches lead to substantial performance gains.

[7] A notable, though rather trivial exception being the estimation of the mean from a sequence of i.i.d. Gaussian random variables. It appears that nonlinearities, such as the matrix inversion we experience here, are what induce the bias/variance conflict.

It is insightful to consider the special occurrence when $\mathbf{S}_{ii}(\omega)$ is rank one:

$$\mathbf{S}_{ii}(\omega) = s(\omega)\mathbf{j}(\omega)\mathbf{j}(\omega)^H \tag{3-21}$$

where $s(\omega)$ is the source spectral density and $\mathbf{j}$ is the interference direction vector. This case arises for example with a single point source interferer coupled with thermal noise. We now find the error to be

$$\text{Optimal MSE} = \frac{1}{2\pi}\int_{-\infty}^{\infty} \left(S_{ss}^{-1}(\omega) + S_{uu}^{-1}(\omega)\left[\mathbf{d}(\omega)^H\mathbf{d}(\omega) - \frac{|\mathbf{d}^H(\omega)\mathbf{j}(\omega)|^2 s(\omega)}{S_{uu}(\omega) + \mathbf{j}(\omega)^H\mathbf{j}(\omega)s(\omega)}\right]\right)^{-1} d\omega \tag{3-22}$$

When the power density in the directional interference $\mathbf{i}$ dominates that of the white interference $\mathbf{u}$ this simplifies to

$$\text{Optimal MSE} = \frac{1}{2\pi}\int_{-\infty}^{\infty} \text{harmonic mean}(S_{ss}, [S_{uu} \div \text{length}(\text{projection}(\mathbf{d} \text{ onto } \perp\mathbf{j}))])\, d\omega \tag{3-23}$$

Note that the directional interference $\mathbf{i}$ is entirely eliminated. Note also that prior to forming the harmonic mean we attenuate the nondirectional interference $\mathbf{u}$ by the length squared of the target direction vector component orthogonal to the directional interference. This is not surprising, in the absence of directional interference, $\mathbf{h}(\omega) \propto \mathbf{d}(\omega)$ and the spatially white noise is attenuated by the norm squared of $\mathbf{d}$ (see [26] or simply take the proper limits in (3-17)). When directional interference is present, the component of $\mathbf{d}$ in its direction is used up before the directivity embedded in $\mathbf{d}$ can be used to excise the known, directional target return from the spatially uncorrelated, nondirectional noise. When directional interference dominates, the error expression no longer includes this term. This implies that this interference source is attenuated to a power level roughly equal to that of the weaker white noise $\mathbf{u}$. Therefore the sensor gain in the direction of the directional interference is on the order of the ratio $S_{uu}/s(\omega)$. This result is often used to estimate the anticipated null depth in an antenna pattern as a function of directional interference power [7].[8] Note that the stronger the directional interference, the deeper the null, achieving an infinite null (perfect suppression) in the absence of thermal noise. (We rederive this result from a different approach in the next section.) As the thermal noise

[8] It is also used to establish connections between eigenvalue spread and the directional interference to thermal noise power ratio. The eigenvalue spread determines convergence rate (in the least mean square algorithm), and word length requirements (for the recursive least squares or sample matrix inversion algorithms) [26].

rises, the null decreases. The jammer is now perceived as less threatening compared to the omnidirectional noise, so the Wiener filter increasingly focuses on suppressing this latter term.

The filter **h** is generally not realizable, at least in the wideband case, since it is noncausal. One can pose the problem of constructing an optimal filter $\mathbf{h}(\omega)$ which is constrained to be causal. This causal Wiener filtering problem is discussed in [39] and it can be solved by means of whitening followed by spectral factorization of the cross spectrum between target spectrum and the whitened data. In the present multivariate case one can generalize the scalar approach, using the notion of whitening filters and factorizations for rational system transfer function matrices as discussed in [16]. We omit details for brevity.

Let us explore the reduction in the MSE resulting from the incremental addition of a single sensor. Suppose we have an N element array and we add one more sensor. Denote the net interference power of this new sensor as η, obtaining a noise power density of $S_{\eta\eta}(\omega)$. Partition the new $N+1$ by 1 direction vector $\tilde{\mathbf{d}}(\omega)$ as $\tilde{\mathbf{d}}(\omega) = [\alpha, \mathbf{d}(\omega)]^T$, and denote the new cross spectral vector $\mathbf{S}_{i+u,\eta}(\omega)$ by β. Then the inverse of $\tilde{\mathbf{S}}(\omega)$, the cross spectral density matrix for the length $N+1$ interference snapshot vector $[\eta, \mathbf{i}+\mathbf{u}]^T$ is given in partitioned form by

$$\tilde{\mathbf{S}}^{-1}(\omega) = \frac{1}{S_{\eta\eta}(\omega) - \beta^H \mathbf{S}_{i+u,i+u}^{-1}(\omega)\beta} \begin{bmatrix} 1 & -\beta^H \mathbf{S}_{i+u,i+u}^{-1}(\omega) \\ \mathbf{S}_{i+u,i+u}^{-1}(\omega)\beta & \mathbf{S}_{i+u,i+u}^{-1}(\omega)(\mathbf{I}(S_{\eta\eta}(\omega) - \beta^H \mathbf{S}_{i+u,i+u}^{-1}(\omega)\beta) + \beta\beta^H \mathbf{S}_{i+u,i+u}^{-1}(\omega)) \end{bmatrix} \tag{3-24}$$

It then follows after some tedious and uninspiring algebra that

$$\tilde{\mathbf{d}}(\omega)^H \tilde{\mathbf{S}}^{-1}(\omega)\tilde{\mathbf{d}}(\omega) = \mathbf{d}(\omega)^H \mathbf{S}_{i+u,i+u}^{-1}(\omega)\mathbf{d}(\omega) + \left[\frac{1}{S_{\eta\eta}(\omega) - \beta^H \mathbf{S}_{i+u,i+u}^{-1}(\omega)\beta} \, |\tilde{\mathbf{d}}(\omega)^H (1, -\beta^H \mathbf{S}_{i+u,i+u}^{-1}(\omega))^H|^2 \right] \tag{3-25}$$

Hence we can express the new error density $\tilde{g}(\omega)$, after some more algebra, as

$$\tilde{g}(\omega)^{-1} = g(\omega)^{-1} + \left[\frac{|\alpha - \mathbf{h}_0^H \beta|^2}{S_{\eta\eta}(\omega) - \beta^H (\mathbf{S}_{ii}(\omega) + S_{uu}\mathbf{I})^{-1}\beta} \right] \tag{3-26}$$

The inverse of a positive definite matrix is itself positive definite [39]. Since cross spectral matrices are nonnegative definite [39], it follows that the denominator in (3-26), which equals $1/(\mathbf{S}^{-1})_{1,1}$, is itself positive (if it exists).

Thus the term in brackets in (3-26) is nonnegative. It follows from (3-17) that the error decreases monotonically as this term grows. Thus to place a new sensor in a good location one selects it to maximize the [ ] term in (3-26). Of course we do not generally know the cross spectral densities involved beforehand, making such a choice rather difficult. Nonetheless (3-26) allows us to assess the utility of incorporating additional sensors as a function of interference scenario.

Note that if the cross correlation β vanishes, the decrease in error by adding a new sensor is given by $|\alpha|^2 S_{\eta\eta}^{-1}$. Hence the less noise power on this new sensor, the better the enhancement of the target estimate. If there is no noise on the new sensor the error becomes zero! This is of course exactly what one expects, but it is assuring that it arises out of our theory.

In a manner analogous to that for sensor augmentation, one can explore the incremental effects due to the presence of an additional (rank one) directional interferer with cross power density:

$$\mathbf{S}_{\text{new interferer}}(\omega) = \xi\xi^H \tag{3-27}$$

One can show that for the new resulting error density $\hat{g}$,

$$\hat{g}(\omega)^{-1} = g(\omega)^{-1} - \frac{\|\mathbf{h}_0(\omega)^H \xi\|^2}{1 + \xi^H \mathbf{S}_{i+u,i+u}^{-1} \xi} \tag{3-28}$$

Thus the new error density grows monotonically with the output power from the original, unnormalized, multichannel Wiener filter $\mathbf{h}_0$ when the new interference is input, and it also grows monotonically with $1/[\xi^H \mathbf{S}_{\mathbf{i+u,i+u}}^{-1} \xi]$, the output power from an MVDR beamformer pointed in the direction of the interference, with the initial interference as input. This latter quantity in some sense measures how much power the array sees from the new interference source, and the former quantity measures how much of the new interference leaks through if the optimal filter which focuses only on the original interference is employed. It is not at all surprising that the error should depend monotonically on these two quantities, again affirming the physical reasonableness of the Wiener theory.

3-2-3 Narrowband Processing

We complete our discussion of Wiener theory with a discussion of narrowband processing. Consider two (omnidirectional) sensors i and k acquiring a single source signal from bearing θ. Assume the source spectrum is flat across the band $\Omega, \Omega + \Delta$. Now suppose we downshift the data on these two channels to baseband (translating down by frequency Ω) and sample the resulting complex waveform at the Nyquist rate of Δ. Then we obtain a discrete time sequence on each channel whose cross spectrum equals

$$S_{ik}(e^{j\omega}) = P \exp\left(j\left(\frac{\omega\Delta}{2\pi} + \Omega\right) \frac{d}{c} \sin(\theta)\right) \tag{3-29}$$

where the constant P is the power density, d is the spacing between sensors i and k, and c is the propagation velocity. Note that for small

$$\mu \equiv \frac{\Delta d \sin(\theta)}{c} \tag{3-30}$$

(relative to Ω), this is approximately frequency independent. Therefore its inverse transform is approximately $r_{i,k}(l) = P \exp(j\Omega(d/c)\sin(\theta))\delta(l)$. We conclude that one can, for small μ, replace the cross spectrum in (3-7) by the covariance matrix $\mathbf{R}_{xx} \equiv E[\mathbf{x}(t)\mathbf{x}(t)^H]$, and thereby replace the (unscaled) filter $\mathbf{h}_0(\omega)$ by a constant (frequency independent) weighting. This is called the narrowband approximation, since one ignores the frequency variability in the effective source spectra. Note that μ can be interpreted as a time-bandwidth product (TBWP). The role of the TBWP in antenna arrays is described further in Chapter 4. The advantage of the narrowband approximation, of course, is that it allows one to replace the N convolution operations in (3-2) by N simple complex multiplications. Furthermore it poses a simplified estimation problem when employed adaptively; $\mathbf{R}_{xx}$ alone needs to be adaptively determined from the data.

Henceforth (save for Section 3-4-5) we assume the narrowband approximation is in effect. Thus power spectral densities will no longer be confronted and all expressions will be in the time domain.

3-2-4 Chaos

We close this discussion with an alternative to the conventional stochastic interference model employed in the Wiener, least squares theory.

Recently, empirical evidence has emerged suggesting that certain forms of radar interference, mainly sea clutter, are in fact *deterministic* [25]. Although clutter looks random, phase space analysis reveals it is deterministic, that is, it can be modeled as the output of a nonlinear, deterministic, dynamical system. The correlation dimension has been measured, and it possesses a fractional dimension and hence is chaotic. If clutter is truly chaotic, not random, this has significant implications for target detection. Performance now depends on our ability to identify the nonlinear dynamical equations, and usual metrics, such as (linear) MSE and SNR become meaningless.

To find the correlation dimension of a set of M N-vectors one first computes

$$\eta(\theta, N) = \frac{1}{M^2 - M} \sum_{i \neq j} u(\|\mathbf{x}_i - \mathbf{x}_j\| - \theta) \tag{3-31}$$

where as usual u is the unit step function. One then looks at the slope of η on a log-log scale as a function of N. If this slope tapers off as N increases, the resulting asymptote is called the correlation dimension. Note that for i.i.d. random variables, $E[\eta(\theta)]$ equals the cumulative distribution function (CDF) for the length of $\mathbf{x}$. If these random variables tend to fill up the space R^N (or C^N) then $\ln(E[\eta(\theta)])$ will increase linearly with N. Thus there is no asymptote unless the data $\mathbf{x}_i$ do not fill up the space, but rather lie on some manifold. If the correlation dimension is fractional the data is chaotic. Obviously, it is well beyond the scope of this text to explore this recent alternative in detail. In fact at present it remains little more than an intriguing curiosity. One can test one's data, by means of histograms and correlation dimension analysis, to determine whether a multivariate Gaussian (or some other PDF), or rather a chaotic model, is appropriate.

3-3 DATA BASED TARGET RECOVERY AND BEARING ESTIMATION

3-3-1 Introduction

In practice, the second order statistics ($\mathbf{S}_{xx}(\omega)$ or $\mathbf{R}_{xx}(\tau)$) are not known a priori. Therefore they must be inferred from the data. This is the focus of the remainder of this chapter. In this section we will study the efficacy of data adaptive estimates of the covariance. We will make highly idealized assumptions which allow us to proceed without introducing probabilistic considerations.[9] The resulting toy problems serve to illustrate the potential power of adaptive schemes, motivating the more mathematically sophisticated (and more realistic) material which follows. First we consider target waveform estimation, and then target bearing estimation, each of which offers its own separate insights.

3-3-2 Data Adaptive Target Reconstruction

Let us consider the problem of data adaptive (i.e., unknown covariance) target estimation in the absence of receiver noise. We will state a key theorem regarding this task.

Theorem 3-1 (Ideal Target Signal Recovery) Suppose we have a narrowband array of N sensors, upon which up to $N-1$ interference sources impinge. We observe $N-1$ snapshot vectors $\mathbf{x}_n$, $n=1,\ldots,N-1$ by sampling the waveforms on the N sensors. These snapshots contain the $N-1$ interferer signals only, no thermal noise, and no target return. We form from these snapshots the matrix

[9] Other than an assumption regarding continuity of the cumulative PDF.

$$\hat{\mathbf{R}} \equiv \mathbf{X}\mathbf{X}^H , \qquad \mathbf{X} \equiv [\mathbf{x}_1 | \mathbf{x}_2 | \cdots | \mathbf{x}_{N-1}] \tag{3-32}$$

and then compute

$$\mathbf{h} \equiv \frac{P_{\text{null}\{\hat{\mathbf{R}}\}}(\mathbf{d})}{\| P_{\text{null}\{\hat{\mathbf{R}}\}}(\mathbf{d}) \|^2} \tag{3-33}$$

where $\mathbf{d}$ is the target steering vector, $\mathbf{P}(\ )$ the projection operator, and null$\{\mathbf{Z}\}$ the nullspace of a matrix $\mathbf{Z}$. We assume the denominator in (3-33) does not vanish. The above snapshots are zero mean, i.i.d., with continuous joint cumulative distribution function. Then, if subsequent snapshots $\mathbf{x}_n$, $n \geqslant N$ contain target returns s_n, the weight vector in (3-33) recovers these returns, that is,

$$y_n \equiv \mathbf{h}^H \mathbf{x}_n = s_n , \qquad n \geqslant N \tag{3-34}$$

with unit probability.

Note we assume that the desired signal is not present when $\hat{\mathbf{R}}$ is formed, but is when the inner product $\mathbf{h}^H\mathbf{x}$ is formed on subsequent snapshots. Many applications do in fact allow desired-signal free looks at the interference; often certain range gates are known to be target free. The theorem states that in the absence of thermal noise one need not merely estimate the target signal in the narrowband case, one can actually recover it exactly, provided $N-1$ target free snapshots are available during weight vector formation.

Proof. To begin assume there are exactly $N-1$ jammers so that the net received interference snapshots have the form

$$\mathbf{i}_n = \sum_{k=1}^{N-1} j_k(n)\mathbf{v}_k \tag{3-35}$$

where $\mathbf{v}_k$ is the direction vector of the kth jammer and $j_k(n)$ is the nth sample of the kth jammer. Note that $\mathbf{x}_n = \mathbf{i}_n$, $n \in [1, N-1]$. Assume the jammers are not perfectly correlated, in which case the ensemble covariance $\mathbf{R}_{ii} \equiv E[\mathbf{i}_n \mathbf{i}_n^H]$ has rank exactly $N-1$. Denote by $\mathbf{q}$ the normalized zero eigenvector of $\mathbf{R}_{ii}$:

$$\mathbf{R}_{ii}\mathbf{q} = \mathbf{0} , \qquad \mathbf{q}^H\mathbf{q} = 1 \tag{3-36}$$

Now form

$$\mathbf{z} = \frac{\mathbf{q}^H\mathbf{d}\mathbf{q}}{|\mathbf{q}^H\mathbf{d}|^2} = \frac{P_{\text{null}\{\mathbf{R}_{ii}\}}(\mathbf{d})}{\| P_{\text{null}\{\mathbf{R}_{ii}\}}(\mathbf{d}) \|^2} = \frac{\mathbf{q}}{\mathbf{d}^H\mathbf{q}} \tag{3-37}$$

Note that by assumption $\mathbf{x}_n = s_n\mathbf{d} + \mathbf{i}_n$, $n > N-1$. Note also that

$$\mathbf{z}^H\mathbf{x}_n = s_n + \mathbf{z}^H\mathbf{i}_n, \qquad n > N-1 \tag{3-38}$$

Now $E[\mathbf{z}^H\mathbf{i}_n] = 0$ and

$$\mathrm{var}(\mathbf{z}^H\mathbf{i}_n) = \mathbf{z}^H\mathbf{R}_{ii}\mathbf{z} = \frac{\mathbf{q}^H\mathbf{R}_{ii}\mathbf{q}}{|\mathbf{q}^H\mathbf{d}|^2} = \mathbf{q}^H\mathbf{0} = 0 \rightarrow \mathbf{z}^H\mathbf{i}_n = 0 \tag{3-39}$$

Thus

$$\mathbf{z}^H\mathbf{x}_n = s_n \tag{3-40}$$

suggesting perfect recovery of the desired target return s_n is possible provided that the projection of the target direction vector onto the null space of $\mathbf{R}_{ii}$ does not vanish. Physically this merely restricts the interference to disparate bearings from that of the target, a modest restriction indeed.

Note the contrast with conventional scalar Wiener filtering where the frequency bands cannot overlap at all to get perfect recovery. This latter restriction is clearly much more severe. The restrictions we placed on the jammers to obtain (3-40) (we assumed there are exactly $N-1$ and that they are not perfectly correlated) merely simplified our discussion; (3-40) holds without them. The only modification is that the null space might then have higher rank and $\mathbf{h}$ is no longer unique.

We wish to prove not only that perfect recovery is feasible but also that it can be obtained adaptively from $N-1$ snapshots using (3-33). It suffices to show that

$$\mathrm{null}\{\mathbf{R}_{ii}\} = \mathrm{null}\{\hat{\mathbf{R}}\} \tag{3-41}$$

From (3-39) it follows that all snapshots in (3-32) are orthogonal to the nullspace of $\mathbf{R}_{ii}$ with unit probability. If we can show that $\hat{\mathbf{R}}$ has rank $N-1$ then (3-41) follows, since the space $\perp\mathbf{q}$ has rank $N-1$. This is readily done using the fact that $N-1$ i.i.d. vectors from a continuous CDF are linearly independent with unit probability (see Appendix A).

Note that if the target is present during the formation of $\hat{\mathbf{R}}$, Theorem 3-1 no longer applies, null($\hat{\mathbf{R}}$) now has no particular significance. It is known that in the more general case of full rank $\mathbf{R}_{ii}$ (such as arises with thermal noise added to the $N-1$ interferers), the absence of the desired target signal during adaptation greatly enhances the convergence rate of $\mathbf{z}^H\mathbf{x}_n$ to s_n [26]. Theorem 3-1 provides a simple algebraic, nonstochastic justification for this observed phenomenon.

3-3-3 Data Adaptive Bearing Recovery

Now we turn to the issue of bearing estimation. In passive systems, bearing is inherently unknown. In active systems such as radar, the target's bearing is known within some tolerance, since it has been illuminated with a directional transmit beam. However active systems can benefit from bearing location also. One approach to interference suppression [5] is to estimate the interference-source bearings and then place beams in these directions. collect the beam outputs and then perform ordinary stochastic MSE estimation of the desired signal. To proceed in this case requires a knowledge of the propagation of all the source signals. This increase in required prior knowledge leads to heightened sensitivity in the estimates to imperfections such as multipath, fading, channel mismatch, etcetera. For simplicity we assume a linear array with far field narrowband interference (jammers). We will find that, under mild regularity conditions, we can infer the bearings of $N-1$ signals with N sensors. The method we describe is known as Root MUSIC. It has many variants [30]. We will only consider the uniform linear array case, in which case we can express the $N-1$ received snapshot vectors as

$$\mathbf{x}_n = \boldsymbol{\Psi}\mathbf{j}_n , \qquad n = 1, \ldots, N-1 ,$$
$$\Psi_{i,k} \equiv \exp\left(-j \frac{d}{c} \omega \sin(\theta_k i)\right) \tag{3-42}$$
$$k \in [1, N-1], \qquad i \in [0, N-1], \qquad \mathbf{j}_n \in \mathscr{C}^{N-1}$$

where as before d is the sensor spacing, c the propagation velocity, θ_i the bearing of the ith signal, and ω the carrier frequency, and $\mathscr{C}^L$ denotes the space of complex L-vectors. This model assumes the presence of $N-1$ directional signals and no receiver noise. $\mathbf{j}_n$ is the vector of source signals at time n, as per (3-35). We will assume that $\sin(\theta_i) \neq \sin(\theta_j) \bmod(2c\pi/d\omega))$, in which case $\boldsymbol{\Psi}$ is full rank by Vandermonde's relation [12]. We also assume the $\mathbf{j}_n$ are i.i.d. Under mild restrictions on the CDF of $\mathbf{j}$ it again follows from Appendix A that the $\mathbf{x}_n$ will be linearly independent. We now present a theorem describing the bearing recovery process:

Theorem 3-2 (Ideal Bearing Recovery) The bearings θ_i can be found, given the snapshots $\mathbf{x}_n \in \mathscr{C}^N$, $n = 1, \ldots, N-1$ as follows:

1. Find $\mathbf{v}_0$, the eigenvector for the zero eigenvalue in $\hat{\mathbf{R}} \equiv \Sigma_{n=1}^{N-1} \mathbf{x}_n \mathbf{x}_n^H$.
2. Form from this eigenvector the polynomial $P(z) \equiv \Sigma_{i=1}^{N} \mathbf{v}_0(i) z^{i-1}$ and find its roots z_i, $i = 1, \ldots, N-1$.
3. Set $\theta_i = \sin^{-1}(c\angle z_i / d\omega)$, $i = 1, \ldots, N-1$.

Note that perfect bearing recovery of $N-1$ signals is possible, with a number of snapshots equal to $N-1$. As in Theorem 3-1 this result has value even in the case when thermal noise is present. Since this noise is spatially white in expected value it adds a scalar diagonal matrix to $\hat{\mathbf{R}}$. This does not affect the eigenvectors of $\hat{\mathbf{R}}$. We conclude that in the presence of thermal noise the recovery process reverts to an (imperfect) estimation process. Appendix F describes some of the statistical behavior of the presence of thermal noise using subspace perturbation arguments. As the number of snapshots grows we move closer to the ensemble covariance. Therefore while $N-1$ snapshots works in the absence of thermal noise, many more may be needed in the general case. (This comment applies to target recovery as well.)

Proof. The zero eigenvector satisfies

$$\mathbf{\Psi J J}^H \mathbf{\Psi}^H \mathbf{v}_0 = \mathbf{0} \tag{3-43}$$

where $\mathbf{J}$ is a matrix whose columns are built from the $\mathbf{j}_n$, and $\mathbf{0} \in \mathscr{C}^{N-1}$ is the zero vector. $\mathbf{\Psi}$ is full rank by assumption, as is, with unit probability (by the i.i.d. assumption) the $N-1$ by $N-1$ matrix $\mathbf{JJ}^H$. Since $\mathbf{\Psi}$ is full rank its column space spans $\mathscr{C}^{N-1}$. Thus no vector when left multiplied by $\mathbf{\Psi}$ yields zero, except the zero vector. Likewise for $\mathbf{JJ}^H$. It follows that $\mathbf{\Psi}^H \mathbf{v}_0 = \mathbf{0} \rightarrow P(\exp(j(d\omega/c)\sin(\theta_i))) = 0,\ i = 1, \ldots, N-1$ where $P(\ )$ is a polynomial built from $\mathbf{v}_0$ as per stage 2 in the theorem. Clearly the roots of this polynomial are $\exp(j(d\omega/c)\sin(\theta_i))$ and stage 3 follows immediately.

Note that the $N-1$ roots of any polynomial which is not identically zero are uniquely defined by the polynomial itself. Thus, since $\mathbf{v}_0$ is unique to within a scale factor, the correct bearings are uniquely obtained, to within aliasing induced by the arcsine function.

Example 3-1. Suppose $N = 3$, $d\omega/c = 2\pi$, and the two snapshots are

$$\mathbf{x}_1 = \begin{bmatrix} 0 \\ 1-j \\ 0 \end{bmatrix}, \qquad \mathbf{x}_2 = \begin{bmatrix} 1 \\ j \\ 1 \end{bmatrix} \tag{3-44}$$

so

$$\hat{\mathbf{R}} = \begin{bmatrix} 1 & -j & 1 \\ j & 3 & j \\ 1 & -j & 1 \end{bmatrix} \tag{3-45}$$

Thus

$$\mathbf{x}_1 = \begin{bmatrix} 1 \\ 0 \\ -1 \end{bmatrix} \tag{3-46}$$

so the roots are ± 1 and the angles are $\angle \mathbf{z}_i = 0, \pi$. It follows that the bearings are 0 and $\sin^{-1}(\frac{1}{2}) = \pi/6$.

Once the bearings are found we can if we so choose recover the source signal vector matrix **J** given the corresponding observation snapshot matrix **X** via the relation

$$\mathbf{X} = \boldsymbol{\Psi}\mathbf{J} \rightarrow \mathbf{J} = \boldsymbol{\Psi}^P\mathbf{X} \tag{3-47}$$

where $\boldsymbol{\Psi}^P$ is the pseudo inverse of $\boldsymbol{\Psi}$, $\boldsymbol{\Psi}^P\boldsymbol{\Psi} = \mathbf{I}_{N-1}$. Given the SVD of $\boldsymbol{\Psi}$, that is, $\boldsymbol{\Psi} = \mathbf{U}_N[\boldsymbol{\Lambda} \,|\, \mathbf{0}]^H\mathbf{V}_{N-1}$, with $\mathbf{U}_N$, $\mathbf{V}_{N-1}$ unitary of size N and $N-1$ respectively, the pseudo inverse is given by $\boldsymbol{\Psi}^P = \mathbf{V}_{N-1}^H[\boldsymbol{\Lambda}^{-1} \,|\, \mathbf{0}]\mathbf{U}_N^H$. The rows of this pseudo inverse then yield the weight vectors needed to recover any desired interference source.

This pseudo inverse interpretation of Theorem 3-2 gives some valuable intuitive insight. We are given the matrix product of an unknown matrix ($\boldsymbol{\Psi}$) and some unknown vectors ($\mathbf{j}_i$). We seek to infer from this product the constituent matrix and the vectors. In general this is of course impossible but the parametric form of $\boldsymbol{\Psi}$ allows us to proceed, albeit in a rather circuitous, nonlinear fashion.

It is sometimes possible to get by with less than the $N-1$ snapshots considered here [4]. Also if fewer than $N-1$ signals impinge on the array one can to some extent use radii to separate noise and signal zeroes in $P(z)$. We omit details for brevity.

Note that in both bearing estimation and target reconstruction the matrix $\hat{\mathbf{R}}$ played a central role. This matrix is a time average of the snapshot dyadic (i.e., outer product) $\mathbf{x}\mathbf{x}^H$, just as the covariance matrix $\mathbf{R}$ is this dyadic's ensemble average. $\hat{\mathbf{R}}$ is known as the *sample covariance matrix* and as we shall see it lies at the core of every adaptive detection and estimation procedure.

3-4 ADAPTIVE ESTIMATION

3-4-1 Formulation

Consider again the relation (3-1). Now suppose we form a sequence of snapshots $\mathbf{x}_n$, $n = 1, \ldots, L$ by sampling $\mathbf{x}(t)$, with the narrowband assumption in force. Then we can express this relation as

$$\boxed{\mathbf{x}_n = s_n\mathbf{d} + \mathbf{n}_n \quad , \qquad \mathbf{n}_n \equiv \mathbf{i}_n + \mathbf{u}_n} \tag{3-48}$$

Recall that the narrowband assumption essentially maintains that the steering vector $\mathbf{d}$ be approximately constant across the bandwidth of the target signal $s(t)$. We absorb both noise components $\mathbf{i}$ (directional interference) and $\mathbf{u}$ (spatially white) into the one vector $\mathbf{n}$ (net noise vector impinging on the array). Henceforth we exclusively consider the narrowband sampled data model in (3-48). Specifically we explore the inference (estimation), and later detection, of the sampled scalar target waveform s_n, $n = 1, \ldots, L$, given the observations $\mathbf{x}_n$, $n = 1, \ldots, L$. Let us introduce the N by L complex matrix

$$\mathbf{X} \equiv [\mathbf{x}_1 | \cdots | \mathbf{x}_L] \tag{3-49}$$

and the 1 by L complex vector

$$\mathbf{s}^H \equiv [s_1, \ldots, s_L] \tag{3-50}$$

Then we can view our estimation problem as that of finding a mapping L taking the data matrix $\mathbf{X}$ into a vector near $\mathbf{s}^H$, that is

$$L: \mathscr{C}^{NL} \rightarrow \mathscr{C}^L, \qquad \text{so that } L(\mathbf{X}) = [\hat{s}_1, \hat{s}_2, \ldots, \hat{s}_L] \approx [s_1, \ldots, s_L] \tag{3-51}$$

We shall assume that the vectors $\mathbf{n}_n$ are i.i.d., zero mean, with common covariance matrix $\mathbf{R_{nn}} = E[\mathbf{n}_n \mathbf{n}_n^H]$, and that $\mathbf{d}$ is known. A number of special cases arise, depending on our prior knowledge regarding the noise $\mathbf{n}_i$, and target s_i, and measure of goodness of approximation. A list of several possibilities regarding prior knowledge is provided below, along with various figures of merit:

1. $\mathbf{R_{n,n}}$ is known and s_n is a stationary random process. Figure of merit: MSE.
2. $\mathbf{R_{n,n}}$ is known and s_n is unknown but deterministic. Figure of merit: MVDR.
3. $\mathbf{R_{n,n}}$ and s_n are unknown and deterministic. Figures of merit: risk (expected loss) minimization, likelihood maximization.
4. $\mathbf{R_{n,n}}$ and s_n are unknown and random. Figures of merit: MSE, maximization of a posteriori probability.

For case 1 we obtain the Wiener solution of Section 3-3-2, if our cost functional is the MSE $E[|\hat{s}_n - s_n|^2]$, and our processing is restricted to be linear. Note that this approach does not require prior knowledge of the probability density functions for $\mathbf{n}$ and/or s, beyond mean and covariance. (For non-Gaussian data, we can obtain an MSE smaller than that provided by the Wiener solution provided (i) we employ nonlinear processing and (ii) we know the PDF of the data.)

For case 2 we obtain the MVDR solution of (3-19) if we seek a linear unbiased estimate of minimum variance. Again we do not require a knowledge of the PDF of $\mathbf{n}$ for MVDR processing.

Cases 1 and 2 are really not adaptive, but are included for completeness. The latter two cases, discussed below, are adaptive, in the sense that they involve prior uncertainty regarding the structure of the interference.

Case 3 differs from 1 in that the interference covariance is now unknown, and things complicate markedly. We cannot employ the MSE criterion directly since the resulting expectations cannot be evaluated. One way around this is to replace ensemble averages with time averages. The justification for this approach is that if ergodicity holds, the two averages converge in the limit. Since in practice we have a finite number of snapshots, this is no longer really a Wiener solution, but some sort of approximation, with no obvious optimality properties.

An alternative solution method to case 3 is to employ the so called risk theory to construct an estimator which minimizes some expected loss (risk) metric [28]. A popular loss metric is the quadratic loss which in our case is $\|\hat{\mathbf{s}}(\mathbf{X}) - \mathbf{s}\|^2$. This approach is intractable, even in the case far simpler than ours where one merely seeks to estimate an unknown mean [28]! Indeed, Anderson [1] shows that if one has a set of L Gaussian random N-vectors with unknown mean vector, the sample mean (which is maximum likelihood and minimum variance unbiased) does not minimize the risk! Other counterexamples, some dealing with covariance estimation, are detailed in Anderson and Muirhead [1, 28]. The bizarre nature of many of the estimates arising from risk theory has spawned a good bit of controversy. By expecting over the loss risk theory puts enormous weight on losses which are high but occur very rarely. This sensitivity to tails of the loss density may account for some of the observed peculiarities arising from the theory, and place some question on the practicality of the results.

A third method for solving case 3 is that of maximum likelihood (ML) estimation. The idea here is to find **s** to maximize the joint PDF of the snapshots evaluated at the observations. In other words we select the unknown covariance, and the vector **s** to make the observed data as likely as possible:

$$\hat{s}_{\mathrm{ML}} = \underset{\mathbf{s}, \mathbf{R_{n,n}} \in \Omega}{\operatorname{argmax}} \left(f(\mathbf{X} \mid \mathbf{s}, \mathbf{R_{n,n}}) \right) \tag{3-52}$$

This has some intuitive appeal and it satisfies certain optimality properties. Under mild regularity conditions in the limit as $L \rightarrow \infty$ it is unbiased, normally distributed, and it minimizes the expected quadratic loss [39]. Another intriguing optimality property of the ML estimate emerges from information theory. One discovers that this estimate approximately minimizes the mean Kullback–Leibler information distance between PDFs constrained to lie in Ω and the true unknown density [22].

The set Ω in (3-52) is the set of joint signal models and covariance matrices one seeks to sup over. It is common to choose the covariance matrix to be positive semidefinite and Hermitian but otherwise arbitrary.

The family of target waveforms must be restricted in some fashion to obtain unique (useful) estimates. We will consider specifics later.

As we will see one can also derive the ML estimate as a special case of 4 considered below.

Note that although we seek s_n we are forced to estimate $\mathbf{R}_{\mathbf{n},\mathbf{n}}$ as part of the ML procedure. Essentially the interference structure is a nuisance parameter which must be dealt with although it is not part of the final estimation objective. Note also that the ML estimate requires a parametric model for the PDF of the data. A common choice is the multivariate normal density.

Case 4 treats the covariance matrix of $\mathbf{n}$ and the statistics of s as themselves random variables.

By ascribing densities to these unknown quantities we can once again speak of expectations, unlike in case 3. If we employ the MSE criterion we obtain the estimate [31],

$$\hat{\mathbf{s}} = E[\mathbf{s} \mid \mathbf{X}] \tag{3.53}$$

where the expectation is over $\mathbf{s}$ and $\mathbf{R}_{\mathbf{n},\mathbf{n}}$. As with risk theory, the pursuit of elegant objective functionals leads to a hopelessly intractable estimator. Furthermore it seems wasteful to integrate over a prior density for the interference covariance, since we can obtain a decent estimate of this parameter from the data.

A tractable, intuitively pleasing alternative is the so called maximum a priori (MAP) estimate. The MAP estimate finds the most probable s_n given $\mathbf{X}$:

$$\hat{\mathbf{s}}_{\text{MAP}} = \underset{\mathbf{s},\mathbf{R}_{\mathbf{n},\mathbf{n}} \in \Omega}{\operatorname{argmax}} \left(f(\mathbf{s}, \mathbf{R}_{\mathbf{n},\mathbf{n}} \mid \mathbf{X}) \right) \tag{3-54}$$

Again we sup over the interference covariance, as well as the target waveform, and its statistics. As in case 3 we require knowledge of the parametric form of the PDF for the data.

We can express (3-54) in the following slightly different form using Baye's theorem:

$$\hat{\mathbf{s}}_{\text{MAP}} = \underset{\mathbf{s},\mathbf{R}_{\mathbf{n},\mathbf{n}} \in \Omega}{\operatorname{argmax}} \left(\frac{f(\mathbf{X} \mid \mathbf{s}, \mathbf{R}_{\mathbf{n},\mathbf{n}}) f(\mathbf{s}, \mathbf{R}_{\mathbf{n},\mathbf{n}})}{f(\mathbf{X})} \right) \tag{3-55}$$

The denominator in (3-55) is independent of the free parameters and therefore can be discarded. If the prior density of the target and interference is uniform then we obtain the ML estimate![10] Finding prior densities for the

[10] One cannot have a uniform density over an infinite support, so we would have to restrict Ω to have finite (but possibly enormous) support.

parameters in MAP estimation is admittedly a nebulous task. To this end we note that a uniform density is maximum entropy in the absence of any prior information other than the support of Ω. If the mean and covariance of **s** and/or $\mathbf{R}_{n,n}$ is known then the Gaussian density would be the maximum entropy solution. This provides some justification for using the former density (which leads to the ML estimate) in the absence of any prior knowledge, and using the latter density if the average prior value of the parameters is known, and some level of uncertainty (variance) can be reasonably attached to them.

There is at least one example where a nonuniform prior leads to interesting results (for details see [36]). If one chooses as prior on $\mathbf{R}_{n,n}$ the function in (3-58) (with **R** now the random variable and **S** replaced by some constant matrix $\mathbf{S}_0$) raised to the power α and suitably re-scaled one obtains the covariance estimate $(\mathbf{S}L + \alpha\mathbf{S}_0)/(L + \alpha)$. This is effectively the recursive least squares estimate [26] with an exponential window, with S_0 the covariance estimate from prior data.

In discussing cases 3 and 4 we should note that there is a raging controversy about whether unknowns should be viewed as random (the Bayesian's creed), or deterministic (the non-Bayesian's creed). A random model seems appropriate when it is difficult to obtain a good estimate of the unknown parameter from the data. Such is not the case here, but it is for instance in handling phase in asynchronous communication. As engineers we are concerned with system design, not merely abstract knowledge, so we can avoid the Bayesian fray by merely trying different approaches to see which work.

While we do not seek to imply it has the final word necessarily, the ML approach offers an attractive compromise between tractability and fundamental optimality. Henceforth, we focus exclusively on ML estimates.

3-4-2 The PDF of the Array Snapshots

We seek the PDF of the N by L matrix of snapshots in (3-49). To this end we assume that the noise is zero mean normal with covariance **R**, and that s_n is deterministic. We prove in Appendix B, Theorem 3-11, that the PDF for a zero mean normal complex random vector **x** is given by

$$f_x(\mathbf{x}) = \frac{1}{\pi^N \det(\mathbf{R})} \exp(-\mathbf{x}^H \mathbf{R}^{-1} \mathbf{x}) \tag{3-56}$$

If **x** has nonzero mean we simply replace **x** by $\mathbf{x} - E[\mathbf{x}]$ in (3-56).

Why assume Gaussian data? (i) The linear averaging due to the various IF bandpass, baseband anti-aliasing, pulse compression, filters, etcetera, tend to yield Gaussian output data (due to the central limit theorem).[11] (ii) If we do not, everything is too hard to analyze! (iii) If $\mathbf{x}$ has a PDF which is a function of $\|\mathbf{x}\|$ and the entries are independent then the vector is Gaussian (Proof: see [31]).

We seek the density for $\mathbf{X}$. By independence we have

$$f(\mathbf{X}) = \prod_n f(\mathbf{x}_n)$$
$$= \left(\frac{1}{\pi}\right)^{NL} \det(\mathbf{R})^{-L} \exp\left(-\sum_n (\mathbf{x}_n - E[\mathbf{x}_n])^H \mathbf{R}^{-1}(\mathbf{x}_n - E[\mathbf{x}_n])\right) \tag{3-57}$$

The covariance in (3-57) is both the covariance of $\mathbf{x}$ and the covariance of $\mathbf{n}$, since s_n is deterministic, and therefore does not affect covariance. Using the fact that $\text{Trace}(AB) = \text{Trace}(BA)$ we obtain

$$\boxed{\begin{aligned} f(\mathbf{X}) &= [(\pi)^{-N} \det(\mathbf{R})^{-1} \exp(-\text{Trace}(\mathbf{R}^{-1}\mathbf{S}))]^L , \\ \mathbf{S} &\equiv \frac{1}{L} (\mathbf{X} - E[\mathbf{X}])(\mathbf{X} - E[\mathbf{X}])^H \end{aligned}} \tag{3-58}$$

This form will be most useful in our subsequent discussions.

3-4-3 Maximum Likelihood Estimates of Covariance

In the ML method we must take the sup over both the covariance and the target waveform. It does not matter which quantity we maximize over first, so let us pick the covariance. As a side benefit this yields the ML estimate of the covariance, which is of interest in its own right in some applications. Rather than maximize the likelihood directly we can minimize $-\log(f(\mathbf{X}))/L$:

$$\mathbf{R}_{\text{ML}} = \text{argmin}(\log(\det(\mathbf{R})) + \text{Trace}(\mathbf{R}^{-1}\mathbf{S})) \tag{3-59}$$

since the logarithm is a monotonic function. Note that by construction $\mathbf{S}$ is positive definite (for $L > N$ with unit probability, see Appendix A) and it therefore possesses a positive definite square root, $\sqrt{\mathbf{S}}$. Using the fact that

$$\det(\mathbf{AB}) = \det(\mathbf{A}) \det(\mathbf{B}) , \qquad \det(\mathbf{A}^{-1}) = 1/\det(\mathbf{A}) \tag{3-60}$$

we can re-express (3-59) as

[11] Actually other density functions do play a role in radar detection, but not in characterizing the snapshot's joint PDF.

$$\mathbf{R}_{\mathrm{ML}} = \operatorname{argmin}(-\log(\det(\boldsymbol{\Psi})) + \operatorname{Trace}(\boldsymbol{\Psi})) , \qquad \boldsymbol{\Psi} \equiv \sqrt{\mathbf{S}}\mathbf{R}^{-1}\sqrt{\mathbf{S}} \tag{3-61}$$

The set of covariances we seek to minimize over is the set of positive semidefinite Hermitian matrices. Therefore minimizing over $\boldsymbol{\Psi}$ is equivalent to minimizing over $\mathbf{R}$. This follows from Sylvester's theorem [12], or a number of other simple arguments. Now we can express $\boldsymbol{\Psi}$ in diagonal form as follows:

$$\boldsymbol{\Psi} = \mathbf{U}\boldsymbol{\Delta}\mathbf{U}^H , \qquad \mathbf{U}\mathbf{U}^H = \mathbf{I} , \qquad \boldsymbol{\Delta} = \operatorname{diag}(\{\lambda_i\}) , \qquad \lambda_i \geqslant 0 \tag{3-62}$$

Since the trace is the sum of the eigenvalues of a matrix and the determinant is the product of the eigenvalues, it follows that we seek those eigenvalues which satisfy

$$\operatorname{argmin}\left(- \sum_n \log(\lambda_n) + \sum_n \lambda_n \right) , \qquad \lambda_n \geqslant 0 \tag{3-63}$$

Clearly we have $\lambda = 1$, $\rightarrow \mathbf{U} = \mathbf{I}$, $\rightarrow \boldsymbol{\Psi} = \mathbf{I}$, and hence

$$\boxed{\mathbf{R}_{\mathrm{ML}} = \mathbf{S} \equiv \frac{1}{L} \sum_{n=1}^{L} (\mathbf{x}_n - E[\mathbf{x}_n])(\mathbf{x}_n - E[\mathbf{x}_n])^H} \tag{3-64}$$

This is an important result. It states that the familiar sum of outer products sample covariance matrix is the maximum likelihood estimate. This is a stronger optimality property than the limiting ergodicity justification often provided for the sum of outer products estimator.

Suppose there is prior knowledge about the form of $\mathbf{R}$. Then the set of positive semidefinite Hermitian matrices is not the proper set to maximize over. We shall now consider three cases of alternate covariance sets: (i) subspace models; (ii) Toeplitz models; and (iii) the extendible set. These sets evolve from the highly tractable but somewhat ad hoc (subspace), to the completely intractable but physically authentic (extendible).

3-4-4 Subspace Structured Covariance Matrices

If a waveform satisfies the narrowband assumption then it yields a rank one array covariance matrix. Furthermore, a broadband waveform approaching from broadside also yields a rank one covariance matrix. This suggests that for multiple narrowband interferers, a low rank model might be appropriate for ML estimation [40]. Further justification for low rank modeling stems from the fact that a broadband signal has an approximately low rank covariance with rank specified by the time bandwidth product [26]. The low rank, that is, subspace, covariance model is the following:

$$\mathbf{R} = \mathbf{U}\boldsymbol{\Psi}\mathbf{U}^H + \delta^2\mathbf{I} \tag{3-65}$$

where $\mathbf{U}$ is an N by M, $M < N$ matrix, $\mathbf{U}^H\mathbf{U} = \mathbf{I}$, and σ^2 is an unknown scalar (representing white receiver noise) and $\boldsymbol{\Psi}$ is an unknown M by M matrix.

It will be convenient to rewrite $\mathbf{R}$ as

$$\mathbf{R} = \sigma^2(\mathbf{I} - \mathbf{U}\mathbf{U}^H) + \mathbf{U}(\Psi + \sigma^2\mathbf{I})\mathbf{U}^H \tag{3-66}$$

In this form we have decomposed $\mathbf{R}$ into the sum of two invariant subspaces. Let us make a change of coordinates by introducing the Hermitian matrix variable $\mathbf{K} = \boldsymbol{\Psi} + \sigma^2\mathbf{I}$. Then after some algebra we find we seek to solve

$$\sup_{\sigma^2,\mathbf{K}} \frac{1}{\det(\sigma^2(\mathbf{I} - \mathbf{U}\mathbf{U}^H) + \mathbf{U}\mathbf{K}\mathbf{U}^H)} \exp(-\mathrm{Tr}(\sigma^{-2}(\mathbf{I} - \mathbf{U}\mathbf{U}^H)\mathbf{S} + \mathbf{U}\mathbf{K}^{-1}\mathbf{U}^H\mathbf{S})) \tag{3-67}$$

Using the fact that the eigenvalues of an invariant sum is the union of the constituent matrices' eigenvalues, we can rewrite (3-67) as

$$\min_{\sigma^2,\mathbf{K}} \sigma^{2(N-M)} \det(\mathbf{K}) \exp(\mathrm{Tr}(\sigma^{-2}(\mathbf{I} - \mathbf{U}\mathbf{U}^H)\mathbf{S} + \mathbf{U}\mathbf{K}^{-1}\mathbf{U}^H\mathbf{S})) \tag{3-68}$$

Minimizing over $\mathbf{K}$ as in Section 3-4-3, we conclude that $\mathbf{K} = \mathbf{U}^H\mathbf{S}\mathbf{U}$. We now arrive at the extremal problem for the scalar variable σ^2:

$$\min_{\sigma^2} \sigma^{2(N-M)} \exp(\mathrm{Tr}(\sigma^{-2}(\mathbf{I} - \mathbf{U}\mathbf{U}^H)\mathbf{S})) \tag{3-69}$$

After some manipulation we conclude that

$$\sigma^2 = \frac{\mathrm{Tr}(\mathbf{S} - \mathbf{U}^H\mathbf{S}\mathbf{U})}{N - M} \tag{3-70}$$

resulting in the final subspace-structured covariance estimate of

$$\boxed{\mathbf{R}_{\mathrm{ML}} = \frac{\mathrm{Tr}(\mathbf{S} - \mathbf{U}^H\mathbf{S}\mathbf{U})}{N - M}(\mathbf{I} - \mathbf{U}\mathbf{U}^H) + \mathbf{U}\mathbf{U}^H\mathbf{S}\mathbf{U}\mathbf{U}^H} \tag{3-71}$$

If $\mathbf{U}$ is unknown then one must sup over it as well. The ML estimate of $\mathbf{U}$ is simply the dominant invariant subspace of $\mathbf{S}$. Details are found in [2].

This discussion on subspace covariance models would be incomplete without a discussion of the following remarkable theorem due to Xu and Kailath [41].

Theorem 3-3 (Fast Subspace Decomposition) Let **R** be an N by N matrix which is Hermitian and which has the form

$$\mathbf{R} = \mathbf{B} + \sigma^2 \mathbf{I} \tag{3-72}$$

where σ^2 is unknown, **I** is the N by N identity, and **B** is Hermitian of rank $d < N - 1$, having distinct nonzero eigenvalues. **B** is unknown as is d. There exists an algorithm which involves a finite, and fixed number of multiplies, adds, and square roots, which will find:

(i) d, and subsequently
(ii) σ^2,
(iii) **B**, and
(iv) eigenvectors in the null space of $\mathbf{R} - \sigma^2 \mathbf{I}$.

Note that once (i) and (ii) are found the rest follows easily, so we focus on these two steps. Theorem 3-3 reveals that if one had available the ensemble covariance matrix the task of extracting spanning vectors for the directional interference (or signal) subspace **B**, or eigenvectors for the thermal noise subspace, is (unlike eigenanalysis in general) *noniterative*! The cost of the finite algorithm, which we construct during the proof, is $O(N^2 d)$. The more practical case when only sample covariance matrices are available is discussed in [42]. This reference also describes further savings available when **R** exhibits Toeplitz structure.

Proof. Let us express **B** as

$$\mathbf{B} = \mathbf{E}\Lambda\mathbf{E}^H , \qquad \Lambda \equiv \text{diag}(\lambda_1, \lambda_2, \ldots, \lambda_d) \tag{3-73}$$

where the columns of **E** equal the eigenvectors of **B**, which we denote by $\mathbf{e}_i$, $i = 1, \ldots, d$. Then we can express **R** as

$$\mathbf{R} = \mathbf{E}(\Lambda + \sigma^2 \mathbf{I})\mathbf{E}^H + \sigma^2 (I - \mathbf{E}\mathbf{E}^H) \tag{3-74}$$

First we generate a vector **f** which can be expressed as

$$\mathbf{f} = \sum_{i=1}^{d} \alpha_i \mathbf{e}_i + \mathbf{v} , \qquad \alpha_i \neq 0 , \qquad \|\mathbf{v}\| \neq 0 , \qquad \mathbf{v} \perp \mathbf{e}_i , \qquad i = 1, \ldots, d \tag{3-75}$$

Note that we can form such a vector with probability 1 by selecting it from any distribution with continuous CDF (see Appendix A). Next we form the N by M matrix,

$$\mathbf{K}_M = [\mathbf{f}, \mathbf{R}\mathbf{f}, \ldots, \mathbf{R}^{M-1}\mathbf{f}] \tag{3-76}$$

that is, the ith column of $\mathbf{K}_M$ is $\mathbf{R}^{i-1}\mathbf{f}$. M must be chosen to satisfy $M > d+1$. If nothing is known about d other than $d < N-1$, then one selects $M = N$. If d is further bounded then one can select M to be smaller, thereby saving operations. We now show that

$$\operatorname{rank}(\mathbf{K}_M) = d + 1 \tag{3-77}$$

Note that since M is free, this implies that the first $d+1$ columns of $\mathbf{K}_M$ are linearly independent, with dependency first arising in the $(d+2)$nd column. This is handy computationally; one can orthogonalize columns from left to right, terminating when a dependent vector is reached. First note that

$$\mathbf{f} = \mathbf{E}\mathbf{a} + \mathbf{v}\,, \qquad \mathbf{a} \equiv [\alpha_1 \quad \cdots \quad \alpha_d]^T \tag{3-78}$$

It follows from (3-74), and the property $\mathbf{E}^H\mathbf{v} = \mathbf{0}$, and some algebra, that

$$\begin{aligned}\mathbf{K}_M = {} & \mathbf{E}\operatorname{diag}(\alpha_1, \ldots, \alpha_d)[\mathbf{1}, (\Lambda + \sigma^2\mathbf{I})\mathbf{1}, \ldots, (\Lambda + \sigma^2\mathbf{I})^{M-1}\mathbf{1}] \\ & + \mathbf{v}(1, \sigma^2, (\sigma^2)^2, \ldots, (\sigma^2)^{M-1})\end{aligned} \tag{3-79}$$

where $\mathbf{1}$ is a length d column vector whose entries are all unity. Now premultiplication by a matrix with orthonormal columns does not affect rank (see [12]). Multiplying by a full rank square diagonal matrix does not affect rank either. Finally the fact that $\mathbf{E} \perp \mathbf{v}$ ensures that the outer product involving $\mathbf{v}$ augments the rank by exactly one. By expanding out the matrix in brackets it follows that

$$\operatorname{rank}(\mathbf{K}_M) = 1 + \operatorname{rank}\left(\begin{bmatrix} 1 & x_1 & x_1^2 & \cdots & x_1^{M-1} \\ 1 & x_2 & x_2^2 & \cdots & x_2^{M-1} \\ \vdots & \vdots & \vdots & \vdots & \vdots \\ 1 & x_d & x_d^2 & \cdots & x_d^{M-1} \end{bmatrix}\right), \quad x_i \equiv (\lambda_i + \sigma^2) \tag{3-80}$$

Note by assumption that the λ_i, and hence the x_i are distinct. The matrix involving various powers of the x_i, which we now denote by $\mathcal{V}$ is known as the *Vandermonde* matrix [12]. One can determine its rank by finding the number of linearly independent vectors $\mathbf{q}$ satisfying $\mathcal{V}\mathbf{q} = \mathbf{0}$. Denote by $P_q(x)$ the polynomial built from $\mathbf{q}$. Then we can express the relation $\mathcal{V}\mathbf{q} = \mathbf{0}$ as $P_q(x) = 0$, $x = x_i$, $i = 1, \ldots, d$. The number of linearly independent polynomials of order $M-1$ which share d (distinct) specified roots is exactly $M-d$. (The vector space of polynomials with k free roots has dimension $k+1$.) Hence the column null space of $\mathcal{V}$ has dimension $M-d$, so the column space has dimension d, that is, rank $(\mathcal{V}) = d$, and part (i) of Theorem 3-3 follows. Now that d is known it remains to find σ^2. To this end we form a unit norm vector $\mathbf{w}$ which has the property that it is orthonormal to $\mathbf{K}_M$. (We can construct $\mathbf{w}$ using information at hand using projection matrices.) Note that $\mathbf{w} \perp \mathbf{K}_M \rightarrow \mathbf{w} \perp \mathbf{B}$. It follows that

$$\mathbf{w}^H R \mathbf{w} = \sigma^2 \tag{3-81}$$

which establishes part (ii) of Theorem 3-3. Once σ^2 is found it is trivial to find $\mathbf{B}$ (and its null space) via $\mathbf{B} = \mathbf{R} - \sigma^2 \mathbf{I}$.

Another popular set of covariances to maximize over is the set of positive semidefinite Toeplitz matrices [9]. This set arises from a uniform linear array of sensors (with identical beampatterns). Unfortunately closed form solutions to the ML estimate do not exist, and one must resort to nonlinear, iterative methods, which only promise local maxima. There is however an elegant theoretical result regarding ML estimation for Toeplitz matrices and related classes. Let Ω denote a family of covariance matrices, with the property that for any $\mathbf{R}_0 \in \Omega$ the variation $\delta \mathbf{R}_0$ is also in Ω. (This is true for Toeplitz models in particular, and linear models generally.) Then the ML estimation problem reduces to the following extremal problem:

$$\mathbf{R}_{\mathrm{ML}} = \underset{\mathbf{R} \in \Omega}{\operatorname{argmin}} \det(\mathbf{R}) \quad \text{subject to} \quad \operatorname{Trace}(\mathbf{R}^{-1}\mathbf{S}) = N \tag{3-82}$$

One can also use (3-82) to derive the ML covariance estimates in (3-64), (3-71).

3-4-5 Extendibility

We now present a very flexible means of constructing the exact set of covariances which a given array can produce. Specifically, we explore the constraints on $\mathbf{R}$ imposed by the array geometry, the operating frequencies of the array, the wavefront propagation velocity and the angular sector in which the interference (or sources) may reside. We will consider the case where each of the N sensor elements used in forming $\mathbf{R}$ has an arbitrary beampattern. We will, for notational expediency alone, consider the case of a linear array. These constraints have not yet been employed in adaptive procedures, but they provide insights which may ultimately prove quite valuable.

Denote by $r_{i,j}$ and $\tau_{i,j}(\theta)$ the array covariance elements and interelement propagation delays respectively. In a linear array the propagation delay $\tau_{i,j}$ simply equals the signed distance between sensor i and sensor j divided by the wavefront propagation velocity c multiplied by $\sin(\theta)$, where θ is the angle of arrival of the source with respect to the normal vector of the array. Thus $\tau_{i,j}$ contains all array geometry and wavefront velocity information. The elements of the covariance matrix by definition equal $E[x_i x_j^*]$. Denote by $S(\theta, \omega)$ the spatial/temporal spectra of the sources which give rise to

$\mathbf{r} = \{r_{i,j}\}$. Denote by $\beta_i(\theta, \omega)$ the transfer function of the ith sensor, that is, antenna element. Observe that for an omnidirectional antenna, β_i is simply a constant. Let $\Gamma = \{\theta, \omega: \theta_0 \leq \theta \leq \theta_1, \omega_0 \leq \omega \leq \omega_1\}$. The bounds on ω correspond to the known operating frequency band of the array (or the known frequency band of the sources/interference), and the bounds on θ arise from possible a priori knowledge on the location of the radiating sources. (More generally Γ might contain disjoint sets of frequency bands and/or angular sectors.) We seek to characterize the set Ω consisting of those $r_{i,j}$, $i, j = 1, \ldots, N$ attainable for given β_i, $\tau_{i,j}$, and Γ. The set Ω is known as the set of *extendible* correlations. This nomenclature is due to Lang and McClellan [23], as is (in essence) the following characterization, derived in Appendix C.

$$\Omega = \left\{ r_{i,j}: \sum_{i,j=1}^{N} p_{i,j} r_{i,j} \geq 0, \forall p_{i,j}: \sum_{i,j=1}^{N} p_{i,j} \beta_i(\theta, \omega) \beta_j^*(\theta, \omega) \exp(j\omega\tau_{i,j}(\theta)) \geq 0 \;\; \omega, \theta \in \Gamma \right\} \tag{3-83}$$

Given this representation one can readily test for the membership of a given $\mathbf{r}$ in Ω by means of linear programming.

The above representation of Ω is independent of the source spatial/temporal power spectra which generates the correlations. These spectra are positive but otherwise arbitrary functions on the support Γ, and are therefore infinite dimensional variables. They have been replaced in (3-83) by the N^2 dual variables $p_{i,j}$, a tremendous simplification.

Extendibility forms the basis of multidimensional maximum entropy spectral estimation [23].

To illustrate the notation of extendibility consider the following toy examples.

Example 3-2. Let $N = 2$, $\theta = \pi/4$, $\omega_0 \leq |\omega| \leq \omega_1$, $\tau_{1,2} = \sqrt{2}\cos(\theta) = 1$,

$$\beta_1(\omega) = \sin(\omega), \qquad \beta_2 = 1, \qquad \mathbf{R}_x \text{ real} \tag{3-84}$$

The set of extendible covariance matrices is given from (3-83) by those entries $r_{i,j}$, $i, j = 1, 2$ which satisfy the following inequality for all $p_{1,1}, p_{2,2}, p_{1,2} = p_{2,1}$:

$$\begin{aligned} p_{1,1} r_{1,1} + p_{2,2} r_{2,2} + p_{1,2}(r_{1,2} + r_{2,1}) \geq 0, \qquad & \forall p_{1,1}, p_{2,2}, p_{1,2} \\ p_{2,2} + p_{1,1}\sin^2(\omega) + p_{1,2}\sin(2\omega) \geq 0, \qquad & \omega \in \Gamma \end{aligned} \tag{3-85}$$

In contrast the set of nonnegative definite matrices $\mathbf{R}$ for $N = 2$ is given by

$$r_{1,1} \geqslant 0\,, \qquad r_{1,1} r_{2,2} \geqslant r_{1,2}^2 \tag{3-86}$$

The projection of the set in (3-86) in the $r_{1,1}r_{2,2}$ plane exactly fills the first quadrant. In contrast, this same projection maps the set in (3-85) into the set given by

$$\langle r_{1,1}, r_{2,2} \rangle = \alpha \langle \cos(\phi), \sin(\phi) \rangle\,, \qquad \phi \in \left[\frac{\pi}{4}, \frac{\pi}{2}\right], \qquad \alpha \geqslant 0 \tag{3-87}$$

when $\omega_0 = 0$, $\omega_1 = \pi$, and the corresponding dual set for the $p_{i,j}$ equals

$$\langle p_{1,1}, p_{2,2} \rangle = \alpha \langle \cos(\phi), \sin(\phi) \rangle\,, \qquad \phi \in \left[0, \frac{3\pi}{4}\right] \tag{3-88}$$

For comparison, the set of $r_{i,j}$ associated with spatially white spectra for this array is

$$\langle r_{1,1}\, r_{2,2}, r_{1,2} \rangle = \frac{\alpha}{4\pi} \left\langle \frac{1}{\omega_1 - \omega_0}, \frac{1}{\omega_1 - \omega_0} + \frac{\sin(2\omega_0) - \sin(2\omega_1)}{2}, \right.$$
$$\left. \cos(2\omega_0) - \cos(2\omega_1) \right\rangle \tag{3-89}$$

which is a small (zero measure) subset of (3-85).

Example 3-3 (Uniform Array). Now modify the above example so that $\beta_i = 1$. This corresponds to a two element linear array, so $\mathbf{R}$ is Toeplitz, that is, $r_{1,1} = r_{2,2}$. The extendible set is now

$$\langle r_{1,1}, r_{1,2} \rangle = \alpha \langle \cos(\phi), \sin(\phi) \rangle, \cos(\omega_1) \leqslant \tan(\phi) \leqslant \cos(\omega_0) \tag{3-90}$$

(assuming $\cos(\omega_0) > 0, \cos(\omega_1) < 0$) which for unrestricted Γ reverts as anticipated to $|r_{1,1}| \geqslant |r_{1,2}|$, the nonnegative definite constraints for a second order Toeplitz matrix.

For example if $\omega_0 = \pi/4$, $\omega_1 = 3/4\pi$, $-35° \leqslant \phi \leqslant 35°$ (approximately). In contrast the positive definite constraints alone yield $-45° \leqslant \phi \leqslant 45°$, and a white noise model, flat across ω_0, ω_1 yields $\phi = 0$.

3-4-6 Adaptive Waveform Estimation

In adaptive estimation the interference covariance is unknown, as is the desired target waveform. After we form an ML covariance matrix estimate we must place this estimate into the likelihood function and maximize over the set of feasible target waveforms. We restrict ourselves to the unstructured (sum of outer product) covariance model in (3-64). Note that the trace term vanishes leaving the determinant

$$\mathbf{s} = \underset{\mathbf{s} \in \Omega_s}{\operatorname{argmin}} (\det((\mathbf{X} - E[\mathbf{X}])(\mathbf{X} - E[\mathbf{X}])^H)) \tag{3-91}$$

Recall from (3-48) that

$$E[\mathbf{X}] = \mathbf{d}\mathbf{s}^H \tag{3-92}$$

To proceed we must consider the class Ω_s of $\mathbf{s}$ over which we optimize. One simple possibility, arising in a single pulse continuous wave (CW) system is that

$$s_n = A \exp(j\omega n + j\phi) \tag{3-93}$$

where any combination of A, ω, ϕ is unknown. This would correspond to the return from a single point source, with cross section proportional to A, Doppler proportional to ω, and range proportional to ϕ (modulo ambiguities). The returns from a high performance radar system, such as a pulsed Doppler radar, tend to be a bit more complicated than this.

Let us then consider the type of processing which takes place in a typical pulse Doppler radar. The radar will send out a train of pulses which illuminate a target. Now suppose we uniformly sample each pulse return. If the target motion is negligible during a train of pulses, the nth sample of each pulse will correspond to the same range (to within the sampling ambiguity, and within a finite aperture), and is therefore called a *range gate*. Now suppose we compute the (discrete) Fourier transform of the sequence of range samples during the train duration, at a collection of frequencies, and that we repeat this operation over each antenna element. Then we will have at our disposal a sequence of doubly indexed snapshot vectors, one index referring to Doppler and the other to range. If we seek nonaccelerating targets then we are looking for targets at a single unknown Doppler, at unknown range. The direction of the target we can take to be known, since we are illuminating a known, fixed direction, during each pulse burst. Other scenarios of course occur, but the above model will serve as an excellent guide. We then model the target as deterministic, but unknown, placing a nonzero mean on the PDF for the received data. A reasonable data model might then be the following:

$$\mathbf{s}^H = g\mathbf{b}^H \tag{3-94}$$

where, for now, $\mathbf{b}$ is a unit vector (one entry of unity and the rest zero) denoting the snapshot interrogated for target presence, and the quantity g is a complex number whose amplitude represents the target's radar cross section. The phase of g contains, in principle, refined range information from within the specified range gate.

We see then that the matrix $\mathbf{X}$ might actually be built in a variety of ways, besides directly sampling the data on each sensor. Suppose the interference is highly variable in frequency, but fairly constant over range. In that case we might build **X** out of a sequence of snapshots, at a given DFT (Doppler) frequency, at differing range gates. Since targets are fairly rare we could then test for targets at different range gates by varying the unit entry location in **b**. We point out that more general signal models have been proposed [13, 17]. In the following discussion we will employ (3-94) as our signal model, that is, as our class Ω_s.

Note that the model in (3-94) also models target returns in a CW system if $b_n = \exp(j\omega n)$. In this case **b** is variable, depending as it does on the free frequency (Doppler) parameter ω.

Substituting (3-94) into (3-91), (3-92) results after minor algebra in

$$\min_{S\in\Omega_s} [\det((\mathbf{X} - g\mathbf{d}\mathbf{b}^H)(\mathbf{X}^H - g^*\mathbf{b}\mathbf{d}^H))] \tag{3-95}$$

where **X** is as defined in (3-49). We can, without loss of generality restrict **b** to have unit Euclidean norm. This simplifies some of the subsequent algebra. Using a trick analogous to completing the square, we can rewrite (3-95) in the following more convenient form:[12]

$$\min_{S\in\Omega_s} [\det(\mathbf{A} + \mathbf{v}\mathbf{v}^H)], \qquad \mathbf{v} \equiv \mathbf{X}\mathbf{b} - \mathbf{d}g, \qquad \mathbf{A} \equiv \mathbf{X}(\mathbf{I} - \mathbf{b}\mathbf{b}^H)\mathbf{X}^H \tag{3-96}$$

We will make use of the following matrix identity known as Woodbury's identity [12, p. 3], valid for any Hermitian matrix **A**, and any appropriately dimensioned column vector **v**, provided the stated inverses exist:

$$(\mathbf{A} \pm \mathbf{v}\mathbf{v}^H)^{-1} = \mathbf{A}^{-1} \mp \frac{\mathbf{A}^{-1}\mathbf{v}\mathbf{v}^H\mathbf{A}^{-1}}{1 \pm \mathbf{v}^H\mathbf{A}^{-1}\mathbf{v}} \tag{3-97}$$

This identity is easily verified by direct algebra. One can explicitly evaluate the determinant of a matrix comprised of a rank one modification to the identity. Using this fact, the following determinant relation follows, with some effort, from (3-97):

$$\det(\mathbf{A} \pm \mathbf{v}\mathbf{v}^H) = \det(\mathbf{A})(1 \pm \mathbf{v}^H\mathbf{A}^{-1}\mathbf{v}) \tag{3-98}$$

Applying (3-98) to the bracketed term in (3-96) leads to

$$\det(\mathbf{X}(\mathbf{I} - \mathbf{b}\mathbf{b}^H)\mathbf{X}^H)(1 + \{(\mathbf{b}^H\mathbf{X}^H - \mathbf{d}^H g^*)(\mathbf{X}(\mathbf{I} - \mathbf{b}\mathbf{b}^H)\mathbf{X}^H)^{-1} \times (\mathbf{X}\mathbf{b} - \mathbf{d}g)\}) \tag{3-99}$$

[12] The terms in the brackets of (3-96) and (3-95) are equal, as can be shown by direct algebraic expansion. Just as completing the squares allows us to solve certain scalar problems by inspection this device allows us to use formulas for rank one modifications to obtain our desired result.

Applying (3-98) once again to the determinant in (3-99) leads to the minimization problem,

$$\min[(1 - \mathbf{b}^H\mathbf{X}^H(\mathbf{X}\mathbf{X}^H)^{-1}\mathbf{X}\mathbf{b})(1 + \{(\mathbf{b}^H\mathbf{X}^H - \mathbf{d}^H g^*) \times (\mathbf{X}(\mathbf{I} - \mathbf{b}\mathbf{b}^H)\mathbf{X}^H)^{-1}(\mathbf{X}\mathbf{b} - \mathbf{d}g)\}] \tag{3-100}$$

It follows from the Cauchy–Schwartz inequality, or a number of other standard techniques, that the number g which minimizes (3-100) is given by

$$g_{\text{ML}} = \frac{\mathbf{d}^H\mathbf{Y}^{-1}\mathbf{X}\mathbf{b}}{\mathbf{d}^H\mathbf{Y}^{-1}\mathbf{d}}, \qquad \mathbf{Y} \equiv (\mathbf{X}(\mathbf{I} - \mathbf{b}\mathbf{b}^H)\mathbf{X}^H) \tag{3-101}$$

$\mathbf{Y}$ will generally be nonsingular (even though $\mathbf{I} - \mathbf{b}\mathbf{b}^H$ is itself singular), provided $L > N$. Note that $\mathbf{Y}$ is the (unscaled) summed product formula covariance estimate in (3-64), modified by projecting off the component in the signal direction $\mathbf{b}$.

If $\mathbf{b}$ is constant then g_{ML}, and hence $\mathbf{s}_{\text{ML}} = g_{\text{ML}}\mathbf{b}$, are fully specified. Otherwise, as in the CW model of (3-93), we need to substitute (3-101) into (3-100) and maximize again to excise the unknown parameters in (3-101). To do this note that the { } term in (3-100) is a weighted norm minimization problem in g. Thus the resulting squared norm will simply be the squared length of the component of the vector $\mathbf{y} = \mathbf{X}\mathbf{b}$ in the direction $\perp\mathbf{d}$ which clearly equals

$$\|\mathbf{y} - (\langle \mathbf{y}, \mathbf{d}\rangle / \|\mathbf{d}\|^2)\mathbf{d}\|^2 = \langle \mathbf{y}, \mathbf{y}\rangle - |\langle \mathbf{y}, \mathbf{d}\rangle|^2 / \langle \mathbf{d}, \mathbf{d}\rangle \tag{3-102}$$

We thus have

$$\min_{\mathbf{b}\in\Omega_b}\left[\left(1 - \frac{1}{L}\,\mathbf{y}^H\mathbf{S}^{-1}\mathbf{y}\right)\left(1 + \left\{\mathbf{y}^H\mathbf{Y}^{-1}\mathbf{y} - \frac{|\mathbf{d}^H\mathbf{Y}^{-1}\mathbf{y}|^2}{\mathbf{d}^H\mathbf{Y}^{-1}\mathbf{d}}\right\}\right)\right], \qquad \mathbf{S} \equiv \frac{1}{L}\,\mathbf{X}\mathbf{X}^H \tag{3-103}$$

where the { } term equals the bracketed term in (3.100) after substituting in the ML estimate for g.

By means of Woodbury's identity and tedious algebra, we ascertain that the ML estimate solves

$$\min_{\mathbf{b}\in\Omega_b}\left[\frac{\mathbf{d}^H\mathbf{S}^{-1}\mathbf{d}\left(1 - \frac{1}{L}\,\mathbf{y}^H\mathbf{S}^{-1}\mathbf{y}\right)}{|\mathbf{d}^H\mathbf{S}^{-1}\mathbf{y}|^2}\right] \tag{3-104}$$

which for fixed $\mathbf{d}$ leads to

$$\min_{\mathbf{b}\in\Omega_b}\left[\frac{1-\frac{1}{L}\,\mathbf{y}^H\mathbf{S}^{-1}\mathbf{y}}{|\mathbf{d}^H\mathbf{S}^{-1}\mathbf{y}|^2}\right] \tag{3-105}$$

We now consider two special cases. Further examples, including a subspace model for **b** are found in [17].

1. *Target-signal free snapshots.* This case corresponds to the post-Doppler processing scheme described earlier, with **b** equal to a unit vector. We do not have any free parameters in **b** and so we are done after (3-101). Without loss of generality, for expediency, we set $i = L$. Partition **X** as $\mathbf{X} = [\mathbf{X}_{\text{free}} \,|\, \mathbf{x}_L]$. Note that $g_{\text{ML}} = \hat{s}_L$, the estimate of the target return at sample L, yielding

$$\boxed{\hat{s}_L = \frac{\mathbf{d}^H\mathbf{S}_{\text{free}}^{-1}\mathbf{x}_L}{\mathbf{d}^H\mathbf{S}_{\text{free}}^{-1}\mathbf{d}}\,, \qquad \mathbf{S}_{\text{free}} \equiv \frac{1}{L-1}\sum_{n=1}^{L-1}\mathbf{x}_n\mathbf{x}_n^H} \tag{3-106}$$

Comparing (3-106) with (3-19), for the narrowband case, we see that $\hat{s}_L$ is the output of the minimum variance distortionless response beamformer with the sample covariance replacing its ensemble value. Thus MVDR is optimal in the ML sense, when the appropriate ML covariance estimate replaces **R**, provided signal free samples are available. (Often the desired signal is very weak and is only discernible after analysis of the nulled output. Thus even when the signal is present its effect may be small.) This result, and the subsequent PDF analysis in Section 3-5-4 is due to Kelly [18].

2. *Estimation of a coherent signal of unknown Doppler.* This is the CW radar case in (3-93), and Ω_b becomes the set of vectors

$$\mathbf{b}^H = \frac{1}{\sqrt{L}}\,[1, \exp(j\omega), \ldots, \exp(j\omega(L-1))]\,, \qquad \omega \in [-\pi, \pi] \tag{3-107}$$

The maximum likelihood frequency estimate is now given by

$$\min_{\omega}\left[\frac{(1-\tilde{\mathbf{y}}^H\tilde{\mathbf{y}})}{|\tilde{\mathbf{d}}^H\tilde{\mathbf{y}}|^2}\right],\ \tilde{\mathbf{d}}^H = \mathbf{d}^H\mathbf{C}\,, \tag{3-108}$$

$$\tilde{\mathbf{y}} = \mathbf{C}^H\mathbf{X}\mathbf{b}\,, \qquad \mathbf{C}\mathbf{C}^H = (\mathbf{X}\mathbf{X}^H)^{-1}$$

The estimate g_{ML} is then found by substituting the vector **b** found in (3-108) into (3-101).

What happens if $g\mathbf{b} \in \mathscr{C}^L$, that is **s** is completely unconstrained? Then we can make the [ ] term in (3-105) vanish, by setting **b** to be any vector in the span of $\mathbf{X}$. Since the [ ] term is nonnegative, this is also a minimum. We find that the ML estimate is nonunique, and does not depend on $\mathbf{R}$! Actually the likelihood function becomes unbounded for this choice of $\mathbf{b}$. We see then that unstructured target signals yield meaningless ML procedures. If we know nothing about our targets then, either in terms of their temporal or spectral support, or parametric form, what do we do? One approach is to jettison ML theory after the sample covariance $\mathbf{S}$ has been formed and plug this estimate into a Wiener theory (or MVDR) based weight vector formula. The presence of a target during the formation of the interference covariance then has deleterious effects, unless the target is extremely weak, or has prior known structure [26].

3-5 PERFORMANCE ANALYSIS

3-5-1 Introduction

We now derive the PDF and various moments of the ML signal estimates and related quantities. First we derive the PDF of various functions of the sample covariance matrix $\mathbf{S}$ found in (3-64). Next we use these properties to derive the PDF of the adaptive SNR, the adaptive weight vectors, and the adaptive target estimates. The statistical properties of the sample covariance are essential for the threshold selection process in adaptive detection, so the developments in this section will be employed heavily in the subsequent, final section which covers detection.

3-5-2 Properties of Sample Covariance Matrices

We now explore the properties of the sample covariance matrix $\mathbf{S}$, which for zero mean data is defined (see (3-64)) as

$$\mathbf{S} \equiv \frac{1}{L} \mathbf{X}\mathbf{X}^H \tag{3-109}$$

where $\mathbf{X}$ is the N by L snapshot matrix of (3-49). We restrict ourselves to the case when $L \geqslant N$ in which case $\mathbf{S}$ is positive definite with unit probability (see Appendix A). It will be convenient to work with $\mathbf{X}\mathbf{X}^H = L\mathbf{S}$ rather than $\mathbf{S}$ directly. $L\mathbf{S}$, is called the Wishart matrix in the statistical literature [28].

We can express $\mathbf{X}$ as follows:

$$\mathbf{X} = \sqrt{\mathbf{R}}\mathbf{W} \tag{3-110}$$

where $\mathbf{W}$ is an N by L matrix comprised of zero mean i.i.d. normal complex random variables with unit variance, which we shall henceforth denote as $N(0, 1)$. (Likewise a normal random variable with mean μ and variance σ^2 we denote by $N(\mu, \sigma^2)$.) $\sqrt{\boldsymbol{R}}$ is the positive definite square root of $\mathbf{R}$,

$\sqrt{\mathbf{R}}\sqrt{\mathbf{R}} = \mathbf{R}$. Let $\underset{d}{=}$ denote equality in distribution. A vector of i.i.d. $N(0, 1)$ entries has a covariance matrix which equals the identity. If we premultiply this random vector by a unitary matrix it follows that the covariance is unaltered, which for normal vectors implies the density is invariant as well. It follows that

$$\mathbf{QW} \underset{d}{=} \mathbf{W} \rightarrow \mathbf{WW}^H \underset{d}{=} \mathbf{QWW}^H\mathbf{Q}, \qquad \forall \mathbf{Q}: \mathbf{Q} = \mathbf{Q}^H, \mathbf{Q}^2 = \mathbf{I} \tag{3-111}$$

The additional constraint in (3-111) that $\mathbf{Q}$ be Hermitian as well as unitary is useful for algebraic reasons.

We will find it useful to determine the PDF of the quadratic form $\mathbf{d}^H\mathbf{S}^{-1}\mathbf{d}$ where $\mathbf{d}$ is any column vector, possibly random, but independent of $\mathbf{S}$. Suppose we pick $\mathbf{Q}$ so that

$$\mathbf{Q}\sqrt{\mathbf{R}}^{-1}\mathbf{d} = \exp(j \arg([\sqrt{\mathbf{R}}^{-1}\mathbf{d}]_{1,1}))\sqrt{\mathbf{d}^H\mathbf{R}^{-1}\mathbf{d}}\,\mathbf{e}_1$$
$$\mathbf{e}_1 \equiv [1 \quad 0 \quad \cdots \quad]^T \tag{3-112}$$

(Such a $\mathbf{Q}$ can always be found for instance via Householder transform schemes [34].) Then,

$$\mathbf{d}^H(L\mathbf{S})^{-1}\mathbf{d} = \mathbf{d}^H\sqrt{\mathbf{R}}^{-1}(\mathbf{WW}^H)^{-1}\sqrt{\mathbf{R}}^{-1}\mathbf{d} \underset{d}{=} \mathbf{d}^H\mathbf{R}^{-1}\mathbf{d}\mathbf{e}_1^H(\mathbf{WW}^H)^{-1}\mathbf{e}_1 \tag{3-113}$$

Next partition $\mathbf{W}$ as

$$\mathbf{W} = \begin{bmatrix} \mathbf{z}_1^H \\ \mathbf{Z}^H \end{bmatrix} \tag{3-114}$$

where $\mathbf{z}_1^H$ is 1 by L and $\mathbf{Z}^H$ is $N - 1$ by L, both with i.i.d. entries which are $N(0, 1)$. The Frobenius relations for the inverse of a partitioned matrix [18, 28] inform us that if a Hermitian matrix $\mathbf{T}$ is partitioned as

$$\mathbf{T} = \begin{bmatrix} a & \mathbf{b}^H \\ \mathbf{b} & \mathbf{C} \end{bmatrix} \tag{3-115}$$

with a scalar, $\mathbf{b}$ a column vector, and $\mathbf{C}$ Hermitian, then the inverse can be partitioned as

$$\mathbf{T}^{-1} = \begin{bmatrix} 1 & -\mathbf{b}^H\mathbf{C}^{-1} \\ -\mathbf{C}^{-1}\mathbf{b} & \alpha^{-1}\mathbf{C}^{-1} + \mathbf{C}^{-1}\mathbf{b}\mathbf{b}^H\mathbf{C}^{-1} \end{bmatrix}\alpha, \qquad \alpha \equiv (a - \mathbf{b}^H\mathbf{C}^{-1}\mathbf{b})^{-1} \tag{3-116}$$

provided the accompanying inverses exist, which they will if $\mathbf{T}$ is positive definite. These relations can easily be verified by showing that indeed $\mathbf{TT}^{-1} = \mathbf{I}$. Applying these relations to $\mathbf{WW}^H$, which is positive definite, we find that

$$\alpha = (\mathbf{z}_1^H \mathbf{z}_1 - \mathbf{z}_1^H \mathbf{Z}(\mathbf{Z}^H \mathbf{Z})^{-1} \mathbf{Z}^H \mathbf{z}_1)^{-1} = [(\mathbf{W}\mathbf{W}^H)^{-1}]_{1,1} - \mathbf{C}^{-1}\mathbf{b}\alpha$$
$$= -(\mathbf{Z}^H \mathbf{Z})^{-1} \mathbf{Z}^H \mathbf{z}_1 \alpha \tag{3-117}$$

Now we can rewrite α as

$$\alpha^{-1} = \mathbf{z}_1^H [\mathbf{I} - \mathbf{P}] \mathbf{z}_1 , \qquad \mathbf{P} = \mathbf{Z}(\mathbf{Z}^H \mathbf{Z})^{-1} \mathbf{Z} \tag{3-118}$$

We note that $\mathbf{P}$ is a projection of rank $N - 1$. $\mathbf{I} - \mathbf{P}$ is then also a projection of rank $L - N + 1$.[13] Note that $\mathbf{P}$ is independent of $\mathbf{z}_1$. The PDF of $\mathbf{z}_1$ (which has i.i.d. $N(0, 1)$ entries) is spherically symmetric. It follows that the PDF of α is invariant to the particular subspace $(\mathbf{I} - \mathbf{P})$ we project onto; only the dimensionality matters. (Varying the form of $\mathbf{P}$ merely varies the solid angle we rotate through prior to truncating the component of $\mathbf{z}_1 \perp \mathbf{P}$.) The simplest subspace is that spanned by the first $L - N + 1$ unit vectors. It follows that

$$\alpha^{-1} \underset{d}{=} \sum_{i=1}^{L-N+1} |\mathbf{z}_1(i)|^2 \underset{d}{=} \chi^2_{L-N+1} \tag{3-119}$$

where χ^2_m is a complex normalized chi-squared random variable [18] with m degrees of freedom, whose PDF is

$$f\chi^2_m(y) = y^{m-1} \exp(-y)/(m-1)! \tag{3-120}$$

Note that we can express the quadratic form in (3-113) as $\mathbf{d}^H \mathbf{R}^{-1} \mathbf{d}\alpha$. We arrive then at the following useful theorem.

Theorem 3-4 (Wishart Inversion Theorem) Let $\mathbf{X}$ be an N by L matrix of L i.i.d. complex Gaussian vectors (i.e., array snapshots), with zero mean and with covariance $\mathbf{R}$. Let $\mathbf{d}$ be a column vector independent of $\mathbf{X}$. Then,

$$\boxed{\mathbf{d}^H (\mathbf{X}\mathbf{X}^H)^{-1} \mathbf{d} \underset{\mathrm{d}}{=} \frac{\mathbf{d}^H R^{-1} \mathbf{d}}{\chi^2_{L-N+1}}} \tag{3-121}$$

where χ^2_m is a complex normalized chi-squared random variable.

Theorem 3-4 tells us a great deal about $\mathbf{S}^{-1} = L(\mathbf{X}\mathbf{X}^H)^{-1}$. First we note that by integration of (3-120) by parts $E[(\chi^2_{L-N+1})^{-1}] = (L - N)^{-1}$. Now consider $E[\mathbf{S}^{-1}]$. Note that

[13] To see that $\mathbf{P}$ is indeed a projection one can perform the QR decomposition, again via Householder schemes, whereupon one discovers that $\mathbf{P}$ is an outer product of rectangular orthonormal matrices [34].

$$\mathbf{d}^H E[\mathbf{S}^{-1}]\mathbf{d} = \frac{\mathbf{d}^H \mathbf{R}^{-1} \mathbf{d} L}{(L-N)}, \qquad \forall \mathbf{d} \tag{3-122}$$

It follows that $\mathbf{d}^H\{E[\mathbf{S}^{-1}] - \mathbf{R}^{-1}L/(L-N)\}\mathbf{d} = 0, \forall \mathbf{d}$. From this we conclude (via Rayleigh's theorem [12] or otherwise) that all eigenvalues of the matrix in brackets vanish. Since the matrix in { } is Hermitian it must then be the zero matrix. This implies that the matrices on both sides of (3-122) must in fact be equal, and we find that

$$E[\mathbf{S}^{-1}] = \frac{\mathbf{R}^{-1}L}{(L-N)} \tag{3-123}$$

In words $\mathbf{S}^{-1}$ is, within a scale factor, an unbiased estimate of the inverse of the true covariance matrix. Matrix inversion maps the set of positive definite matrices onto itself. Thus substituting $\mathbf{R}^{-1}$ for $\mathbf{R}$ in the ML formulation merely corresponds to a change of variables, and it follows (from, e.g., the implicit function theorem [28]) that the ML estimate of $\mathbf{R}^{-1}$ is simply $\mathbf{S}^{-1}$.

Estimators of $\mathbf{R}^{-1}$ which minimize various risk functions are considered in [1, 28]. Certain risk functions result in the ML estimate. Other risks yield alternative, sometimes quite counterintuitive, estimators.

We will find more applications of (3-121) shortly. First we furnish two more theorems regarding Wishart matrices.

Theorem 3-5 (Cholesky Factor Density) Let $\mathbf{X}$ be an N by L matrix whose columns $\mathbf{x}_n$ are i.i.d. zero mean complex normal random vectors (snapshots) with common (nonsingular) covariance $\mathbf{R}$. Let $\mathbf{C}$ be a lower triangular matrix, with positive real diagonal entries, such that $\mathbf{CC}^H = \mathbf{R}$. Such a matrix is called the Cholesky factor of $\mathbf{R}$. Let $\mathbf{B}$ be the Cholesky factor of $\mathbf{XX}^H$, that is, $\mathbf{BB}^H = \mathbf{XX}^H$, with $\mathbf{B}$ lower triangular. Then $\mathbf{B}$ has the form $\mathbf{B} = \mathbf{C}\xi$, where ξ is a lower triangular matrix, with independent entries, distributed as follows:

$$\xi \underset{\mathrm{d}}{=} \begin{bmatrix} \chi_L & 0 & \cdots & \cdots & 0 \\ n & \chi_{L-1} & 0 & \cdots & 0 \\ n & n & \chi_{L-2} & \ddots & \vdots \\ \vdots & & \ddots & \ddots & 0 \\ n & \cdots & \cdots & n & \chi_{L-N+1} \end{bmatrix} \tag{3-124}$$

where n is a unit normal ($N(0, 1)$) variate, and χ_m is a complex chi variate with m degrees of freedom.

Theorem 3-5 is quite useful in simulations. It is far more efficient to generate $\mathbf{S}$ via (3-124) than it is to form L random snapshots and compute (3-109) explicitly.

Sketch of proof. First note that we can express $\mathbf{X}$ as $\mathbf{X} = \mathbf{C}\mathbf{W}$ where as before $\mathbf{W}$ is an N by L matrix of $N(0, 1)$ variates. Thus it suffices to show that ξ is the Cholesky factor of the whitened sample matrix $\mathbf{W}\mathbf{W}^H$.

A complete proof of Theorem 3-5 is found in [28]. We provide a heuristic proof below. More precisely, we show that the marginal densities for the entries of both sides of the matrix equation in (3-124) are equal. To complete the proof one must show that the joint densities agree as well. It is trivial to verify that the diagonal entries of $\xi\xi^H$ are all χ_L^2, which indeed matches the density for the diagonals of $\mathbf{W}\mathbf{W}^H$. How about the off diagonals? Let us look at the i, j elements of $\xi\xi^H$, $i > j$. These elements can be expressed as

$$[\xi\xi^H]_{i,j} = \sum_{k=1}^{j-1} n_k \tilde{n}_k^* + \chi_{L+1-j} \tilde{n}_j^* \tag{3-125}$$

where the terms n_k and $\tilde{n}_k$ are independent $N(0, 1)$ variates. (These are just the same n which appear in the ith and jth rows of (3-124), but we place subscripts and tildes on them to indicate that they are distinct, independent variables.) Now let us condition on the $\tilde{n}$ variates. We then have the normal variate $N(0, \Sigma_{k=1}^{j-1} |n_k|^2 + \chi_{L+1-j}^2) \underset{d}{=} N(0, \chi_L^2)$. It follows that $[\xi\xi^H]_{i,j} \underset{d}{=} u\chi_L$, where u is some $N(0, 1)$ variate independent of χ_L. One can similarly show that $[\mathbf{W}\mathbf{W}^H]_{i,j} \underset{d}{=} u\chi_L$, $i \neq j$ as well, which implies the marginal density equivalence$[\xi\xi^H]_{i,j} \underset{d}{=} [\mathbf{W}\mathbf{W}^H]_{i,j}$, $\forall i, j$. QED

It is actually possible to prove Theorem 3-4 from Theorem 3-5 (an exercise for the reader) but many of the steps leading up to the latter are important in their own right, and so we included them.

Now for one final important theorem, which builds on the previous two.

Theorem 3-6 (Inverse Wishart Independence) Let $\mathbf{W}$ be an N by L matrix of $N(0, 1)$ variates, with $\mathbf{z}_1, \mathbf{Z}$ as per (3-114). Then

$$\boxed{\delta \equiv \mathbf{z}_1^H \mathbf{Z}(\mathbf{Z}^H \mathbf{Z})^{-2} \mathbf{Z}^H \mathbf{z}_1 = \frac{\chi_{N-1}^2}{\chi_{L-N+2}^2}} \tag{3-126}$$

where the numerator and denominator chi squared variables are independent. Furthermore δ is independent of α, as defined in (3-117).

Theorem 3-6 vastly simplifies a surprising array of PDF calculations involving various functions of $\mathbf{S}^{-1}$.

Proof. To begin we compute the QR decomposition of $\mathbf{Z}$, that is we express $\mathbf{Z}$ as $\mathbf{Z} = \mathbf{Q_q R_r}$, where $\mathbf{R_r}$ is $N-1$ by $N-1$ and lower triangular, and $\mathbf{Q_q}$ is L by $N-1$, with $\mathbf{Q_q}^H \mathbf{Q_q} = \mathbf{I}$. We can form this decomposition using Householder schemes as discussed in [12, 34]. After some algebra we then find that we can express δ as

$$\delta = \tilde{\mathbf{z}}_1^H \mathbf{R_r}^{-H} \mathbf{R_r}^{-1} \tilde{\mathbf{z}}_1 , \qquad \tilde{\mathbf{z}}_1 \equiv \mathbf{Q_q}^H \mathbf{z}_1 \tag{3-127}$$

Note also that since $\mathbf{Z}$ has i.i.d. $N(0, 1)$ entries, $\mathbf{Z} \underset{d}{=} \mathbf{QZ}$ for any $N-1$ by $N-1$ unitary $\mathbf{Q}$, selected independently of $\mathbf{Q}$. Now pick $\mathbf{Q}$ so that $\mathbf{Q}\tilde{\mathbf{z}}_1 = \exp(j\theta)\sqrt{(\tilde{\mathbf{z}}_1^H \tilde{\mathbf{z}}_1)}[0 \quad \cdots \quad 0, 1]^T$ (again via Householder manipulation). Then by direct matrix multiplication we discover that

$$\delta \underset{d}{=} \tilde{\mathbf{z}}_1^H \mathbf{Q} \mathbf{R_r}^{-H} \mathbf{R_r}^{-1} \mathbf{Q} \tilde{\mathbf{z}}_1 \underset{d}{=} \tilde{\mathbf{z}}_1^H \tilde{\mathbf{z}}_1 (\mathbf{R}_{\mathbf{r}N-1,N-1})^{-2} \tag{3-128}$$

Note that $\tilde{\mathbf{z}}_1$ is a length $N-1$ column vector of i.i.d. $N(0, 1)$ variates, independent of $\mathbf{Q_q}$ (and therefore $\mathbf{Z}$), and hence $\tilde{\mathbf{z}}_1^H \tilde{\mathbf{z}}_1 \underset{d}{=} \chi^2_{N-1}$ independent of $\mathbf{R_r}$. Note also from Theorem 3-5, with N replaced by $N-1$, that $(\mathbf{R}_{\mathbf{r}N-1,N-1})^{-2} \underset{d}{=} 1/\chi^2_{L-N+2}$. Relation (3-126) follows. It remains to establish independence of δ with α. From (3-118), α can be expressed as $\mathbf{z}_1^H(\mathbf{I} - \mathbf{Q_q Q_q}^H)\mathbf{z}_1$. α is obviously independent of $(\mathbf{R}_{\mathbf{r}N-1,N-1})^{-2}$ since it is independent of $\mathbf{Z}$, which formed $\mathbf{R_r}$. We only need to show that it is independent of $\tilde{\mathbf{z}}_1$. We seek then to establish the independence of $\mathbf{z}_1^H \mathbf{Q_q Q_q}^H \mathbf{Q_q Q_q}^H \mathbf{z}_1$ (see (3-127)) and $\mathbf{z}_1^H(\mathbf{I} - \mathbf{Q_q Q_q}^H)(\mathbf{I} - \mathbf{Q_q Q_q}^H)\mathbf{z}_1$. Clearly these are independent if $\mathbf{z}_1^H \mathbf{Q_q Q_q}^H$ and $\mathbf{z}_1^H(\mathbf{I} - \mathbf{Q_q Q_q}^H)$ are independent. By normality (and the zero mean assumption) it suffices to verify that these two normal vectors have a vanishing cross correlation matrix, which we do below:

$$\begin{aligned} E[\mathbf{Q_q Q_q}^H \mathbf{z}_1 \mathbf{z}_1^H (\mathbf{I} - \mathbf{Q_q Q_q}^H)] &= \mathbf{Q_q Q_q}^H E[\mathbf{z}_1 \mathbf{z}_1^H](\mathbf{I} - \mathbf{Q_q Q_q}^H) \\ &= \mathbf{Q_q Q_q}^H (\mathbf{I})(\mathbf{I} - \mathbf{Q_q Q_q}^H) = 0 \end{aligned} \tag{3-129}$$

It will be useful for later discussion to introduce the quantity

$$\rho \equiv (1 + \delta)^{-1} = \frac{1}{1 + \mathbf{z}_1^H \mathbf{Z}(\mathbf{Z}^H \mathbf{Z})^{-2} \mathbf{Z}^H \mathbf{z}_1} \tag{3-130}$$

This quantity obeys the beta distribution,

$$f_\rho(y) = \frac{L!}{(N-2)!}\, y^{L-N+1}(1-y)^{N-2} \tag{3-131}$$

To establish (3-131) one maps $\chi^2_{L-N+2}, \chi^2_{N-1}$ into the pair of variables $D = \chi^2_{L-N+2} + \chi^2_{N-1}$ and ρ. Next one computes the Jacobian of this mapping and integrates over D in the joint PDF [2].

The beta variate ρ is well known in the statistical literature and it was apparently introduced in the context of array signal processing by Capon and Goodman in the late 1960s [8]. Note that ρ varies between zero and one. It is often referred to as the *loss factor* for reasons that will soon be apparent.

3-5-3 The PDF, Mean, and Variance of the Adaptive SNR

Suppose we form the narrowband adaptive MVDR weight vector, which from (3-19), (3-106) has the form

$$\boxed{\mathbf{w}_{\text{adaptive}} = \frac{\mathbf{S}^{-1}\mathbf{d}}{\mathbf{d}^H\mathbf{S}^{-1}\mathbf{d}}} \tag{3-132}$$

when the signal s_n is not present during the formation of the interference covariance estimate $\mathbf{S}$. Recall that the nth snapshot vector formed from sampling the sensor outputs satisfies the relation (3-48) where $\mathbf{n}_n$ is the net interference in the data at time n. Now the output signal to interference ratio, also termed the SNR, is given by

$$\begin{aligned}\text{SNR} &\equiv \frac{E[|\text{output due to desired signal alone}|^2]}{E[|\text{output due to interference alone}|^2]} \\ &= \frac{E[|s_n\mathbf{w}^H\mathbf{d}|^2]}{E[|\mathbf{w}^H\mathbf{n}_n|^2]} = \frac{E[|s_n|^2]|\mathbf{w}^H\mathbf{d}|^2}{\mathbf{w}^H\mathbf{R}\mathbf{w}}\end{aligned} \tag{3-133}$$

The expectations in both numerator and denominator are over the snapshot $\mathbf{x}_L$ to which the weights are being applied, with $\mathbf{w}$ held fixed. Specifically, the numerator expectation is over s_L (the desired signal component of $\mathbf{x}_L$) and the denominator expectation is over $\mathbf{n}_L$ (the interference, i.e., noise component of $\mathbf{x}_L$). Note that $\mathbf{x}_L$ is not among the snapshots used to form the covariance estimate $\mathbf{S}$.

The relation in (3-133) is actually valid regardless of how the weight vector $\mathbf{w}$ is selected.[14] It will be convenient to introduce the notion of the normalized SNR, SNR^{norm}, which is given by

$$\boxed{\text{SNR}^{\text{norm}} = \frac{\text{SNR}}{E[|s_n|^2]} = \frac{|\mathbf{w}^H\mathbf{d}|^2}{\mathbf{w}^H\mathbf{R}\mathbf{w}}} \tag{3-134}$$

[14] Provided it is independent of $\mathbf{x}_L$.

This normalized SNR is the SNR for a unit variance signal. For clairvoyant weights (those formed with knowledge of **R**) the normalized SNR is given by employing the clairvoyant, nonadaptive, MVDR weights

$$\mathbf{w}_{\text{clair}} = \frac{\mathbf{R}^{-1}\mathbf{d}}{\mathbf{d}^H\mathbf{R}^{-1}\mathbf{d}} \tag{3-135}$$

in (3-133), which yields the simple relation

$$\text{SNR}_{\text{clair}}^{\text{norm}} = \mathbf{d}^H\mathbf{R}^{-1}\mathbf{d} \tag{3-136}$$

Note that in the scalar case this quantity equals the squared desired-signal array response over the noise power which is indeed the output SNR for a unit-variance input.

For a **w** formed adaptively from the data as in (3-132) the output SNR becomes

$$\text{SNR}_{\text{adaptive}}^{\text{norm}} = \frac{|\mathbf{d}^H\mathbf{S}^{-1}\mathbf{d}|^2}{\mathbf{d}^H\mathbf{S}^{-1}\mathbf{R}\mathbf{S}^{-1}\mathbf{d}} \tag{3-137}$$

This normalized SNR is a random variable. We will now compute its PDF, first derived by Reed, Mallet, and Brennan in 1974 [32]. As before we note that we can express **S** as $\mathbf{S} = \sqrt{\mathbf{R}}\mathbf{W}\mathbf{W}^H\sqrt{\mathbf{R}}/L$, with **W** a matrix of whitened snapshots as before. We also note again that we can premultiply **W** by any Hermitian unitary matrix that is independent of **W** (see (3-111)) and still preserve its statistics. We then have, after some algebra

$$\text{SNR}_{\text{adaptive}}^{\text{norm}} = \frac{|\mathbf{p}^H\mathbf{Q}(\mathbf{W}\mathbf{W}^H)^{-1}\mathbf{Q}\mathbf{p}|^2}{\mathbf{p}^H\mathbf{Q}(\mathbf{W}\mathbf{W}^H)^{-2}\mathbf{Q}\mathbf{p}}\,[\mathbf{d}^H\mathbf{R}^{-1}\mathbf{d}], \qquad \mathbf{p} \equiv \frac{\sqrt{\mathbf{R}}^{-1}\mathbf{d}}{\sqrt{\mathbf{d}^H\mathbf{R}^{-1}\mathbf{d}}} \tag{3-138}$$

Note that the term in [ ] equals the normalized clairvoyant SNR with prior knowledge of the covariance **R** as per (3-136). The rest of the right hand side expresses the statistical oscillations due to the fact that we are estimating **R**. Also note that **p** has unit norm. We can then Householder-ize, that is, we can pick **Q** so as to reflect coordinates in such a way that $\mathbf{Q}\mathbf{p} = \exp(j\theta)\mathbf{e}_1$. We then discover that

$$\text{SNR}_{\text{adaptive}}^{\text{norm}} = \text{SNR}_{\text{clair}}^{\text{norm}} \frac{[(\mathbf{W}\mathbf{W}^H)^{-1}]_{1,1}}{[(\mathbf{W}\mathbf{W}^H)^{-2}]_{1,1}} \tag{3-139}$$

The numerator is simply the quantity α arising in (3-117). By means of the matrix partitioning in (3-116) we can expand out the numerator in (3-139), finding after a bit of algebra

$$\boxed{\mathrm{SNR}_{\mathrm{adaptive}}^{\mathrm{norm}} = \mathrm{SNR}_{\mathrm{clair}}^{\mathrm{norm}} \frac{1}{1 + \mathbf{z}_1^H \mathbf{Z}(\mathbf{Z}^H \mathbf{Z})^{-2} \mathbf{Z}^H \mathbf{z}_1} = \mathrm{SNR}_{\mathrm{clair}}^{\mathrm{norm}} \rho} \qquad (3\text{-}140)$$

We conclude that the loss in $\mathrm{SNR}^{\mathrm{norm}}$ due to adaptively determining the covariance matrix of the interference is simply the beta distributed random variable ρ! ($\rho = 1$ implies no loss, $\rho = 0$ implies total loss.) Note that this quantity depends only on the dimensionality of the problem, mainly the number of snapshots and array elements. Note that as N increases the expected value of ρ decreases, and as L increases the expected value increases as well. This is as one would expect. Equation (3-140) allows one to determine, in a stationary environment, the number of snapshots required to keep the adaptive loss below any prescribed value with any specified probability. Alternatively, one can evaluate the average loss in SNR when adaptive weights are employed. To this end one must compute the expected value of ρ. Using the expression in (3-131) we can readily use direct integration to compute the expectation, and hence from (3-140) we find

$$\boxed{E[\mathrm{SNR}_{\mathrm{adaptive}}^{\mathrm{norm}}] = \mathrm{SNR}_{\mathrm{clair}}^{\mathrm{norm}} \frac{L - N + 3}{L + 2}} \qquad (3\text{-}141)$$

Note that for large N this expression depends only on the ratio of N to L. For $L = 2N$ the expected SNR loss is about $2 \approx 3\,\mathrm{dB}$. For $L = 5N$ the expected SNR loss is about $5/4 \approx 1\,\mathrm{dB}$. Although (3-141) provides less complete information than (3-140) it is very popular in radar design due to its simplicity. Note that we must have many more snapshots than array elements to obtain tolerable losses, a fact that is confirmed by the form of ρ described in (3-126), (3-130). The variance of ρ can also be readily computed, and we find that

$$\boxed{\mathrm{var}(\mathrm{SNR}_{\mathrm{adaptive}}^{\mathrm{norm}}) = \mathrm{SNR}_{\mathrm{clair}}^{\mathrm{norm}2} \frac{(L - N + 3)(N - 1)}{(L + 2)^2 (L + 3)}} \qquad (3\text{-}142)$$

For large N, L the variance goes as the ratio N/L, an appealing result.

3-5-4 The PDF, Mean, and MSE for Target Estimates

Now suppose we compute the MVDR weight vector in (3-132), with s_n zero during formation of $\mathbf{S}$, and then apply it to an additional snapshot, $n \geq L$, which contained a nonzero s_n. We saw in (3-106) that this resulted in an ML estimate for the unknown s_n. What is the PDF of $\hat{s}_{n\ \text{adaptive}} \equiv \mathbf{w}^H_{\text{adaptive}}\mathbf{x}_n$, the resulting estimate of s_n? Note that $\mathbf{w}_{\text{adaptive}}$ and $\mathbf{x}_n$ are both random variables. Thus the estimate of s_n is degraded by both random fluctuations in estimating $\mathbf{w}_{\text{clair}}$, and the presence of the noise component $\mathbf{n}_n$ in the snapshot containing the true s_n. Minor algebra yields

$$\hat{s}_{n\ \text{adaptive}} = s_n + \frac{\mathbf{d}^H\mathbf{S}^{-1}\mathbf{n}_n}{\mathbf{d}^H\mathbf{S}^{-1}\mathbf{d}} \tag{3-143}$$

Applying the (by now) familiar whitening and unitary transformations we find that

$$\hat{s}_{n\ \text{adaptive}} = s_n + \frac{\mathbf{e}_1^H(\mathbf{W}\mathbf{W}^H)^{-1}\sqrt{\mathbf{R}}^{-1}\mathbf{n}_n}{\sqrt{\mathbf{d}^H\mathbf{R}^{-1}\mathbf{d}}\alpha} \tag{3-144}$$

Now condition on $\mathbf{W}$ and note that we then have a scalar normal variate for the second right hand term in (3-144). By computing its variance we then arrive at

$$\boxed{\hat{s}_{n\ \text{adaptive}} = s_n + \frac{u}{\sqrt{\rho\mathbf{d}^H\mathbf{R}^{-1}\mathbf{d}}} = s_n + \frac{u}{\sqrt{\text{SNR}^{\text{norm}}_{\text{adaptive}}}}} \tag{3-145}$$

where u is $N(0, 1)$ and ρ is the loss factor. Note that this estimate is unbiased, with a variance independent of s_n, increasing with N, decreasing with L, and decreasing with the normalized clairvoyant SNR.

If we know the interference covariance $\mathbf{R}$ a priori then we can compute the clairvoyant weights in (3-135), but we still cannot determine s_n exactly, because of the presence of the interference $\mathbf{n}_n$ in the snapshot containing s_n. Now that we know the weights, the only random component in our estimate of s_n is the snapshot interference $\mathbf{n}_n$.

After some algebra one finds that $\hat{s}_{n\ \text{clair}}$, the resulting clairvoyant estimate of s_n, is given by

$$\hat{s}_{n\ \text{clair}} = s_n + \frac{u}{\sqrt{\mathbf{d}^H\mathbf{R}^{-1}\mathbf{d}}} = s_n + \frac{u}{\sqrt{\text{SNR}^{\text{norm}}_{\text{clair}}}} \tag{3-146}$$

which is identical to (3-145), except the loss factor ρ has vanished! When $\mathbf{R}$ must be estimated from the data, the resulting SNR expression, and the signal estimate PDF, differ from their clairvoyant counterparts only through a single random variable whose form is invariant to the true covariance and the target steering vector. This is an incredible (and fortuitous) result!

The variance of s_n depends on the product of the clairvoyant SNR and the loss factor. As N increases the SNR rises (see Section 3-2), and ρ falls, leading to a tradeoff that is scenario dependent. As more sensors are used the Wiener solution (and hence the clairvoyant SNR) improve. However more sensors require more data to learn from, to keep the adaptive loss factor sufficiently near unity. Equation (3-145) is useful in exploring this tradeoff in an application dependent fashion. To this end, one can use (3-131) to find the MSE of $\hat{s}_{n\ \text{adaptive}}$. The MSE equals the variance since $\hat{s}_n$ is unbiased. One can then compare this adaptive MSE with the clairvoyant MSE when the true weight $\mathbf{w}_{\text{clair}}$ is known. Tedious algebra yields the following result:

$$\boxed{\mathrm{MSE}(\hat{s}_{n\ \text{adaptive}}) = \mathrm{MSE}(\hat{s}_{n\ \text{clair}})\left[1 + \frac{2(N-1)^2}{(L-N+2)(2(L-N)+3)}\right]} \tag{3-147}$$

with

$$MSE(\hat{s}_{n\ \text{clair}}) = \frac{1}{\mathrm{SNR}_{\text{clair}}^{\text{norm}}} \tag{3-148}$$

Note that for large N, L the loss incurred (excess MSE) for adaptive processing goes as $1 + ((L/N) - 1)^{-2}$, which asymptotes to unity for a large ratio of snapshot quantity to array size.

It is interesting to compare the clairvoyant MSE with the MSE obtained by conventional fixed phased array beamforming where $\mathbf{w} = \mathbf{d}/\|\mathbf{d}\|$. In this case the MSE is simply $\mathbf{d}^H\mathbf{R}\mathbf{d}$. The ratio of this conventional MSE to the clairvoyant adaptive MSE is then simply

$$\frac{\mathrm{MSE}(\hat{s}_{n\ \text{fixed}})}{\mathrm{MSE}(\hat{s}_{n\ \text{clair}})} = \frac{(\mathbf{d}^H\mathbf{R}\mathbf{d})(\mathbf{d}^H\mathbf{R}^{-1}\mathbf{d})}{(\mathbf{d}^H\mathbf{d})^2} \tag{3-149}$$

One can show that, for fixed $\mathbf{R}$, this product of quadratic forms is bounded below by unity, and above by the extremal eigenvalues of $\mathbf{R}$:

$$1 \leqslant \frac{(\mathbf{d}^H\mathbf{R}\mathbf{d})(\mathbf{d}^H\mathbf{R}^{-1}\mathbf{d})}{(\mathbf{d}^H\mathbf{d})^2} \leqslant \frac{1}{2} + x + \frac{1}{x},\ x \equiv \frac{\lambda_{\max}(R)}{\lambda_{\min}(R)} \tag{3-150}$$

To establish (3-150) one employs duality methods to argue that only the extremal eigenspaces contribute to the upperbound [36]. The bound in (3-150) was apparently first derived by Kantorovich, in a different context, in 1948.

3-5-5 The PDF of the Adaptive Weights

We now consider the (marginal) density of the adaptive weight vector [35]. This expression is useful for assessing the dynamic range requirements in implementing these weights. $\hat{w}_i$, the ith entry of the adaptive MVDR weight vector can be expressed as

$$\hat{w}_i = \frac{\mathbf{e}_i^H \mathbf{S}^{-1} \mathbf{d}}{\mathbf{d}^H \mathbf{S}^{-1} \mathbf{d}} \tag{3-151}$$

where $\mathbf{e}_i$ is the ith unit vector. We again whiten and reflect to obtain

$$\hat{w}_i \underset{\mathrm{d}}{=} \frac{\mathbf{e}_i^H \sqrt{\mathbf{R}}^{-1} \mathbf{Q} \exp(j\theta)(\mathbf{W}\mathbf{W}^H)^{-1} \mathbf{e}_1}{\sqrt{\mathbf{d}^H \mathbf{R}^{-1} \mathbf{d}}(\mathbf{W}\mathbf{W}^H)^{-1}_{1,1}},$$

$$\mathbf{Q}\sqrt{\mathbf{R}}^{-1}\mathbf{d} = \exp(j\theta)\sqrt{\mathbf{d}^H \mathbf{R}^{-1} \mathbf{d}}\,\mathbf{e}_1 \tag{3-152}$$

with $\mathbf{W}$ as defined previously. After some algebra, and by means of the partitioning in (3-114), (3-117) we find that

$$\hat{w}_i \underset{\mathrm{d}}{=} w_i - [\mathbf{p}^H (\mathbf{Z}^H \mathbf{Z})^{-1} \mathbf{Z}^H \mathbf{Z}_1], \quad \frac{\mathbf{e}_i^H \sqrt{\mathbf{R}}^{-1} \mathbf{Q} \exp(j\theta)}{\sqrt{\mathbf{d}^H \mathbf{R}^{-1} \mathbf{d}}} = \begin{bmatrix} w_i^* \\ p \end{bmatrix}^H \tag{3-153}$$

where w_i is the ensemble, clairvoyant MVDR weight. Note that

$$\mathbf{p}^H \mathbf{p} = \mathbf{R}_{i,i}^{-1} / (\mathbf{d}^H \mathbf{R}^{-1} \mathbf{d}) - |w_i|^2 \tag{3-154}$$

Conditioning on $\mathbf{Z}$ we obtain a normal density for the term in brackets. Computing its variance we then obtain the following relation, which remains valid in the absence of conditioning:

$$\hat{w}_i \underset{\mathrm{d}}{=} w_i + u\sqrt{\mathbf{p}^H (\mathbf{Z}^H \mathbf{Z})^{-1} \mathbf{p}} \tag{3-155}$$

where u is $N(0, 1)$ and we are justified in removing the minus sign by the relation $u \underset{\mathrm{d}}{=} -u$. By Theorem 3-4 and (3-135) we find that

$$\boxed{\hat{w}_i \underset{\mathrm{d}}{=} w_i + \sqrt{\mathbf{R}_{i,i}^{-1} / (\mathbf{d}^H \mathbf{R}^{-1} \mathbf{d}) - |w_i|^2}\; \frac{u}{\chi_{L-N+2}}} \tag{3-156}$$

The ratio of a complex normal to a complex chi-squared random variable, appearing in the right hand side in (3-156), is the complex version of the ordinary Student's t statistic.

The above expressions for the SNR, signal estimates, and weights, have been expressed as the sum and/or product of scalar deterministic quantities and various combinations of $N(0, 1)$, ρ (i.e., beta), and χ^2 variates. One can obtain the PDFs for these quantities by conditioning and integration involving conditional densities [17].

3-6 ADAPTIVE DETECTION

3-6-1 Introduction

A search radar scans a broad surveillance area seeking possible targets. These targets are characterized by their range (distance to the radar), Doppler (frequency shift from transmitted pulse), azimuth and elevation angle, and cross section (amplitude). Many or all of these parameters are unknown a priori. With the exception of the gain, which is proportional to the radar cross section, all these parameters enter nonlinearly into the received waveform equation. Because of the enormous difficulties surrounding the solution of a detection problem containing nonlinearly occurring target parameters, combined with the potential presence of multiple targets, it is customary to pose a single detection problem at fixed Doppler, angle, and range. One then must solve a large number of detection problems, forming a dense mesh over all angle, frequency, and distance parameters. Even with these latter factors held fixed, the detection problem remains composite, that is, there are free variables remaining under both the null hypothesis H_0 (target absence), and the alternate hypothesis H_1 (target presence). The parameter set, Ω_0, under H_0 contains the unknown interference covariance matrix $\mathbf{R}_{\mathbf{n},\mathbf{n}}$. The parameter set, Ω_1, under H_1 contains the unknown interference covariance matrix $\mathbf{R}_{\mathbf{n},\mathbf{n}}$ plus the target waveform gain.

In radar we do not know a priori the odds of target presence, so unlike in, say, digital communications, we lack the requisite information necessary for minimizing the probability of error. The best we can do is simultaneously retain a low probability of false alarm (PFA), and a high probability of detection (PD). In the absence of free parameters (noncomposite, or simple, detection) the best detector is the likelihood ratio test,

$$\frac{f_{H_0}(\{\mathbf{x}_n\})}{f_{H_1}(\{\mathbf{x}_n\})} \underset{\text{“}H_1\text{”}}{\overset{\text{“}H_0\text{”}}{\gtrless}} \tau \tag{3-157}$$

that is, we say H_0, written "H_0", if the ratio exceeds the constant threshold τ and we say H_1 otherwise, where $f_{H_k}(\{\mathbf{x}_n\})$ denotes the PDF of the observed snapshots $\mathbf{x}_n$ under hypothesis H_k, $k = 0, 1$. We can vary the threshold τ to flexibly trade off the PD, $P(\text{“}H_1\text{”} \mid H_1)$, and PFA, $P(\text{“}H_1\text{”} \mid H_0)$.

It is extremely rare that a best detector exists (in the sense of uniformly maximizing PD for fixed PFA) for a composite detection problem. Probably no adaptive radar detection problem of practical interest has a solution which is uniformly optimal for all physically meaningful values of the unknown free parameters. A technique that usually works quite well for composite detection is the so called generalized likelihood ratio test (GLRT), which mimics (3-157) as follows:

$$\frac{\sup_{\Omega_0} f_{H_0}(\{\mathbf{x}_n\})}{\sup_{\Omega_1} f_{H_1}(\{\mathbf{x}_n\})} \underset{``H_1"}{\overset{``H_0"}{\gtrless}} \tau \tag{3-158}$$

This test in many cases satisfies a weaker version of optimality. Loosely speaking it is the best detector, in the sense that it trades off PD verses PFA in an optimal manner subject to natural symmetries in how it handles the data (so called uniformly most powerful maximal invariance). See [2, 28, 37] for details.

After a target has been detected at a given location and frequency it is tracked over time, that is, from sweep to sweep. If the time evolution does not correspond to a physically realizable object (e.g., perhaps it suddenly vanishes, or maneuvers at an impossible rate), we discover it is not a true target after all but a false detection. This separation of false ghost target tracks from valid target tracks requires substantial post processing, and so the number of initial false detections, or false hits, must be kept low to avoid overwhelming the available computational resources. Furthermore if there are too many target tracks one comes up against another problem which no amount of computation can resolve, mainly the nonrecoverable intermingling of multiple tracks. Since many detection decisions are made per unit time it follows that radars must operate with very low PFAs. While a PFA of a few percent is adequate, and common, in other applications, such as experimental design and data analysis, PFAs of 10^{-6} or even 10^{-9} are common in radar.

It is highly desirable to have a constant false alarm rate (CFAR) detector, that is a decision rule whose PFA is independent of the free parameters in Ω_0.[15] A CFAR rule allows us to regulate the effort spent by the post processor in processing false hits. In practice the GLRT often renders CFAR decision rules. Obviously it would be even better if one could construct a CPD detector, that is, one with a fixed detection probability, but this is out of the question since the number of targets, and their gains, is inherently unknown a priori!

3-6-2 Matched Filter Detection

Consider again the snapshot model in (3-48). Suppose we seek to decide if at time n, $s_n = 0$ (H_0) or $s_n \neq 0$ (H_1). Suppose we know $\mathbf{R} = E[\mathbf{n}_n \mathbf{n}_n^H]$, so the only unknown is s_n. Rarely in practice is $\mathbf{R}$ known a priori; it must almost always be measured. However one can in some cases collect a lot of data known to be target free, so one can form a good estimate of $\mathbf{R}$, and then treat it is a given ensemble parameter, not a time average. Although ad hoc, this method is popular and works fairly well. It appears throughout various chapters in the present text. Several generalizations of this ad hoc procedure

[15] The PFA of the GLRT test is always free of the target parameters which enter into Ω_1.

are described in [13]. There are now no unknowns in f_{H_0}. Maximizing over the variable s_n in f_{H_1} we find that the ML estimate of the target is $\hat{s}_n = \mathbf{d}^H\mathbf{R}^{-1}\mathbf{x}_n/\mathbf{d}^H\mathbf{R}^{-1}\mathbf{d}$, and the resulting GLRT ratio from (3-158) can be expressed as

$$\boxed{\frac{|\mathbf{d}^H\mathbf{R}^{-1}\mathbf{x}_n|^2}{\mathbf{d}^H\mathbf{R}^{-1}\mathbf{d}} \underset{\text{``}H_0\text{''}}{\overset{\text{``}H_1\text{''}}{\gtrless}} -\ln \tau} \tag{3-159}$$

To obtain (3-159) we applied the monotonic (and therefore inequality preserving) function ln( ) to both sides of the inequality in (3-158).

This approach is called matched filter detection because one computes the inner product of the data with the signal steering vector, suitably whitened to account for noise color. This is precisely the spatial dual of the matched filter procedure which is so common in time series [39].

Now let us turn our attention to the more realistic problem where **R** is unknown. One simple approach is to replace **R** in (3-159) by the sample covariance matrix **S**. This approach was studied by Kelly [19], and he referred to it as the "adaptive matched filter".[16] What happens if one formulates the GLRT test, treating **R** as an unknown to be maximized over as well as s_n? One might expect that one simply obtains the adaptive matched filter. This is not the case! In fact the resulting GLRT detector, first derived by Kelly in 1985 [18] and discussed next, has been found to outperform this latter approach at low to moderate SNR.

3-6-3 The Kelly Test

In the following discussion we will consider the case where L snapshots are available, the last of which possibly contains a target signal s_L. Our goal is to determine whether or not $s_L = 0$. (If it is not zero we get an estimate for it, mainly (3-106), as part of the GLRT procedure.) Recall from Section 3-4-6 that this corresponds to the signal model in (3-94) with $\mathbf{b} = \mathbf{e}_L$. One can apply a coordinate transformation to map any arbitrary **b** into a unit vector (times a scale factor) so our results are actually applicable (with the obvious transformations) for arbitrary **b** [17]. Our goal then is to determine which of the following hypothesis is true:

$$\begin{aligned} H_0&: \ \mathbf{X} = \mathbf{N}, \\ H_1&: \ \mathbf{X} = s_L \mathbf{d}\mathbf{e}_L^H + \mathbf{N}, \end{aligned} \qquad \mathbf{N} = [\mathbf{n}_1 \ \cdots \ \mathbf{n}_L] \tag{3-160}$$

[16] One might argue that most matched filters are in fact adaptive matched filters since **R** is virtually always measured (estimated) somewhere along the way. The novelty of [19] is that this fact is confronted up front and the corresponding test, and subsequent PFA and PD, are derived.

where $\mathbf{X}$ is as per (3-49), and the $\mathbf{n}_i$ are i.i.d. normal with common covariance $\mathbf{R}$. We take the target direction vector $\mathbf{d}$ as known, with s_L an unknown complex variable. Note that the snapshots $\mathbf{x}_n$ are independent, but not identically distributed, since they have varying means. As we saw earlier (again, see (3-106)), the ML estimate of s_L is found by applying the adaptive MVDR weights, formed from the first $L-1$ columns of $\mathbf{X}$, to $\mathbf{x}_L$. We must then plug that estimate into f_{H_1}, the density function for H_1, along with the sample covariance, the latter having also been plugged into f_{H_0}. The exponential terms in both densities vanish, leaving us with a ratio of determinants (raised to a power which, by evoking monotonicity, we can ignore). The H_1 determinant, which from (3-158) appears in the denominator, is given by (3-99). The argument of the H_0 determinant is merely $\mathbf{X}\mathbf{X}^H$. By means of (3-98) and (3-103), coupled with minor algebra we find that the GLRT ratio becomes

$$\frac{\sup_{\Omega_0} f_{H_0}(\{\mathbf{x}_n\})}{\sup_{\Omega_1} f_{H_n}(\{\mathbf{x}_n\})} = 1 - \frac{|\mathbf{d}^H \mathbf{S}_{\text{free}}^{-1} \mathbf{x}_L|^2}{\mathbf{d}^H \mathbf{S}_{\text{free}}^{-1} \mathbf{d} \left[1 + \dfrac{1}{L-1} \mathbf{x}_L^H \mathbf{S}_{\text{free}}^{-1} \mathbf{x}_L \right]} \tag{3-161}$$

and so the final decision rule simplifies to

$$\boxed{\frac{|\mathbf{d}^H \mathbf{S}_{\text{free}}^{-1} \mathbf{x}_L|^2}{\mathbf{d}^H \mathbf{S}_{\text{free}}^{-1} \mathbf{d} \left[1 + \dfrac{1}{L-1} \mathbf{x}_L^H \mathbf{S}_{\text{free}}^{-1} \mathbf{x}_L \right]} \underset{\text{``}H_0\text{''}}{\overset{\text{``}H_1\text{''}}{\gtrless}} \tau'} \tag{3-162}$$

with $\mathbf{S}_{\text{free}}$ the sample covariance of the target free snapshots as in (3-106). Note that in the absence of the term in brackets this decision rule is the same as (3-159) with a sample covariance matrix replacing the ensemble average $\mathbf{R}$. Thus without the [ ] term we get the so called adaptive matched filter. This extra term, which would be hard to guess based on nonadaptive theory, leads to improved detectibility at low SNR. The term in brackets is the ratio of the determinant of the unscaled sample matrix $\mathbf{X}\mathbf{X}^H$ over the determinant of the unscaled sample matrix $(L-1)\mathbf{S}_{\text{free}}$, formed in the absence of the target bearing snapshot $\mathbf{x}_L$. If $\mathbf{x}_L$ looks very different from its predecessors then this term will be small, and a detection will be declared even if the adaptive matched filter output is small. This explains intuitively the wisdom furnished by the likelihood theory. Of course for large numbers of snapshots ($L \to \infty$), the [ ] term tends to unity and the Kelly test reverts to the adaptive matched filter. Kelly considered a variety of generalizations of this test, based on more elaborate signal models [17].

3-6-4 Detection and False Alarm Probabilities

The decision rules in (3-159), (3-162) are not complete until we specify the fixed constants τ, τ'. These in turn are chosen by specifying the desired PFA, which we shall find is scenario independent (CFAR). The PD in contrast depends on the (unnormalized) SNR. This quantity in turn depends on array element location, transmit power, antenna pattern, etcetera. Therefore the PD plays a crucial role in the overall radar system design and it must be computed, at least numerically.

Denote by ζ the detection statistic formed from the observations, and subsequently compared to a fixed threshold. (ζ is the left hand side in (3-159) and (3-162).) ζ is often referred to as the test statistic of the detector. ζ is a random variable whose PDF depends on which of the two hypotheses is true. The PFA equals unity minus the CDF of ζ_{H_0} evaluated at the detector threshold. Likewise the PD equals unity minus the CDF of ζ_{H_1} evaluated at the detector threshold. Therefore to compute the PFA and PD we must find the distributions of ζ_{H_i}.

Consider first the matched filter. Using familiar manipulations we find

$$\zeta^{\mathrm{MF}} = \frac{|\mathbf{d}^H\mathbf{R}^{-1}\mathbf{x}_L|^2}{\mathbf{d}^H\mathbf{R}^{-1}\mathbf{d}} \underset{\mathrm{d}}{=} \frac{|\mathbf{d}^H\sqrt{\mathbf{R}}^{-1}\mathbf{u}|^2}{\mathbf{d}^H\mathbf{R}^{-1}\mathbf{d}} \tag{3-163}$$

where $\mathbf{u}$ has independent normal entries, with unit variance, with $E[\mathbf{u}] = \sqrt{\mathbf{R}}^{-1}s_L\mathbf{d}$. The inner product of such a normal vector and a deterministic vector is a normal scalar, u, whose covariance is the norm squared of the deterministic vector. Therefore

$$\zeta^{\mathrm{MF}} \underset{\mathrm{d}}{=} \frac{\|\mathbf{d}^H\sqrt{\mathbf{R}}^{-1}\|^2}{\mathbf{d}^H\mathbf{R}^{-1}\mathbf{d}} |u|^2 \tag{3-164}$$

where

$$u \underset{\mathrm{d}}{=} \begin{cases} N(\sqrt{\mathbf{d}^H\mathbf{R}^{-1}\mathbf{d}}|s_L|, 1), & \text{under } H_1 \\ N(0, 1), & \text{under } H_0 \end{cases} \tag{3-165}$$

In establishing (3-165) we used the invariance $u \underset{\mathrm{d}}{=} \exp(j\theta)u$ (see Appendix B). Since $\zeta_{H_0}^{\mathrm{MF}}$ does not involve any specifics of the desired signal (mainly $s_L, \mathbf{d}$) or the interference (mainly $\mathbf{R}$) we conclude that the matched filter is indeed CFAR. Note that under H_1 the square of the mean of u equals the unnormalized $\mathrm{SNR}_{\mathrm{clair}}$.

From (3-120) a single degree of freedom complex chi squared variate is an exponential random variable. Hence upon simple integration we discover that $\mathrm{PFA}_{\mathrm{ML}} = \tau$. The integral arising in the evaluation of the PD is simple but not elementary [18], and will not be further considered here.

Now consider the Kelly test. We again freely apply the ubiquitous whitening and unitary transforms:

$$\zeta^{\mathrm{LR}} \underset{\mathrm{d}}{=} \frac{|\mathbf{e}_1^H(\mathbf{W}\mathbf{W}^H)^{-1}\mathbf{u}|^2}{((\mathbf{W}\mathbf{W}^H)^{-1})_{1,1}[1+\mathbf{u}^H(\mathbf{W}\mathbf{W}^H)^{-1}\mathbf{u}]} \tag{3-166}$$

The vector $\mathbf{u}$ has independent entries, $N(0,1)$, save the first which is $N(\sqrt{\mathrm{SNR}_{\mathrm{clair}}}, 1)$. $\mathbf{S}_{\mathrm{free}}$ is comprised of only the first $L-1$ (target-free) snapshots. $\mathbf{W}$ and all ensuing variables are hence formed from one less snapshot than in previous discussions. This has the effect of replacing the prior L by $L-1$ in the current case. For notational expediency we resist introducing a whole new set of variables to reflect this minor change. Now for the trick which allows us to proceed. Note that (3-116) can be re-expressed as

$$\mathbf{T}^{-1} = \alpha\begin{bmatrix} 1 \\ -\mathbf{C}^{-1}\mathbf{b} \end{bmatrix}[1, -\mathbf{b}^H\mathbf{C}^{-1}] + \begin{bmatrix} 0 & \mathbf{0} \\ \mathbf{0} & \mathbf{C}^{-1} \end{bmatrix} \tag{3-167}$$

Let us partition the N vector $\mathbf{u}$ into the scalar u' (first entry) and the vector $\mathbf{u}'$ (subsequent $N-1$ entries). From (3-167), (3-117) we then obtain

$$\zeta^{\mathrm{LR}} \underset{\mathrm{d}}{=} \frac{\alpha^2|(1, -\mathbf{b}^H\mathbf{C}^{-1})\mathbf{u}|^2}{\alpha\left[1 + \alpha\mathbf{u}^H\begin{bmatrix} 1 \\ -\mathbf{C}^{-1}\mathbf{b} \end{bmatrix}[1, -\mathbf{b}^H\mathbf{C}^{-1}]\mathbf{u} + \mathbf{u}'^H\mathbf{C}^{-1}\mathbf{u}'\right]} \tag{3-168}$$

$$\zeta^{\mathrm{LR}} \underset{\mathrm{d}}{=} \left(1 + \frac{1+\mathbf{u}'^H\mathbf{C}^{-1}\mathbf{u}'}{\alpha|(1, -\mathbf{b}^H\mathbf{C}^{-1})\mathbf{u}|^2}\right)^{-1} \tag{3-169}$$

$$[(\zeta^{\mathrm{LR}})^{-1} - 1]^{-1} \underset{\mathrm{d}}{=} \frac{\alpha|(1, -\mathbf{z}_1^H\mathbf{Z}(\mathbf{Z}^H\mathbf{Z})^{-1})\mathbf{u}|^2}{1+\mathbf{u}'^H(\mathbf{Z}^H\mathbf{Z})^{-1}\mathbf{u}'} \tag{3-170}$$

Conditioning on $\mathbf{u}'$ and $\mathbf{Z}$ in the numerator one again obtains a conditionally normal variate, yielding

$$[(\zeta^{\mathrm{LR}})^{-1} - 1]^{-1} \underset{\mathrm{d}}{=} \frac{\alpha(\rho^{-1}|\tilde{u}|^2)}{\rho^{-1}}, \qquad \rho^{-1} \equiv [1+\mathbf{u}'^H(\mathbf{Z}^H\mathbf{Z})^{-1}\mathbf{u}'] \tag{3-171}$$

where

$$\tilde{u} \underset{\mathrm{d}}{=} \begin{cases} N(0,1), & \text{under } H_0 \\ N(\sqrt{\rho\mathrm{SNR}_{\mathrm{clair}}}, 1) = N(\sqrt{\mathrm{SNR}_{\mathrm{adaptive}}}, 1), & \text{under } H_1 \end{cases} \tag{3-172}$$

independent of $\mathbf{Z}, \mathbf{u}'$. The quadratic forms in $\mathbf{Z}$ vanish from top and bottom, and from Theorem 3-4 we find that the Kelly test is equal in distribution to the test

$$\frac{|\tilde{u}|^2}{\chi^2_{L-N}} \underset{\text{“}H_0\text{”}}{\overset{\text{“}H_1\text{”}}{\gtrless}} \tilde{\tau}, \qquad \text{where } \tilde{\tau} \equiv \frac{\tau'}{1-\tau'} \tag{3-173}$$

The CFAR structure of the Kelly test is now obvious. We can now readily compute the PFA as a function of $\tilde{\tau}$:

$$\mathrm{PFA}_{\mathrm{LR}} = 1 - \int_0^\infty P\left(\frac{|\tilde{u}|^2}{y} \leqslant \tilde{\tau} \mid \chi^2_{L-N} = y\right) f_{\chi^2_{L-N}}(y)\, dy = (1+\tau)^{(N-L)} \tag{3-174}$$

For fixed $\tilde{\tau}$ the PFA increases with N and decreases with L.

The PD calculations involve conditioning as well but they cannot be expressed in closed form [18].

Equations (3-173), (3-172) provide valuable insight into the adaptive detection process. First we note that there is a χ^2 term appearing in the denominator of the adaptive test which does not appear in the matched filter test. This random term yields a CFAR loss; the variation due to this random term forces us to raise the threshold over that of the matched filter to retain the same PFA. By monotonicity of the CDF, this in turn lowers the PD.

The extra loss in detection performance over the nonadaptive matched filter arises from the fact that the test statistic u now has a squared mean given by the adaptive SNR in contrast to the clairvoyant SNR. This lowers the discrimination between ζ_{H_0} and ζ_{H_1}, again lowering the PD.

The simple expressions for the Kelly test in (3-173), (3-172) allow one to explore how the test behaves as a function of various parameters. Numerous examples are provided in [18, 19].

It is extraordinary that the effects of the necessity to adaptively infer R can be absorbed into two scenario independent (but dimensionally dependent) random variables χ^2 and ρ!

3-6-5 Extensions

A distinctive feature of the adaptive detection presented here is that the absence of prior knowledge of the interference covariance $\mathbf{R}$ is confronted directly, and the effect of the requirement to estimate $\mathbf{R}$ is analyzed via the PDF of the parameter estimates, and the detector's PD and PFA.

A number of recent papers have proceeded along these lines. Bose has shown that the adaptive matched filter and the Kelly test form a spanning set for any viable detector. The Kelly test is shown to be nearly locally most powerful [2]. Reed and Yu recently independently derived the GLRT test of Kelly, in a different coordinate system, and have applied it to radar imaging [32], with excellent results. Tufts and Kirsteins have analytically explored the gain in normalized SNR for a rank one subspace model [38], and shown that significant PD gains are possible when this model is valid. Van Veen and Lee [40] have explored the use of subspace models as a means of enhancing detection even in cases where the model is not strictly accurate. Fuhrman and Robey [11] have explored the detection performance enhancement when the unstructured covariance estimate is replaced by various structured estimates. The performance can improve dramatically, but the computational cost is high, and the CFAR property is often sacrificed. Generalizations of the GLRT test have been explored by Kalson [14]. Kalson has shown that the GLRT test and the adaptive matched filter belong to a parametric class of detectors. The class allows for some interesting tradeoffs of sidelobe suppression verses PD, for fixed PFA. Kelly [20] and Monticiollo [27] have explored extensions of the GLRT for cases where the snapshots are non-stationary. A strict GLRT test now becomes intractable, but Monticiollo has shown in simulations that some suboptimal approximations lead to a good decision rule in the presence of fluctuating interference. Dlugos and Scholtz [10] have explored the use of GLRT procedures for signal classification. The analysis now complicates because the binary decision considered here is replace by a K decision, $K > 2$, detector. One must now partition the detector output into K disjoint sets, and a complete characterization of the receiver is much more complicated than the ROC curve (PD verses PFA plot) which describes a binary rule. Onn and Steinhardt have considered the problem of sinusoidal detection in the presence of noise of unknown structure correlated both temporally and spatially [29]. Our work is based on a localized expansion of the Cramer spectrum in a basis of prolate functions. One interesting structured covariance model is that whereby all eigenvalues are known to exceed a threshold. (The threshold would then correspond to the ambient, white, thermal noise level.) An exact ML estimate in this case is tractable (one merely truncates all eigenvalues below a threshold). One can approximately achieve this effect by diagonal loading. The detection behavior of this structured covariance estimate has been explored by Kalson in [15].

Radar is a field that is highly interdisciplinary. Improvements in power supplies, microwave devices, VLSI, etcetera have a synergistic effect on overall performance. However, components alone do not make a system. The design and analysis of contemporary high performance radars requires the understanding and application of sophisticated mathematics, in the analysis and design of the baseband signal processing and elsewhere. We hope that this book in general, and this chapter in particular, will promote the successful deployment of advanced adaptive detection and estimation concepts in emerging radars.

ACKNOWLEDGMENTS

The contents of this chapter began as a set of lecture notes for a graduate course at Cornell University on multisensor signal processing. I have taught this course each of the last four years. I have benefited immensely from numerous curious, thoughtful, probing students who have challenged me to repeatedly return to first principles. This chapter is far more self-contained than it would have been without their prodding. Special thanks are due to my PhD students, Sandip Bose, Oli Jonsson, Ruth Onn, and Sing So. Many of the theorems were discussed in great detail in group seminars with these students, often resulting in simplifications and/or enhanced clarity. The material regarding adaptive detection and maximum likelihood estimation is largely an outgrowth of the seminal work by E. J. Kelly of MIT Lincoln Laboratory. Much of my understanding of this subject stems from conversations with Dr. Kelly. Finally, I gratefully acknowledge the support of the Air Force Office of Scientific Research, with personal thanks to Dr. John Sjogren, grant officer.

APPENDIX A RANDOM MATRICES

Suppose the n by m real matrix $\mathbf{X}$, $m \leq n$ has random entries. We can think of $\mathbf{X}$ as a wound up vector, which, when unraveled, we denote by ξ. This vector, which lies in $\mathcal{R}^{nm}$, has a corresponding joint CDF $p(\xi)$ which we assume is continuous in $\mathcal{R}^{nm}$. (For instance the ubiquitous multivariate normal CDF is continuous.) What is prob($\mathbf{X}$ has rank $<m$)? We have:

$$\text{prob}(\mathbf{X} \text{ has rank } <m) = \int_{(\mathbf{X}|\det(\mathbf{X}^T\mathbf{X})=0)} p(\xi)\, d\xi \tag{A-1}$$

We claim that the set $\det(\mathbf{X}^T\mathbf{X}) = 0$ has zero measure in $\mathcal{R}^{nm}$. This can be demonstrated using abstract arguments based on real analyticity but we will use a more direct approach. Denote the matrix we get when $x_{1,1}$ vanishes as $\mathbf{X}^0$. Then we have: $\mathbf{X}^0 + x_{1,1}\mathbf{E}_1^n\mathbf{E}_1^{mH} = \mathbf{X}$ where $\mathbf{E}_1^k$ denotes the unit vector in k-space. It follows that the resulting determinant

$$\det(\mathbf{X}^{0T}\mathbf{X}^0 + x_{1,1}^2\mathbf{E}_1^m\mathbf{E}_1^{mT} + \mathbf{X}^{0T}\mathbf{E}_1^n\mathbf{E}_1^{mT}x_{1,1} + (\mathbf{X}^{0T}\mathbf{E}_1^n\mathbf{E}_1^{mT}x_{1,1})^T)$$

is a quadratic form in $x_{1,1}$ which vanishes for at most two distinct values of $x_{1,1}$. Therefore the region of integration in (A-1) has measure zero in $\mathcal{R}^{nm}$ implying, by continuity, that the resultant probability is also zero. It follows that

$$\text{prob}(\text{rank}(\mathbf{X}^T\mathbf{X}) = \text{rank}(\mathbf{X}\mathbf{X}^T) = m) = 1 \tag{A-2}$$

or more generally for arbitrary m, n we have established the following theorem.

Theorem 7 (Wishart Invertibility) Let $\mathbf{X}$ be an n by m random matrix whose (possibly correlated) elements have continuous CDF. Then

$$\boxed{\text{prob}(\text{rank}(\mathbf{X}^T\mathbf{X}) = \text{rank}(\mathbf{X}\mathbf{X}^T) = \min(m, n)) = 1} \tag{A-3}$$

The result readily extends to complex matrices. Weaker conditions on the CDF of $\mathbf{X}$ under which (A-3) holds are discussed in [28].

APPENDIX B COMPLEX NORMAL DISTRIBUTION

We make heavy use of the complex multivariate Gaussian distribution. This distribution is central to antenna array theory, but a derivation of it is not readily available in the literature. (The complex multivariate normal density was first derived apparently by Woodin in 1956 [3].) Therefore we provide such a derivation here.

Let us begin by a review of linear mappings. Such mappings will be used repeatedly in subsequent sections.

Theorem 3-8 Under linear transformation a parallelogram maps into another parallelogram.

Proof. A line segment in N dimensional Euclidean space, denoted by $\mathcal{R}^N$, can be expressed as

$$\mathbf{x}_1\alpha + (1-\alpha)\mathbf{x}_2\,, \qquad \alpha \in [0, 1] \tag{B-1}$$

where $\mathbf{x}_1$ and $\mathbf{x}_2$ are the segment endpoints. A linear transformation, represented by the matrix $\mathbf{A}$, yields a new line segment,

$$\mathbf{A}\mathbf{x}_1\alpha + (1-\alpha)\mathbf{A}\mathbf{x}_2\,, \qquad \alpha \in [0, 1] \tag{B-2}$$

Thus the boundary of a parallelogram maps into a set of line segments. By continuity of linear transforms, vertices map into vertices, and interiors map into interiors. It follows that the image of a parallelogram is a rhombus. It remains to show that parallel lines remain parallel after mapping through $\mathbf{A}$. Two line segments are parallel if their equations are respectively of the form

$$\mathbf{z}_1\alpha + \mathbf{z}_2 \quad \text{and} \quad \mathbf{z}_1\alpha + \mathbf{z}_3 \tag{B-3}$$

for some $\mathbf{z}_1$, $\mathbf{z}_2$, $\mathbf{z}_3$. This general form clearly remains invariant under matrix multiplication, establishing the theorem.

Now let us consider linear mappings of more general objects.

Theorem 3-9 Suppose we have a solid object Ω (e.g., a hypersphere and its interior) in Euclidean N space. Now suppose we map this object into another object Ω' by means of the linear transform $y = \mathbf{A}x$ where $\mathbf{A}$ is an N by N matrix. In other words every point in Ω, viewed as a vector, is premultiplied by $\mathbf{A}$, to obtain the resulting set Ω'. Then the volume of the new set satisfies

$$\text{volume}(\Omega') = \text{volume}(\Omega)|\det(\mathbf{A})| \tag{B-4}$$

Proof. Via the singular value decomposition of $\mathbf{A}$, we obtain a sequence of sets:

$$\begin{aligned}
\Omega_1 &\equiv \{\mathbf{t}_1 \in \mathscr{R}^N \mid \mathbf{t}_1 = \mathbf{V}\mathbf{x}, \text{ for some } \mathbf{x} \in \Omega\} \\
\Omega_2 &\equiv \{\mathbf{t}_2 \in \mathscr{R}^N \mid \mathbf{t}_2 = \mathbf{S}\mathbf{t}_1, \text{ for some } \mathbf{t}_1 \in \Omega_1\} \\
\Omega' &\equiv \{\mathbf{t}_3 \in \mathscr{R}^N \mid \mathbf{t}_3 = \mathbf{U}\mathbf{t}_2, \text{ for some } \mathbf{t}_2 \in \Omega_2\}
\end{aligned} \tag{B-5}$$

where

$$\mathbf{A} = \mathbf{USV}\,, \qquad \mathbf{UU}^T = \mathbf{VV}^T = \mathbf{I}\,, \qquad \mathbf{S} = \text{diag}(s_n)\,, \qquad s_n > 0 \tag{B-6}$$

First we note that $\text{volume}(\Omega_1) = \text{volume}(\Omega)$ and $\text{volume}(\Omega_2) = \text{volume}(\Omega')$. This follows from the familiar fact that orthogonal transforms, such as $\mathbf{V}$ and $\mathbf{U}$, can always be expressed as a product of rotations and/or reflections [12], and are therefore volume-invariant mappings. Next suppose we parcel Ω_1 into hypercubes with sides of length ε. Each hypercube in Ω_1 maps into a distinct hypercube in Ω_2 of volume $\varepsilon^N \Pi_n s_n$. (This follows from (B-1), (B-2), and the related discussion.) Summing, or integrating as $\varepsilon \to 0$, over all hypercubes we conclude that

$$\begin{aligned}
\frac{\text{volume}(\Omega_2)}{\text{volume}(\Omega_1)} = \frac{\text{volume}(\Omega')}{\text{volume}(\Omega)} &= \det(\mathbf{S}) = \det(\mathbf{A})\,\text{sign}(\det(\mathbf{U})\det(\mathbf{V})) \\
&= |\det(\mathbf{A})|
\end{aligned} \tag{B-7}$$

Now we are equipped to handle multivariate normal distributions. Consider a real-valued white unit variance zero mean Gaussian vector $\mathbf{x} \in \mathscr{R}^N$. Its PDF is given by

$$f_x(\mathbf{x}) = \prod_n f_{x_n}(x_n) = \left(\frac{1}{\sqrt{2\pi}}\right)^N \exp\left(-\frac{1}{2}\mathbf{x}^T\mathbf{x}\right) \tag{B-8}$$

A proof that this PDF integrates to unity is trivial using polar coordinates. Now suppose we want correlated Gaussian RVs $\mathbf{y}$ with a given $\mathbf{R} \equiv E[\mathbf{y}\mathbf{y}^T]$? Recall that $\exists \sqrt{\mathbf{R}} \mid \sqrt{\mathbf{R}}\sqrt{\mathbf{R}} = \mathbf{R}$. Then let $\mathbf{y} = \sqrt{\mathbf{R}}\mathbf{x} \rightarrow E[\mathbf{y}\mathbf{y}^T] = \sqrt{\mathbf{R}} I \sqrt{\mathbf{R}} = \mathbf{R}$. We have found how to construct the new correlated RVs $\mathbf{y}$ from the initial uncorrelated $\mathbf{x}$. Now we need to find their density $f_y(\mathbf{y})$. Let Ω' be a small solid object containing the point $\mathbf{y}$. Note that

$$\frac{P(\mathbf{y} \in \Omega')}{\text{volume}(\Omega')} \approx f_y(\mathbf{y}) \tag{B-9}$$

with equality in the limit as the volume vanishes. Now we have

$$P(\mathbf{y} \in \Omega') = P(\mathbf{x} \in \Omega) \approx f_x(R^{-1/2}\mathbf{y}) \text{ volume}(\Omega) \tag{B-10}$$

where

$$\Omega \equiv \{\mathbf{x} \in \mathcal{R}^N \mid \mathbf{x} = \mathbf{R}^{-1/2}\mathbf{y}, \text{ for some } \mathbf{y} \in \Omega'\} \tag{B-11}$$

So in the limit

$$f_y(\mathbf{y}) = f_x(\mathbf{R}^{-1/2}\mathbf{y}) \frac{\text{volume}(\Omega)}{\text{volume}(\Omega')}$$

$$f_y(\mathbf{y}) = \frac{f_x(\sqrt{\mathbf{R}}^{-1}\mathbf{y})}{\sqrt{\det(\mathbf{R})}} \tag{B-12}$$

In establishing (B-12) we used Theorem 3-9 combined with the fact that the determinant and power operators commute, that is $\det(\mathbf{A})^p = \det(\mathbf{A}^p)$. With the help of (B-8) we obtain the following theorem.

Theorem 3-10 The joint PDF of a real-valued correlated Gaussian random vector with zero mean is

$$f_y(\mathbf{y}) = \frac{1}{\sqrt{(2\pi)^N \det(\mathbf{R})}} \exp\left(-\frac{1}{2}\mathbf{y}^T\mathbf{R}^{-1}\mathbf{y}\right) \tag{B-13}$$

When $\mathbf{R} = \sigma^2\mathbf{I}$, $\det(\mathbf{R}) = \sigma^{2N}$, and (B-13) reverts to the familiar i.i.d. PDF form with variance σ^2. For a nonzero mean density we simply substitute $\mathbf{y}$ by $\mathbf{y} - E[\mathbf{y}]$. Since $\mathbf{R}$ is positive definite we know that $\det(\mathbf{R})$ is nonnegative, rendering a real-valued and positive density. Note that the PDF is only valid if $\mathbf{R}$ is nonsingular. If $\mathbf{R}$ is singular a subset of the elements of $\mathbf{y}$ are determined exactly from the rest, and one can express the PDF of the truly random part of $\mathbf{y}$ using a lower dimensional quadratic form.

We are interested in the density of complex normal random variables $\mathbf{x}$. Let $\mathbf{x} = \mathbf{x}_r + j\mathbf{x}_n$. We shall make the physically meaningful, widely used, and simplifying assumption that $\mathbf{x}$ is circularly symmetric. In other words we assume that if we replace $\mathbf{x}$ by $\exp(j\theta)\mathbf{x}$ for all θ the statistics of $\mathbf{x}$ do not change.[17] Let $\mathbf{R} = E[\mathbf{x}\mathbf{x}^H] = \mathbf{R}_r + j\mathbf{R}_i$, with $\mathbf{R}_r, \mathbf{R}_i$ real. Note that $(\mathbf{R}_i)$ $\mathbf{R}_r$ is (skew) symmetric. What does circular symmetry imply about $\mathbf{x}_r, \mathbf{x}_i$? By assumption of normality it suffices to study their second order statistics. Let $\mathbf{x}_\updownarrow \equiv [\mathbf{x}_r \quad \mathbf{x}_i]^T$. Then in partitioned form the covariance of $\mathbf{x}_\updownarrow$ is

$$E[\mathbf{x}_\updownarrow \mathbf{x}_\updownarrow^T] = \begin{bmatrix} \mathbf{A} & \mathbf{B} \\ \mathbf{B}^T & \mathbf{C} \end{bmatrix} \tag{B-14}$$

Multiplying $\mathbf{x}$ by $\exp(j\theta)$ corresponds to premultiplying $\mathbf{x}_\updownarrow$ by a two by two rotation matrix, with rotation angle θ. Now compute the resulting covariance matrix, which by the symmetry-imposed invariance must equal the right hand side of (B-14) for all θ. We then find, from the case $\theta = \pi/2$, that $\mathbf{A} = \mathbf{C}$. It follows in turn that $\mathbf{B}^T = -\mathbf{B}$. We conclude that

$$E[\mathbf{x}_\updownarrow \mathbf{x}_\updownarrow^T] = \frac{1}{2}\begin{bmatrix} \mathbf{R}_r & -\mathbf{R}_i \\ \mathbf{R}_i & \mathbf{R}_r \end{bmatrix} \tag{B-15}$$

The determinant of the above matrix can be evaluated by first applying the following operations (which correspond to multiplication by unit diagonal triangular matrices and are hence determinant invariant). First multiply the second (block) row by $j\mathbf{I}$ and add it to the first (block) row. Next multiply the first column by $-j\mathbf{I}$ and add it to the second column. The result is a block lower triangular matrix, with diagonals blocks $\mathbf{R}_r + j\mathbf{R}_i$ and $\mathbf{R}_r - j\mathbf{R}_i$. The determinant of a block triangular matrix is the product of the determinants of the diagonal blocks. We note from the (skew) symmetry of $(\mathbf{R}_i)\mathbf{R}_r$ that the second block equals $\mathbf{R}^*$, the conjugate of $\mathbf{R}$. Addition and multiplication both commute with conjugation. Since the determinant is a multinomial operator it follows that $\det(\mathbf{R})^* = \det(\mathbf{R}^*)$. Since the determinant of $\mathbf{R}$ is real we conclude that the determinant of the matrix in brackets in (B-15) is merely $\det(\mathbf{R})^2$! The 1/2 becomes a 2^{-N} when placed outside the determinant. Setting $\mathbf{y} = \mathbf{x}_\updownarrow$ in (B-13), and replacing N by $2N$ in Theorem 3-10, we see that the scale factor multiplying the exponent in the density function of $f(\mathbf{x}_\updownarrow)$ is then simply $\pi^{-N}\det(\mathbf{R})^{-1}$. It remains to simplify the quadratic form involving the inverse of the covariance matrix in (B-15). The inverse enjoys the same circular symmetry invariance, and therefore the same block structure, as the original covariance. After some algebra we then find

$$\frac{\mathbf{x}_\updownarrow^T\left(\frac{1}{2}\begin{bmatrix} \mathbf{R}_r & -\mathbf{R}_i \\ \mathbf{R}_i & \mathbf{R}_r \end{bmatrix}\right)^{-1}\mathbf{x}_\updownarrow}{2} = \mathbf{x}_r^T\,\Delta\mathbf{x}_r + \mathbf{x}_i^T\,\Delta\mathbf{x}_i^T - 2\mathbf{x}_i^T\Diamond\mathbf{x}_r$$

$$\Delta \equiv (\mathbf{R}_r + \mathbf{R}_i\mathbf{R}_r^{-1}\mathbf{R}_i)^{-1}, \qquad \Diamond \equiv (\mathbf{R}_i + \mathbf{R}_r\mathbf{R}_i^{-1}\mathbf{R}_r)^{-1} \tag{B-16}$$

[17] For single sideband signals (signals with no symmetry about the center of the band), the waveforms on both the in-phase and quadrature channels have the same statistics, so the assumption is reasonable.

Note that $\mathbf{R}^{-1} = \boldsymbol{\Delta} + j\Diamond$. It follows that the quadratic form in (B-16) equals $\mathbf{x}^H\mathbf{R}^{-1}\mathbf{x}$. We have arrived at the following theorem.

Theorem 3-11 Multivariate Complex Normal Density) The density of the zero mean circularly symmetric complex normal N-vector $\mathbf{x}$ with covariance $E[\mathbf{x}\mathbf{x}^H] = \mathbf{R}$ is

$$\boxed{f_x(\mathbf{x}) = \frac{1}{\pi^N \det(\mathbf{R})} \exp(-\mathbf{x}^H\mathbf{R}^{-1}\mathbf{x})} \tag{B-17}$$

Note that the form of the PDF for complex vectors is (within a 2^N scaling) the square of the PDF for real vectors!

APPENDIX C EXTENDIBILITY

We now prove (3-83). By use of Parseval's theorem applied to cross correlations [39], Ω can be expressed as follows:

$$\Omega = \left\{ r_{i,j}: r_{i,j} = \frac{1}{2\pi} \int_\Gamma S(\theta, \omega)\beta_i(\theta, \omega)\beta_j^*(\theta, \omega) \right.$$
$$\left. \exp(j\omega\tau_{i,j}(\theta))\, d\theta\, d\omega, \text{ for some } S(\theta, \omega) \geqslant 0 \right\}. \tag{C-1}$$

The positivity constraint on $S(\theta, \omega)$ corresponds geometrically to a cone that is a convex set. The mapping of $S(\theta, \omega)$ into $r_{i,j}$ is a linear transformation. Since linear transformations preserve convexity it follows that Ω is convex. This is a very useful result and it allows us to employ the following powerful theorem to represent Ω.

Theorem 3-12 (Hahn Banach Theorem) Let Ω be a convex set in $\mathscr{R}^L$. Then $\mathbf{q} \equiv [q_1 \quad \cdots \quad q_L] \in \Omega$ iff every hyperplane which intersects $\mathbf{q}$ also intersects Ω, that is,

$$\Omega = \{\mathbf{q}: \forall \mathbf{p}, \gamma \text{ for which } \mathbf{p}^T\mathbf{q} = \gamma,\ \mathbf{p}^T\mathbf{r} = \gamma, \text{ for some } \mathbf{r} \in \Omega\} \tag{C-2}$$

For a readable discussion of this theorem see [24]. Our treatment of this theorem, and of extendibility in general, is tutorial and is not intended to be a full rigorous treatment. For example we have not specified the function space to which $S(\theta, \omega)$ belongs in (C-1), and our later handling of impulse functions disregards technicalities surrounding their use. Missing details are available in [23].

To characterize Ω it now suffices to characterize those hyperplanes which intersect it. Let $p_{i,j}$, $i, j = 1, \ldots, N$ be such that $p_{i,j} = p_{j,i}^*$. Then consider the real-valued hyperplane given by

$$\sum_{i,j=1}^{N} p_{i,j} r_{i,j} = \sum_{i=1}^{N} r_{i,i} p_{i,i} + 2\,\mathrm{Re}\Big(\sum_{i>j}^{N} p_{i,j} r_{i,j}\Big) = \gamma \tag{C-3}$$

We seek to determine those values of γ, $\mathbf{p}$ for which this intersects the set in (C-1). We require that one element of the array be omnidirectional so that one of the $r_{i,i}$, say r_{11}, equals the net power impinging on the array. Note that

$$\frac{1}{2\pi} \int_{\Gamma} S(\theta, \omega)\, d\theta\, d\omega = r_{11} \tag{C-4}$$

Substituting the expression for $r_{i,j}$ in (C-1) into (C-3) we find that there is an intersection iff there exists a solution to

$$\frac{1}{2\pi} \int_{\Gamma} \left\{ \frac{S(\theta, \omega)}{\frac{1}{2}\pi \int_{\Gamma} S(\theta', \omega')\, d\theta'\, d\omega'} \right\} \left[\sum_{i,j=1}^{N} p_{i,j} \beta_i(\theta, \omega) \beta_j^*(\theta, \omega) \times \exp(j\omega\tau_{i,j}(\theta)) \right] d\theta\, d\omega = \frac{\gamma}{r_{11}} \tag{C-5}$$

Observe that the square bracketed quantity is real and the integral is largest or smallest when the { } term is a (normalized) impulse function located at the maximum or minimum of this quantity. We need only concern ourselves with the minimum, since the two extrema are transposed when $p_{i,j}$ is replaced by $-p_{i,j}$. This leads to the following characterization of Ω:

$$\Omega = \left\{ r_{i,j}\colon \sum_{i,j=1}^{N} r_{i,j} p_{i,j} / r_{11} \geq \min_{\Gamma} \left[\sum_{i,j=1}^{N} p_{i,j} \beta_i(\theta, \omega) \beta_j^*(\theta, \omega) \times \exp(j\omega\tau_{i,j}(\theta)) \right] \right\} \tag{C-6}$$

We can introduce the slack variable κ which allows us to rewrite the inequality in (C-6) as

$$\sum_{i,j=1}^{N} p_{i,j} r_{i,j} - \kappa r_{11} \geq 0\,, \qquad \forall \kappa,\ p_{i,j}$$

$$\sum_{i,j=1}^{N} p_{i,j} \beta_i(\theta, \omega) \beta_j^*(\theta, \omega) \exp(j\omega\tau_{i,j}(\theta)) - \kappa \geq 0\,, \qquad \forall \theta, \omega \in \Gamma \tag{C-7}$$

Because r_{11} is omnidirectional, the corresponding $\beta_i \beta_j^* \exp(j\omega\tau_{i,j})$ term is unity. (We always assume without loss of generality that omnidirectional antennas have unit gain.) Thus κ can be absorbed into p_{11} in both inequalities, yielding the following desired fundamental result [23].

Theorem 3-13 (Extendibility Theorem) The set of covariances which can be formed from an array of sensors, whose ith sensor has pattern $\beta_i(\theta, \omega)$, whose interelement delay is $\tau_{i,j}(\theta)$, and whose angular/spectral support is Γ is given by

$$\Omega = \left\{ r_{i,j}: \sum_{i,j=1}^{N} p_{i,j} r_{i,j} \geq 0, \forall p_{i,j}: \sum_{i,j=1}^{N} p_{i,j} \beta_i(\theta, \omega) \beta_j^*(\theta, \omega) \times \exp(j\omega\tau_{i,j}(\theta)) \geq 0 \ \omega, \theta \in \Gamma \right\} \tag{C-8}$$

APPENDIX D THE MVDR WEIGHT VECTOR

We now find the solution to the MVDR minimization problem in (3-18), repeated below (suppressing the ω dependence):

$$\min_{\mathbf{d}^H \mathbf{w}=1} (\mathbf{w}^H \mathbf{S} \mathbf{w}) \tag{D-1}$$

where $\mathbf{S}$ is the cross spectral matrix of the interference, which equals the interference covariance matrix in the narrowband case. Let us make the change of variables $\tilde{\mathbf{d}} = \sqrt{\mathbf{S}^{-1}}\mathbf{d}$, $\tilde{\mathbf{w}} = \sqrt{\mathbf{S}}\mathbf{w}$. Then we seek to solve

$$\min_{\tilde{\mathbf{d}}^H \tilde{\mathbf{w}}=1} [\tilde{\mathbf{w}}^H \tilde{\mathbf{w}}] \tag{D-2}$$

Now the constraint is a hyperplane which is nearest the origin at $\tilde{\mathbf{w}} \propto \tilde{\mathbf{d}}$, and the quantity to be minimized is simply the distance to the origin (i.e., the norm) squared. Therefore the solution is $\tilde{\mathbf{w}} \propto \sqrt{\mathbf{S}^{-1}}\mathbf{d} \rightarrow$

$$\mathbf{w} = \frac{\mathbf{S}^{-1}\mathbf{d}}{\mathbf{d}^H \mathbf{S}^{-1}\mathbf{d}} \tag{D-3}$$

and the resulting minimum is just the norm squared of $\tilde{\mathbf{w}}$, which is $(\mathbf{d}^H \mathbf{S}^{-1}\mathbf{d})^{-1}$.

APPENDIX E SPECTRAL REPRESENTATION OF STATIONARY PROCESSES

We now establish (3-4). Let $w(t)$ denote a real-valued white noise process, with unit power spectral density. If we pass $w(t)$ through a filter with transfer function $\sqrt{S_{xx}(\omega)}$ then the output will have a power spectrum matching $x(t)$. Let us, merely for convenience, suppose that we can represent $y(\omega)$ as

$y(\omega) = H(\omega)x(\omega)$, for some filter $H(\omega)$. (Note that $S_{yx}(\omega) = H(\omega)S_{xx}(\omega)$.) Then we can express (3-4) as

$$\begin{aligned} E[x(\omega)y^*(\omega')] &= S_{xx}(\omega)H^*(\omega)E[w(\omega)w(\omega')^*] \\ &= S_{xy}(\omega)\frac{1}{4\pi^2}\int_{-\infty}^{\infty}\int_{-\infty}^{\infty} E[w(t)w(t+\tau)]\exp(-j\omega t) \qquad \text{(E-1)} \\ &\quad \times \exp(-j\omega'(t+\tau))^*\,\mathrm{d}t\,\mathrm{d}\tau \\ &= S_{xy}(\omega)\frac{1}{4\pi^2}\int_{-\infty}^{\infty}\int_{-\infty}^{\infty} \delta(\tau)\exp(-j(\omega-\omega')t - j\omega'\tau)\,dt\,d\tau \\ &= S_{xy}(\omega)2\pi\delta(\omega-\omega') \end{aligned}$$

APPENDIX F SUBSPACE PERTURBATIONS

We now explore the effects of thermal noise on the directional-interference (jammer) subspace. This analysis reveals that the subspace perturbation is a fairly simple quantity that is a conditionally complex noncentral beta distributed random variable with noncentrality parameter (which occurs in the bottom chi-squared term) that is a weighted sum of chi squared variates times the jammer to receiver noise power. Therefore we find that the effects of thermal noise disipates as the interference power to receiver noise ratio rises, exactly as simulations suggest.

The error measure we employ is the generalized cosine of the angle between the true subspace and its noisy version.

Let us now introduce some notation. Let $\mathbf{X}$ be an N by M matrix whose columns are equal to the M snapshot vectors $\mathbf{x}_i$, $i = 1, \ldots, M$. We assume the data consists of M jammers (narrowband), whose steering vectors (assumed linear independent) are spanned by the N by M matrix $\mathbf{U}$ with orthonormal columns. (We will call such a matrix rectangular-unitary.) We can express the component of the snapshots $\mathbf{x}_i$ which arrives from these jammers at time index i as $\mathbf{Ub}_i$, $i = 1, \ldots, M$. Note that $\mathbf{b}_i$ is a vector of length M. Since $\mathbf{U}$ only spans the true steering vectors, and its columns do not equal the jammer steering vectors, it cannot be said that $\mathbf{b}_i$ equals the voltage values from the M jammers at time i. Rather $\mathbf{b}_i$ contains the *virtual* jammer voltage values in the space whose basis is $\mathbf{U}$. Regardless of what (unitary) basis we are in the eigenvalues of the jammers are preserved, a property we will employ later. Let $\mathbf{B}$ be an M by M matrix whose columns are equal to the virtual jammer signals $\mathbf{b}_i$. We will assume that the jamming is stationary, zero mean, and normally distributed, during the observation of the M snapshots, with a common covariance of $\mathbf{R}_j$. Now we are equipped to write down a usable expression for $\mathbf{X}$:

$$\mathbf{X} = \mathbf{UB} + \mathbf{W} \qquad \text{(F-1)}$$

where **W** is an N by M matrix (as is **X**) whose NM elements are i.i.d. and complex normal with variance equal to that of the receiver, which we denote by σ^2.

The cosine between the span of the rectangular-unitary matrix **U** and the span of **X** is defined by

$$\cos^2(\theta) = \frac{\|\mathbf{U}^H\mathbf{X}\|^2}{\|\mathbf{X}\|^2} \tag{F-2}$$

where the norm here is the Frobenius norm,

$$\|\mathbf{T}\|^2_{\text{Fro}} = \text{Trace}(\mathbf{T}\mathbf{T}^H) = \text{Trace}(\mathbf{T}^H\mathbf{T}) = \sum_{i,j} |\mathbf{T}_{i,j}|^2 , \quad \text{for any matrix } \mathbf{T} \tag{F-3}$$

Now let us compute the singular value decomposition of **B**, $\mathbf{B} = \mathbf{U_b}\mathbf{D_b}\mathbf{V_b}^H$, where all three matrices are M by M as is **B**, and the first and last are unitary and the middle diagonal, with diagonal entries d_i. Note that $\|\mathbf{D}_b\| = \|\mathbf{B}\|$.

One can easily show the following five properties of projections:

(i) $\|AB\| = \|B\|$ for all rectangular unitary A possessing at least as many rows as columns (so that we are projecting up not down in dimension);
(ii) $\|A + B\|^2 = \|A\|^2 + \|B\|^2$ for A, B orthogonal (A and B are said to be orthogonal if $A^H B = 0$);
(iii) $(I - \mathbf{U}\mathbf{U}^H)B$ and $\mathbf{U}\mathbf{U}^H A$ are orthogonal, for any appropriately dimensioned A, B; and
(iv) $\mathbf{W} = (I - \mathbf{U}\mathbf{U}^H)\mathbf{W} + \mathbf{U}\mathbf{U}^H\mathbf{W}$ which is trivial, but from which follows, along with (ii), (i) that finally, and not so trivially,

(v) $\|\mathbf{W}\|^2 = \|(I - \mathbf{U}\mathbf{U}^H)\mathbf{W}\|^2 + \|\mathbf{U}^H\mathbf{W}\|^2$.

We can use these properties to establish that

$$\cos^2(\theta) = \frac{\|\mathbf{D_b} + \mathbf{U_b}^H\mathbf{U}^H\mathbf{W}\mathbf{V}_b\|^2_{\text{Fro}}}{\|\mathbf{D_b} + \mathbf{U_b}^H\mathbf{U}^H\mathbf{W}\mathbf{V}_b^H\|^2_{\text{Fro}} + [\|(\mathbf{I} - \mathbf{U}\mathbf{U}^H)\mathbf{W}\|^2_{\text{Fro}}]} \tag{F-4}$$

The term in brackets in (F-4) is σ^2 times a $(N - M)M$ degree normalized complex central chi squared distributed variate, which we denote as $\chi^2_{M(N-M)}$.

To see this we note that

$$\|(\mathbf{I} - \mathbf{U}\mathbf{U}^H)\mathbf{W}\|^2_{\text{Fro}} = \|(\mathbf{U} \,|\, \mathbf{U}_\perp)^H(\mathbf{I} - \mathbf{U}\mathbf{U}^H)\mathbf{W}\|^2 = \|\mathbf{U}_\perp^H\mathbf{W}\|^2 \tag{F-5}$$

where $\mathbf{U}_\perp$ is any rectangular unitary matrix, full rank, of size $N - M$ by M, orthogonal to $\mathbf{U}$. The last matrix product in (F-5) is of size M by $N - M$. Its columns are i.i.d., since those of $\mathbf{W}$ are, and premultiplication cannot change that (for example weight vectors cannot yield a correlated scalar process if the initial snapshots are temporally white). The covariance of each column of $\mathbf{W}$ is the identity, so the covariance of each column of $\mathbf{U}_{\perp}^H\mathbf{W}$ is merely $\mathbf{U}_{\perp}^H E[\mathbf{W}\mathbf{W}^H](\mathbf{U}_{\perp}) = I$ where I is the identity matrix of size $N - M$ by $N - M$. Thus we have, upon division by σ^2 the Frobenius norm (the quadrature sum) of a matrix of i.i.d. zero mean unit normal variates, and the result follows.

To proceed we will momentarily condition on the matrix $\mathbf{B}$, and hence on its SVD. We will find that only the diagonal, central matrix in the SVD remains after simplifying the conditional expressions. We now show that $\|\mathbf{D}_b + \mathbf{U}_b^H\mathbf{U}^H\mathbf{W}\mathbf{V}_b^H\|^2$ is equal in distribution to the norm squared of an M by M matrix with entries, $x_{i,j}$, that are conditionally independent complex normal, of mean zero off the diagonal and mean d_i on the diagonal, and with variance σ^2, each of which is independent of the $\chi^2_{NM-M^2}$ variate described above. The normality follows from the fact that deterministic affine mappings preserve normality. (Since $\mathbf{U_b}, \mathbf{V_b}, \mathbf{D_b}$ are conditioned upon, and since they are independent of $\mathbf{W}$, they behave for now as deterministic mappings.) The conditionally jointly independent nature of the $x_{i,j}$ follows from the orthogonality of $\mathbf{U}, \mathbf{U_b}, \mathbf{V_b}$ by extending the argument of the last paragraph. The conditional independence of the $x_{i,j}$ with the $\chi^2_{NM-M^2}$ variate follows from the fact that the former are built from $\mathbf{U}^H\mathbf{W}$, and the latter from $(I - \mathbf{U}\mathbf{U}^H)\mathbf{W}$, and these two quantities are independent based on their normality and orthogonality. We also note that the $\chi^2_{NM-M^2}$ variate is independent of $\mathbf{D}_b$, since $\mathbf{W}$ (from which the former is built) is independent of $\mathbf{B}$ (from which the latter is built). We note that the conditioning by $\mathbf{U}_b, \mathbf{V}_b$ does not affect the statistics of the Forbenius norm of x_{ij}, and therefore they do not affect the PDF of $\cos^2(\theta)$. Hence the influence of $\mathbf{B}$ on $\cos^2(\theta)$ is felt only through the d_i.

After some algebra, we now obtain

$$\cos^2(\theta) = \left(1 + \frac{\sigma^2\chi^2_{M(N-M)}}{\Sigma_{i,j}|x_{i,j}|^2}\right)^{-1} \tag{F-6}$$

Now if we view the entries $x_{i,j}$ as one long vector we can apply a unitary mapping to them, which renders the PDF invariant, and which maps all the noncentrality terms d_i into a single entry. This mapping is dependent on the d_i, but it is deterministic for the moment since we are conditioning on $\mathbf{B}$. We then have the following relation:

$$\cos^2(\theta) = \left(1 + \frac{\chi^2_{M(N-M)}}{\chi^2_{M^2-1} + |n + (\|D_b\|/\sigma)|^2}\right)^{-1} \tag{F-7}$$

where n is complex unit normal, and it along with the two normalized chi squared terms are mutually independent. Note that in all there are exactly NM degrees of freedom represented in this expression, the same as in $\mathbf{W}$. Since $\mathbf{B}$ is held fixed this is the net number of conditional random degrees of freedom available, so all of them are accounted for. We are not yet done because $\mathbf{D}_b$ is itself random. Since it arises from the jamming matrix $\mathbf{B}$ it is independent of the other terms in our expression. We now lift the conditioning on the d_i. The chi squared variate only depends on the norm of the noncentrality parameter, not on how it is distributed across the $x_{i,j}$, hence only the norm of $\mathbf{D}_b$ influences the density of $\cos^2(\theta)$. This is true even though the matrices $\mathbf{U}_b$, $\mathbf{V}_b$ are not necessarily independent of the d_i.

Geometrically the invariance to conditioning that we observe results from the spherical symmetry of the jointly independent equivariant normal PDF itself. If we rotate (and/or reflect) this PDF by any stochastic rotation that is independent of the constituent normal variates we render the PDF invariant provided the mean vector is zero. The invariance of the norm remains true even if the mean vector is nonzero, even if the rotation is a function of the mean vector. (Rotation only influences how the mean vector is partitioned onto the constituent coordinate axes, its norm cannot be influenced.) Spherical symmetry is a powerful structure and it alone allows for the vast simplifications that we experience here.

Notice that we can whiten $\mathbf{B}$ via the relation $\mathbf{B} = \sqrt{\mathbf{R}_j}\tilde{B}$ where $\tilde{B}$ is M by M with unit normal complex entries. It follows that

$$\|\mathbf{D}_b\|^2 = \text{Trace}(\mathbf{B}\mathbf{B}^H) = \text{Trace}(\sqrt{\mathbf{R}_j}\tilde{B}\tilde{B}^H\sqrt{\mathbf{R}_j}) = \text{Trace}(\tilde{B}^H\mathbf{R}_j\tilde{B}) \tag{F-8}$$

Now diagonalize $\mathbf{R}_j$, that is, construct a unitary matrix $\mathbf{U}_j$ of size M by M and an M by M diagonal matrix of eigenvalues, $\mathbf{\Lambda}$, so that

$$\mathbf{R}_j = \mathbf{U}_j\mathbf{\Lambda}\mathbf{U}_j^H \tag{F-9}$$

Then, again using the equality of distribution of $\tilde{B}$ to deterministic unitary mappings, we can rewrite $\|\mathbf{D}_b\|^2$ as

$$\|\mathbf{D}_b\|^2 = \sum_{i=1}^{M} \mathbf{w}_i^H\mathbf{\Lambda}\mathbf{w}_i\,, \qquad \mathbf{w}_i,\ i=1,\ldots,M, \text{ are i.i.d. normal with identity covariance} \tag{F-10}$$

or simplifying,

$$\|\mathbf{D}_b\|^2 = \sum_{i=1}^{M} \lambda_i(\chi_M^2)_i \tag{F-11}$$

where the $(\chi_M^2)_i$ terms are M i.i.d. chi squared variates, each of degree M. (To avoid confusion we use a subscript to differentiate them since they all have the same degrees of freedom.) The weighting makes the PDF of this sum very hard to express in any useful form. Note that $\|\mathbf{D}_b\|^2$ has exactly M^2 degrees of freedom, the same as **B**, so all terms have been accounted for!

Note that the eigenvalues of $\mathbf{R}_j$ equal the nonzero eigenvalues of $\mathbf{U}\mathbf{R}_j\mathbf{U}^H$, the covariance of the jammers in the element domain. If these eigenvalues are equal, that is the jammer covariance has the form $P_j^2\mathbf{U}\mathbf{U}^H$ (F-11) collapses into a single, ordinary central $\chi^2_{M^2}$ term. We now have our desired formula:

$$\cos^2(\theta) = \left(1 + \frac{\chi^2_{NM-M^2}}{\chi^2_{M^2-1} + |n + \sqrt{\Sigma_{i=1}^M \lambda_i(\chi^2_M)_i}/\sigma|^2}\right)^{-1} \tag{F-12}$$

where the $M + 2$ chi squared terms and the normal variate n are independent of each other.

REFERENCES

[1] T. W. Anderson (1984) *Introduction to Multivariate Statistical Analysis*, 2nd edition, Wiley, New York.

[2] S. Bose and A. Steinhardt (1992) Adaptive detection with arrays: Insights using a maximal invariant framework, IEEE Trans. on Signal Processing, to appear.

[3] D. Brillinger (1981) *Time Series: Data Analysis and Theory*, Holden Day, San Francisco, CA.

[4] Y. Bressler and A. Macovski (1986) "On the number of signals resolvable by a uniform linear array", *IEEE Trans. Acoust., Speech, Signal Process.*, Vol. 34, No. 6, 1612–1628.

[5] E. Brookner and J. Howells (1986) "Adaptive-adaptive array processing", *Proc. IEEE*, Vol. 74, No. 4, 602–604.

[6] J. Burg, D. Luenburger, and D. Wegner (1982) "Estimation of structured covariance matrices", *Proc. IEEE*, Vol. 70, No. 9, 963–974.

[7] B. Carlson (1988) "Covariance matrix estimation errors and diagonal loading in adaptive arrays", *IEEE Trans. Aerosp. Electron. Systems*, Vol. 24, No. 4, 397–401.

[8] J. Capon and N.R. Goodman (1970) "Probability distributions for estimators of the frequency wavenumber spectrum", *Proc. IEEE*, Vol. 58, No. 10, 1785–1786.

[9] A. Dembo, C.L. Mallows, and L. Shepp (1989) "Embedding Nonnegative definite matrices in nonnegative definite circulants", *IEEE Trans. Inform. Theory*, Vol. 35, No. 6, pp. 1206–1212.

[10] D. Dlugos and R. Scholtz (1989) "Acquisition of a spread spectrum signal by an adaptive array", *IEEE Trans. Acoust., Speech, Signal Process.*, Vol. 37, No. 8, pp. 1253–1270.

[11] D. Fuhrmann (1991) "Application of Toeplitz Covariance Estimation to Adaptive Beamforming and Detection", *IEEE Trans. Acoust., Speech, Signal Process*, Vol. 39, No. 10, p. 2194–2198.

[12] G. Golub and C. Van Loan (1983) *Matrix Computations*, John Hopkins Press, Baltimore, MD.

[13] S. Haykin (Ed.) (1985) *Array Signal Processing*, Prentice-Hall, New York.

[14] S. Kalson (1991) "An adaptive array detector with mismatched signal rejection", *IEEE Aerosp. Electron. Systems*, Vol. 28, No. 1, pp. 25–32, Jan. 1992.

[15] S. Kalson (1990) "An adaptive detection algorithm with diagonal loading", In *Proc. IEEE ICASSP Conf.*

[16] T. Kailath (1980) *Linear Systems*, Prentice Hall, Englewood Cliffs, NJ.

[17] E. Kelly and K. Forsyth (1989) "Adaptive detection and parameter estimation for multidimensional signal models", MIT Lincoln Laboratory report, TR848.

[18] E. J. Kelly (1986) "An adaptive detection algorithm", *IEEE Trans. Aerosp. Electron. Systems*, **22**, 115–127.

[19] E. J. Kelly (1985) "Adaptive detection in nonstationary interference: Parts I, II", Technical report 724, MIT Lincoln Laboratory.

[20] E. J. Kelly (1987) "Adaptive detection in nonstationary interference: Part III", Technical report 761, MIT Lincoln Laboratory.

[21] E. J. Kelly and M. Levin (1964) "Signal parameter estimation for seismic arrays", MIT Lincoln Lab, TR-339.

[22] S. Kullbach (1959) *Information Theory and Statistics*, Wiley, New York.

[23] S. Lang and J. McClellan (1983) "Multidimensional MEM Spectral Estimation", *IEEE Trans. Acoust., Speech, Signal Process.*, **31**, 1349–1359.

[24] D. Luenburger (1969) *Optimization by Vector Space Methods*, Wiley, New York.

[25] H. Leung and S. Haykin (1990) "Is there a radar clutter attractor?", *Appl. Phys. Lett.*, Vol. 53, pp. 173–175.

[26] R. Monzingo and T. Miller (1980) *Introduction to Adaptive Arrays*, Wiley, New York.

[27] P. Monticiollo, E. Kelly, and J. Proakis, "An analysis of a noncoherent adaptive detection scheme", *IEEE Aerosp. Electron. Systems*, Vol. 28, No. 1, pp. 12–24, Jan. 1992.

[28] M. Muirhead (1985) *Introduction to Multivariate Analysis*, Prentice Hall, Englewood Cliffs, NJ.

[29] R. Onn and A. Steinhardt (1992) "The application of multiwindow detection to arrays", *IEEE Int. Conf. on Acoust., Speech, and Signal Proc.*, March 1992, San Francisco, Conf. proceedings, to appear.

[30] S. Pillai (1989) *Array Signal Processing*, Springer-Verlag, Berlin.

[31] A. Papoulis (1991) *Probability, Random Variables, and Stochastic Processes*, 3rd edition, McGraw-Hill, New York.

[32] I. Reed, S. Mallet, and L. Brennan (1974) "Rapid convergence rate in adaptive arrays", *IEEE Trans. Aerosp. Electron. Systems*, 853–863.

[33] I. Reed and X. Yu (1990) "Adaptive multiband CFAR detection of an optical pattern with unknown spectral distribution", *IEEE Trans. Signal Process.*, pp. 1760–1770.

[34] A. Steinhardt (1988) "Householder transforms in signal processing", *IEEE Acoust., Speech, Signal Process Mag.*, Vol. 5, No. 3, 4–1.

[35] A. Steinhardt (1991) "The PDF of adaptive beamforming weights", *IEEE Trans. Acoust., Speech, Signal Process.*, Vol. 39, No. 5, pp. 1232–1235.

[36] A. Steinhardt (1991) "Adaptive detection for sensor arrays", unpublished notes, Cornell University.

[37] L. L. Scharf (1991) *Statistical Signal Processing*, Addison-Wesley, New York.

[38] D. Tufts and I. Kirsteins (1989) "On the pdf of the SNR in an improved adaptive detector", in *IEEE ICASSP*, 1985, pp. 572–575; see also (1989) "Rapidly adaptive nulling of interference", in *Proc. GRETSI Conf.*, Lecture Notes in Control and Information Sciences, Springer-Verlag, Berlin.

[39] H. Van Trees (1968) *Detection, Estimation, and Modulation Theory*, Vol. I, Wiley, New York.

[40] B. D. VanVeen and C. Lee (1990) "Adaptive detection in subspaces", ASSP Spectrum Estimation Workshop. Rochester, NY.

[41] G. Xu and T. Kailath (1993) "Fast signal-subspace decomposition Part I: ideal covariances matrices", *IEEE Trans. Signal Process.*, to appear.

[42] G. Xu and T. Kailath (1993) "Fast signal-subspace decomposition Part II: sample covariances matrices", *IEEE Trans. Signal Process.*, to appear.

4

MINIMUM VARIANCE BEAMFORMING

BARRY VAN VEEN
Department of Electrical and Computer Engineering,
University of Wisconsin-Madison,
Madison, USA

4-1 INTRODUCTION

A *beamformer* is a signal processor used in conjunction with a set of antennas that are spatially separated. The beamformer output is simply a weighted combination of the outputs of the set of antennas. The set of antennas used with a beamformer is commonly referred to as an *array*. Usually the goal of beamforming is spatial filtering, that is, separation of signals which have similar temporal frequency content but originate from different spatial locations. The spatial samples of the propagating field obtained by the array are necessary for spatial filtering in the same manner that temporal samples of a waveform are necessary for temporal filtering. In radar, the spatial filtering capability of a beamformer facilitates rejection of interference sources in hostile environments and aids in the suppression of clutter. Beamforming also extends the effective width of the electromagnetic spectrum, since it permits separation of signals based on spatial diversity as well as frequency content. In an *adaptive beamformer* the weights are adjusted or adapted in response to the data received at the antennas to optimize the beamformer's spatial response.

A single dish antenna has a nonomnidirectional gain pattern and is thus a spatial filter. However, its gain pattern or spatial response is inherently

Adaptive Radar Detection and Estimation, Edited by Simon Haykin and Allan Steinhardt.
ISBN 0-471-54468-X

nonadaptive, since changing it requires changing the physical attributes of the dish. Clearly, this is a formidable task. In contrast, the spatial response of a beamformer is determined by the weights used to combine the individual antenna outputs. Changing the beamformer's spatial response is as simple as changing the weights. Although beamformers having fixed weights, termed nonadaptive beamformers, are common, the discussion in this chapter is limited to adaptive beamformers. The weights in an adaptive beamformer are controlled by an *adaptive algorithm*. The adaptive algorithm changes the weights in response to changes in the received data to optimize criteria associated with the beamformer output such as mean squared error or signal to noise ratio (SNR).

The class of adaptive beamformers discussed in this chapter are termed *minimum variance* beamformers. Minimum variance beamformers choose their weights to minimize the beamformer's output power or variance. This criterion is usually used in conjunction with constraints that ensure preservation of the desired signal. The relationship between the minimum variance criterion and other criteria such as minimum mean squared error (Wiener filtering in Chapter 3) and maximum SNR are established in certain cases. A signal processing perspective is taken throughout the chapter. Hardware and deployed system considerations are not addressed except as motivation in a general sense for certain signal processing problems which arise in beamforming. Thus, most of the discussion which follows is not specifically limited to radar, but is also applicable to other beamforming problems, including those involving acoustic wave propagation, such as sonar or speech.

Historically, the sidelobe canceller [1, 26] was perhaps the earliest adaptive beamformer. An omnidirectional, low gain auxiliary antenna is used to estimate interference that leaks through the sidelobes of a high gain primary antenna. The estimate of the sidelobe interference is subtracted from the primary antenna output to cancel it; hence the name sidelobe canceller. This concept was extended to include multiple auxiliary antennas (see Fig. 4-6) and then to arbitrary weighted combinations of multiple antennas, without distinction between primary and auxiliary antennas. At present, arrays composed of thousands of individual antennas are not unusual [8, 51]. Arrays with large numbers of antennas generally possess a high degree of spatial selectivity and provide a large signal gain relative to spatially white noise.

Several textbooks [9, 27, 34], tutorial papers [17, 44], special journal issues (IEEE Transactions on Antennas and Propagation 1976, 1986, Journal of Ocean Engineering 1987), and a bibliography [33] have been devoted to adaptive beamforming. This chapter will be somewhat tutorial in that it begins from first principles but will emphasize current issues and recent research. Some of the material is new. The discussion is limited to adaptive beamforming; the spatial spectrum estimation problem is not covered nor is the nonadaptive beamformer design problem, except to the extent of its role in adaptive beamforming.

The following section introduces notation and mathematical definitions. A key concept in Section 4-2 is the beamformer's spatial response. Section 4-3 discusses the sidelobe canceller, examining basic assumptions and limitations. Many of the limitations of the sidelobe canceller are overcome through the use of constraints, the topic of Section 4-4. The emphasis in Section 4-4 is on linear weight vector constraints. A general approach to converting the linearly constrained beamformer to unconstrained form, termed the generalized sidelobe canceller (GSC), is given. Section 4-5 briefly discusses the use of quadratic constraints in minimum variance beamforming. In Section 4-6 we introduce partially adaptive beamformers. A partially adaptive beamformer uses a subset of the available adaptive degrees of freedom to improve convergence characteristics while reducing the real-time computational burden. Section 4-7 presents an analysis of interference cancellation capabilities for partially and fully adaptive linearly constrained minimum variance beamformers. Adaptive algorithms are surveyed in Section 4-8 with an emphasis on algorithms for the GSC. A statistical analysis of the convergence of the beamformer output power and mean squared error to their steady state values is given in Section 4-9 for a particular class of adaptive algorithms. This analysis illustrates the improvement in adaptive convergence that is obtained by partially adaptive beamforming. A lattice-like modular structure for implementing the GSC is described in Section 4-10. This structure is ideally suited for implementing multiple beamformers using different numbers of adaptive degrees of freedom or a single beamformer in which the number of adaptive degrees of freedom is varied. Partially adaptive beamformer design, that is, selection of which degrees of freedom to utilize adaptively, is the topic of Section 4-11. Section 4-12 describes the signal cancellation phenomena that occurs in minimum variance beamformers when the desired signal is correlated with the interference. The chapter concludes with a summary.

4-2 BASIC DEFINITIONS AND CONCEPTS

Matrix and vector notation is used throughout this chapter to represent various quantities associated with beamformers. Thus, the reader is assumed to have a basic background in linear algebra. Vectors are denoted by boldface lower case symbols, for example, $\mathbf{w}$, and are assumed to be column vectors. Matrices are denoted by boldface upper case symbols, for example $\mathbf{C}$. In the figures, vector valued data paths are indicated by heavy lines with the dimension indicated by the symbol below the slash mark, /. Superscripts $*$, H and -1 represent complex conjugate, complex conjugate transpose and matrix inverse, respectively. Note that in radar applications beamformers are often complex valued systems due to the presence of both in-phase and quadrature data.

Historically, radar systems have been considered narrowband, although broadband systems are becoming more common. One definition of narrowband is in the context of fractional bandwidth, defined as the signal bandwidth as a percentage of the carrier frequency. For example, a 1 MHz signal bandwidth at a carrier frequency of 500 MHz has a fractional bandwidth of 0.2%. It is reasonable to define signals with fractional bandwidths much less than 1% as narrowband and those with fractional bandwidths much greater than 1% as broadband. An alternate definition considers signals to be narrowband if their time bandwidth product (TBWP) is much less than one, where time refers to the time it takes the signal to traverse the array. This definition is more appropriate considering the issues discussed in this chapter, although it is also more complicated and less intuitive. Note that the time it takes a signal to traverse the array is a function of its direction relative to the array, so a signal's apparent bandwidth is also a function of direction. A signal that is received by all antennas in the array simultaneously has zero TBWP, regardless of its temporal frequency content, and is narrowband according to this definition while a signal with a very narrow actual bandwidth could be considered broadband if it arrives from a direction for which the travel time across the array is very large. The rationale for this differentiation between narrowband and broadband lies in beamformer properties and limitations.

Figure 4-1 depicts a beamformer typically used for processing narrowband signals. The antenna outputs are assumed to have been sampled and are represented in discrete time for notational convenience and consistency; in practice beamformers are implemented in both continuous and discrete time. The presence of receiver electronics, demodulation circuitry, and (possible) sampling is implicit and will not be discussed here. The antenna outputs are weighted and then summed to compute the beamformer output. The output, $y(n)$, is given by

$$y(n) = \sum_{p=1}^{N_s} w_p^* x_p(n) \tag{4-1}$$

where w_p and $x_p(n)$ are the weight and data, respectively, in the pth antenna

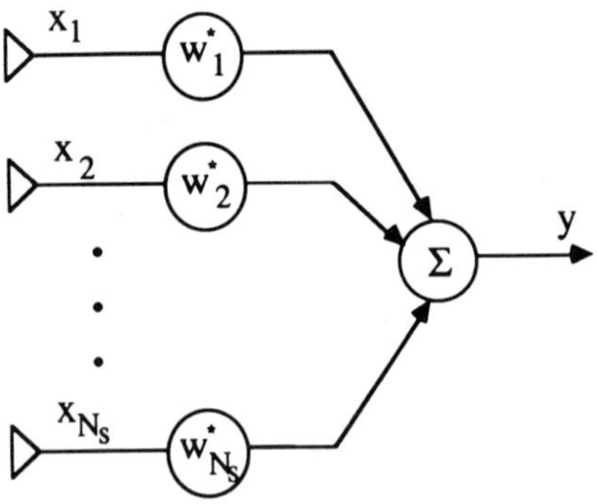

Figure 4-1 A beamformer usually used for narrowband signals.

channel and N_s is the number of antennas. It is conventional to multiply the data by the complex conjugates of the weights to simplify notation.

A beamforming structure commonly utilized to process broadband signals is depicted in Figure 4-2. In this case there are tap delay lines in each channel. The output, $y(n)$, is expressed as

$$y(n) = \sum_{p=1}^{N_s} \sum_{k=0}^{N_t-1} w_{p,k}^* x_p(n-k) \tag{4-2}$$

where N_t is the number of taps in each of the N_s channels and $w_{p,k}$ is the weight applied to the kth tap of the pth channel. Note that in this case, time, as defined for the TBWP, represents the total time it takes a signal to travel across the array and the delay lines. Thus, with this structure a signal that arrives at all antennas simultaneously may be classified as narrowband or broadband, depending only on its bandwidth. Both (4-2) and (4-1) are compactly written as the inner product

$$y(n) = \mathbf{w}^H \mathbf{x}(n) \tag{4-3}$$

where $\mathbf{w}$ is the vector of beamformer weights and $\mathbf{x}(n)$ is a similarly defined vector of data. Let the dimension of $\mathbf{w}$ and $\mathbf{x}(n)$ be N where $N = N_s$ in the narrowband case (4-1) and $N = N_t N_s$ in the broadband case (4-2).

Beamformer response is defined analogously to the frequency response of

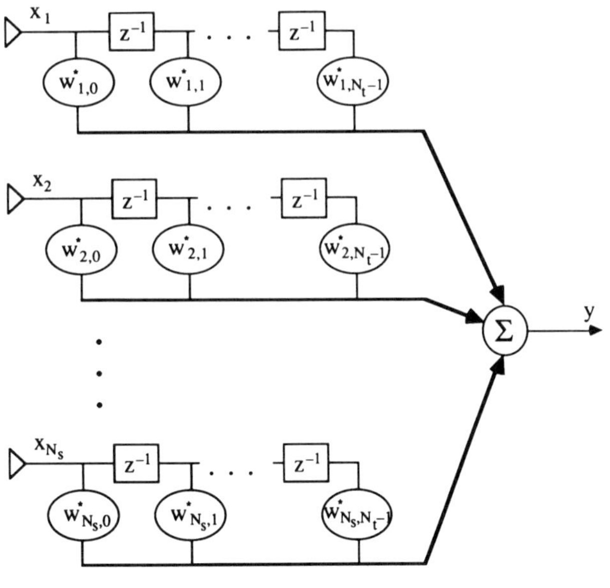

Figure 4-2 A beamformer containing FIR filters in each sensor channel. This structure is generally used when broadband signals are of interest.

a linear system. Recall that if the input to a linear system is a complex sinusoid, the output is a sinusoid of the same frequency. The system's frequency response is the amplitude and phase change experienced by the sinusoid as a function of frequency. The response of the beamformer is defined as the amplitude and phase change presented to a complex plane wave as a function of frequency and location. Location is in general a three-dimensional quantity; here we will only consider direction of arrival and ignore range.

Assume that the array spatially samples a propagating wave of frequency ω and direction θ, defined with respect to a suitable reference direction, as depicted in Fig. 4-3. For convenience let the phase due to propagation be referenced to zero at the first sensor and let $a_p(\theta, \omega)$ be the response of the pth antenna as a function of direction and frequency. This implies that $x_1(n) = a_1(\theta, \omega)e^{j\omega n}$ and $x_p(n) = a_p(\theta, \omega)e^{j\omega[n-\Delta_p(\theta)]}$, $2 \leq p \leq N_s$, where $\Delta_p(\theta)$ represents the time delay due to propagation from the first to the pth antenna. Substituting into (4-2) gives

$$\begin{aligned} y(n) &= e^{j\omega n} \sum_{p=1}^{N_s} \sum_{k=0}^{N_t-1} w_{p,k}^* a_p(\theta, \omega) e^{-j\omega[\Delta_p(\theta)+k]} \\ &= e^{j\omega n} r(\theta, \omega) \end{aligned} \tag{4-4}$$

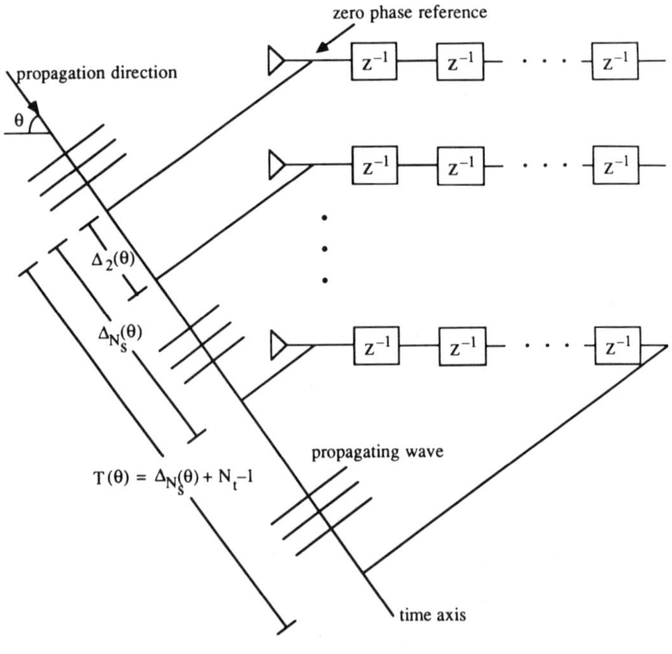

Figure 4-3 Manner in which the array samples a spatially propagating wave.

where $\Delta_1(\theta) = 0$. The term $r(\theta, \omega)$ is defined to be the beamformer response and can be expressed in vector form as

$$r(\theta, \omega) = \mathbf{w}^H \mathbf{d}(\theta, \omega) \tag{4-5a}$$

The elements of $\mathbf{d}(\theta, \omega)$ correspond to the complex exponentials $a_p(\theta, \omega)e^{-j\omega[\Delta_p(\theta)+k]}$. Assuming distortionless omnidirectional antennas, $a_p(\theta, \omega) = 1$, $\mathbf{d}(\theta, \omega)$ is written

$$\mathbf{d}(\theta, \omega) = [1 \quad e^{-j\omega\tau_2(\theta)} \quad e^{-j\omega\tau_3(\theta)} \quad \cdots \quad e^{-j\omega\tau_N(\theta)}]^H \tag{4-5b}$$

where the $\tau_i(\theta)$, $1 \leq i \leq N$, are the time delays due to propagation and any tap delays from the zero phase reference to the point at which the ith weight is applied. In (4-5b) $\tau_1(\theta)$ is zero because the first element of $\mathbf{d}(\theta, \omega)$ is the zero phase reference. The vector $\mathbf{d}(\theta, \omega)$ is termed the *array response vector*. It is also known as the source location vector or direction vector. The time it takes a signal to travel across the array and delay lines is given by

$$T(\theta) = \max_i\{\tau_i(\theta)\} - \min_i\{\tau_i(\theta), 0\} \tag{4-6}$$

The *beampattern* is defined as the magnitude squared of $r(\theta, \omega)$. An example beampattern is depicted in Fig. 4-4 as a function of the sine of θ at a single frequency ω_0. The beamformer is of the narrowband configuration illustrated in Fig. 4-1 and the twenty antenna elements are arranged in a linear equispaced geometry with an intersensor spacing of one-half wavelength. Direction θ is defined with respect to the normal to the line on which the antennas lie. This implies that the ith element of $\mathbf{d}(\theta, \omega_0)$ is $e^{-j\pi(i-1)\sin\theta}$.

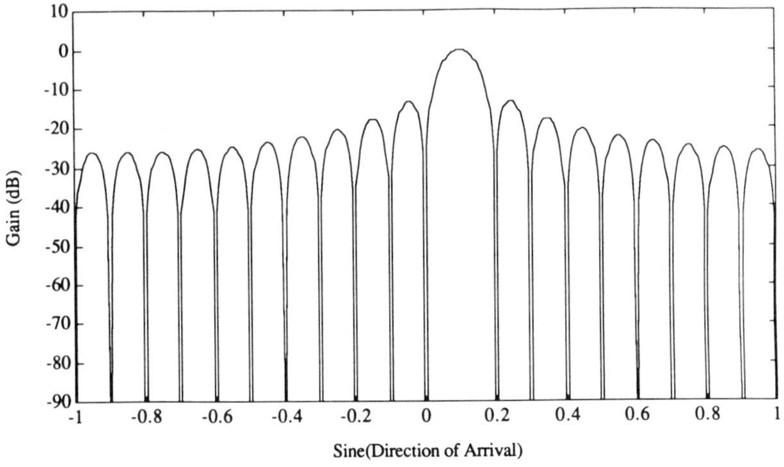

Figure 4-4 An example beampattern.

The dependence on ω_0 disappears because the sensor spacing is given in terms of the wavelength, which is inversely proportional to ω_0. The beampattern in Fig. 4-4 is obtained with weight vector $\mathbf{w}$ set equal to $0.05\mathbf{d}(\arcsin(0.1), \omega_0)$. This results in $|r(\theta, \omega_0)|^2 = 0.05 \sin^2[10\pi(\sin\theta - 0.1)] / \sin^2[\pi(\sin\theta - 0.1)/2]$.

Note that (4-5a) indicates that each weight in $\mathbf{w}$ affects both the temporal and spatial response of the beamformer. Historically, use of FIR filters has been viewed as providing frequency dependent weights in each channel. This interpretation is accurate but somewhat incomplete since the coefficients in each filter also influence the spatial filtering characteristics of the beamformer. As a multi-input single output system, the spatial and temporal filtering that occurs is a result of mutual interaction between spatial and temporal sampling.

A vector space interpretation of beamforming is used throughout this chapter. This geometric point of view is useful both in beamformer design and analysis. The inner product of the N dimensional vectors $\mathbf{w}$ and $\mathbf{d}(\theta, \omega)$ determines the response $r(\theta, \omega)$. For example, if for some (θ, ω) the angle between $\mathbf{w}$ and $\mathbf{d}(\theta, \omega)$ is 90° (i.e., if $\mathbf{w}$ is orthogonal to $\mathbf{d}(\theta, \omega)$), then the response is zero. If the angle is close to 0°, then the response magnitude will be relatively large. The ability to discriminate between sources at different locations and/or frequencies, say (θ_1, ω_1) and (θ_2, ω_2), is determined by the angle between their array response vectors, $\mathbf{d}(\theta_1, \omega_1)$ and $\mathbf{d}(\theta_2, \omega_2)$ [10]. In a general sense, the efficiency of the array geometry is determined by the dimension of the span of $\{\mathbf{d}(\theta, \omega), \theta \in \Theta, \omega \in \Omega\}$, where Θ and Ω represent the range of directions and frequencies of interest, respectively. In order to take advantage of all N degrees of freedom in the weight vector, $\{\mathbf{d}(\theta, \omega), \theta \in \Theta, \omega \in \Omega\}$ must span N dimensional space. If $\{\mathbf{d}(\theta, \omega), \theta = \Theta, \omega \in \Omega\}$ only spans an $M < N$ dimensional space, then only M of the N available degrees of freedom in $\mathbf{w}$ determine the response over Θ and Ω.

Aliasing is possible since we are sampling in both space and time. Essentially, aliasing implies nonuniqueness of array response vectors, that is, $\mathbf{d}(\theta_1, \omega_1) = \mathbf{d}(\theta_2, \omega_2)$ for some $(\theta_1, \omega_1) \neq (\theta_2, \omega_2)$. If $\omega_1 = \omega_2$, then this corresponds to pure spatial aliasing, an ambiguity in source directions, and can occur if the antennas are spaced too far apart. A well known example where spatial aliasing occurs is with linear equispaced array geometries that have antennas spaced at greater than one-half wavelength [34]. If $\mathbf{d}(\theta_1, \omega_1) = \mathbf{d}(\theta_2, \omega_2)$ for $\omega_1 \neq \omega_2$ and $\theta_1 \neq \theta_2$ with a general array geometry, then a source at one direction and one frequency cannot be distinguished from a source at another direction and frequency. This phenomena occurs with linear equispaced array geometries whenever $\omega_1 \sin\theta_1 = \omega_2 \sin\theta_2$, provided there are no tap delay lines in any of the antenna channels. The array geometry affects $\mathbf{d}(\theta, \omega)$ and thus has a significant impact on the beamformer's response. It is logical to attempt optimization of array geometry; however, the complicated manner by which the array geometry enters into the response equation makes this a very difficult problem. Array geometry is not further addressed in this chapter.

All data received by the array, for example, signal interference, noise, is assumed zero mean. Thus, the beamformer's output power is given by $E\{|y(n)|^2\} = \mathbf{w}^H E\{\mathbf{x}(n)\mathbf{x}^H(n)\}\mathbf{w}$. The data covariance matrix $\mathbf{R}_x = E\{\mathbf{x}(n)\mathbf{x}^H(n)\}$ is independent of time if the data is wide sense stationary. Although nonstationary data is often encountered and is one of the motivations for adaptive beamforming, the wide sense stationarity assumption is used in assessing steady state and optimum performance. *Array gain* is a commonly used performance measure; it is defined as the ratio of the output signal to noise ratio to the input signal to noise ratio. In this chapter array gain is only used with beamformers constrained to pass the signal with unit gain. Under this condition it simplifies to the ratio of the noise power at a single antenna to that at the beamformer output. This definition assumes that all antennas receive the same interference and noise power. Let $\mathbf{R}_n$ denote the covariance matrix associated with the interference and noise component of the data. The interference and noise power at the beamformer output is $\mathbf{w}^H\mathbf{R}_n\mathbf{w}$ so the array gain is given by

$$\text{Array gain} = 10 \log\left(\frac{[\mathbf{R}_n]_{1,1}}{\mathbf{w}^H\mathbf{R}_n\mathbf{w}}\right)$$

The twenty antenna linearly equispaced narrowband beamformer used in the previous example is utilized here to illustrate adaptive beamformer performance. The weight vector $\mathbf{w}$ is chosen to minimize the interference output power subject to the constraint that the response be unity (this criterion for choosing $\mathbf{w}$ is described in Section 4-4) in direction $\theta = \arcsin(0.1)$. The interference environment consists of eight interferers located at direction sines of -0.85, -0.55, -0.25, 0.05, 0.25, 0.45, 0.65, and 0.85 in uncorrelated noise. The interferers are of power 60 dB relative to the uncorrelated noise. The beampattern is depicted in Fig. 4-5; note the deep nulls (>90 dB) in the interferer directions. The array gain in this example is 68.16 dB, that is, the signal to noise ratio at the beamformer output is improved by 68 dB as a result of adaptive beamforming.

The data covariance matrix resulting from sampling a spatially propagating field is expressed in terms of the array response vector and the power spectral density $S(\theta, \omega)$ as

$$\mathbf{R}_x = \frac{1}{2\pi}\int_{\theta\in\Theta}\int_{\omega\in\Omega} S(\theta, \omega)\mathbf{d}(\theta, \omega)\mathbf{d}^H(\theta, \omega)\, d\omega\, d\theta \tag{4-7a}$$

This representation for $\mathbf{R}_x$ assumes the data at each frequency and direction is uncorrelated with the data at all other directions and frequencies. A similar, but more complicated representation allows inclusion of any such correlation. In practice $\mathbf{x}$ inevitably also contains uncorrelated or white noise due to receiver electronics, quantization, etcetera, and the expression for $\mathbf{R}_x$ becomes

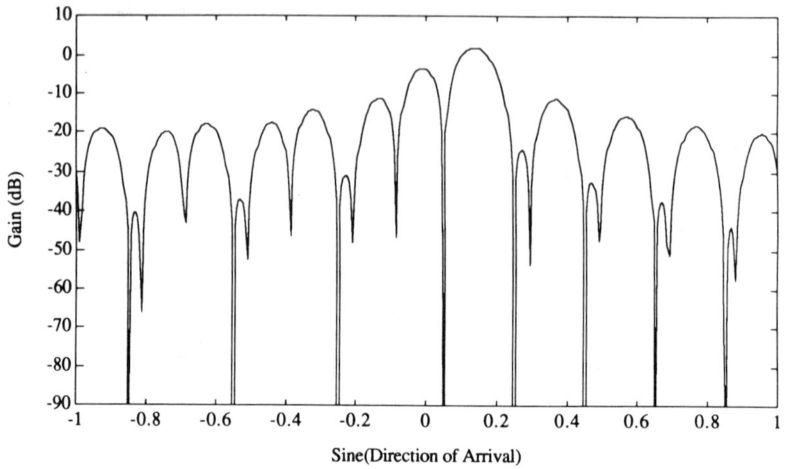

Figure 4-5 Beampattern of an adaptive beamformer. Interferers arrive from direction sines 0.85, 0.65, 0.45, 0.25, 0.05, −0.25, −0.55, and −0.85.

$$\mathbf{R}_x = \frac{1}{2\pi} \int_{\theta \in \Theta} \int_{\omega \in \Omega} S(\theta, \omega)\mathbf{d}(\theta, \omega)\mathbf{d}^H(\theta, \omega)\, d\omega\, d\theta + \sigma_n^2 \mathbf{I} \tag{4-7b}$$

where σ_n^2 is the white noise power. It is common to assume that $S(\theta, \omega)$ represents a sum of sources localized to points in space. The covariance matrix $\mathbf{R}_s$ resulting from a single point source located at θ_s with temporal power spectral density $S(\omega)$ is

$$\mathbf{R}_s = \frac{1}{2\pi} \int_{\omega \in \Omega} S(\omega)\mathbf{d}(\theta_s, \omega)\mathbf{d}^H(\theta_s, \omega)\, d\omega \tag{4-8}$$

The rank of $\mathbf{R}_s$ determines whether a source is considered narrowband or broadband from the beamformer's perspective. We consider a signal to be narrowband if it has an approximately rank one covariance matrix. The rank of $\mathbf{R}_s$ is a function of the TBWP for the source [5; 9, Chap. 3]. If the TBWP is much less than one, $\mathbf{R}_s$ is approximately rank one while if the TBWP is greater than one, $\mathbf{R}_s$ has rank greater than one. Thus, the covariance matrix for a narrowband signal is represented as

$$\mathbf{R}_s = \sigma_s^2 \mathbf{d}(\theta_s, \omega_0)\mathbf{d}^H(\theta_s, \omega_0) \tag{4-9}$$

where ω_0 is the signal frequency and θ_s the signal direction. As discussed later in this chapter, signals having rank one representations require a single degree of freedom for constraint design or adaptive cancellation.

In practice the true covariance matrix $\mathbf{R}_x$ is usually unknown and is estimated from the data. Covariance matrix estimation is often performed either implicitly or explicitly by an adaptive algorithm. This topic is discussed in greater detail in Section 4-8.

4-3 SIDELOBE CANCELLERS

This section introduces the multiple sidelobe canceller (MSC) and discusses the signal cancellation phenomena that can occur if the desired signal is present in the auxiliary data. The interference cancellation capabilities of the MSC are examined in Section 4-8.

An MSC is depicted in Fig. 4-6. The main channel or primary antenna has a high gain in the signal direction and receives both the desired signal and interference signals. The interference and noise is received through the sidelobes of the primary antenna. The auxiliary channels utilize low gain, omnidirectional antennas and are assumed to contain only interference and noise. Conditions under which it is reasonable to assume that the signal is absent from the auxiliary channels are discussed later in this section. A weighted combination of the auxiliary channels is used to estimate the interference leaking through the sidelobes of the primary antenna. The primary antenna can be either a single, high gain device such as a dish antenna or a nonadaptive beamformer, that is, a fixed weighted combination of multiple antennas outputs.

Let $\mathbf{x}_a$ and $\mathbf{w}_a$ represent the auxiliary data and weight vectors, respectively, and y_m the output of the primary antenna. The weights are chosen to minimize the output power

$$P_o = \min_{\mathbf{w}_a} E\{|y_m - \mathbf{w}_a^H \mathbf{x}_a|^2\} \tag{4-10}$$

yielding

$$\mathbf{w}_a = \mathbf{R}_a^{-1} \mathbf{r}_{ma} \tag{4-11}$$

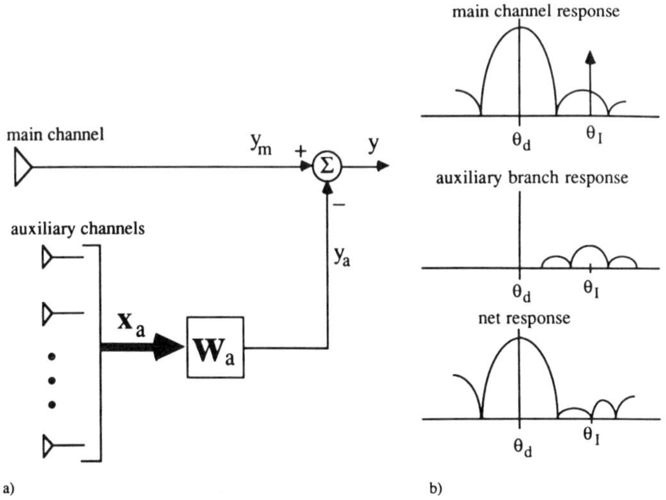

Figure 4-6 (a) Block diagram of a multiple sidelobe canceller (MSC). (b) The MSC cancels interference arriving through the sidelobes of the main channel or primary antenna.

where $\mathbf{R}_a = E\{\mathbf{x}_a \mathbf{x}_a^H\}$ and $\mathbf{r}_{ma} = E\{\mathbf{x}_a y_m^*\}$. Substituting for $\mathbf{w}_a$, the output power is expressed as

$$P_0 = \sigma_m^2 - \mathbf{r}_{ma}^H \mathbf{R}_a^{-1} \mathbf{r}_{ma} \tag{4-12}$$

with $\sigma_m^2 = E\{|y_m|^2\}$.

The assumption that the auxiliary data does not contain the signal is necessary because $\mathbf{w}_a$ is chosen to minimize output power. If the signal is present in the auxiliary data, then the power minimization process may result in signal cancellation. In reality the signal is almost always present in the auxiliary data. However, in many problems of interest the signal is very weak relative to the interference and thermal noise and all the degrees of freedom in $\mathbf{w}_a$ are used to cancel the interference, since it is the dominant contributor to the output power. In other problems it may be known that the signal is absent at specific times. In this case $\mathbf{w}_a$ is determined using the data arriving at times for which the signal is known to be absent. Note that signal cancellation also occurs if the interference is correlated with the signal since the weights will attempt to minimize power by cancelling the component of the signal that is correlated with the interference. This topic is discussed in greater detail in Section 4-12.

In order to illustrate the problematic effect of signal presence in the auxiliary data, consider the following scenario. A sidelobe canceller with M unit gain omnidirectional auxiliaries receives a narrowband signal $s(n)$ from direction θ_s, a narrowband interferer $i(n)$ from direction θ_i, and uncorrelated noise. We have

$$\begin{aligned} y_m(n) &= s(n)g_s + i(n)g_i + w(n) \\ \mathbf{x}_a(n) &= s(n)\mathbf{d}_s + i(n)\mathbf{d}_i + \mathbf{n}(n) \end{aligned} \tag{4-13}$$

where $w(n)$ is the uncorrelated noise received by the primary antenna, $\mathbf{n}(n)$ is the uncorrelated noise received by the auxiliaries, g_s and g_i are the complex gains of the primary antenna to the signal and interferer, respectively, and $\mathbf{d}_s$ and $\mathbf{d}_i$ are the array response vectors of the auxiliary array corresponding to the signal and interference directions. The noises $w(n)$ and $\mathbf{n}(n)$ are assumed to be uncorrelated, as are the interferer and signal. This implies

$$\begin{aligned} \mathbf{R}_a &= \sigma_s^2 \mathbf{d}_s \mathbf{d}_s^H + \sigma_i^2 \mathbf{d}_i \mathbf{d}_i^H + \sigma_n^2 \mathbf{I} \\ \mathbf{r}_{ma} &= \sigma_s^2 \mathbf{d}_s g_s^* + \sigma_i^2 \mathbf{d}_i g_i^* \end{aligned} \tag{4-14}$$

where σ_s^2, σ_i^2, and σ_n^2 are the signal, interferer, and uncorrelated noise powers, respectively. The signal output power, P_s, is

$$P_s = E\{|s(n)g_s - s(n)\mathbf{w}_a^H\mathbf{d}_s|^2\}$$
$$= \sigma_s^2|g_s - \mathbf{w}_a^H\mathbf{d}_s|^2 \tag{4-15}$$

The term $|g_s - \mathbf{w}_a^H\mathbf{d}_s|^2$ represents the magnitude squared of the beamformer response to the signal and is concisely referred to as the signal power gain. Using (4-14) in (4-11), applying the matrix inversion lemma, and substituting into (4-15) we determine that the signal power gain is given by

$$|g_s - \mathbf{w}_a^H\mathbf{d}_s|^2 = \left[\frac{1 + p_{in}M}{1 + p_{in}M + p_{sn}[M + p_{in}(M^2 - |\rho|^2)]}\right]^2$$
$$\times \left\{|g_s|^2 - 2\,\frac{p_{in}}{1 + p_{in}M}\,\mathrm{Re}[g_i g_s^* \rho^*] + \left(\frac{p_{in}}{1 + p_{in}M}\right)^2 |g_i|^2|\rho|^2\right\} \tag{4-16}$$

where $p_{sn} = \sigma_s^2/\sigma_n^2$, $p_{in} = \sigma_i^2/\sigma_n^2$, and $\rho = \mathbf{d}_s^H\mathbf{d}_i$. Note that since the auxiliary antennas are omnidirectional and have unit gain, each element of $\mathbf{d}_s$ and $\mathbf{d}_i$ has unit magnitude and thus $|\rho| \leq M$. In general $|\rho|$ decreases as $|\theta_s - \theta_i|$ increases.

There are two factors in (4-16) which affect the signal power gain. The term in brackets $[*]$ is unity when the signal is absent since then $p_{sn} = 0$. Thus, the term in braces $\{\cdot\}$ represents the power gain in the signal direction due to cancellation of the interferer. If the gain of the primary antenna to the signal, $|g_s|$, is large relative to its gain to the interferer, $|g_i|$, then the dominant component of the term in braces is $|g_s|^2$. This is easily verified using the fact that $|\rho| \leq M$ and

$$\frac{p_{in}}{1 + p_{in}M} < \frac{1}{M}$$

$|g_s|/|g_i|$ is the ratio of main lobe to sidelobe gain for the primary antenna; it is reasonable to assume this is significantly greater than one. Therefore, $|g_s|^2$ is the power gain to the signal if the signal is absent or weak relative to the noise. It is also the power gain if no auxiliaries are used ($M = 0$).

The bracketed term in (4-16) is always less than one for $p_{sn} \neq 0$, thus it represents a loss in signal gain associated with the presence of the signal. The existence of p_{sn} in the denominator indicates that the signal power gain decreases as the signal power increases. Note also that the effect of p_{sn} increases as $|\rho|$ decreases. If $M = 1$, then $|\rho|$ must be 1 and the denominator is $1 + p_{in} + p_{sn}$; the signal and interference are equally weighted. As we shall see later, one adaptive weight can cancel a single narrowband interferer; as p_{sn} increases, the weight best minimizes output power by attempting to cancel the signal rather than the interferer. For $M > 1$, there are excess degrees of freedom beyond that required for cancelling a single interferer. These additional degrees of freedom are used to minimize output power by cancelling the signal, hence the additional emphasis of p_{sn} by $p_{in}(M^2 - |\rho|^2)$.

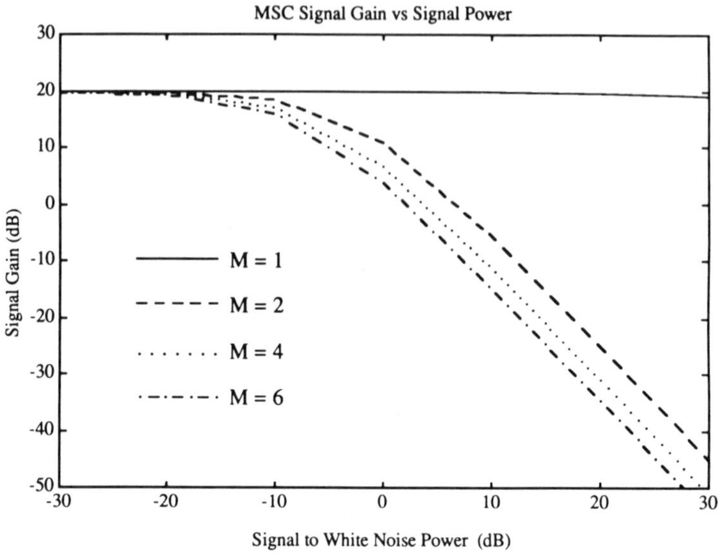

Figure 4-7 MSC signal gain (4-16) as a function of the signal to white noise power in dB for several values of M.

Figure 4-7 depicts (4-16) as a function of p_{sn} for $M = 1$, 2, 4, and 6. We have assumed $p_{in} = 10^4$, $|g_s|^2 = 100$, $|g_i|^2 = 1$, $|\rho| = M/3$ (except for $M = 1$, in which case $|\rho|$ must be 1), and $\mathrm{Re}\{g_i g_s^* \rho^*\} = 2M$. With these parameters the term in braces is approximately $96 = 19.8$ dB, which is the signal gain for small p_{sn}. The signal gain decreases dramatically as M increases from one to two. The extra degree of freedom is used by the MSC to cancel the signal. Additional weights result in further signal cancellation, although the signal cancellation is not significant until p_{sn} begins to approach unity $(10\log_{10}(p_{sn}) = 0)$. Negative values of $10\log_{10}(p_{sn})$ correspond to very weak signals while positive values of $10\log_{10}(p_{sn})$ correspond to strong signals. If the beamformer output is used for signal detection, then reliable detection may be possible even if the value $10\log_{10}(p_{sn})$ is negative because the post beamforming detection processing will enhance the signal relative to the noise. However, a single system must be capable of detecting both weak and strong signals. Excess degrees of freedom in the MSC cause the greatest loss in signal power gain for positive $10\log_{10}(p_{sn})$. This loss may result in strong signals being undetectable.

4-4 LINEARLY CONSTRAINED BEAMFORMING

Constraints on the beamformer weights are often utilized to control its response to the desired signal. Usually the constraints are designed to prevent signal distortion and/or cancellation. In this section we limit our discussion to linear constraints; quadratic constraints are discussed briefly in Section 4-5. This section begins by reviewing three types of linear constraints: point, derivative, and eigenvector constraints. The linearly constrained minimum variance (LCMV) criterion for choosing the weights is then discussed. The generalized sidelobe canceller (GSC) implementation is developed and the section concludes with an example.

4-4-1 Constraint Design

Linear constraints on the weight vector **w** are expressed in the form

$$\mathbf{C}^H\mathbf{w} = \mathbf{f} \tag{4-17}$$

where **C**, termed the constraint matrix, and **f**, the response vector, have constant elements. If there are K linear constraints, then **C** is N by K and **f** is K by 1. Each column of **C** and the corresponding element of **f** represent a single constraint. We assume that **C** is chosen to have linearly independent columns so that (4-17) represents K linearly independent equations in N unknowns. If K equals N, then **w** is uniquely determined by the constraints. If K is greater than N, there may be no **w** that satisfies the constraints. We assume that K is less than N. Each constraint uses up one degree of freedom in **w**.

As an example, suppose we wish to constrain the beamformer response at direction θ_s and frequency ω_s to be unity. We require

$$\mathbf{w}^H\mathbf{d}(\theta_s, \omega_s) = 1 \tag{4-18}$$

which implies $\mathbf{C} = \mathbf{d}(\theta_s, \omega_s)$ and $\mathbf{f} = 1$. This is an example of a point constraint, since the beamformer response is constrained at a single point in direction and frequency. In general, a point constraint is of the form $\mathbf{w}^H\mathbf{d}(\theta, \omega) = r$, where r is the desired (complex) response in direction θ at frequency ω. If we know that an interferer is located at direction θ_i and frequency ω_s, then we may wish to force the beamformer to have zero response to the interferer: $\mathbf{w}^H\mathbf{d}(\theta_i, \omega_s) = 0$. Combining this with the signal response constraint in (4-18) we obtain multiple point constraints

$$\mathbf{w}^H[\mathbf{d}(\theta_s, \omega_s) \quad \mathbf{d}(\theta_i, \omega_s)] = [1 \quad 0] \tag{4-19}$$

Here $K = 2$, $\mathbf{C} = [\mathbf{d}(\theta_s, \omega_s) \quad \mathbf{d}(\theta_i, \omega_s)]$, and $\mathbf{f}^H = [1 \quad 0]$.

Point constraints are most often used in narrowband beamforming since the response often needs to be constrained only at a single frequency. Broadband systems require the response to be constrained over a band of frequencies. In this situation, as we shall see shortly, point constraints are inefficient. The presence of mismatch or uncertainty in both narrowband and broadband problems can also require the response to be constrained over a region instead of at a single point. The term mismatch implies that the actual $\mathbf{d}(\theta, \omega)$ does not match the assumed $\mathbf{d}(\theta, \omega)$. For example, if the direction of a signal is not known exactly, then the response must be constrained over a range of directions to ensure that the beamformer does not distort the signal. Other types of mismatch arise due to unknown antenna gain and phase characteristics. Several philosophies can be used to constrain response over ranges of unknown parameters. In this section we focus on direction and frequency, although the techniques also apply to other kinds of mismatch.

One philosophy employs multiple point constraints. To illustrate, suppose the signal direction is known to be in a small region $\theta_s - \delta$ to $\theta_s + \delta$. We could set up constraints

$$\mathbf{w}^H[\mathbf{d}(\theta_s - \delta, \omega_s) \quad \mathbf{d}(\theta_s, \omega_s) \quad \mathbf{d}(\theta_s + \delta, \omega_s)] = [1 \quad 1 \quad 1]$$

to approximately control the response over the region. Note that multiple point constraints only control the response at the constraint points. If the constraint points densely sample the region to be controlled, then we expect to have good control over the region. However, it is generally undesirable to use large numbers of constraints since each constraint uses one degree of freedom in $\mathbf{w}$. Degrees of freedom used by the constraints cannot be used to cancel interference. Also, as the spacing between constraint points decreases, the individual constraints become more linearly dependent.

Another approach constrains derivatives of the response to be zero at a point of direction or frequency. This prevents rapid changes in the response near the constraint point. For example, in the example of the previous paragraph we would require

$$\mathbf{w}^H\mathbf{d}(\theta_s, \omega_s) = 1$$

and

$$\frac{\partial}{\partial\theta}\, \mathbf{w}^H\mathbf{d}(\theta, \omega_s)\bigg|_{\theta=\theta_s} = 0 \qquad (4\text{-}20)$$

Additional flatness is enforced on the response in the vicinity of θ_s by constraining higher order derivatives also to be zero. Note that since the response is a linear function of $\mathbf{w}$, derivative constraints are also linear; (4-20) can be written in the form

$$\mathbf{w}^H\left(\frac{\partial}{\partial\theta}\, \mathbf{d}(\theta, \omega_s)\bigg|_{\theta=\theta_s}\right) = 0$$

The primary disadvantage of derivative constraints is that they only indirectly control response. It is not evident how many derivatives must be constrained to achieve a specified level of response control over a given region. Additional discussion of derivative constraints is given in [2, 6, 14, 27].

The last class of constraints to be discussed in this section is sometimes termed eigenvector constraints. Eigenvector constraints are based on a least squares approximation to the desired response over the region of interest and are optimum in the sense that the total squared error between the desired and actual response is minimized for a given number of constraints. Thus, they provide more direct control over the beamformer response than derivative constraints. Discussion of the relationship between eigenvector and point constraints follows the development of eigenvector constraints.

Let $r_d(\theta, \omega)$ be the desired beamformer response over the region of direction $\theta \in \Theta$ and frequency $\omega \in \Omega$. The total squared error between the desired and actual response is given by

$$e^2 = \int_{\theta \in \Theta} \int_{\omega \in \Omega} |r_d(\theta, \omega) - \mathbf{w}^H \mathbf{d}(\theta, \omega)|^2 \, d\omega \, d\theta \tag{4-21}$$

If desired, a nonnegative weighting function can be inserted in the integrand to emphasize the error at certain values of ω and θ. Expanding the square, (4-21) is rewritten as

$$e^2 = \mathbf{w}^H \mathbf{Q} \mathbf{w} - \mathbf{w}^H \mathbf{p} - \mathbf{p}^H \mathbf{w} + \sigma_r^2 \tag{4-22}$$

where

$$\mathbf{Q} = \int_{\theta \in \Theta} \int_{\omega \in \Omega} \mathbf{d}(\theta, \omega) \mathbf{d}^H(\theta, \omega) \, d\omega \, d\theta \tag{4-23a}$$

$$\mathbf{p} = \int_{\theta \in \Theta} \int_{\omega \in \Omega} \mathbf{d}(\theta, \omega) r_d^*(\theta, \omega) \, d\omega \, d\theta \tag{4-23b}$$

$$\sigma_t^2 = \int_{\theta \in \Theta} \int_{\omega \in \Omega} |r_d(\theta, \omega)|^2 \, d\omega \, d\theta \tag{4-23c}$$

$\mathbf{Q}$ is a nonnegative definite Hermitian matrix and hence has the eigenvector decomposition

$$\mathbf{Q} = \mathbf{E} \boldsymbol{\Lambda} \mathbf{E}^H = \sum_{i=1}^{N} \lambda_i \mathbf{e}_i \mathbf{e}_i^H \tag{4-24}$$

where $\boldsymbol{\Lambda}$ is a diagonal matrix of eigenvalues λ_i, $i = 1, 2, \ldots, N$ that satisfy $\lambda_1 \geqslant \lambda_2 \geqslant \cdots \geqslant \lambda_N \geqslant 0$ and the ith column of $\mathbf{E}$ is the eigenvector $\mathbf{e}_i$. The orthonormality of the eigenvectors, $\mathbf{E}\mathbf{E}^H = \mathbf{I}$, implies that $\mathbf{p} = \mathbf{E}\mathbf{E}^H \mathbf{p}$. Thus, (4-22) may be rewritten in the form

$$e^2 = \sigma_r^2 + \sum_{i=1}^{K} \lambda_i |\mathbf{e}_i^H \mathbf{w}|^2 - \mathbf{w}^H \mathbf{e}_i \mathbf{e}_i^H \mathbf{p} - \mathbf{p}^H \mathbf{e}_i \mathbf{e}_i^H \mathbf{w}$$
$$+ \sum_{i=K+1}^{N} \lambda_i |\mathbf{e}_i^H \mathbf{w}|^2 - \mathbf{w}^H \mathbf{e}_i \mathbf{e}_i^H \mathbf{p} - \mathbf{p}^H \mathbf{e}_i \mathbf{e}_i^H \mathbf{w} \tag{4-25}$$

Complete the square for the sum from one to K to obtain

$$e^2 = \sigma_r^2 + \sum_{i=1}^{K} \lambda_i |\mathbf{e}_i^H \mathbf{w} - \lambda_i^{-1} \mathbf{e}_i^H \mathbf{p}|^2 - \lambda_i^{-1} \mathbf{p}^H \mathbf{e}_i \mathbf{e}_i^H \mathbf{p}$$
$$+ \sum_{i=K+1}^{N} \lambda_i |\mathbf{e}_i^H \mathbf{w}|^2 - \mathbf{w}^H \mathbf{e}_i \mathbf{e}_i^H \mathbf{p} - \mathbf{p}^H \mathbf{e}_i \mathbf{e}_i^H \mathbf{w} \tag{4-26}$$

The error involving the first K terms is minimized by constraining the weights $\mathbf{w}$ to satisfy $\mathbf{e}_i^H \mathbf{w} = \lambda_i^{-1} \mathbf{e}_i^H \mathbf{p}$, $i = 1, 2, \ldots, K$. In this set of constraints the columns of $\mathbf{C}$ correspond to the K largest eigenvalues of $\mathbf{Q}$, hence the name eigenvector constraints. If $\mathbf{w}$ is constrained in this manner, then the error is

$$e^2 = \sigma_r^2 - \sum_{i=1}^{K} \lambda_i^{-1} \mathbf{p}^H \mathbf{e}_i \mathbf{e}_i^H \mathbf{p} + \sum_{i=K+1}^{N} \lambda_i |\mathbf{e}_i^H \mathbf{w}|^2 - \mathbf{w}^H \mathbf{e}_i \mathbf{e}_i^H \mathbf{p} - \mathbf{p}^H \mathbf{e}_i \mathbf{e}_i^H \mathbf{w} \tag{4-27}$$

σ_r^2 and the first sum (4-27) are independent of $\mathbf{w}$ and thus only depend on the number of constraints K. The second sum is a function of $\mathbf{w}$; however, if K is chosen such that the λ_i, $i = K+1, K+2, \ldots, N$ are very small, then this sum is insignificant and e^2 is approximately independent of $\mathbf{w}$. To see this, first note that

$$\lambda_i = \mathbf{e}_i^H \mathbf{Q} \mathbf{e}_i = \int_{\theta \in \Theta} \int_{\omega \in \Omega} |\mathbf{d}^H(\theta, \omega) \mathbf{e}_i|^2 \, d\omega \, d\theta \tag{4-28}$$

Now consider

$$|\mathbf{p}^H \mathbf{e}_i \mathbf{e}_i^H \mathbf{w}|^2 = |\mathbf{e}_i^H \mathbf{w}|^2 \left| \int_{\theta \in \Theta} \int_{\omega \in \Omega} r_d(\theta, \omega) \mathbf{d}^H(\theta, \omega) \mathbf{e}_i \, d\omega \, d\theta \right|^2 \tag{4-29}$$

and apply the Cauchy–Schwarz inequality to obtain

$$|\mathbf{p}^H \mathbf{e}_i \mathbf{e}_i^H \mathbf{w}|^2 \leq c |\mathbf{e}_i^H \mathbf{w}|^2 \int_{\theta \in \Theta} \int_{\omega \in \Omega} |r_d(\theta, \omega)|^2 |\mathbf{d}^H(\theta, \omega) \mathbf{e}_i|^2 \, d\omega \, d\theta \tag{4-30a}$$

where

$$c = \int_{\theta \in \Theta} \int_{\omega \in \Omega} d\omega \, d\theta \tag{4-30b}$$

In most cases of interest we have $|r_d(\theta, \omega)|^2 \leq 1$, so substituting (4-28) in (4-30a) we obtain

$$|\mathbf{p}^H \mathbf{e}_i \mathbf{e}_i^H \mathbf{w}|^2 \leq c|\mathbf{e}_i^H \mathbf{w}|^2 \lambda_i \tag{4-31}$$

Thus, if the λ_i, $i = K+1, K+2, \ldots, N$ are very small, then the terms in (4-27) that depend on **w** can be ignored.

This suggests that the number of constraints, K, should be chosen as the number of significant eigenvalues of **Q** to obtain small values for e^2. Note that **Q** corresponds to the covariance matrix for a source with power spectral density $S(\theta, \omega) = 2\pi$ on Θ and Ω (see (4-7a)). In general one must compute the eigenvalues of **Q** to determine K, however, if Θ corresponds to a single point of direction then the number of significant eigenvalues is determined by the TBWP [5]. If the response is constrained at a single point of direction and frequency, say θ_s and ω_s, then the eigenvector constraints correspond to the point constraint of (4-18) since $\mathbf{Q} = \mathbf{d}(\theta_s, \omega_s)\mathbf{d}^H(\theta_s, \omega_s)$ has only one nonzero eigenvalue and $\mathbf{e}_1 = \mathbf{d}(\theta_s, \omega_s)/[\mathbf{d}^H(\theta_s, \omega_s)\mathbf{d}(\theta_s, \omega_s)]^{1/2}$.

The general relationship between point constraints and eigenvector constraints is now examined. Consider the desired set of point constraints

$$\mathbf{A}\mathbf{w} = \mathbf{r}_d \tag{4-32}$$

where $\mathbf{A}^H = [\mathbf{d}(\theta_1, \omega_1) \quad \cdots \quad \mathbf{d}(\theta_i, \omega_j) \quad \cdots \quad \mathbf{d}(\theta_g, \omega_h)]$, $\mathbf{r}_d^H = [r_d(\theta_1, \omega_1) \quad \cdots \quad r_d(\theta_i, \omega_j) \quad \cdots \quad r_d(\theta_g, \omega_h)]$ and θ_i, ω_j, $i = 1, 2, \ldots, g$, $j = 1, 2, \ldots, h$ densely sample Θ and Ω. Assuming $gh > N$, (4-32) is an overdetermined system of linear equations and in general cannot be satisfied by any **w**. The squared error, $e^2 = |\mathbf{A}\mathbf{w} - \mathbf{r}_d|^2$, is given by

$$e^2 = \mathbf{w}^H\mathbf{A}^H\mathbf{A}\mathbf{w} - \mathbf{r}_d^H\mathbf{A}\mathbf{w} - \mathbf{w}^H\mathbf{A}^H\mathbf{r}_d + \mathbf{r}_d^H\mathbf{r}_d \tag{4-33}$$

Noting that $\mathbf{A}^H\mathbf{A}$, $\mathbf{A}^H\mathbf{r}_d$, and $\mathbf{r}_d^H\mathbf{r}_d$ are numerical approximations to **Q**, **p**, and σ_r^2, respectively, we see that (4-33) and (4-22) are equivalent. Thus, eigenvector constraints approximately minimize the squared error e^2 associated with an overdetermined system of point constraints. Direct use of point constraints requires $gh \leq N$ and results in exactly zero error at the constraint points. However, the error is uncontrolled between constraint points. In contrast, eigenvector constraints minimize the error over the entire region.

4-4-2 Linearly Constrained Minimum Variance Criterion

The linearly constrained minimum variance (LCMV) criterion for choosing **w** is commonly used in both narrowband and broadband beamforming. It is based on minimizing the output power or variance, $E\{|y(n)|^2\} = \mathbf{w}^H E\{\mathbf{x}(n)\mathbf{x}^H(n)\}\mathbf{w}$, while constraining **w** to present a desired response to the signal of interest. Variance minimization effectively minimizes the interference and noise power at the beamformer output while the constraints preserve the signal of interest. Thus, we solve

$$P_0 = \min_{\mathbf{w}} \mathbf{w}^H \mathbf{R}_x \mathbf{w} \quad \text{subject to} \quad \mathbf{C}^H \mathbf{w} = \mathbf{f} \tag{4-34}$$

where $\mathbf{R}_x = E\{\mathbf{x}(n)\mathbf{x}^H(n)\}$. The solution (4-34) is easily obtained using Lagrange multipliers (the solution to (4-34) is derived in Appendix D of Chapter 3 for the case of a single linear constraint, see also [16]) and is given by

$$\mathbf{w} = \mathbf{R}_x^{-1}\mathbf{C}(\mathbf{C}^H\mathbf{R}_x^{-1}\mathbf{C})^{-1}\mathbf{f} \tag{4-35}$$

The inverses exist since we assume that $\mathbf{C}$ is full rank and $\mathbf{R}_x$ is positive definite because the data always contain an uncorrelated noise component. Substituting (4-35) into the expression for the output variance (equivalent to the output power) yields

$$P_0 = \mathbf{f}^H(\mathbf{C}^H\mathbf{R}_x^{-1}\mathbf{C})^{-1}\mathbf{f} \tag{4-36}$$

An example illustrating LCMV beamformer performance is given later in this section.

The minimum variance distortionless response (MVDR) beamformer is a special case of the LCMV beamformer that uses the specific set of constraints given in (4-18), that is, $\mathbf{C} = \mathbf{d}(\theta_s, \omega_s)$ and $\mathbf{f} = 1$. The MVDR beamformer is inherently narrowband since the response is only constrained at a single frequency, ω_s. Note that in the MVDR case (4-35) and (4-36) simplify somewhat to

$$\mathbf{w} = \frac{1}{\mathbf{d}^H(\theta_s, \omega_s)\mathbf{R}_x^{-1}\mathbf{d}(\theta_s, \omega_s)}\mathbf{R}_x^{-1}\mathbf{d}(\theta_s, \omega_s) \tag{4-37}$$

and

$$P_0 = \frac{1}{\mathbf{d}^H(\theta_s, \omega_s)\mathbf{R}_x^{-1}\mathbf{d}(\theta_s, \omega_s)} \tag{4-38}$$

If the signal and interference are uncorrelated and the signal component of $\mathbf{R}_x$ is that of a narrowband signal, $\mathbf{R}_s = \sigma_s^2\mathbf{d}(\theta_s, \omega_s)\mathbf{d}^H(\theta_s, \omega_s)$ (see (4-9)), then (4-37) also maximizes the signal to noise ratio (SNR) at the beamformer output. To see this, express $\mathbf{R}_x$ as the sum of signal and interference/noise terms: $\mathbf{R}_x = \sigma_s^2\mathbf{d}(\theta_s, \omega_s)\mathbf{d}^H(\theta_s, \omega_s) + \mathbf{R}_n$, substitute into (4-37) and apply the matrix inversion lemma to obtain

$$\mathbf{w} = \alpha\mathbf{R}_n^{-1}\mathbf{d}(\theta_s, \omega_s) \tag{4-39}$$

where the scalar $\alpha = (\mathbf{d}^H(\theta_s, \omega_s)\mathbf{R}_n\mathbf{d}(\theta_s, \omega_s))^{-1}$. The SNR is the ratio of quadratic forms

$$\text{SNR} = \frac{\mathbf{w}^H \sigma_s^2 \mathbf{d}(\theta_s, \omega_s)\mathbf{d}^H(\theta_s, \omega_s)\mathbf{w}}{\mathbf{w}^H \mathbf{R}_n \mathbf{w}} \tag{4-40}$$

and is maximized by the solution $\mathbf{w}$ corresponding to the largest generalized eigenvalue γ in the generalized eigenproblem

$$\sigma_s^2 \mathbf{d}(\theta_s, \omega_s)\mathbf{d}^H(\theta_s, \omega_s)\mathbf{w} = \gamma \mathbf{R}_n \mathbf{w} \tag{4-41}$$

$\sigma_s^2 \mathbf{d}(\theta_s, \omega_s)\mathbf{d}^H(\theta_s, \omega_s)$ is a rank one matrix and $\mathbf{R}_n$ is positive definite so there is only one nonzero (and positive) generalized eigenvalue. It is straightforward to verify by substitution that (4-39) is a generalized eigenvector corresponding to the only positive generalized eigenvalue. Furthermore, the largest generalized eigenvalue is the maximum SNR.

In general the LCMV beamformer does not maximize the SNR, that is, $\mathbf{w}$ does not satisfy

$$\max_{\mathbf{w}} \frac{\mathbf{w}^H \mathbf{R}_s \mathbf{w}}{\mathbf{w}^H \mathbf{R}_n \mathbf{w}} \tag{4-42}$$

The SNR is often maximized by tolerating some loss of signal power in return for a proportionally larger reduction in noise output power. Thus, the maximum SNR weight vector $\mathbf{w}$ generally introduces distortion of the desired signal. In contrast, the constraints on $\mathbf{w}$ in the LCMV beamformer restrict the distortion of the desired signal. Maximization of SNR is an intuitively satisfying criterion for choosing $\mathbf{w}$. However, like the Wiener theory discussed in Chapter 3, maximization of SNR for cases where $\mathbf{R}_s$ is not rank one is often not practical since it requires knowledge of both $\mathbf{R}_s$ and $\mathbf{R}_n$. The LCMV criterion is a practical and often very effective criterion for choosing $\mathbf{w}$.

4-4-3 Generalized Sidelobe Canceller

The generalized sidelobe canceller (GSC) is a decomposition of the LCMV weights into constrained and unconstrained components. This decomposition provides both an alternate implementation for the LCMV beamformer and insight into LCMV beamformer characteristics. Both of these aspects of the GSC are exploited in this chapter. Essentially, the GSC corresponds to the method proposed by Hanson and Lawson [24] for solving linearly constrained quadratic minimization problems using an orthogonal basis for the null space of the constraint matrix. This concept was applied in beamforming by Applebaum and Chapman [2] and by Griffiths and Jim [23], who coined the term GSC.

Let the columns of $\mathbf{C}_n$, an N by $N-K$ matrix, represent a basis for the orthogonal complement of the space spanned by the columns of the constraint matrix $\mathbf{C}$. Together the columns of $\mathbf{C}$ and $\mathbf{C}_n$ span the entire N

dimensional space so we can express an arbitrary weight vector $\mathbf{w}$ in terms of this set of basis vectors as $\mathbf{w} = \mathbf{C}\mathbf{v} - \mathbf{C}_n\mathbf{w}_n$. Here $\mathbf{v}$ (K by 1) and $-\mathbf{w}_n$ ($N - K$ by 1) represent the components of $\mathbf{w}$ in the space spanned by the columns of $\mathbf{C}$ and $\mathbf{C}_n$, respectively. $\mathbf{C}$ and $\mathbf{C}_n$ have orthogonal columns, so applying the constraint to $\mathbf{w}$ gives $\mathbf{C}^H\mathbf{C}\mathbf{v} = \mathbf{f}$. The solution for $\mathbf{v}$ is $\mathbf{v} = (\mathbf{C}^H\mathbf{C})^{-1}\mathbf{f}$. Thus, the GSC representation of $\mathbf{w}$ is

$$\mathbf{w} = \mathbf{w}_q - \mathbf{C}_n\mathbf{w}_n \tag{4-43}$$

where $\mathbf{w}_q = \mathbf{C}(\mathbf{C}^H\mathbf{C})^{-1}\mathbf{f}$. The weight vector $\mathbf{w}_q$ represents a nonadaptive beamformer that satisfies the constraints. Note that the constraint does not affect $\mathbf{w}_n$; $\mathbf{w}_n$ embodies the available degrees of freedom in $\mathbf{w}$. Equation (4-43) is represented in block diagram form in Fig. 4-8.

Suppose the data $\mathbf{x}$ is also decomposed in terms of the spaces spanned by the columns of $\mathbf{C}$ and $\mathbf{C}_n$, that is, $\mathbf{x} = \mathbf{C}\mathbf{x}_c + \mathbf{C}_n\mathbf{x}_n$. Substituting (4-43) into $y = \mathbf{w}^H\mathbf{x}$ and noting that $\mathbf{w}_q^H\mathbf{C}_n = \mathbf{0}$ gives

$$y = \mathbf{w}_q^H\mathbf{C}\mathbf{x}_c - \mathbf{w}_n^H\mathbf{C}_n^H\mathbf{C}_n\mathbf{x}_n$$

Thus, $\mathbf{w}_q$ only affects the component of the data in the space spanned by $\mathbf{C}$ while $\mathbf{w}_n$ only affects the component of the data in the orthogonal complement of the space spanned by the columns of $\mathbf{C}$. $\mathbf{C}_n$ blocks the component of $\mathbf{x}$ that lies in the space spanned by the columns of $\mathbf{C}$. $\mathbf{C}$ is usually chosen to preserve the desired signal which implies that the desired signal lies in the space spanned by the columns of $\mathbf{C}$. $\mathbf{C}_n$ is often termed the *signal blocking matrix* because its columns are orthogonal to the space in which the signal lies.

The LCMV problem (4-34) is expressed in terms of the GSC as the unconstrained minimization problem

$$P_0 = \min_{\mathbf{w}_n}(\mathbf{w}_q - \mathbf{C}_n\mathbf{w}_n)^H\mathbf{R}_x(w_q - \mathbf{C}_n\mathbf{w}_n) \tag{4-44a}$$

or

$$P_0 = \min_{\mathbf{w}_n}(\mathbf{w}_q^H\mathbf{R}_x\mathbf{w}_q - \mathbf{w}_q^H\mathbf{R}_x\mathbf{C}_n\mathbf{w}_n - \mathbf{w}_n^H\mathbf{C}_n^H\mathbf{R}_x\mathbf{w}_q + \mathbf{w}_n^H\mathbf{C}_n^H\mathbf{R}_x\mathbf{C}_n\mathbf{w}_n) \tag{4-44b}$$

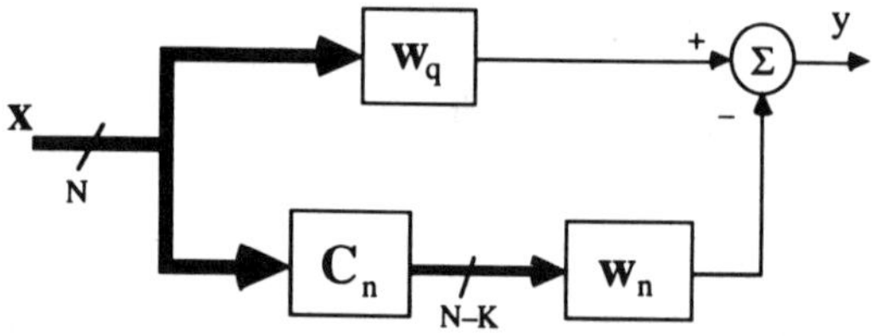

Figure 4-8 Block diagram for a generalized sidelobe canceller (GSC).

Completing the square we have

$$P_0 = \min_{\mathbf{w}_n}[\mathbf{w}_n - (\mathbf{C}_n^H\mathbf{R}_x\mathbf{C}_n)^{-1}\mathbf{C}_n^H\mathbf{R}_x\mathbf{w}_q]^H\mathbf{C}_n^H\mathbf{R}_x\mathbf{C}_n[\mathbf{w}_n - (\mathbf{C}_n^H\mathbf{R}_x\mathbf{C}_n)^{-1}\mathbf{C}_n^H\mathbf{R}_x\mathbf{w}_q]$$
$$+ \mathbf{w}_q^H\mathbf{R}_x\mathbf{w}_q - \mathbf{w}_q^H\mathbf{R}_x\mathbf{C}_n(\mathbf{C}_n^H\mathbf{R}_x\mathbf{C}_n)^{-1}\mathbf{C}_n^H\mathbf{R}_x\mathbf{w}_q \qquad (4\text{-}45)$$

The matrix $\mathbf{C}_n$ is full rank and $\mathbf{R}_x$ is positive definite so $\mathbf{C}_n^H\mathbf{R}_x\mathbf{C}_n$ is positive definite and the inverse exists. Thus, (4-45) is minimized by

$$\mathbf{w}_n = (\mathbf{C}_n^H\mathbf{R}_x\mathbf{C}_n)^{-1}\mathbf{C}_n^H\mathbf{R}_x\mathbf{w}_q \qquad (4\text{-}46)$$

The minimum output power is obtained using this set of weights and is given by

$$P_0 = \mathbf{w}_q^H\mathbf{R}_x\mathbf{w}_q - \mathbf{w}_q^H\mathbf{R}_x\mathbf{C}_n(\mathbf{C}_n^H\mathbf{R}_x\mathbf{C}_n)^{-1}\mathbf{C}_n^H\mathbf{R}_x\mathbf{w}_q \qquad (4\text{-}47)$$

Suppose $\mathbf{x} = \mathbf{s} + \mathbf{n}$ where $\mathbf{s}$ is the component of the data due to the desired signal and $\mathbf{n}$ is the component due to interference and noise. Assume that the constraints are chosen to preserve the desired signal, that is, $\mathbf{s}$ lies in the space spanned by the columns of $\mathbf{C}$ so $\mathbf{C}_n^H\mathbf{s} = \mathbf{0}$. This implies that the beamformer output is

$$y = \mathbf{w}_q^H\mathbf{s} + (\mathbf{w}_q - \mathbf{C}_n\mathbf{w}_n)^H\mathbf{n} \qquad (4\text{-}48)$$

If we further assume that the signal and interference are uncorrelated, then the output power or variance is expressed as

$$E\{|y|^2\} = \mathbf{w}_q^H\mathbf{R}_s\mathbf{w}_q + \mathbf{w}^H\mathbf{R}_n\mathbf{w} \qquad (4\text{-}49)$$

where $\mathbf{w}^H\mathbf{R}_n\mathbf{w}$ is the interference and noise component of the output power. Therefore, minimizing the total output power subject to $\mathbf{C}^H\mathbf{w} = \mathbf{f}$ is equivalent to minimizing the interference and noise output power subject to the same constraints. Substitute the GSC representation for $\mathbf{w}$ into (4-49) with $\mathbf{w}_n$ given by (4-46) and utilize the identity $\mathbf{C}_n^H\mathbf{R}_s = \mathbf{0}$ to express the minimum output power P_0 as a sum of signal, P_s, and interference and noise, P_n, output powers where

$$P_s = \mathbf{w}_q^H\mathbf{R}_s\mathbf{w}_q \qquad (4\text{-}50a)$$

$$P_n = \mathbf{w}_q^H\mathbf{R}_n\mathbf{w}_q - \mathbf{w}_q^H\mathbf{R}_n\mathbf{C}_n(\mathbf{C}_n^H\mathbf{R}_n\mathbf{C}_n)^{-1}\mathbf{C}_n^H\mathbf{R}_n\mathbf{w}_q \qquad (4\text{-}50b)$$

The case where the signal and interference are correlated is discussed in Section 4-12.

The mean squared error (MSE) between the desired signal and the beamformer output is easily derived using the GSC representation. We assume the constraints are chosen so that the beamformer output in the absence of interference and noise is equal to the desired signal, y_d. That is, $y_d = \mathbf{w}_q^H \mathbf{s}$. Define the MSE, e, as

$$e = E\{|y_d - y|^2\} \tag{4-51}$$

Substitute (4-48) into (4-51) to obtain

$$\begin{aligned} e &= E\{|\mathbf{w}^H \mathbf{n}|^2\} \\ &= \mathbf{w}^H \mathbf{R}_n \mathbf{w} \end{aligned} \tag{4-52}$$

Replacing $\mathbf{w}$ by its GSC representation shows that $e = P_n$. Comparison of (4-52) and (4-49) indicates that the LCMV criterion is equivalent to a minimum MSE criterion provided the constraints are chosen to preserve the desired signal.

The weight vector $\mathbf{w}_q$ is often termed the *quiescent* weight vector because it is the optimum LCMV weight vector when the environment is quiet, that is, characterized by uncorrelated noise so that $\mathbf{R}_x = \sigma_n^2 \mathbf{I}$. Substituting $\mathbf{R}_x = \sigma_n^2 \mathbf{I}$ into (4-46) we obtain $\mathbf{w}_n = \mathbf{0}$ as a result of the orthogonality of the columns of $\mathbf{C}_n$ and $\mathbf{w}_q$, that is, $\mathbf{C}_n^H \mathbf{w}_q = \mathbf{0}$, so in this case $\mathbf{w} = \mathbf{w}_q$. If a signal is present and the noise covariance is $\mathbf{R}_n = \sigma_n^2 \mathbf{I}$ so that $\mathbf{R}_x = \mathbf{R}_s + \sigma_n^2 \mathbf{I}$, then we still have $\mathbf{w}_n = \mathbf{0}$ since $\mathbf{C}_n^H \mathbf{R}_x \mathbf{w}_q = \mathbf{0}$. Thus, for $\mathbf{R}_n = \sigma_n^2 \mathbf{I}$ we have $\mathbf{w} = \mathbf{w}_q$ and $P_n = \sigma_n^2 \mathbf{w}_q^H \mathbf{w}_q$. The noise output power is thus proportional to the square of the norm of $\mathbf{w}_q$. Recall that $\mathbf{w}_q = \mathbf{C}(\mathbf{C}^H \mathbf{C})^{-1} \mathbf{f}$. If the constraint matrix $\mathbf{C}$ is ill-conditioned, then the norm of $\mathbf{w}_q$ can become very large. For example, a large norm results if two columns of $\mathbf{C}$ are almost linearly dependent but the corresponding elements of $\mathbf{f}$ are different, such that the constraints are almost inconsistent. Consider the set of two point constraints in (4-19). If θ_s and θ_i are very close to each other, then $\mathbf{d}(\theta_s, \omega_s)$ and $\mathbf{d}(\theta_i, \omega_s)$ are almost linearly dependent and $\mathbf{w}_q$ has a large norm because the projection of $\mathbf{w}_q$ onto the space spanned by $\mathbf{d}(\theta_s, \omega_s)$ is required to be unity while that onto the nearly identical space spanned by $\mathbf{d}(\theta_i, \omega_s)$ is required to be zero.

The overall GSC weights are equivalent for all matrices $\mathbf{C}_n$ whose columns span the orthogonal complement of the space spanned by the columns of $\mathbf{C}$. To show this let GSC_1 and GSC_2 be represented by $\mathbf{w}_q - \mathbf{C}_{n1}\mathbf{w}_{n1}$ and $\mathbf{w}_q - \mathbf{C}_{n2}\mathbf{w}_{n2}$, respectively, where $\mathbf{C}_{n1} = \mathbf{C}_{n2}\mathbf{S}$ with $\mathbf{S}$ an arbitrary nonsingular matrix. The vector $\mathbf{w}_{n1}$ is given by

$$\begin{aligned} w_{n1} &= (\mathbf{C}_{n1}^H \mathbf{R}_x \mathbf{C}_{n1})^{-1} \mathbf{C}_{n1}^H \mathbf{R}_x \mathbf{w}_q \\ &= (\mathbf{S}^H \mathbf{C}_{n2}^H \mathbf{R}_x \mathbf{C}_{n2} \mathbf{S})^{-1} \mathbf{S}^H \mathbf{C}_{n2}^H \mathbf{R}_x \mathbf{w}_q \\ &= \mathbf{S}^{-1} (\mathbf{C}_{n2}^H \mathbf{R}_x \mathbf{C}_{n2})^{-1} \mathbf{C}_{n2}^H \mathbf{R}_x \mathbf{w}_q \\ &= \mathbf{S}^{-1} \mathbf{w}_{n2} \end{aligned} \tag{4-53}$$

Thus, $\mathbf{C}_{n1}\mathbf{w}_{n1} = \mathbf{C}_{n2}\mathbf{w}_{n2}$. It follows that P_0, P_s, and P_n are also independent of $\mathbf{S}$. The equivalence of all $\mathbf{C}_n$ matrices whose columns span the same space provides freedom that can be used to optimize other criteria associated with the GSC, such as computational complexity or adaptive algorithm performance.

Multiple sidelobe cancellers can be represented with a structure similar to the GSC. Initially let us assume that the primary antenna is formed as a linear combination of antenna outputs using a weight vector $\mathbf{w}_p$. The auxiliary channel data $\mathbf{x}_a$ is represented by applying a selection matrix, $\mathbf{T}_a$, to the array data $\mathbf{x}$: $\mathbf{x}_a = \mathbf{T}_a^H\mathbf{x}$. $\mathbf{T}_a$ is an N by M matrix having a single unity element in each column and row with remaining elements all zero. The rows of $\mathbf{T}_a$ having unity elements correspond to the antennas in the array which serve as auxiliaries. Thus, this MSC is represented by the overall weight vector $\mathbf{w} = \mathbf{w}_p - \mathbf{T}_a\mathbf{w}_a$ and $\mathbf{w}_a$ is chosen according to

$$\min_{\mathbf{w}_a}(\mathbf{w}_p - \mathbf{T}_a\mathbf{w}_a)^H\mathbf{R}_x(\mathbf{w}_p - \mathbf{T}_a\mathbf{w}_a) \tag{4-54}$$

Comparison of (4-54) and (4-44a) indicates that $\mathbf{w}_p$ corresponds to $\mathbf{w}_q$, $\mathbf{T}_a$ to $\mathbf{C}_n$, and $\mathbf{w}_a$ to $\mathbf{w}_n$. Note that the MSC satisfies $\mathbf{w}_p^H\mathbf{T}_a = \mathbf{0}$ only if the set of antennas used to form the primary antenna and the set used as auxiliaries are disjoint. MSCs that employ a single high gain antenna for the primary channel are represented by modeling the primary antenna response as the response of a nonadaptive beamformer. The array of antennas modeling the primary antenna and the auxiliary antennas are combined into a single array which generates a data vector $\mathbf{x}$. In this case $\mathbf{w}_p$ contains zeros in the positions corresponding to the auxiliary antennas while $\mathbf{T}_a$ contains zeros in the rows associated with the antennas used to model the primary antenna and we have $\mathbf{w}_p^H\mathbf{T}_a = \mathbf{0}$.

With the exception of the following section, the analysis in the rest of this chapter focuses on LCMV beamformers and utilizes the GSC representation. Much of it is also applicable to the MSC using the relationship between the GSC and MSC previously described.

4-4-4 Examples

We conclude this section with two examples that illustrate LCMV beamformer performance. An array of fifteen antennas equally spaced along a line at one-half wavelength of the highest frequency is employed. Each antenna channel contains six tap FIR filters resulting in a total of ninety weights in the beamformer. The tap spacing is chosen in accordance with the Nyquist sampling rate so that frequency is normalized on the interval $-\pi$ to π radians. The signal of interest and the interference occupy the band $3\pi/5 \leq |\omega| \leq 4\pi/5$, corresponding to a 20% bandwidth signal. The interferer power spectral densities are constant over this band. The data and

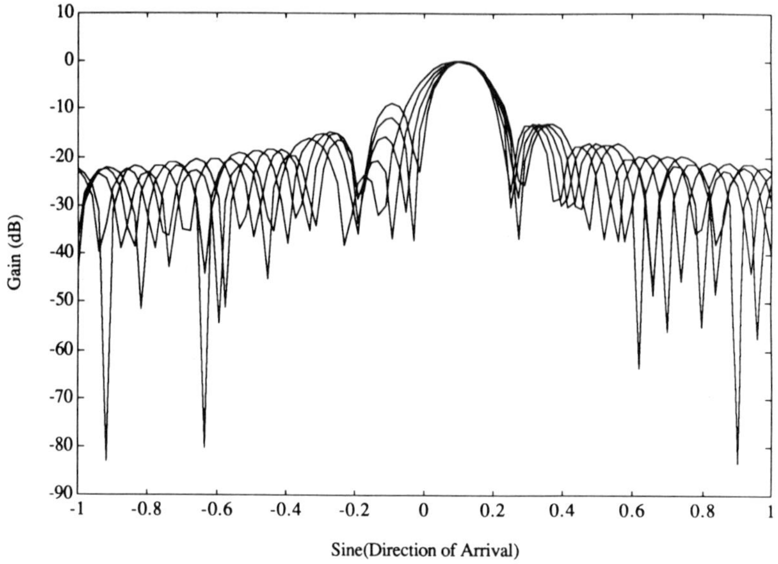

Figure 4-9 Magnitude response of $\mathbf{w}_q$ at five frequencies on $3\pi/5 \leqslant \omega \leqslant 4\pi/5$ as a function of direction sine.

beamformer weights are real. Direction is measured in terms of the sine of the angle relative to a line perpendicular to the array axis, henceforth termed the direction sine. The signal of interest has a direction sine of 0.1. The beamformer is implemented in GSC form and eleven eigenvector constraints are employed to ensure unit gain and linear phase response to the signal of interest over the frequency band $3\pi/5 \leqslant |\omega| \leqslant 4\pi/5$. Thus, this beamformer has seventy-nine adaptive degrees of freedom.

The series of curves in Figure 4-9 depict the magnitude response of $\mathbf{w}_q$ in dB, $20 \log|\mathbf{w}_q^H \mathbf{d}(\theta, \omega)|$, as a function of direction sine at five evenly spaced frequencies in the interval $3\pi/5 \leqslant \omega \leqslant 4\pi/5$. Note that all curves coincide at 0 dB in the desired signal direction (direction sine of 0.1) indicating that $\mathbf{w}_q$ satisfies the unity gain constraint. The gain and phase response of $\mathbf{w}_q$ in the desired signal direction sine is displayed as a function of frequency in Fig. 4-10. This figure indicates that the constraints result in unity gain (0 dB) and linear phase for frequencies between $3\pi/5$ (~1.89) and $4\pi/5$ (~2.51). The effectiveness of the signal blocking matrix $\mathbf{C}_n$ at blocking the signal is evident from the plot of the gain of each column in the signal direction sine at five evenly spaced frequencies in the interval $3\pi/5 \leqslant \omega \leqslant 4\pi/5$ given in Fig. 4-11. All columns of $\mathbf{C}_n$ attenuate the signal by greater than 100 dB; the greatest attenuation is more than 300 dB.

The interference cancellation capability of the LCMV beamformer is illustrated in Fig. 4-12. The beamformer gain in dB is plotted as a function of direction sine for five frequencies in the interval $3\pi/5 \leqslant \omega \leqslant 4\pi/5$

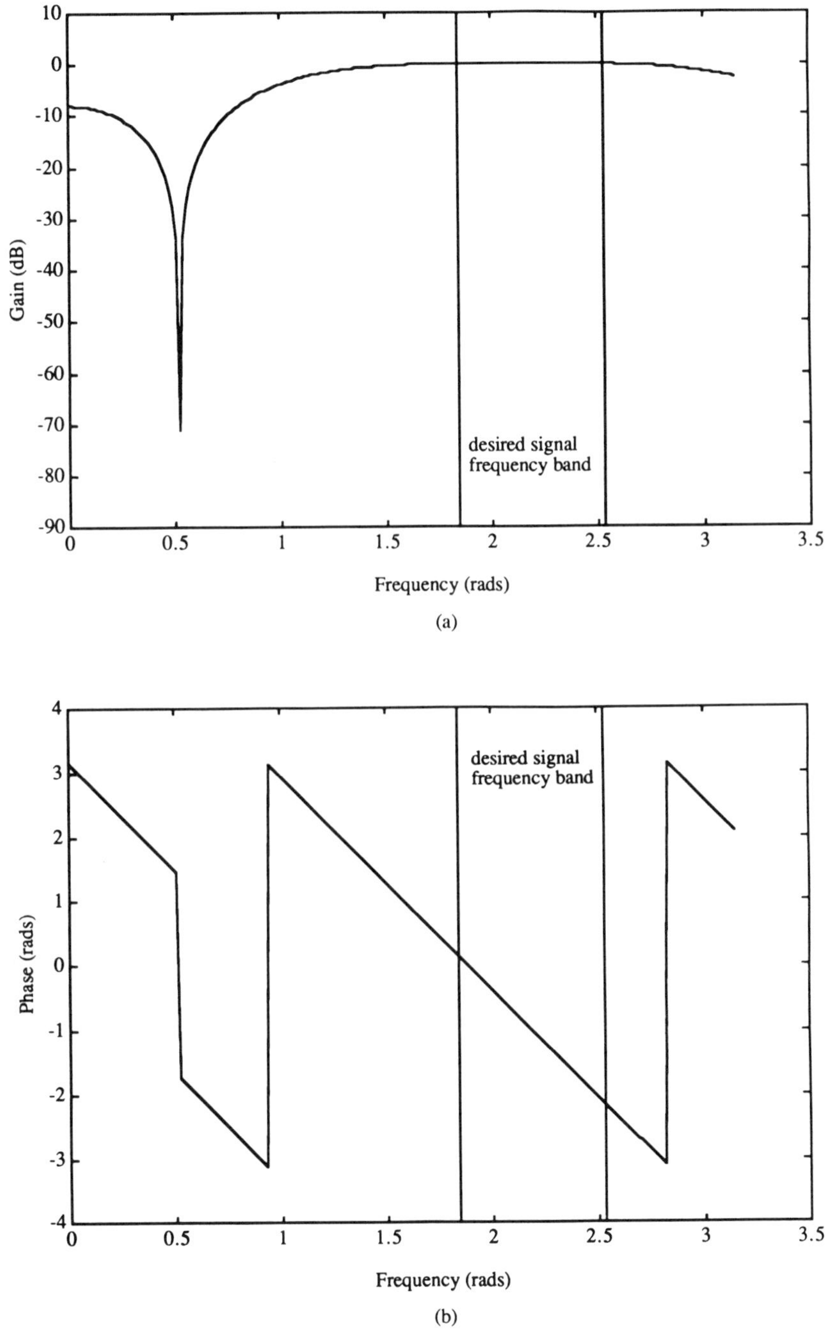

Figure 4-10 (a) Magnitude response of $\mathbf{w}_q$ as a function of frequency in the desired signal direction, direction sine = 0.1. (b) Phase response of $\mathbf{w}_q$ as a function of frequency in the desired signal direction, direction sine = 0.1.

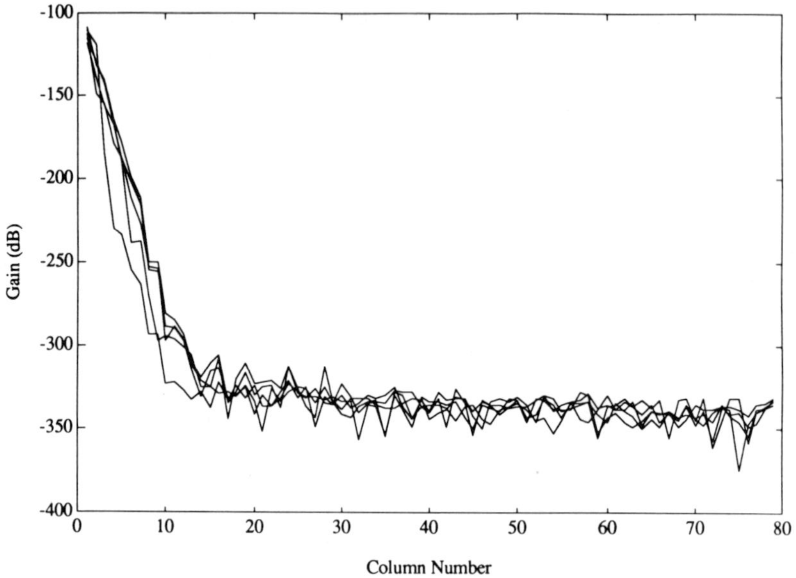

Figure 4-11 Gain of each column of $\mathbf{C}_n$ in the desired signal direction at five frequencies on $3\pi/5 \leqslant \omega \leqslant 4\pi/5$.

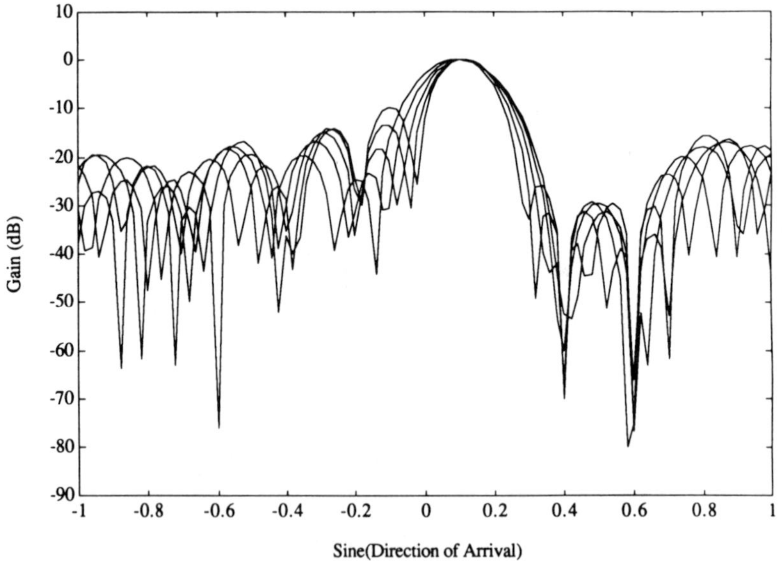

Figure 4-12 LCMV beamformer magnitude response as a function of direction sine at five frequencies on $3\pi/5 \leqslant \omega \leqslant 4\pi/5$. Interference environment consists of two interferers in white noise. The interferer powers are 30 dB and 40 dB relative to the white noise and arrive at direction sines of 0.4 and 0.6 respectively.

assuming the environment consists of two interferers and uncorrelated noise. The interferers are located at direction sines of 0.4 and 0.6 with powers of 30 dB and 40 dB relative to the white noise level, respectively. The beamformer maintains unit gain at the desired signal direction sine while severely attenuating the interferers. The weights are chosen to minimize output power subject to the signal preservation constraint so the stronger interferer receives greater attenuation than the weaker interferer. In this example the array gain is 33.7 dB. The array gain associated with $\mathbf{w}_q$ is 2.9 dB. Thus, LCMV beamforming provides almost 31 dB in additional noise reduction.

As a second example of interference cancellation, the LCMV beamformer cancellation capabilities are evaluated in an environment consisting of five interferers: a 40 dB interferer with raised cosine power spectral density, $1+\cos(10\omega - 7\pi/10)$ on $3\pi/5 \leqslant |\omega| \leqslant 4\pi/5$ at direction sine 0.4, a 30 dB interferer with constant power spectral density on $3\pi/5 \leqslant |\omega| \leqslant 4\pi/5$ at direction sine 0.85, a 40 dB interferer with constant power spectral density on $7\pi/10 \leqslant |\omega| \leqslant 4\pi/5$ at direction sine -0.2, a 50 dB interferer with constant power spectral density on $79\pi/100 \leqslant |\omega| \leqslant 4\pi/5$ at direction sine 0.7, and a 40 dB interferer with constant power spectral density on $7\pi/10 \leqslant |\omega| \leqslant 71\pi/100$ at direction sine -0.15 in uncorrelated noise. The interferer power levels are given relative to the uncorrelated noise level. The LCMV beamformer array gain is computed to be 38.34 dB. The array gain associated with $\mathbf{w}_q$ is 3.36 dB so LCMV beamforming provides almost 35 dB in additional noise reduction.

4-5 QUADRATICALLY CONSTRAINED BEAMFORMING

Quadratic constraints on the weight vector can also be used to control beamformer response. They are capable of providing response control which cannot be achieved using linear constraints, but in general are more difficult to implement. Quadratic constraints have been suggested for controlling the beamformer's response to the desired signal [15], making the beamformer robust with respect to uncertainties in source direction, antenna location and response errors [11, 27], and improving the signal to noise ratio [47]. In this section we discuss quadratically constrained minimum variance criteria for choosing the beamformer weights and review several types of quadratic constraints.

The eigenvector constraints of Section 4-4 were derived by finding a set of linear constraints on $\mathbf{w}$ that minimized the mean squared response error (4-21). A set of quadratic constraints on $\mathbf{w}$ is obtained by directly constraining the mean squared error. That is, using (4-22), we require

$$\mathbf{w}^H\mathbf{Q}\mathbf{w} - \mathbf{w}^H\mathbf{p} - \mathbf{p}^H\mathbf{w} + \sigma_r^2 \leqslant e_m \tag{4-55}$$

where e_m is the maximum tolerable mean squared response error. It is convenient to express (4-55) in the form

$$(\mathbf{w} - \mathbf{Q}^{-1}\mathbf{p})^H \mathbf{Q}(\mathbf{w} - \mathbf{Q}^{-1}\mathbf{p}) \leq e_o \tag{4-56}$$

where $e_o = e_m - \sigma_r^2 + \mathbf{p}^H\mathbf{Q}^{-1}\mathbf{p}$. Provided Θ or Ω in (4-23a) define a region of nonzero width, $\mathbf{Q}$ is guaranteed to be nonsingular, although it may be so ill-conditioned that it appears numerically singular. The quadratic constraint (4-56) is more general than the eigenvector constraints of Section 4-4 since it directly controls the mean squared error and does not rely on some of the eigenvalues of $\mathbf{Q}$ being very small. The cost of this generality is a considerable increase in the computational complexity associated with the determination of $\mathbf{w}$. Quadratic constraints of this type are discussed in [15, 47].

Another common use of quadratic constraints is to decrease the sensitivity of the beamformer to signal mismatch, that is, differences between the true and assumed signal characteristics. Examples of errors which result in mismatch include gain and phase errors in the antennas and receiver electronics, differences between the actual and assumed signal direction, and errors in the assumed sensor locations. Mismatch usually results in cancellation of some or all of the signal because the beamformer treats the signal as interference. For example, if the beamformer is constrained to pass signals from one direction and the signal of interest arrives from a different direction, then the signal appears as an interferer.

The gain of the array to uncorrelated noise is a commonly used measure of the robustness of the beamformer to errors that are uncorrelated from sensor to sensor [11]. Thus, the sensitivity of the beamformer to mismatch is controlled by constraining the white noise gain

$$\mathbf{w}^H\mathbf{w} \leq e_0 \tag{4-57}$$

This quadratic constraint is usually used in conjunction with other constraints. A thorough bibliography on quadratic constraints is given in [11]. Hudson [27] contains a detailed discussion of constraints on the norm of $\mathbf{w}$.

A general form for quadratic constraints is $(\mathbf{w} - \mathbf{w}_o)^H \mathbf{Q}_o(\mathbf{w} - \mathbf{w}_o) \leq e_o$ where $\mathbf{Q}_o$ is a positive semidefinite Hermitian matrix, $\mathbf{w}_o$ is a known vector, and e_o is a nonnegative constant. For example, in the quadratic constraint of (4-56), $\mathbf{Q}_o$ and $\mathbf{w}_o$ correspond to $\mathbf{Q}$ and $\mathbf{Q}^{-1}\mathbf{p}$, respectively. In the absence of other constraints the quadratically constrained minimum variance problem is

$$\min_{\mathbf{w}} \mathbf{w}^H\mathbf{R}_x\mathbf{w} \quad \text{subject to} \quad (\mathbf{w} - \mathbf{w}_o)^H\mathbf{Q}_o(\mathbf{w} - \mathbf{w}_o) \leq e_o \tag{4-58}$$

The solution to (4-58) is obtained via Lagrange multiplier techniques and is given by

$$\mathbf{w} = (\mathbf{R}_x + \lambda\mathbf{Q}_o)^{-1}\lambda\mathbf{Q}_o\mathbf{w}_o \tag{4-59}$$

where λ is the Lagrange multiplier. λ must be chosen so that $\mathbf{w}$ satisfies the constraint $(\mathbf{w}-\mathbf{w}_o)^H\mathbf{Q}_o(\mathbf{w}-\mathbf{w}_o) \leq e_o$. It can be shown that λ and $(\mathbf{w}-\mathbf{w}_o)^H\mathbf{Q}_o(\mathbf{w}-\mathbf{w}_o)$ are inversely related so the constraint is satisfied with any λ greater than the one for which $(\mathbf{w}-\mathbf{w}_o)^H\mathbf{Q}_o(\mathbf{w}-\mathbf{w}_o) = e_o$. However, $\mathbf{w}^H\mathbf{R}_x\mathbf{w}$ is an increasing function of λ so simultaneous satisfaction of the constraint and minimization of $\mathbf{w}^H\mathbf{R}_x\mathbf{w}$ implies we choose λ such that $(\mathbf{w}-\mathbf{w}_o)^H\mathbf{Q}_o(\mathbf{w}-\mathbf{w}_o) = e_o$, or

$$\mathbf{w}_o^H[(\mathbf{R}_x+\lambda\mathbf{Q}_o)^{-1}\lambda\mathbf{Q}_o-\mathbf{I}]^H\mathbf{Q}_o[(\mathbf{R}_x+\lambda\mathbf{Q}_o)^{-1}\lambda\mathbf{Q}_o-\mathbf{I}]\mathbf{w}_o = e_o \tag{4-60}$$

Equation (4-60) can be simplified considerably by simultaneously diagonalizing $\mathbf{R}_x$ and $\mathbf{Q}_o$; nevertheless λ cannot be directly expressed as a function of e_o and must be solved for using an iterative algorithm, resulting in additional computational burden. The one to one correspondence between e_o and λ suggests that this additional computational burden could be avoided by choosing a value for λ and tolerating the resulting value for e_o. e_o is dependent on $\mathbf{R}_x$ if λ is held fixed, so this approach is only satisfactory in situations where variation in e_o for different signal/interference scenarios is acceptable.

In some problems both linear and quadratic constraints are utilized (see e.g., [11]). That is, $\mathbf{w}$ is chosen to satisfy

$$\min_{\mathbf{w}} \mathbf{w}^H\mathbf{R}_x\mathbf{w} \quad \text{subject to} \quad \mathbf{C}^H\mathbf{w}=\mathbf{f} \text{ and } (\mathbf{w}-\mathbf{w}_o)^H\mathbf{Q}_o(\mathbf{w}-\mathbf{w}_o) \leq e_o \tag{4-61}$$

It is possible that for sufficiently small values of e_o there is no $\mathbf{w}$ which satisfies both constraints. This occurs when the linear variety defined by $\mathbf{C}^H\mathbf{w}=\mathbf{f}$ does not intersect the ellipsoidal volume defined by $(\mathbf{w}-\mathbf{w}_o)^H\mathbf{Q}_o(\mathbf{w}-\mathbf{w}_o) \leq e_o$. The linear constraint is eliminated using the GSC decomposition $\mathbf{w}=\mathbf{w}_q-\mathbf{C}_n\mathbf{w}_n$ to rewrite (4-61) as

$$\min_{\mathbf{w}_n}(\mathbf{w}_q-\mathbf{C}_n\mathbf{w}_n)^H\mathbf{R}_x(\mathbf{w}_q-\mathbf{C}_n\mathbf{w}_n) \quad \text{subject to}$$

$$(\mathbf{C}_n\mathbf{w}_n-\mathbf{w}_{oq})^H\mathbf{Q}_o(\mathbf{C}_n\mathbf{w}_n-\mathbf{w}_{oq}) \leq e_o \tag{4-62}$$

where $\mathbf{w}_{oq}=\mathbf{w}_q-\mathbf{w}_o$. Again using Lagrange multiplier techniques we find that the solution, as a function of the Lagrange multiplier λ, is expressed as

$$\mathbf{w}_n = [\mathbf{C}_n^H(\mathbf{R}_x+\lambda\mathbf{Q}_o)\mathbf{C}_n]^{-1}\mathbf{C}_n^H(\mathbf{R}_x\mathbf{w}_q+\lambda\mathbf{Q}_o\mathbf{w}_{oq}) \tag{4-63}$$

λ is chosen as the smallest positive number which satisfies the constraint in (4-62) after $\mathbf{w}_n$ is replaced by (4-63).

Adaptive algorithms which incorporate quadratic constraints are given in [11, 12, 13, 37, 47].

4-6 PARTIALLY ADAPTIVE BEAMFORMING CONCEPTS

The purpose of this section is to provide qualitative and quantitative definitions of partially adaptive beamformers and examine their basic properties. Thus far in this chapter we have defined different types of minimum variance beamformers: multiple sidelobe cancellers, LCMV beamformers, and quadratically constrained minimum variance beamformers. A partially adaptive beamformer is the most general form of the LCMV beamformer and is introduced now so that we can make use of it in the performance analyses contained in the following sections. These analyses also provide strong motivation for the use of partially adaptive beamformers.

In a partially adaptive beamformer only a portion of the available degrees of freedom are used adaptively, that is, allowed to change in response to the received data. The principal benefits associated with reducing the number of adaptive degrees of freedom are reduced computational burden and improved adaptive convergence rate. As we shall see in Section 4-8, the computational cost of adaptive algorithms is generally either directly proportional to the number of adaptive weights, or proportional to the square or cube of the number of adaptive weights. It is often mandatory that the number of adaptive weights be reduced with large antenna arrays because of the adaptive algorithm's computational requirements. Adaptive convergence is discussed in Section 4-9, where we show that the number of data vectors needed for the weights to converge to their optimal values is also proportional to the number of adaptive weights. Thus, in some applications adaptive response requirements dictate reductions in the number of adaptive weights. The primary disadvantage of reducing the number of adaptive weights is a degradation in the beamformer's steady state interference cancellation capability. This degradation is a function of which adaptive degrees of freedom are utilized and is the motivation for the partially adaptive beamformer design problem discussed in Section 4-11.

The GSC decomposition of the beamformer's weight vector $\mathbf{w}$ is $\mathbf{w} = \mathbf{w}_q - \mathbf{C}_n\mathbf{w}_n$. Recall that $\mathbf{C}_n$ is an N by $N - K$ dimensioned full rank matrix whose columns are orthogonal to the columns of the constraint matrix $\mathbf{C}$. The $N - K$ dimensional weight vector $\mathbf{w}_n$ embodies the available degrees of freedom because it is independent of the constraint $\mathbf{C}^H\mathbf{w} = \mathbf{f}$. The number of adaptive weights is reduced by forcing $\mathbf{w}_n$ to adapt in an M $(M < N - K)$ dimensional subspace of the $N - K$ dimensional space spanned by the columns of $\mathbf{C}_n$. Formally, we restrict $\mathbf{w}_n$ to lie in the M dimensional space spanned by the columns of an $N - K$ by M dimensioned matrix $\mathbf{T}$. That is, let $\mathbf{w}_n = \mathbf{T}\mathbf{w}_a$ where $\mathbf{w}_a$ is an M dimensional vector that represents the available M degrees of freedom. Now $\mathbf{w} = \mathbf{w}_q - \mathbf{C}_n\mathbf{T}\mathbf{w}_a$, or to simplify notation somewhat, $\mathbf{w} = \mathbf{w}_q - \mathbf{T}_a\mathbf{w}_a$ where $\mathbf{T}_a = \mathbf{C}_n\mathbf{T}$, is an N by M dimensioned matrix. A block diagram illustrating this decomposition is given in Fig. 4-13.

The identical notation was purposely used at the end of Section 4-4 to

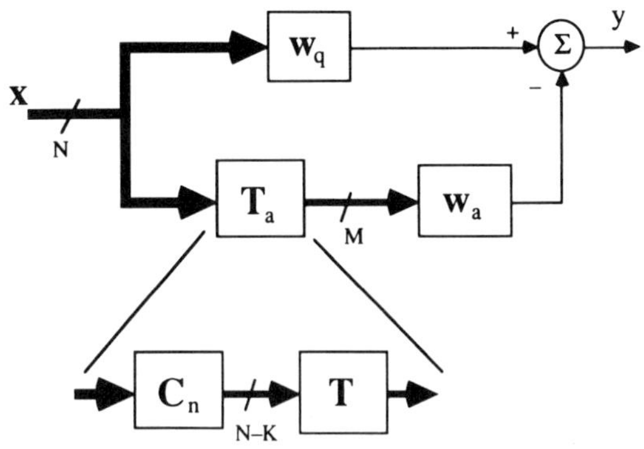

Figure 4-13 Block diagram of a partially adaptive GSC.

represent the MSC ($\mathbf{T}_a$ selected the auxiliary channel data and $\mathbf{w}_a$ represented the adaptive weights) because the MSC is a special case of a partially adaptive beamformer. This notation can also represent beamformers that operate on the output of a beamforming preprocessing network. If a preprocessing network is used, $\mathbf{w}_q$ represents the beam directed to the signal while $\mathbf{T}_a$ represents the beams that are used adaptively. As a result of the generality of this representation the analysis in this chapter has broad applicability. It is limited only in cases where specific properties of the GSC are invoked, such as $\mathbf{w}_q^H \mathbf{T}_a = \mathbf{0}$, that may not apply to more general representations.

Reducing the number of adaptive weights is equivalent to increasing the number of constraints on $\mathbf{w}$. Define an N by $N - K - M$ dimensioned full rank matrix $\mathbf{V}$ whose columns are orthogonal to the space spanned by the columns of both $\mathbf{C}$ and $\mathbf{T}_a$, that is $\mathbf{V}^H \mathbf{C} = \mathbf{0}$ and $\mathbf{V}^H \mathbf{T}_a = \mathbf{0}$. Noting that $\mathbf{V}^H \mathbf{w} = \mathbf{V}^H (\mathbf{w}_q - \mathbf{T}_a \mathbf{w}_a) = \mathbf{0}$, we see that the partially adaptive beamformer is equivalent to a fully adaptive beamformer with the augmented constraint set $\mathbf{C}_a^H \mathbf{w} = \mathbf{f}_a$ where $\mathbf{C}_a = [\mathbf{C} \quad \mathbf{V}]$ and $\mathbf{f}_a^H = [\mathbf{f}^H \quad \mathbf{0}]$.

An unconstrained minimization problem analogous to (4-44a) is solved to obtain $\mathbf{w}_a$. In this case

$$P_o = \min_{\mathbf{w}_a} (\mathbf{w}_q - \mathbf{T}_a \mathbf{w}_a)^H \mathbf{R}_x (\mathbf{w}_q - \mathbf{T}_a \mathbf{w}_a) \tag{4-64}$$

and

$$\mathbf{w}_a = (\mathbf{T}_a^H \mathbf{R}_x \mathbf{T}_a)^{-1} \mathbf{T}_a^H \mathbf{R}_x \mathbf{w}_q \tag{4-65}$$

The matrix to be inverted in (4-65) is of dimension M by M. The fully adaptive beamformer in (4-46) required an $N - K$ by $N - K$ matrix inverse. Similar to (4-47), the output power is

$$P_o = \mathbf{w}_q^H \mathbf{R}_x \mathbf{w}_q - \mathbf{w}_q^H \mathbf{R}_x \mathbf{T}_a (\mathbf{T}_a^H \mathbf{R}_x \mathbf{T}_a)^{-1} \mathbf{T}_a^H \mathbf{R}_x \mathbf{w}_q \tag{4-66}$$

Assuming that the signal and noise are uncorrelated, $\mathbf{R}_x = \mathbf{R}_s + \mathbf{R}_n$, and $\mathbf{C}_n^H \mathbf{R}_s = \mathbf{0}$,

$$P_s = \mathbf{w}_q^H \mathbf{R}_s \mathbf{w}_q \tag{4-67a}$$

$$P_n = \mathbf{w}_q^H \mathbf{R}_n \mathbf{w}_q - \mathbf{w}_q^H \mathbf{R}_n \mathbf{T}_a (\mathbf{T}_a^H \mathbf{R}_n \mathbf{T}_a)^{-1} \mathbf{T}_a^H \mathbf{R}_n \mathbf{w}_q \tag{4-67b}$$

Thus, the signal output power is independent of $\mathbf{T}_a$. $\mathbf{R}_n$ is Hermitian and positive definite so it has a square root decomposition $\mathbf{R}_n = \mathbf{R}_n^{H/2} \mathbf{R}_n^{1/2}$. Let $\mathbf{v} = \mathbf{R}_n^{1/2} \mathbf{w}_q$ and rewrite (4-67b) as

$$P_n = \mathbf{v}^H (\mathbf{I} - \mathbf{R}_n^{1/2} \mathbf{T}_a (\mathbf{T}_a^H \mathbf{R}_n^{H/2} \mathbf{R}_n^{1/2} \mathbf{T}_a)^{-1} \mathbf{T}_a^H \mathbf{R}_n^{H/2}) \mathbf{v} \tag{4-68}$$

This shows that P_n is the norm squared of the projection of $\mathbf{v}$ onto the orthogonal complement of the space spanned by the columns of $\mathbf{R}_n^{1/2} \mathbf{T}_a$. In the fully adaptive case P_n is the norm squared of the projection of $\mathbf{v}$ onto the orthogonal complement of the space spanned by the columns of $\mathbf{R}_n^{1/2} \mathbf{C}_n$. The space spanned by the columns of $\mathbf{R}_n^{1/2} \mathbf{T}_a$ is a subspace of the space spanned by the columns of $\mathbf{R}_n^{1/2} \mathbf{C}_n$; hence, the partially adaptive noise output power must be greater than or at best equal to the fully adaptive noise output power. The goal of partially adaptive beamformer design is to choose $\mathbf{T}_a$ to minimize the increase in output noise power.

4-7 INTERFERENCE CANCELLATION ANALYSIS

In this section we study the interference cancellation capabilities of both narrowband and broadband LCMV beamformers assuming the data covariance matrix is known. The analysis also assumes the desired signal and interference are uncorrelated. We begin with an intuitive discussion of the principles of interference cancellation and then proceed to derive exact expressions. These results illustrate the fundamental difference between narrowband and broadband interference cancellation and provide insight into the partially adaptive beamformer design problem. The results are developed assuming the most general form for the interference and noise environment. Common special cases are then examined utilizing the general results. The analysis given in this section is an extension of that given in [45].

4-7-1 Response Matching

The interference and noise power at the beamformer output is $P_n = \mathbf{w}^H \mathbf{R}_n \mathbf{w}$ where $\mathbf{R}_n$ is the covariance matrix associated with the noise and interference. Suppose the interference is represented by power spectral density

$S(\theta, \omega)$ over the range of directions $\theta \in \Theta$ and frequencies $\omega \in \Omega$ and the noise is white with power σ_n^2. Mathematically we have

$$\mathbf{R}_n = \frac{1}{2\pi} \int_{\theta \in \Theta} \int_{\omega \in \Omega} S(\theta, \omega)\mathbf{d}(\theta, \omega)\mathbf{d}^H(\theta, \omega)\, d\omega\, d\theta + \sigma_n^2 \mathbf{I}$$
$$= \mathbf{R}_I + \sigma_n^2 \mathbf{I} \tag{4-69}$$

The interference often consists of several (d) interfering point sources arriving from distinct directions so that

$$S(\theta, \omega) = \sum_{i=1}^{d} S(\theta_i, \omega)\delta(\theta - \theta_i) \tag{4-70a}$$

Here $S(\theta_i, \omega)$ is the power spectral density of the source from direction θ_i. If the interference consists of d narrowband point sources, then

$$S(\theta, \omega) = \sum_{i=1}^{d} 2\pi S(\theta_i, \omega_o)\delta(\theta - \theta_i)\delta(\omega - \omega_o) \tag{4-70b}$$

The factor 2π permits the broadband analysis given below to be applied to the narrowband case by simply replacing integrals by sums. Both of the cases represented by (4-70a) and (4-70b) are considered later in this section. The more general case represented in (4-69) allows representation of interferers that are distributed in space as well as isotropic background noise.

Substitute the GSC representation for $\mathbf{w}$ and the representation (4-69) for $\mathbf{R}_n$ into the expression $P_n = \mathbf{w}^H\mathbf{R}_n\mathbf{w}$ to obtain

$$P_n = \frac{1}{2\pi} \int_{\theta \in \Theta} \int_{\omega \in \Omega} S(\theta, \omega)|\mathbf{w}_q^H\mathbf{d}(\theta, \omega) - \mathbf{w}_a^H\mathbf{T}_a^H\mathbf{d}(\theta, \omega)|^2\, d\omega\, d\theta$$
$$+ \sigma_n^2\mathbf{w}_q^H\mathbf{w}_q + \sigma_n^2\mathbf{w}_a^H\mathbf{w}_a \tag{4-71}$$

In obtaining (4-71) we have used the orthogonality of $\mathbf{w}_q$ and $\mathbf{T}_a$ ($\mathbf{w}_q^H\mathbf{T}_a = \mathbf{0}$) and assumed $\mathbf{T}_a^H\mathbf{T}_a = \mathbf{I}$. $\mathbf{T}_a$ can be assumed to have orthonormal columns because the GSC is invariant to nonsingular transformations applied to $\mathbf{T}_a$ on the right. This invariance is established for the fully adaptive GSC in Section 4-4, it also holds for a partially adaptive GSC. It is the space spanned by the columns of $\mathbf{T}_a$ that is important, not the individual columns.

Define $q(\theta, \omega) = \mathbf{w}_q^H\mathbf{d}(\theta, \omega)$ as the response associated with $\mathbf{w}_q$ and $\mathbf{a}(\theta, \omega) = \mathbf{T}_a^H\mathbf{d}(\theta, \omega)$ as a column vector whose elements represent the responses associated with the columns of $\mathbf{T}_a$. P_n is now expressed as

$$P_n = \frac{1}{2\pi} \int_{\theta \in \Theta} \int_{\omega \in \Omega} S(\theta, \omega)|q(\theta, \omega) - \mathbf{w}_a^H\mathbf{a}(\theta, \omega)|^2\, d\omega\, d\theta$$
$$+ \sigma_n^2\mathbf{w}_q^H\mathbf{w}_q + \sigma_n^2\mathbf{w}_a^H\mathbf{w}_a \tag{4-72}$$

$\mathbf{w}_a^H \mathbf{a}(\theta, \omega)$, the response of the GSCs adaptive branch, is a linear combination of responses of the columns of $\mathbf{T}_a$. The interference output power depends on how well the response of the adaptive branch matches the response of the nonadaptive branch over Θ and Ω. $S(\theta, \omega)$ serves as a weighting function to emphasize frequencies and directions having large interference energy over those having weak (or zero) interference energy. The contribution of the white noise to the output is dependent on the norm squared of $\mathbf{w}_q$ and $\mathbf{w}_a$.

Choosing $\mathbf{w}_a$ to minimize P_n is equivalent to choosing $\mathbf{w}_a$ to minimize the weighted squared error between the adaptive and nonadaptive branch responses over Ω and Θ subject to a norm constraint. That is, (4-64) corresponds to

$$\min_{\mathbf{w}_a} \frac{1}{2\pi} \int_{\theta \in \Theta} \int_{\omega \in \Omega} S(\theta, \omega) | q(\theta, \omega) - \mathbf{w}_a^H \mathbf{a}(\theta, \omega)|^2 \, d\omega \, d\theta$$
$$\text{subject to} \quad \mathbf{w}_a^H \mathbf{w}_a \leqslant \delta \tag{4-73}$$

If we solve (4-73) using the method of Lagrange multipliers, we find that the Lagrange multiplier corresponds to σ_n^2 and that δ and σ_n^2 are inversely related. Of course, this is also evident from (4-72)—the importance of the norm of $\mathbf{w}_a$ increases as σ_n^2 increases. As σ_n^2 increases, δ is effectively reduced, which generally increases the value of the weighted squared error in (4-73). In the absence of a norm constraint, (4-73) becomes a standard least squares problems and the minimum value of the weighted squared error is determined by the projection of $q(\theta, \omega)$ onto the orthogonal complement of the space spanned by $\mathbf{a}(\theta, \omega)$ for $\omega \in \Omega$, $\theta \in \Theta$. This is shown explicitly later in the section.

4-7-2 Response Matching Examples

In order to fix these ideas, consider the case in which d narrowband interferers are of frequency ω_0 and are located at distinct directions θ_i, $i = 1, 2, \ldots, d$. The interference/noise covariance matrix $\mathbf{R}_n$ is now

$$\mathbf{R}_n = \sum_{i=1}^{d} S(\theta_i, \omega_o)\mathbf{d}(\theta_i, \omega_o)\mathbf{d}^H(\theta_i, \omega_o) + \sigma_n^2 \mathbf{I} \tag{4-74}$$

and P_n is expressed as

$$P_n = \sigma_n^2 \mathbf{w}_q^H \mathbf{w}_q + \|\mathbf{q}^H - \mathbf{w}_a^H \mathbf{A}\|^2 + \sigma_n^2 \mathbf{w}_a^H \mathbf{w}_a \tag{4-75}$$

where

$$\mathbf{q}^H = [S^{1/2}(\theta_1, \omega_o) q(\theta_1, \omega_o) \quad S^{1/2}(\theta_2, \omega_o) q(\theta_2, \omega_o) \quad \cdots \quad S^{1/2}(\theta_d, \omega_o) \times q(\theta_d, \omega_o)] \tag{4-76a}$$

$$\mathbf{A} = [S^{1/2}(\theta_1, \omega_o)\mathbf{a}(\theta_1, \omega_o) \quad S^{1/2}(\theta_2, \omega_o)\mathbf{a}(\theta_2, \omega_o) \quad \cdots \quad S^{1/2}(\theta_d, \omega_o) \times \mathbf{a}(\theta_d, \omega_o)] \tag{4-76b}$$

If we choose $\mathbf{w}_a$ so that $\mathbf{w}_a^H\mathbf{A} = \mathbf{q}^H$, then the output power associated with the interferers is zero. The equation $\mathbf{w}_a^H\mathbf{A} = \mathbf{q}^H$ represents d linear equations in M unknowns. These equations are linearly independent for almost any $\mathbf{T}_a$ if $\mathbf{d}(\theta_i, \omega_o)$, $i = 1, 2, \ldots, d$ are linearly independent.* Thus, $\mathbf{w}_a$ can be chosen to satisfy $\mathbf{w}_a^H\mathbf{A} = \mathbf{q}^H$ provided $M \geqslant d$. This is why it is often said that a beamformer requires one adaptive degree of freedom per interferer to achieve perfect cancellation. Note that this reasoning assumes the norm of $\mathbf{w}_a$ is effectively unconstrained, a valid assumption provided that the interferer powers are significantly greater than σ_n^2. If σ_n^2 is significant relative to the interferer powers, then the norm of $\mathbf{w}_a$ is constrained and the interferers will not be completely cancelled in general. Furthermore, the level of cancellation is dependent on $\mathbf{T}_a$ if σ_n^2 is significant, since $\mathbf{T}_a$ affects the norm of $\mathbf{w}_a$. Perfect cancellation of interferers requires response matching. The response of the adaptive branch, $\mathbf{w}_a^H\mathbf{a}(\theta, \omega)$, must match the response of the quiescent branch, $q(\theta, \omega)$, at frequency ω_0 in the interferer directions θ_i, $i = 1, 2, \ldots, d$.

Perfect cancellation of broadband interferers cannot usually be achieved even if σ_n^2 is zero. The response matching criterion for the ith interferer is $q(\theta_i, \omega) = \mathbf{w}_a^H\mathbf{a}(\theta_i, \omega)$, $\omega \in \Omega$. This equation cannot generally be satisfied for all ω. It can be solved in a least squares sense, or at a set of M or fewer distinct frequencies. Here the level of cancellation and number of adaptive degrees of freedom required for good cancellation are greatly influenced by $\mathbf{T}_a$, since the elements of $\mathbf{a}(\theta_i, \omega)$ represent the response of the columns of $\mathbf{T}_a$ as a function of frequency in direction θ_i. If the function $q(\theta_i, \omega)$ lies very close to the space spanned by the elements of the function $\mathbf{a}(\theta_i, \omega)$, $\omega \in \Omega$, then the adaptive branch can approximately match the response of $\mathbf{w}_q$ with the proper choice for $\mathbf{w}_a$ and good cancellation is obtained, even if M is one. On the contrary, if $q(\theta_i, \omega)$ is far from the space spanned by the elements of $\mathbf{a}(\theta_i, \omega)$, $\omega \in \Omega$, then cancellation performance will be poor, even for large M.

4-7-3 A Quantitative Analysis of Interference Cancellation

Thus far in this section we have studied P_n as a function of $\mathbf{w}_a$ to develop an intuitive feel for the interference cancellation process. We now substitute the optimum solution for $\mathbf{w}_a$ to obtain exact expressions by starting with the equations for $\mathbf{w}_a$ and P_n given in (4-65) and (4-67b). Expressing the individual terms in (4-65) and (4-67b) in the notation of this section gives

* The $\mathbf{d}(\theta_i, \omega_o)$, $i = 1, 2, \ldots, d$ will be linearly independent if the antenna locations are chosen to prevent aliasing. This implies that the $\mathbf{a}(\theta_i, \omega_o)$, $i = 1, 2, \ldots, d$ are almost always linearly independent. Exceptions include cases where $\mathbf{T}^H\mathbf{d}(\theta_i, \omega_o) = 0$.

$$\mathbf{T}_a^H \mathbf{R}_n \mathbf{T}_a = \frac{1}{2\pi} \int_{\theta \in \Theta} \int_{\omega \in \Omega} S(\theta, \omega)\mathbf{a}(\theta, \omega)\mathbf{a}^H(\theta, \omega)\, d\omega\, d\theta + \sigma_n^2 \mathbf{I}$$
$$= \mathbf{R}_a + \sigma_n^2 \mathbf{I} \tag{4-77}$$

$$\mathbf{w}_q^H \mathbf{R}_n \mathbf{w}_q = \frac{1}{2\pi} \int_{\theta \in \Theta} \int_{\omega \in \Omega} S(\theta, \omega)|q(\theta, \omega)|^2\, d\omega\, d\theta + \sigma_n^2 \mathbf{w}_q^H \mathbf{w}_q$$
$$= \mathbf{w}_q^H \mathbf{R}_I \mathbf{w}_q + \sigma_n^2 \mathbf{w}_q^H \mathbf{w}_q \tag{4-78}$$

$$\mathbf{w}_q^H \mathbf{R}_n \mathbf{T}_a = \frac{1}{2\pi} \int_{\theta \in \Theta} \int_{\omega \in \Omega} S(\theta, \omega) q(\theta, \omega)\mathbf{a}^H(\theta, \omega)\, d\omega\, d\theta \tag{4-79}$$

where $\mathbf{R}_I$ is implicitly defined in (4-69). In obtaining (4-77) and (4-79) we have used $\mathbf{T}_a^H \mathbf{T}_a = \mathbf{I}$ and $\mathbf{w}_q^H \mathbf{T}_a = \mathbf{0}$. $\mathbf{R}_a$ represents the covariance matrix of the interference at the output of $\mathbf{T}_a$. Let $\mathbf{R}_a$ have eigendecomposition $\mathbf{V\Lambda V}^H$ where the columns of $\mathbf{V}$ are orthonormal eigenvectors and $\mathbf{\Lambda}$ is a diagonal matrix with eigenvalues λ_i satisfying $\lambda_1 \geq \lambda_2 \geq \cdots \geq \lambda_p > \lambda_{p+1} \approx \cdots \approx \lambda_M \approx 0$ as diagonal elements. Here p is the effective rank of $\mathbf{R}_a$. Note that $\mathbf{R}_a$ is mathematically full rank if the region defined by either Θ or Ω has nonzero width. The value p is defined such that λ_i, $i = p+1, p+2, \ldots, M$ are very small compared to σ_n^2 and the trace of $\mathbf{R}_a$. The following analysis assumes $\lambda_i = 0$, $i = p+1, p+2, \ldots, M$.

Let $\mathbf{V}_p$ be an M by p matrix composed of the first p columns of $\mathbf{V}$ and $\mathbf{\Lambda}_p$ a p by p diagonal matrix with the p nonzero eigenvalues of $\mathbf{R}_a$ as diagonal elements. Thus, $\mathbf{R}_a = \mathbf{V}_p \mathbf{\Lambda}_p \mathbf{V}_p^H$. The columns of $\mathbf{V}_p$ are an orthonormal basis for the space spanned by the columns of $\mathbf{R}_a$ and $\mathbf{R}_a$ is a sum (integral) of outer products involving $S^{1/2}(\theta, \omega)\mathbf{a}(\theta, \omega)$, $\theta \in \Theta$, $\omega \in \Omega$. This implies that $S^{1/2}(\theta, \omega)\mathbf{a}(\theta, \omega)$, $\theta \in \Theta$, $\omega \in \Omega$ lies in the space spanned by the columns of $\mathbf{V}_p$ and that p dimensional vectors $\boldsymbol{\alpha}(\theta, \omega)$, satisfying

$$S^{1/2}(\theta, \omega)\mathbf{a}(\theta, \omega) = \mathbf{V}_p \mathbf{\Lambda}_p^{1/2} \boldsymbol{\alpha}(\theta, \omega) \tag{4-80a}$$

can be defined. That is,

$$\boldsymbol{\alpha}(\theta, \omega) = S^{1/2}(\theta, \omega)\mathbf{\Lambda}_p^{-1/2} \mathbf{V}_p^H \mathbf{a}(\theta, \omega) \tag{4-80b}$$

Each of the elements of $S^{1/2}(\theta, \omega)\mathbf{a}(\theta, \omega)$ can be expressed as linear combination of the elements of $\boldsymbol{\alpha}(\theta, \omega)$ in (4-80a). Thus, the elements of $\boldsymbol{\alpha}(\theta, \omega)$ are a basis for the space spanned by the elements of $S^{1/2}(\theta, \omega)\mathbf{a}(\theta, \omega)$ over Θ and Ω. $\boldsymbol{\alpha}(\theta, \omega)$ represents an orthonormal basis since

$$
\begin{aligned}
&\frac{1}{2\pi}\int_{\theta\in\Theta}\int_{\omega\in\Omega}\boldsymbol{\alpha}(\theta,\omega)\boldsymbol{\alpha}^H(\theta,\omega)\,d\omega\,d\theta \\
&\quad=\boldsymbol{\Lambda}_p^{-1/2}\mathbf{V}_p^H\frac{1}{2\pi}\int_{\theta\in\Theta}\int_{\omega\in\Omega}S(\theta,\omega)\mathbf{a}(\theta,\omega)\mathbf{a}^H(\theta,\omega)\,d\omega\,d\theta\mathbf{V}_p\boldsymbol{\Lambda}_p^{-1/2} \\
&\quad=\boldsymbol{\Lambda}_p^{-1/2}\mathbf{V}_p^H\mathbf{R}_a\mathbf{V}_p\boldsymbol{\Lambda}_p^{-1/2} \\
&\quad=\mathbf{I}
\end{aligned}
\tag{4-81}
$$

The elements of $\mathbf{a}(\theta, \omega)$ represent the responses of the columns of $\mathbf{T}_a$; multiplication by $S^{1/2}(\theta, \omega)$ weights the response by the square root of the interference power spectral density. We say that $\boldsymbol{\alpha}(\theta, \omega)$ is an orthonormal basis for the interference weighted response of the columns of $\mathbf{T}_a$.

Substitute (4-80a) into (4-79) to obtain

$$
\mathbf{w}_q^H\mathbf{R}_n\mathbf{T}_a=\frac{1}{2\pi}\int_{\theta\in\Theta}\int_{\omega\in\Omega}S^{1/2}(\theta,\omega)q(\theta,\omega)\boldsymbol{\alpha}^H(\theta,\omega)\,d\omega\,d\theta\boldsymbol{\Lambda}_p^{1/2}\mathbf{V}_p^H
\tag{4-82}
$$

Define

$$
\mathbf{h}=\frac{1}{2\pi}\int_{\theta\in\Theta}\int_{\omega\in\Omega}S^{1/2}(\theta,\omega)q^*(\theta,\omega)\boldsymbol{\alpha}(\theta,\omega)\,d\omega\,d\theta
\tag{4-83}
$$

so that

$$
\mathbf{w}_q^H\mathbf{R}_n\mathbf{T}_a=\mathbf{h}^H\boldsymbol{\Lambda}_p^{1/2}\mathbf{V}_p^H
\tag{4-84}
$$

The elements of $\mathbf{h}$ are the inner product between the interference weighted response of $\mathbf{w}_q$, $S^{1/2}(\theta, \omega)q(\theta, \omega)$, and the orthonormal basis for the interference weighted response of the columns of $\mathbf{T}_a$, $\boldsymbol{\alpha}(\theta, \omega)$.

Now $\mathbf{R}_a$ has eigendecomposition $\mathbf{V}\boldsymbol{\Lambda}\mathbf{V}^H$, so $\mathbf{T}_a^H\mathbf{R}_n\mathbf{T}_a$ has eigendecomposition $\mathbf{V}(\boldsymbol{\Lambda} + \sigma_n^2\mathbf{I})\mathbf{V}^H$. This implies

$$
(\mathbf{T}_a^H\mathbf{R}_n\mathbf{T}_a)^{-1}=\mathbf{V}(\boldsymbol{\Lambda}+\sigma_n^2\mathbf{I})^{-1}\mathbf{V}^H
\tag{4-85}
$$

Substituting (4-84) and (4-85) into the expression for the adaptive weight vector $\mathbf{w}_a$, (4-65), gives

$$
\begin{aligned}
\mathbf{w}_a&=\mathbf{V}(\boldsymbol{\Lambda}+\sigma_n^2\mathbf{I})^{-1}\mathbf{V}^H\mathbf{V}_p\boldsymbol{\Lambda}_p^{1/2}\mathbf{h} \\
&=\mathbf{V}_p\boldsymbol{\Gamma}\mathbf{h}
\end{aligned}
\tag{4-86}
$$

where $\boldsymbol{\Gamma}$ is a p by p diagonal matrix having ith diagonal element $\lambda_i^{1/2}(\lambda_i + \sigma_n^2)^{-1}$. Note that $\mathbf{w}_a$ lies entirely in the p dimensional subspace spanned by the columns of $\mathbf{V}_p$. The columns of $\mathbf{V}_p$ are a basis for the space spanned by the columns of $\mathbf{R}_a$; hence $\mathbf{w}_a$ and the interference at the output of $\mathbf{T}_a$ lie in the same space.

Now consider expressing P_n (4-67b) using (4-78), (4-84), and (4-85)

$$P_n = \sigma_n^2 \mathbf{w}_q^H \mathbf{w}_q + \mathbf{w}_q^H \mathbf{R}_I \mathbf{w}_q - \mathbf{h}^H \mathbf{\Lambda}_p^{1/2} \mathbf{V}_p^H \mathbf{V}(\mathbf{\Lambda} + \sigma_n^2 \mathbf{I})^{-1} \mathbf{V}^H \mathbf{V}_p \mathbf{\Lambda}_p^{1/2} \mathbf{h}$$

$$= \sigma_n^2 \mathbf{w}_q^H \mathbf{w}_q + \mathbf{w}_q^H \mathbf{R}_I \mathbf{w}_q - \sum_{i=1}^{p} |[\mathbf{h}]_i|^2 \frac{\lambda_i}{\lambda_i + \sigma_n^2} \tag{4-87}$$

where $[\mathbf{h}]_i$ denotes the ith element of the vector $\mathbf{h}$. This expression is simplified somewhat using a geometric interpretation for $|[\mathbf{h}]_i|^2$. Define the cosine squared of the angle ϕ_i between the functions $S^{1/2}(\theta, \omega)q(\theta, \omega)$ and $[\boldsymbol{\alpha}(\theta, \omega)]_i$ as

$$\cos^2\phi_i =$$

$$\frac{\left| \frac{1}{2\pi} \int_{\theta \in \Theta} \int_{\omega \in \Omega} S^{1/2}(\theta, \omega) q^*(\theta, \omega) [\boldsymbol{\alpha}(\theta, \omega)]_i \, d\omega \, d\theta \right|^2}{\left(\frac{1}{2\pi} \int_{\theta \in \Theta} \int_{\omega \in \Omega} S(\theta, \omega) |q(\theta, \omega)|^2 \, d\omega \, d\theta \right) \left(\frac{1}{2\pi} \int_{\theta \in \Theta} \int_{\omega \in \Omega} |[\boldsymbol{\alpha}(\theta, \omega)]_i|^2 \, d\omega \, d\theta \right)}$$

$$= \frac{|[\mathbf{h}]_i|^2}{\mathbf{w}_q^H \mathbf{R}_I \mathbf{w}_q} \tag{4-88}$$

The second line in (4-88) follows from the orthonormality of the $[\boldsymbol{\alpha}(\theta, \omega)]_i$ derived in (4-81). The value $\cos^2\phi_i$ is the square of the fraction of the function $S^{1/2}(\theta, \omega)q(\theta, \omega)$ that lies in the space spanned by the function $[\boldsymbol{\alpha}(\theta, \omega)]_i$. Equation (4-88) implies that $|[\mathbf{h}]_i|^2 = \mathbf{w}_q^H \mathbf{R}_I \mathbf{w}_q \cos^2\phi_i$; substituting this expression into (4-87) yields

$$P_n = \sigma_n^2 \mathbf{w}_q^H \mathbf{w}_q + \gamma \mathbf{w}_q^H \mathbf{R}_I \mathbf{w}_q \tag{4-89}$$

where γ is given by

$$\gamma = \left(1 - \sum_{i=1}^{p} \frac{\lambda_i}{\lambda_i + \sigma_n^2} \cos^2\phi_i \right) \tag{4-90}$$

If the number of adaptive degrees of freedom, M, is zero, that is, we employ a nonadaptive beamformer with weight vector $\mathbf{w}_q$, then $P_n = \sigma_n^2 \mathbf{w}_q^H \mathbf{w}_q + \mathbf{w}_q^H \mathbf{R}_I \mathbf{w}_q$. Thus, γ represents the net effect of interference cancellation on the beamformer interference and noise output power. The value of γ is bounded above by unity because all the terms in the sum are nonnegative and bounded below by zero because

$$\frac{\lambda_i}{\lambda_i + \sigma_n^2} < 1$$

and

$$\sum_{i=1}^{p} \cos^2\phi_i \leq 1$$

Note that the contribution of the ith column of $\mathbf{V}$ to adaptive cancellation is dependent on both the size of λ_i relative to σ_n^2 and $\cos^2\phi_i$. Eigenvectors corresponding to large λ_i do not necessarily provide the largest contribution to adaptive cancellation.

It is tempting to conclude from (4-89) that $\gamma \mathbf{w}_q^H \mathbf{R}_I \mathbf{w}_q$ is the interference component of the output power. However, comparison with (4-72) indicates that it represents the sum of the interference power and an enhanced noise power because $\mathbf{w}_a$ is nonzero. Algebraic manipulation indicates that the components of P_n associated with the white noise, P_w, and interference, P_I are

$$P_w = \sigma_n^2 \left(\mathbf{w}_q^H \mathbf{w}_q + \mathbf{w}_q^H \mathbf{R}_I \mathbf{w}_q \sum_{i=1}^{P} \frac{\lambda_i}{(\lambda_i + \sigma_n^2)^2} \cos^2\phi_i \right) \tag{4-91}$$

$$P_I = \mathbf{w}_q^H \mathbf{R}_I \mathbf{w}_q \left(1 - \sum_{i=1}^{p} \frac{\lambda_i^2 + 2\lambda_i \sigma_n^2}{(\lambda_i + \sigma_n^2)^2} \cos^2\phi_i \right) \tag{4-92}$$

In terms of (4-72), P_w represents $\sigma_n^2 \mathbf{w}_q^H \mathbf{w}_q + \sigma_n^2 \mathbf{w}_a^H \mathbf{w}_a$ while P_I represents the average weighted squared error between the adaptive and nonadaptive branch responses over Θ and Ω.

The value of γ is only dependent on the size of the λ_i relative to σ_n^2, $\cos^2\phi_i$, and p. The eigenvalue λ_i is the power of the interference associated with the ith mode of $\mathbf{R}_a$, that is, the ith column of $\mathbf{V}$. If the interference is weak relative to the white noise, then all the λ_i are small relative to σ_n^2 and γ is approximately one, so there is no reduction in P_n due to adaptive processing. In this case $\mathbf{w}_a$ is approximately the zero vector. At the other extreme, suppose σ_n^2 approaches zero so that $\lambda_i \gg \sigma_n^2$, $i = 1, 2, \ldots, p$. Now γ is dependent only on $\cos^2\phi_i$, $i = 1, 2, \ldots, p$

$$\gamma \approx \left(1 - \sum_{i=1}^{p} \cos^2\phi_i \right) \tag{4-93}$$

and $P_n \approx P_I$ is expressed as

$$P_n \approx \mathbf{w}_q^H \mathbf{R}_I \mathbf{w}_q \left(1 - \sum_{i=1}^{p} \cos^2\phi_i \right) \tag{4-94}$$

The noise power P_n in (4-94) corresponds to the minimum average weighted response error in (4-72). Here $1 - \gamma$ represents the square of the total fraction of $S^{1/2}(\theta, \omega)q(\theta, \omega)$ that lies in the space spanned by the elements of $\boldsymbol{\alpha}(\theta, \omega)$. The elements of $\boldsymbol{\alpha}(\theta, \omega)$ are a basis for the interference weighted response of the columns of $\mathbf{T}_a$; hence, γ is a measure of how closely the response of the columns of $\mathbf{T}_a$ matches the response of $\mathbf{w}_q$ with $S^{1/2}(\theta, \omega)$ as a weighting factor.

4-7-4 Examples

Before concluding this section we apply this analysis to several special cases. Consider first the case in which the interferers are narrowband of frequency ω_0 and are located at distinct directions θ_i, $i = 1, 2, \ldots, d$ so that $\mathbf{R}_n$ is given by (4-74). The integrals in the above analysis are now replaced by sums. We have $\mathbf{R}_a = \mathbf{A}\mathbf{A}^H$ where $\mathbf{A}$ is defined in (4-76b). The rank of $\mathbf{R}_a$, p, is equal to the lessor of the number of adaptive weights, M, or interferers, d (this also assumes that the antenna locations are chosen to prevent aliasing and thus $\mathbf{A}$ is full rank). Define

$$\begin{aligned}\mathbf{B} &= \frac{1}{\sqrt{2\pi}}[\boldsymbol{\alpha}(\theta_1, \omega_o) \quad \boldsymbol{\alpha}(\theta_2, \omega_o) \quad \cdots \quad \boldsymbol{\alpha}(\theta_d, \omega_o)] \\ &= \boldsymbol{\Lambda}_p^{-1/2}\mathbf{V}_p^H\mathbf{A} \qquad (4\text{-}95\text{a})\end{aligned}$$

Note that the factor of 2π in (4-70b) implies that (4-80b) is given by

$$\boldsymbol{\alpha}(\theta_i, \omega_o) = (2\pi)^{1/2}S^{1/2}(\theta_i, \omega_o)\boldsymbol{\Lambda}_p^{-1/2}\mathbf{V}_p^H\mathbf{a}(\theta_i, \omega_o) \qquad (4\text{-}95\text{b})$$

Equation (4-81) implies that $\mathbf{B}\mathbf{B}^H = \mathbf{I}$. $\mathbf{h}$ defined in (4-83) is given by $\mathbf{h} = \mathbf{B}\mathbf{q}$ where $\mathbf{q}$ is defined in (4-76a). We also have $\mathbf{w}_q^H\mathbf{R}_I\mathbf{w}_q = \mathbf{q}^H\mathbf{q}$. Note that

$$\begin{aligned}\sum_{i=1}^{p} \cos^2\phi_i &= \frac{\mathbf{h}^H\mathbf{h}}{\mathbf{q}^H\mathbf{q}} \\ &= \frac{\mathbf{q}^H\mathbf{B}^H\mathbf{B}\mathbf{q}}{\mathbf{q}^H\mathbf{q}} \qquad (4\text{-}96)\end{aligned}$$

However, $\mathbf{B}^H\mathbf{B} = \mathbf{A}^H\mathbf{V}_p\boldsymbol{\Lambda}_p^{-1}\mathbf{V}_p^H\mathbf{A} = \mathbf{A}^H(\mathbf{A}\mathbf{A}^H)^{-1}\mathbf{A}$ is a projection matrix that projects vectors onto the space spanned by the rows of $\mathbf{A}$. It is also evident from the interpretation of the interference term in (4-75) as a system of d linear equations in M unknowns that the level of interference cancellation is dependent on the projection of $\mathbf{q}^H$ onto the space spanned by the rows of $\mathbf{A}$.

If $M \geq d$, then $\mathbf{A}$ is rank d and $\mathbf{q}^H$ must lie entirely in the space spanned by the rows of $\mathbf{A}$. Thus, $\mathbf{q}^H\mathbf{B}^H\mathbf{B} = \mathbf{q}^H$ implying

$$\sum_{i=1}^{p} \cos^2\phi_i = 1 \qquad (4\text{-}97)$$

and complete interference cancellation is obtained provided σ_n^2 is sufficiently small. This indicates once again that perfect interference cancellation is obtained independent of $\mathbf{T}_a$ with one adaptive weight per interferer in the absence of white noise. Note, however, that complete interference cancellation can be obtained even if $M < d$, provided a $\mathbf{T}_a$ exists such that $\mathbf{q}$ lies in the space spanned by the rows of $\mathbf{A}$.

Additional insight is obtained by assuming that $M = d$ and that the ith column of $\mathbf{T}_a$ has unit response to the ith interferer and zero response to all other interferers, that is

$$[\mathbf{T}_a]_i^H \mathbf{d}(\theta_j, \omega_o) = [\mathbf{a}(\theta_j, \omega_o)]_i = \begin{cases} 1, & i = j \\ 0, & i \neq j \end{cases} \tag{4-98}$$

This assumption is usually unrealistic in practice, but leads to a very clean interpretation of the expressions derived in this chapter. Note that (4-98) implies that $\mathbf{R}_a$ is now a diagonal matrix

$$\mathbf{R}_a = \text{diag}\{S(\theta_1, \omega_o), S(\theta_2, \omega_o), \ldots, S(\theta_d, \omega_o)\} \tag{4-99}$$

so $\lambda_i = S(\theta_i, \omega_o)$, $i = 1, 2, \ldots, d$ and $\mathbf{V} = \mathbf{I}$. Substitution of these values into (4-95b) gives

$$[\boldsymbol{\alpha}(\theta_j, \omega_o)]_i = \begin{cases} \sqrt{2\pi}, & i = j \\ 0, & i \neq j \end{cases} \tag{4-100}$$

which implies from (4-95a) and $\mathbf{h} = \mathbf{B}\mathbf{q}$ that

$$[\mathbf{h}]_i = S^{1/2}(\theta_i, \omega_o)\mathbf{d}^H(\theta_i, \omega_o)\mathbf{w}_q, \qquad i = 1, 2, \ldots, d \tag{4-101}$$

Therefore, using (4-86), the adaptive weight associated with the ith column of $\mathbf{T}_a$ is

$$[\mathbf{w}_a]_i = \frac{S(\theta_i, \omega_o)}{S(\theta_i, \omega_o) + \sigma_n^2} \mathbf{d}^H(\theta_i, \omega_o)\mathbf{w}_q \tag{4-102}$$

The gain presented by the beamformer to the ith interferer can be represented as the product of two terms

$$|(\mathbf{w}_q - \mathbf{T}_a\mathbf{w}_a)^H \mathbf{d}(\theta_i, \omega_o)|^2 = |\mathbf{w}_q^H \mathbf{d}(\theta_i, \omega_o)|^2 \left(\frac{\sigma_n^2}{S(\theta_i, \omega_o) + \sigma_n^2} \right)^2 \tag{4-103}$$

The first term is the gain of the nonadaptive beamformer $\mathbf{w}_q$ to the interferer. The second term is always less than one; it represents the reduction in gain associated with adaptive processing. If the interferer power $S(\theta_i, \omega_o)$ is much greater than σ_n^2, then the gain to the ith interferer is proportional to the inverse of that interferer's power. Substitute (4-101) into (4-88) to obtain

$$\cos^2\phi_i = \frac{S(\theta_i, \omega_o)|\mathbf{w}_q^H \mathbf{d}(\theta_i, \omega_o)|^2}{\sum_{j=1}^{d} S(\theta_j, \omega_o)|\mathbf{w}_q^H \mathbf{d}(\theta_j, \omega_o)|^2} \tag{4-104}$$

In this example $\cos^2\phi_i$ represents the ratio of the ith interferer's power at the output of $\mathbf{w}_q$ to the total interference power at the output of $\mathbf{w}_q$. The adaptive cancellation factor γ defined in (4-90) is

$$\gamma = 1 - \sum_{i=1}^{d} \frac{S(\theta_i, \omega_o)}{S(\theta_i, \omega_o) + \sigma_n^2} \cos^2\phi_i \tag{4-105}$$

If the interferer powers are significantly larger than the white noise power, then γ is approximately zero since $\Sigma_{i=1}^{p} \cos^2\phi_i = 1$. Note that the contribution of the ith column of $\mathbf{T}_a$ to interference and noise cancellation is dependent on $S(\theta_i, \omega_o)$, σ_n^2, and $\cos^2\phi_i$, which is dependent on the gain of $\mathbf{w}_q$ to the ith interferer.

Now suppose the interference consists of a single broadband interferer located at direction θ_o with $S(\theta_o, \omega) = 1$, $\omega \in \Omega$ so that

$$\mathbf{R}_n = \frac{1}{2\pi} \int_{\omega \in \Omega} \mathbf{d}(\theta_o, \omega)\mathbf{d}^H(\theta_o, \omega)\, d\omega + \sigma_n^2 \mathbf{I} \tag{4-106}$$

Note that $q(\theta_o, \omega)$ represents the frequency response of $\mathbf{w}_q$ in direction θ_o and the elements of $\mathbf{a}(\theta_o, \omega)$ represent the frequency responses of the individual columns of $\mathbf{T}_a$ in direction θ_o. The elements of $\boldsymbol{\alpha}(\theta_o, \omega)$ are therefore an orthonormal basis for the frequency responses of the columns of $\mathbf{T}_a$. The value $|[\mathbf{h}]_i|^2$ is the norm squared of the projection of the frequency response of $\mathbf{w}_q$ in direction θ_o onto the ith basis for the frequency response of the columns of $\mathbf{T}_a$. The quantity $\cos^2\phi_i$ measures how closely the shape of the frequency response of $\mathbf{w}_q$ matches the shape of the ith basis for the frequency response of the columns of $\mathbf{T}_a$. Thus, $\Sigma_{i=1}^{p} \cos^2\phi_i$ is a measure of how well $q(\theta_0, \omega)$ can be described as a linear combination of the elements of $\mathbf{a}(\theta_o, \omega)$.

As a final example, we use the results obtained in this section to numerically analyze the interference cancellation performance of the example given at the end of Section 4-4. Recall that the desired signal direction sine is 0.1 and the interference environment consists of broadband interferers at direction sines of 0.4 and 0.6 having 30 dB and 40 dB powers, respectively, relative to the white noise. The number of adaptive degrees of freedom, M, is 79. Figure 4-14 depicts $\cos^2\phi_i$ and $\lambda_i/(\lambda_i + \sigma_n^2)$ in dB for $i = 1, 2, \ldots, 26$. Of the 79 eigenvalues, the smallest 53 are insignificant so we set $p = 26$ in (4-90) and obtain $\gamma = 8.593 \times 10^{-5}$. Now $\mathbf{w}_q^H \mathbf{R}_I \mathbf{w}_q$ is 7099 and $\sigma_n^2 \mathbf{w}_q^H \mathbf{w}_q = 5.23$ so using (4.89) we obtain $P_n = 5.84$ which corresponds to an array gain of 33.7 dB. This is identical to the array gain obtained in Section 4-4 even though we have only considered the first 26 eigenvalues as

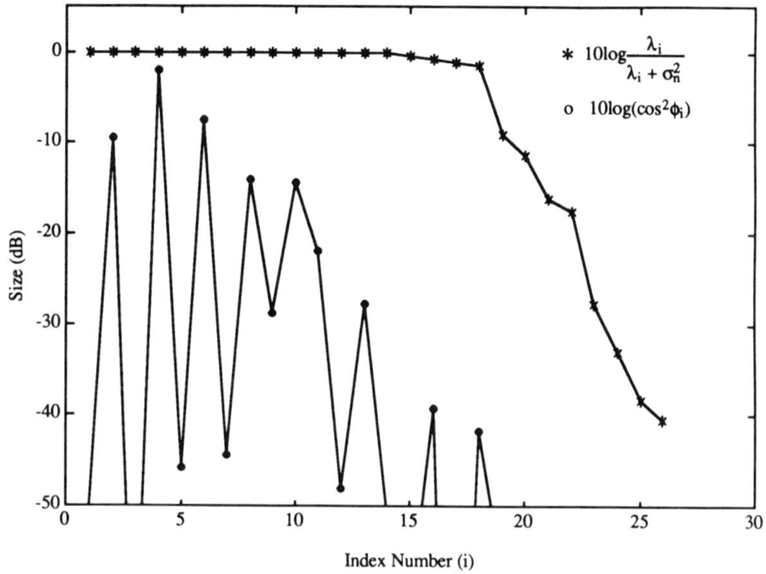

Figure 4-14 Relative importance of terms in the expression for the adaptive cancellation factor γ.

significant. Using (4-91) and (4-92) we have $P_w = 5.76$ and $P_I = 0.08$. If we only use the nine adaptive degrees of freedom that satisfy $\cos^2\phi_i > -40$ dB (modes 2, 4, 6, 8, 9, 10, 11, 13, and 16), then (4-90) gives $\gamma = 2.310 \times 10^{-4}$ and $P_n = 6.87$. Thus, reducing the number of adaptive weights from seventy-nine to nine only results in a noise power increase of 0.7 dB ($10\log(6.87/5.84)$).

4-8 ADAPTIVE ALGORITHMS

The weights for the beamformers considered in this chapter are a function of the data covariance matrix $\mathbf{R}_x$. Thus far $\mathbf{R}_x$ is assumed known, however, in practice $\mathbf{R}_x$ is unknown and all one has available are the data vectors $\mathbf{x}(n)$. The primary purpose of adaptive algorithms is to find an optimum set of weights using the data $\mathbf{x}(n)$ instead of $\mathbf{R}_x$. Since adaptive algorithms determine the weights using the received data, the weights automatically adjust to changes in the environment in order to maintain interference suppression. This section presents an overview of several adaptive algorithms that are appropriate for use with the GSC. The texts by Monzingo and Miller [34], Hudson [27], Haykin [25], and Compton [9] are excellent sources of more detailed information.

Most standard adaptive algorithms are directly applicable to minimum variance beamformers implemented in GSC form. The weight vector $\mathbf{w}$ in a partially adaptive GSC is given by $\mathbf{w} = \mathbf{w}_q - \mathbf{T}_a\mathbf{w}_a$. Define $y_d(n) = \mathbf{w}_q^H\mathbf{x}(n)$ as a desired signal and a data vector $\mathbf{u}(n) = \mathbf{T}_a^H\mathbf{x}(n)$ so that the beamformer output is written

$$y(n) = y_d(n) - \mathbf{w}_a^H\mathbf{u}(n) \tag{4-107}$$

Equation (4-107) represents a standard adaptive filter configuration: a weighted combination of the data $\mathbf{u}(n)$ is used to estimate the desired signal $y_d(n)$. The adaptive algorithm chooses the weight vector $\mathbf{w}_a$ to minimize some function of $|y(n)|^2$. Examples of adaptive algorithms that are applicable to the form (4-107) include LMS, recursive least squares, and sample covariance matrix inversion. One class of adaptive algorithms that cannot be used in the adaptive beamforming problem are the fast recursive least squares algorithms. This is because the elements of the data vector $\mathbf{u}(n)$ do not possess a time shift relationship as occurs in FIR filter applications, that is, the ith element of $\mathbf{u}(n)$ is not related to the $(i-1)$st element of $\mathbf{u}(n)$ through a fixed time delay.

The computational complexity of adaptive algorithms varies, but is always dependent on the number of adaptive weights, M. For example, recursive least squares algorithms are generally of $O(M^2)$ complexity while sample covariance matrix inversion algorithms are of $O(M^3)$. Adaptive algorithm computation cost can be prohibitive for large numbers of adaptive weights and high data rates.

Adaptive algorithms which either directly or indirectly estimate the covariance matrix are of particular interest for radar applications due to their convergence properties. Gradient based algorithms such as LMS have limited applicability because their convergence characteristics are strongly dependent on the eigenvalue spread of the covariance matrix. A common estimate of $\mathbf{R}_x$ is the sample covariance matrix. Given L zero mean data vectors, $\mathbf{x}(n)$, $n = 1, 2, \ldots, L$, the sample covariance estimate is

$$\hat{\mathbf{R}}_x = \frac{1}{L}\sum_{n=1}^{L}\mathbf{x}(n)\mathbf{x}^H(n) \tag{4-108}$$

The sample covariance matrix estimate is the maximum likelihood estimate of $\mathbf{R}_x$ given no prior structural constraints [21]. The adaptive weight vector $\mathbf{w}_a$ is estimated by substituting $\hat{\mathbf{R}}_x$ in place of $\mathbf{R}_x$ in (4-65) (or (4-46)). Standard numerical methods for solving systems of linear equations, for example, Cholesky factorization [20], are used to compute the weights. Note that the data's random nature implies that the weight vector is a random variable. Statistical properties of the beamformer are analyzed in Section 4-9 based on the statistics of the sample covariance estimator.

Exponential and rectangular time windows are often used to update the covariance matrix estimate. With an exponential window the covariance estimate at time n, $\hat{\mathbf{R}}_x(n)$, is computed using the data at time n, and the estimate at time $n-1$, $\hat{\mathbf{R}}_x(n-1)$, as

$$\hat{\mathbf{R}}_x(n) = \lambda \hat{\mathbf{R}}_x(n-1) + \mathbf{x}(n)\mathbf{x}^H(n) \tag{4-109}$$

where the exponential weighting factor, λ, is a positive constant close to but less than unity. In the case of a rectangular window of length L we have

$$\hat{\mathbf{R}}_x(n) = \hat{R}_x(n-1) + \mathbf{x}(n)\mathbf{x}^H(n) - \mathbf{x}(n-L)\mathbf{x}^H(n-L) \tag{4-110}$$

Recursive least squares algorithms employ the matrix inversion lemma to update the weights at time n in terms of the weights at time $n-1$. This results in an order M savings in computational complexity compared to recomputing the matrix inverse at each time.

Numerical problems can occur when the adaptive weights are obtained in terms of the sample covariance matrix since the dynamic range of the elements of the covariance matrix is twice that of the data. This has led to the development of algorithms which solve for the weights directly in terms of the data. Define the N by L data matrix $\mathbf{X} = [\mathbf{x}(1) \quad \mathbf{x}(2) \quad \cdots \quad \mathbf{x}(L)]$. The solution of

$$[\mathbf{T}_a^H \hat{\mathbf{R}}_x \mathbf{T}_a]\mathbf{w}_a = \mathbf{T}_a^H \hat{\mathbf{R}}_x \mathbf{w}_q \tag{4-111}$$

is theoretically equivalent to the solution of the least squares problem

$$\min_{\mathbf{w}_a} |\mathbf{X}^H \mathbf{T}_a \mathbf{w}_a - \mathbf{X}^H \mathbf{w}_q|_2^2 \tag{4-112}$$

since $\hat{\mathbf{R}}_x = L^{-1}\mathbf{X}\mathbf{X}^H$. Assuming (4-112) is solved without forming $\mathbf{X}\mathbf{X}^H$, the solution to (4-112) will be less sensitive to numerical error as a result of the reduced dynamic range. Orthogonalization methods, such as QR and singular value decompositions, are applied to $\mathbf{X}^H\mathbf{T}_a$ to solve for $\mathbf{w}_a$ (see [20, 31]). Exponential and rectangular windowing can be used to update $\mathbf{X}$ as new data is available. As in the covariance matrix case, considerable computational savings are obtained by using the previous weights and new data to compute the weights. Algorithms for updating the weights based on the old weights and the new data using QR decomposition are given by Haykin [25] and Rader and Steinhardt [39].

4-9 ADAPTIVE CONVERGENCE USING THE SAMPLE COVARIANCE MATRIX

The statistics of the mean squared error (MSE) between the beamformer output and the desired signal are analyzed in this section assuming the sample covariance matrix $\hat{\mathbf{R}}_x$ is used to estimate $\mathbf{R}_x$. The statistics of the output power are also derived. These statistics are functions of the number of data vectors, L; thus they indicate the rate at which the MSE and output power converge to their steady state values. The rate of convergence is strongly dependent on the number of adaptive degrees of freedom—the convergence rate increases as the number of degrees of freedom decrease. Similar convergence characteristics can be established for other adaptive algorithms, such as LMS, but are not developed here.

Several investigators have studied the convergence characteristics of adaptive beamformers that utilize the sample covariance matrix estimate of $\mathbf{R}_x$. In a well known paper, Reed *et al.* [40] derive the distribution of a normalized output SNR assuming the beamformer weights are based on signal free data vectors. Capon and Goodman [7] derive the distribution of the output power for a minimum variance beamformer subject to a single linear constraint. Monzingo and Miller [34] treat adaptive convergence of output SNR and MSE for several configurations of beamformers commonly used in narrowband processing; several of these are equivalent to a minimum variance beamformer subject to a single linear constraint. Additional related work is found in [3, 19, 29, 30] and Chapter 3 of this book. The analysis in this section follows that of Van Veen [48].

4-9-1 The Sample Covariance Matrix and MSE

Define the data matrix $\mathbf{X} = [\mathbf{x}(1) \quad \mathbf{x}(2) \quad \cdots \quad \mathbf{x}(L)]$ and let $\mathbf{X} = \mathbf{X}_s + \mathbf{X}_n$ where $\mathbf{X}_s$ is the data due to the desired signal and $\mathbf{X}_n$ the data due to the noise. Now (4-108) becomes

$$\hat{\mathbf{R}}_x = \frac{1}{L}\mathbf{X}\mathbf{X}^H$$
$$= \frac{1}{L}[\mathbf{X}_s\mathbf{X}_s^H + \mathbf{X}_s\mathbf{X}_n^H + \mathbf{X}_n\mathbf{X}_s^H + \mathbf{X}_n\mathbf{X}_n^H] \tag{4-113}$$

Substituting for $\mathbf{R}_x$ in (4-65) we obtain

$$\mathbf{w}_a = (\mathbf{T}_a^H\mathbf{X}\mathbf{X}^H\mathbf{T}_a)^{-1}\mathbf{T}_a^H\mathbf{X}\mathbf{X}^H\mathbf{w}_q \tag{4-114}$$

Assume the constraints are chosen to preserve the desired signal, that is, the columns of $\mathbf{X}_s$ lie in the space spanned by the columns of $\mathbf{C}$. This implies that $\mathbf{T}_a^H\mathbf{X}_s = \mathbf{0}$. Expanding $\mathbf{X}\mathbf{X}^H$ as in (4-113) and using $\mathbf{T}_a^H\mathbf{X}_s = \mathbf{0}$ we obtain

$$\mathbf{w}_a = (\mathbf{T}_a^H\mathbf{X}_n\mathbf{X}_n^H\mathbf{T}_a)^{-1}\mathbf{T}_a^H\mathbf{X}_n(\mathbf{X}_s^H + \mathbf{X}_n^H)\mathbf{w}_q \tag{4-115}$$

Let $\mathbf{s} = [s(1) \quad s(2) \quad \cdots \quad s(L)]^H$ be the column vector containing samples of the complex conjugate of the desired signal at times $n = 1, 2, \ldots, L$ and assume that the constraints are chosen to pass the desired signal without distortion, for example, with unit gain and linear phase. This implies that $\mathbf{w}_q^H\mathbf{X}_s = \mathbf{s}^H$, that is, in the absence of noise ($\mathbf{X}_n = \mathbf{0}$) the beamformer output is the desired signal. Let $\mathbf{y}^H = [y(1) \quad y(2) \quad \cdots \quad y(L)]$ represent the beamformer output vector and note that $\mathbf{y}^H = \mathbf{w}^H\mathbf{X} = (\mathbf{w}_q - \mathbf{T}_a\mathbf{w}_a)^H\mathbf{X}$. Define the sample MSE $\hat{e}$ as

$$\begin{aligned}\hat{e} &= \frac{1}{L}\,|\mathbf{s}^H - \mathbf{y}^H|^2 \\ &= \frac{1}{L}\,|\mathbf{s}^H - \mathbf{w}^H\mathbf{X}|^2 \end{aligned} \tag{4-116}$$

Using $\mathbf{T}_a^H\mathbf{X}_s = \mathbf{0}$ and $\mathbf{w}_q^H\mathbf{X}_s = \mathbf{s}^H$ yields

$$\begin{aligned}\hat{e} &= \frac{1}{L}\,|\mathbf{w}^H\mathbf{X}_n|^2 \\ &= \frac{1}{L}\,\mathbf{w}^H\mathbf{X}_n\mathbf{X}_n^H\mathbf{w} \end{aligned} \tag{4-117}$$

Substituting the GSC representation for $\mathbf{w}$, $\mathbf{w}_q - \mathbf{T}_a\mathbf{w}_a$, into (4-117) results in

$$\hat{e} = \frac{1}{L}\,[\mathbf{w}_q^H\mathbf{X}_n\mathbf{X}_n^H\mathbf{w}_q - \mathbf{w}_q^H\mathbf{X}_n\mathbf{X}_n^H\mathbf{T}_a\mathbf{w}_a - \mathbf{w}_a^H\mathbf{T}_a^H\mathbf{X}_n\mathbf{X}_n^H\mathbf{w}_q + \mathbf{w}_a^H\mathbf{T}_a^H\mathbf{X}_n\mathbf{X}_n^H\mathbf{T}_a\mathbf{w}_a] \tag{4-118}$$

Now replace $\mathbf{w}_a$ by the expression given in (4-115) and simplify to obtain

$$\begin{aligned}\hat{e} = \frac{1}{L}\,[&\mathbf{w}_q^H\mathbf{X}_n\mathbf{X}_n^H\mathbf{w}_q - \mathbf{w}_q^H\mathbf{X}_n\mathbf{X}_n^H\mathbf{T}_a(\mathbf{T}_a^H\mathbf{X}_n\mathbf{X}_n^H\mathbf{T}_a)^{-1}\mathbf{T}_a^H\mathbf{X}_n\mathbf{X}_n^H\mathbf{w}_q \\ &+ \mathbf{w}_q^H\mathbf{X}_s\mathbf{X}_n^H\mathbf{T}_a(\mathbf{T}_a^H\mathbf{X}_n\mathbf{X}_n^H\mathbf{T}_a)^{-1}\mathbf{T}_a^H\mathbf{X}_n\mathbf{X}_s^H\mathbf{w}_q] \end{aligned} \tag{4-119}$$

Define

$$\hat{e}_n = \frac{1}{L}\,[\mathbf{w}_q^H\mathbf{X}_n\mathbf{X}_n^H\mathbf{w}_q - \mathbf{w}_q^H\mathbf{X}_n\mathbf{X}_n^H\mathbf{T}_a(\mathbf{T}_a^H\mathbf{X}_n\mathbf{X}_n^H\mathbf{T}_a)^{-1}\mathbf{T}_a^H\mathbf{X}_n\mathbf{X}_n^H\mathbf{w}_q] \tag{4-120}$$

and

$$\begin{aligned}\hat{e}_s &= \frac{1}{L}\,[\mathbf{w}_q^H\mathbf{X}_s\mathbf{X}_n^H\mathbf{T}_a(\mathbf{T}_a^H\mathbf{X}_n\mathbf{X}_n^H\mathbf{T}_a)^{-1}\mathbf{T}_a^H\mathbf{X}_n\mathbf{X}_s^H\mathbf{w}_q] \\ &= \frac{1}{L}\,[\mathbf{s}^H\mathbf{X}_n^H\mathbf{T}_a(\mathbf{T}_a^H\mathbf{X}_n\mathbf{X}_n^H\mathbf{T}_a)^{-1}\mathbf{T}_a^H\mathbf{X}_n\mathbf{s}] \end{aligned} \tag{4-121}$$

so that $\hat{e} = \hat{e}_n + \hat{e}_s$. $\hat{e}_n$ is the sample MSE due to the noise. That is, $\hat{e}_n$ is the sample MSE in the absence of the signal ($\mathbf{X}_s = \mathbf{0}$). $\hat{e}_s$ is an additional MSE that results from the presence of the signal. Note that $\hat{e}_n$ is equivalent to P_n in (4-67b) if we substitute $\hat{\mathbf{R}}_n = L^{-1}\mathbf{X}_n\mathbf{X}_n^H$ for $\mathbf{R}_n$. This indicates, as noted in Section 4-4, that P_n is the MSE if the covariance matrix is known.

The expression for the output power is obtained by substituting $L^{-1}\mathbf{X}\mathbf{X}^H$ for $\mathbf{R}_x$ in (4-66)

$$\hat{P}_o = \frac{1}{L} [\mathbf{w}_q^H \mathbf{X}\mathbf{X}^H \mathbf{w}_q - \mathbf{w}_q^H \mathbf{X}\mathbf{X}^H \mathbf{T}_a (\mathbf{T}_a^H \mathbf{X}\mathbf{X}^H \mathbf{T}_a)^{-1} \mathbf{T}_a^H \mathbf{X}\mathbf{X}^H \mathbf{w}_q] \quad (4\text{-}122)$$

Note that this expression cannot be written as a sum of signal and noise powers $\hat{P}_s$ and $\hat{P}_n$ analogous to (4-67a,b) because of the cross terms $\mathbf{X}_s\mathbf{X}_n^H$ and $\mathbf{X}_n\mathbf{X}_s^H$ in the expansion for $\hat{\mathbf{R}}_x$.

4-9-2 Probability Distributions

We now derive the probability distributions of $\hat{P}_o$, $\hat{e}_n$, and $\hat{e}_s$. The columns of $\mathbf{X}$ and $\mathbf{X}_n$ are assumed to be independent and identically Gaussian distributed with zero mean. Under these assumptions $\mathbf{X}\mathbf{X}^H$ and $\mathbf{X}_n\mathbf{X}_n^H$ are complex Wishart distributed [21], with distribution denoted as $W(L, N; \mathbf{R}_x)$ and $W(L, N; \mathbf{R}_n)$ respectively. Here L is the number of columns in $\mathbf{X}$ ($\mathbf{X}_n$), N is the number of rows, and $\mathbf{R}_x$ ($\mathbf{R}_n$) is the covariance matrix associated with the columns of $\mathbf{X}$ ($\mathbf{X}_n$). Muirhead [36] is used as a reference for most of the properties of the Wishart distribution needed below. Muirhead [36] only considers real random variables; however, it is straightforward to extend these properties to the complex case (e.g., see [30, Appendix 2]).

The following properties of the Wishart distribution are needed. Let the matrix $\hat{\mathbf{B}}$ be distributed as $W(L, N; \mathbf{B})$ where $\mathbf{B}$ is nonsingular and let $\mathbf{D}$ be a constant N by M ($M \leq N$) full rank matrix.

1. $\hat{\mathbf{B}}$ is nonsingular with probability one if $L \geq N$.
2. $D^H\hat{\mathbf{B}}\mathbf{D}$ is distributed as $W(L, M; \mathbf{D}^H\mathbf{B}\mathbf{D})$.
3. $(\mathbf{D}^H\hat{\mathbf{B}}^{-1}\mathbf{D})^{-1}$ is distributed as $W(L - N + M, M; (\mathbf{D}^H\mathbf{B}^{-1}\mathbf{D})^{-1})$.
4. A complex random variable distributed as $W(P, 1; 1)$ is equivalent to a complex chi squared random variable with P complex degrees of freedom ($2P$ real degrees of freedom).

First consider $\hat{P}_o$. Define the $M + 1$ by $M + 1$ matrices $\mathbf{A}$ and $\hat{\mathbf{A}}$ as

$$\mathbf{A} = [\mathbf{w}_q \quad \mathbf{T}_a]^H \mathbf{R}_x [\mathbf{w}_q \quad \mathbf{T}_a]$$

$$\hat{\mathbf{A}} = [\mathbf{w}_q \quad \mathbf{T}_a]^H \mathbf{X}\mathbf{X}^H [\mathbf{w}_q \quad \mathbf{T}_a]$$

$$= \begin{bmatrix} \mathbf{w}_q^H\mathbf{X}\mathbf{X}^H\mathbf{w}_q & \mathbf{w}_q^H\mathbf{X}\mathbf{X}^H\mathbf{T}_a \\ \mathbf{T}_a^H\mathbf{X}\mathbf{X}^H\mathbf{w}_q & \mathbf{T}_a^H\mathbf{X}\mathbf{X}^H\mathbf{T}_a \end{bmatrix} \quad (4\text{-}123)$$

$\hat{\mathbf{A}}$ is nonsingular with probability 1 provided $L \geq M + 1$. Use the identity for the inverse of a partitioned matrix [28] to express the element in the first row and column of $\hat{\mathbf{A}}^{-1}$ as

$$[\hat{\mathbf{A}}^{-1}]_{1,1} = [\mathbf{w}_q^H \mathbf{X}\mathbf{X}^H \mathbf{w}_q - \mathbf{w}_q^H \mathbf{X}\mathbf{X}^H \mathbf{T}_a (\mathbf{T}_a^H \mathbf{X}\mathbf{X}^H \mathbf{T}_a)^{-1} \mathbf{T}_a^H \mathbf{X}\mathbf{X}^H \mathbf{w}_q]^{-1} \quad (4\text{-}124)$$

This implies that $\hat{P}_o$ is L^{-1} times the inverse of (4-124)

$$\hat{P}_o = \frac{1}{L} [[\hat{\mathbf{A}}^{-1}]_{1,1}]^{-1}$$
$$= \frac{1}{L} [\mathbf{u}_1^H \hat{\mathbf{A}}^{-1} \mathbf{u}_1]^{-1} \tag{4-125}$$

where $\mathbf{u}_1^H = [1 \quad 0 \quad 0 \quad \cdots \quad 0]$. Similarly, (4-66) is written as $P_o = [\mathbf{u}_1^H \mathbf{A}^{-1} \mathbf{u}_1]^{-1}$.

Using property 2 we see that $\hat{\mathbf{A}}$ is distributed as $W(L, M+1; \mathbf{A})$. Now apply property 3 to show that $L\hat{P}_o$ is distributed as $W(L - M, 1; [\mathbf{u}_1^H \mathbf{A}^{-1} \mathbf{u}_1]^{-1}) = W(L - M, 1; P_o)$. $L\hat{P}_o / P_o$ is distributed as $W(L - M, 1; 1)$, which, according to property 4, is a chi squared random variable with $L - M$ complex degrees of freedom. Thus, the mean of $\hat{P}_o$ is given by

$$E\{\hat{P}_o\} = \frac{L-M}{L} P_o \tag{4-126}$$

The factor $(L - M)/L$ represents the loss in output power due to estimation of $\mathbf{R}_x$. At least $L = 2M$ data vectors are required for the mean of $\hat{P}_o$ to converge within 3 dB of P_o.

Now consider the distribution of $\hat{e}_n$, the MSE due to the noise. This distribution is easily obtained by exploiting the similarities between the expressions for $\hat{e}_n$ and $\hat{P}_o$. Note that the expression for $\hat{P}_o$ (4-122) is identical to the expression for $\hat{e}_n$ (4-120) if $\mathbf{X}$ is replaced by $\mathbf{X}_n$. Furthermore, the expression for P_o (4-66) is equivalent to the expression for P_n (4-67b) if $\mathbf{R}_x$ is replaced by $\mathbf{R}_n$. Thus, $L\hat{e}_n$ is distributed as $W(L - M, 1; P_n)$. The mean of $\hat{e}_n$ is

$$E\{\hat{e}_n\} = \frac{L-M}{L} P_n \tag{4-127}$$

As with the output power, the factor $(L - M)/L$ determines the adaptive convergence of the mean sample noise MSE to the steady state value. As L increases, $E\{\hat{e}_n\}$ increases until it reaches the steady state value of P_n.

Lastly, the distribution of $\hat{e}_s$ is derived. First the conditional distribution of $\hat{e}_s$ given $\mathbf{s}$ is obtained. Define $\mathbf{T}_a^H \mathbf{X}_n = \mathbf{V}$ so

$$\hat{e}_s = \frac{1}{L} [\mathbf{s}^H \mathbf{V}^H (\mathbf{V}\mathbf{V}^H)^{-1} \mathbf{V}\mathbf{s}] \tag{4-128}$$

Let $E\{\mathbf{V}\mathbf{V}^H\} = L\mathbf{T}_a^H \mathbf{R}_n \mathbf{T}_a$ have Cholesky factorization $\mathbf{M}\mathbf{M}^H$ and rewrite $\hat{e}_s$ as

$$\hat{e}_s = \frac{1}{L} [\mathbf{s}^H \mathbf{V}^H \mathbf{M}^{-H} \mathbf{M}^H (\mathbf{V}\mathbf{V}^H)^{-1} \mathbf{M}\mathbf{M}^{-1} \mathbf{V}\mathbf{s}]$$
$$= \frac{1}{L} [\mathbf{s}^H \mathbf{V}^H \mathbf{M}^{-H} (\mathbf{M}^{-1} \mathbf{V}\mathbf{V}^H \mathbf{M}^{-H})^{-1} \mathbf{M}^{-1} \mathbf{V}\mathbf{s}]$$
$$= \frac{1}{L} [\mathbf{s}^H \mathbf{U}^H (\mathbf{U}\mathbf{U}^H)^{-1} \mathbf{U}\mathbf{s}] \tag{4-129}$$

The columns of $\mathbf{U} = \mathbf{M}^{-1}\mathbf{V}$ are independent identically distributed normal random vectors with covariance $E\{\mathbf{U}\mathbf{U}^H\} = \mathbf{M}^{-1}E\{\mathbf{V}\mathbf{V}^H\}\mathbf{M}^{-H} = \mathbf{I}$. This implies that $\hat{e}_s$ is independent of the noise covariance $\mathbf{R}_n$ and the elements of $\mathbf{T}_a$. Let $a = (\mathbf{s}^H\mathbf{s})^{1/2}$ so that $\mathbf{s} = a\mathbf{s}_o$ where $\mathbf{s}_o^H\mathbf{s}_o = 1$. Define the vector $\mathbf{q} = \mathbf{U}\mathbf{s}_o$. Given $\mathbf{s}$, $\mathbf{q}$ is a Gaussian random vector with mean $E\{\mathbf{q}\} = \mathbf{0}$ and covariance $E\{\mathbf{q}\mathbf{q}^H\} = \mathbf{I}$. The functional form of the distribution of $\mathbf{q}$ does not depend on $\mathbf{s}_o$, hence, $\mathbf{q}$ is independent of $\mathbf{s}$. Substituting the definition for $\mathbf{q}$ into (4-129) we obtain

$$\begin{aligned} \hat{e}_s &= \frac{a^2}{L}\,[\mathbf{q}^H(\mathbf{U}\mathbf{U}^H)^{-1}\mathbf{q}] \\ &= \frac{a^2}{L}\,\rho \end{aligned} \tag{4-130}$$

where the random variable ρ is defined as $\rho = \mathbf{q}^H(\mathbf{U}\mathbf{U}^H)^{-1}\mathbf{q}$. Since both $\mathbf{q}$ and $\mathbf{U}$ are independent of $\mathbf{s}$, ρ is independent of $\mathbf{s}$.

It is now established that ρ is a complex beta distributed random variable. Define an L by L dimensional unitary matrix $\mathbf{H} = [\mathbf{G} \quad \mathbf{s}_o]$, that is, $\mathbf{G}^H\mathbf{G} = \mathbf{I}$ and $\mathbf{G}^H\mathbf{s}_o = \mathbf{0}$. Let $\mathbf{B} = \mathbf{U}\mathbf{H} = [\mathbf{Z} \quad \mathbf{q}]$ where $\mathbf{Z} = \mathbf{U}\mathbf{G}$ is a M by $L-1$ dimensional matrix. Now, $\mathbf{U}\mathbf{U}^H = \mathbf{B}\mathbf{B}^H = \mathbf{Z}\mathbf{Z}^H + \mathbf{q}\mathbf{q}^H$ so

$$\rho = [\mathbf{q}^H(\mathbf{Z}\mathbf{Z}^H + \mathbf{q}\mathbf{q}^H)^{-1}\mathbf{q}] \tag{4-131}$$

If $L > M$, then $\mathbf{Z}\mathbf{Z}^H$ is nonsingular with probability 1. Applying the matrix inversion lemma to (4-131) yields

$$\begin{aligned} \rho &= \mathbf{q}^H(\mathbf{Z}\mathbf{Z}^H)^{-1}\mathbf{q} - \frac{\mathbf{q}^H(\mathbf{Z}\mathbf{Z}^H)^{-1}\mathbf{q}\mathbf{q}^H(\mathbf{Z}\mathbf{Z}^H)^{-1}\mathbf{q}}{1 + \mathbf{q}^H(\mathbf{Z}\mathbf{Z}^H)^{-1}\mathbf{q}} \\ &= \frac{1}{1 + \dfrac{1}{\mathbf{q}^H(\mathbf{Z}\mathbf{Z}^H)^{-1}\mathbf{q}}} \end{aligned} \tag{4-132}$$

Define $\alpha = \mathbf{q}^H\mathbf{q}$ and $\gamma = \mathbf{q}^H\mathbf{q}/\mathbf{q}^H(\mathbf{Z}\mathbf{Z}^H)^{-1}\mathbf{q}$ so that

$$\rho = \frac{1}{1 + \gamma/\alpha} \tag{4-133}$$

Recall that $\mathbf{q}$ is a zero mean Gaussian random variable with identity covariance. Thus, α is a complex chi squared random variable with M complex degrees of freedom. We now show that γ is also a complex chi squared random variable and is independent of α. First note that the columns of $\mathbf{B}$ are independent and identically Gaussian distributed since $\mathbf{B} = \mathbf{U}\mathbf{H}$ is obtained from $\mathbf{U}$ via a unitary transformation (see, e.g., [30, Appendix A, part F]). Thus, $\mathbf{q}$ and the columns of $\mathbf{Z}$ are independent and identically Gaussian distributed. The complex version of Theorem 3.2.12 in

[36] is therefore applicable and implies that γ is a complex chi squared random variable with $L - M$ degrees of freedom and is independent of $\mathbf{q}$.

The variable γ is independent of $\mathbf{q}$ and is therefore independent of α. Thus, γ/α is a ratio of two complex chi squared random variables and is complex F distributed. A simple change of variables applied to the F distribution indicates that ρ is complex beta distributed

$$p(\rho) = \frac{(L-1)!}{(L-M-1)!(M-1)!}\, \rho^{M-1}(1-\rho)^{L-M-1} \tag{4-134}$$

The distribution of ρ is dependent only on the number of data vectors in the covariance matrix estimate, L, and the number of adaptive degrees of freedom, M, in the beamformer.

The distribution of $\hat{e}_s$ given $\mathbf{s}$ is now obtained as

$$p(\hat{e}_s|s) = \frac{L(L-1)!}{a^2(L-M-1)!(M-1)!}\left(\frac{L\hat{e}_s}{a^2}\right)^{M-1}\left(1 - \frac{L\hat{e}_s}{a^2}\right)^{L-M-1} \tag{4-135}$$

The dependence of $\hat{e}_s$ on $\mathbf{s}$ is only through $a^2 = \mathbf{s}^H\mathbf{s}$; in principle, the distribution of $\hat{e}_s$ is obtained by integrating the product of $p(\hat{e}_s|\mathbf{s})$ and the distribution of a^2 over $\mathbf{s}$. However, the moments of $\hat{e}_s$ provide greater insight into adaptive beamformer performance than the distribution. It is straightforward in this case to obtain the moments of $\hat{e}_s$ without computing its distribution because ρ is independent of $\mathbf{s}$ and thereby a^2. The mean of $\hat{e}_s$ is

$$\begin{aligned} E\{\hat{e}_s\} &= L^{-1}E\{a^2\}E\{\rho\} \\ &= \sigma_s^2\,\frac{M}{L} \end{aligned} \tag{4-136}$$

where $\sigma_s^2 = L^{-1}E\{\mathbf{s}^H\mathbf{s}\} = E\{s(n)s^*(n)\}$ is the variance or power of the signal. The mean sample MSE due to the signal presence is directly proportional to the signal power and the number of adaptive degrees of freedom, but is inversely proportional to the number of data vectors used to estimate $\mathbf{R}_x$. Note that strong signals result in large MSE. The mean sample signal MSE decreases as L increases.

Using (4-127) and (4-136) we see that the expected value of the sample MSE is given by

$$E\{\hat{e}\} = \left(1 - \frac{M}{L}\right)P_n + \frac{M}{L}\,\sigma_s^2 \tag{4-137}$$

The mean sample MSE depends on P_n, σ_s^2, and the ratio of M to L. Reducing M affects $E\{\hat{e}\}$ in two ways. Recall from Section 4-6 that P_n generally increases as M is reduced. If P_n increases, then $E\{\hat{e}\}$ increases.

However, as M is reduced, the number of data vectors L required for $E\{\hat{e}\}$ to converge to its steady state value decreases. P_n is a function of the elements of $\mathbf{T}_a$ while M only represents the number of columns in $\mathbf{T}_a$. These results suggest that M should be kept as small as possible while the elements of $\mathbf{T}_a$ should be chosen to optimize P_n. This approach to partially adaptive beamformer design is discussed in Section 4-11. Finally, note that the beamformer output is not defined until $L \geqslant M$. Thus, reducing M lowers the number of data vectors required to compute the beamformer output.

4-10 A MODULAR STRUCTURE FOR BEAMFORMER IMPLEMENTATION

The results of previous sections indicate that adaptive beamformer performance is dependent on the number of adaptive weights employed. In general, a large number of adaptive weights is desired to maximize steady state interference cancellation capability while a small number is desired to maximize the adaptive convergence rate and minimize computational complexity. These tradeoffs suggest that it is advantageous to simultaneously implement beamformers with differing numbers of adaptive weights or perhaps dynamically change the number of adaptive weights in a single beamformer. The modular structure described in this section is ideally suited to these tasks. Analogous to a lattice filter, the structure consists of a cascade of adaptive modules, each of which represents one (or several) adaptive degree of freedom. A weighted combination of each module's output implements an adaptive beamformer with the number of adaptive weights given by the cumulative number of adaptive degrees of freedom at that point in the cascade of modules. If a module implements P adaptive degrees of freedom, then a set of P linear equations must be solved to obtain the module weights. Thus, the modular structure distributes the computational burden of matrix inversion over a series of lower order problems. Matrix inversion is not required if all modules have $P = 1$ degree of freedom.

The optimal partially adaptive GSC weight vector $\mathbf{w} = \mathbf{w}_q - \mathbf{T}_a\mathbf{w}_a$ is expressed using (4-65) as $\mathbf{w} = (\mathbf{I} - \mathbf{T}_a\mathbf{G})\mathbf{w}_q$, where

$$\mathbf{G} = (\mathbf{T}_a^H\mathbf{R}_x\mathbf{T}_a)^{-1}\mathbf{T}_a^H\mathbf{R}_x \tag{4-138}$$

$\mathbf{G}$ represents the solution to the optimization problem

$$\min_{\mathbf{G}} E\{|(\mathbf{I} - \mathbf{T}_a\mathbf{G})^H\mathbf{x}|^2\} \tag{4-139a}$$

or equivalently

$$\min_{\mathbf{G}} \operatorname{tr}(\mathbf{I} - \mathbf{T}_a\mathbf{G})^H\mathbf{R}_x(\mathbf{I} - \mathbf{T}_a\mathbf{G}) \tag{4-139b}$$

The data vector $\mathbf{z} = (\mathbf{I} - \mathbf{T}_a\mathbf{G})^H\mathbf{x}$ can be viewed as the output of a bank of beamformers with weight vectors given by the columns of $(\mathbf{I} - \mathbf{T}_a\mathbf{G})$; thus, $\mathbf{G}$ minimizes the sum of the powers at the outputs of the beamformers in this bank, $E\{\mathbf{z}^H\mathbf{z}\} = \text{tr}\, E\{\mathbf{z}\mathbf{z}^H\}$. The partially adaptive GSC output is a linear combination of the elements of $\mathbf{z}$: $y = \mathbf{w}_q^H\mathbf{z}$. This representation for the GSC is depicted in Fig. 4-15. The output power is $E\{|y|^2\} = \mathbf{w}_q^H\mathbf{R}_z\mathbf{w}_q$ where $\mathbf{R}_z = \mathbf{R}_x - \mathbf{R}_x\mathbf{T}_a(\mathbf{T}_a^H\mathbf{R}_x\mathbf{T}_a)^{-1}\mathbf{T}_a^H\mathbf{R}_x$. Using the decomposition $\mathbf{R}_x = \mathbf{R}_x^{1/2}\mathbf{R}_x^{H/2}$, we write $\mathbf{R}_z$ as

$$\mathbf{R}_z = \mathbf{R}_x^{1/2}[\mathbf{I} - \mathbf{R}_x^{H/2}\mathbf{T}_a(\mathbf{T}_a^H\mathbf{R}_x\mathbf{T}_a)^{-1}\mathbf{T}_a^H\mathbf{R}_x^{1/2}]\mathbf{R}_x^{H/2} \tag{4-140}$$

The matrix in brackets implements a projection onto the complement of the space spanned by the columns of $\mathbf{R}_x^{H/2}\mathbf{T}_a$; hence, $\mathbf{z}$ does not contain any of the components of $\mathbf{x}$ which lie in the space spanned by columns of $\mathbf{R}_x^{H/2}\mathbf{T}_a$.

Let $\mathbf{T}_a$ be partitioned as $\mathbf{T}_a = [\mathbf{H}_1 \quad \mathbf{H}_2 \quad \cdots \quad \mathbf{H}_J]$ where $\mathbf{H}_i$ has N rows and P_i columns such that P_i, $i = 1, 2, \ldots, J$ sum to M. The integer P_i corresponds to the number of degrees of freedom in the ith module. Define $\mathbf{z}_i = (\mathbf{I} - \mathbf{H}_i\mathbf{G}_i)^H\mathbf{z}_{i-1}$, $i = 1, 2, \ldots, J$, with $\mathbf{z}_0 = \mathbf{x}$. It is shown by Liu and Van Veen [32], using the formulas for the inverses of block matrices, that

$$(\mathbf{I} - \mathbf{H}_1\mathbf{G}_1)(\mathbf{I} - \mathbf{H}_2\mathbf{G}_2)\cdots(\mathbf{I} - \mathbf{H}_J\mathbf{G}_J) = \mathbf{I} - \mathbf{T}_a\mathbf{G} \tag{4-141}$$

where $\mathbf{G}_i$ solves

$$\min_{\mathbf{G}_i} E\{|(\mathbf{I} - \mathbf{H}_i\mathbf{G}_i)^H\mathbf{z}_{i-1}|^2\}, \qquad i = 1, 2, \ldots, J \tag{4-142}$$

That is,

$$\mathbf{G}_i = (\mathbf{H}_i^H\mathbf{R}_{z_{i-1}}\mathbf{H}_i)^{-1}\mathbf{H}_i^H\mathbf{R}_{z_{i-1}} \tag{4-143}$$

Analogous to (4-140), it can be shown that

$$\mathbf{R}_{z_i} = \mathbf{R}_{z_{i-1}}^{1/2}[\mathbf{I} - \mathbf{R}_{z_{i-1}}^{H/2}\mathbf{H}_i(\mathbf{H}_i^H\mathbf{R}_{z_{i-1}}\mathbf{H}_i)^{-1}\mathbf{H}_i^H\mathbf{R}_{z_{i-1}}^{1/2}]\mathbf{R}_{z_{i-1}}^{H/2} \tag{4-144}$$

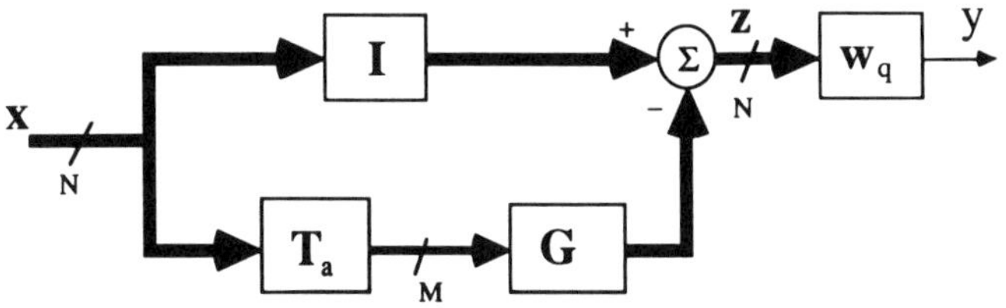

Figure 4-15 Block diagram illustrating separation of adaptive structure and the fixed beamformer $\mathbf{w}_q$.

The ith module removes the components of the input data which lie in the space spanned by the columns of $\mathbf{R}_{z_{i-1}}^{H/2}\mathbf{H}_i$.

This decomposition implies that the GSC can be implemented as a cascade of adaptive modules followed by a nonadaptive beamformer as depicted in Fig. 4-16. The dimension of the matrix inverse required in the ith module is given by P_i, the number of columns in $\mathbf{H}_i$. If each P_i is one, then the M dimensional matrix inverse required in the direct GSC implementation is reduced to a sequence of M scalar inverses in the modular implementation.

The proof of equivalence between the direct and modular implementations is applicable when the covariance matrix is replaced by an estimated covariance matrix. This implies that the direct and modular beamformer outputs are identical when adaptive algorithms based on covariance matrix estimates, such as recursive least squares and sample covariance matrix inversion, are used to update the adaptive weights. Thus, the statistics of the direct and modular beamformer outputs are also identical.

The adaptive components in the ith module do not depend on the adaptive components of the jth module if $j > i$. Thus, modules can be added or removed to dynamically change the number of adaptive degrees of freedom without recomputing the weights in the modules that remain. This is analogous to the order recursive property of lattice filters. If the level of interference cancellation is insufficient, the number of adaptive degrees of freedom can be increased without recomputing all the weights. Alternatively, the number of adaptive degrees of freedom can be reduced to improve convergence properties. Extending this idea, it is evident that the modular GSC structure can simultaneously implement beamformers of differing numbers of adaptive degrees of freedom with only a slight increase in computational burden by passing the data at the output of each module through $\mathbf{w}_q$. That is, if y_i is the output of a beamformer based on the adaptive degrees of freedom represented by $[\mathbf{H}_1 \quad \mathbf{H}_2 \quad \cdots \quad \mathbf{H}_i]$, then $y_i = \mathbf{w}_q^H\mathbf{z}_i$. In this manner, one can simultaneously implement beamformers with rapid convergence (y_i for small numbers of adaptive degrees of freedom) and good interference cancellation (y_i for large numbers of adaptive degrees of freedom).

Lastly, the modular GSC decomposition decouples the fixed beamformer $\mathbf{w}_q$ from the adaptive weight determination process. This implies that one can simultaneously implement beamformers with different $\mathbf{w}_q$ using the

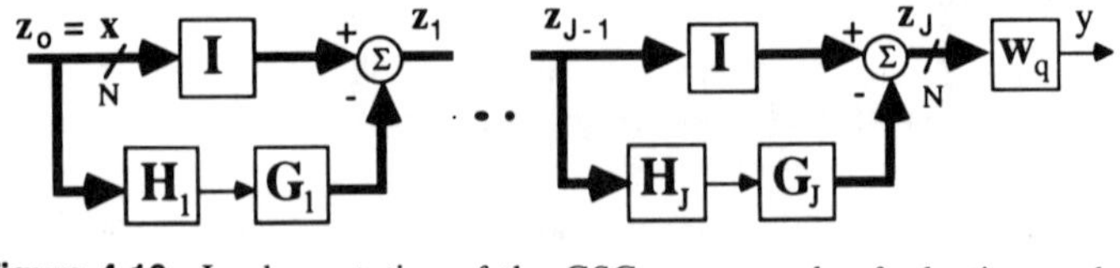

Figure 4-16 Implementation of the GSC as a cascade of adaptive modules.

same set of adaptive weights. Note however, that the only components of $\mathbf{w}_q$ which can be modified are those that lie in the space orthogonal to the columns of $\mathbf{C}$ and $\mathbf{T}_a$. The components in the space spanned by the columns of $\mathbf{C}$ must be fixed to satisfy the constraints on the beamformer weights while the beamformer output is invariant to those in the space spanned by $\mathbf{T}_a$ [22]. Procedures for choosing $\mathbf{w}_q$ to approximate a desired quiescent response and obtain improved interference cancellation are discussed in [46].

4-11 PARTIALLY ADAPTIVE BEAMFORMER DESIGN

In previous sections it has been shown that for a given environment the adaptive convergence rate and computational complexity are dependent only on the number of adaptive weights. However, the interference cancellation capability is dependent on which degrees of freedom, represented by $\mathbf{T}_a$, are utilized. This section addresses the problem of designing beamformers that are capable of good interference cancellation using small numbers of adaptive weights. The section begins with a discussion of traditional philosophies and their limitations. A design method based on the eigenstructure of the interference covariance matrix is then reviewed. This is followed by discussion of an optimization based approach, in which the adaptive degrees of freedom are chosen to minimize the interference and noise output power over a likely set of interference scenarios. The section concludes with a set of simulations illustrating the performance of eigenstructure and optimization based designs.

Early work in partially adaptive beamforming was reported by Chapman [8] and Morgan [35]. Chapman studied the performance obtained using simple sums and differences of the antenna outputs as inputs to an adaptive beamformer. Morgan reported on the performance of an MSC that utilized a subset of the available auxiliary channels adaptively. Another early idea was that of using $\mathbf{T}_a$ to direct beams at interferers. The term beam in this context refers to a set of weights that produce a response of unit gain in one direction and minimize the gain in other directions. The adaptive weights scale the contributions of the interferers to the beam outputs appropriately so that they can be subtracted from the interference contribution to the signal beam output. This approach is very effective if the interferers are narrowband, located at discrete points in space, and are strong relative to the background noise.

The majority of the early work assumed narrowband interferers located at points in space. It was known, as was shown in Section 4-7, that at high interference to noise levels complete cancellation is obtained with one adaptive weight per interferer independent of $\mathbf{T}_a$. This fact suggests that the interference cancellation performance is not highly sensitive to the choice of $\mathbf{T}_a$. In contrast, broadband interference cancellation performance is very sensitive to the choice for $\mathbf{T}_a$. Beam and MSC based partially adaptive approaches typically require large numbers of adaptive degrees of freedom to achieve cancellation of broadband interference.

One approach to the broadband problem is to convert it to a sequence of narrowband problems by passing the antenna outputs through a bank of narrowband filters or by taking a discrete Fourier transform (DFT) of the antenna outputs. The data in each frequency bin is processed independently of other frequency bins. This technique is commonly used in sonar systems [38]. The narrowband assumption is valid provided the product of the frequency bin bandwidth and the temporal aperture, $T(\theta)$, defined in (4-6), is much less than one for all θ of interest. Arrays with large numbers of antennas typically have large $T(\theta)$ and thus require large order narrowband filters or very long DFTs at the output of each antenna for the narrowband assumption to hold. Note that the data is assumed stationary over the filter or DFT length, an assumption that is violated in rapidly changing environments when large order filters or DFTs are required. Large order narrowband filters or DFTs also result in a large number of frequency bins. One adaptive degree of freedom is required per interferer per frequency bin; this often leads to considerably more adaptive degrees of freedom than are required with a direct broadband approach.

4-11-1 Eigenstructure Based Design

Two approaches to the design of $\mathbf{T}_a$ for broadband interference are now examined. The first is motivated by the interference cancellation analysis in Section 4-7. Consider repeating the analysis for the fully adaptive case, that is, replacing $\mathbf{T}_a$ by $\mathbf{C}_n$ and $\mathbf{w}_a$ by $\mathbf{w}_n$. Recall that $\mathbf{w}_n$ is the $N - K$ dimensional fully adaptive weight vector and that $\mathbf{T}_a = \mathbf{C}_n\mathbf{T}$. Now $\mathbf{V}_p\mathbf{\Lambda}_p\mathbf{V}_p^H$ is the eigendecomposition of $\mathbf{C}_n^H\mathbf{R}_I\mathbf{C}_n$ and (4-86) becomes

$$\mathbf{w}_n = \mathbf{V}_p\mathbf{\Gamma}\mathbf{h} \tag{4-145}$$

where $\mathbf{h}$ is appropriately redefined. This implies that $\mathbf{w}_n$ lies in the p dimensional subspace spanned by the columns of $\mathbf{V}_p$. Thus, the number of adaptive degrees of freedom can be reduced from $N - K$ to p with no loss in performance by choosing $\mathbf{T}$ to span the same space as $\mathbf{V}_p$.

The problem with this approach is that it requires knowledge of $\mathbf{R}_I$. In general $\mathbf{R}_I$ is not known a priori. Estimation of $\mathbf{R}_I$ may be possible under some circumstances [4, 18]. However, the additional difficulty of estimating $\mathbf{R}_I$ and solving the eigenproblem to obtain $\mathbf{V}_p$ may offset any potential performance gain or complexity reduction associated with reducing the number of adaptive weights. Alternatively, suppose we choose $\mathbf{T}$ such that the columns of $\mathbf{V}_p$ associated with any likely $\mathbf{R}_I$ lie in the space spanned by the columns of $\mathbf{T}$. This avoids the need to know $\mathbf{R}_I$ a priori or estimate it although the number of adaptive weights required is increased. Let Θ_I

represent the set of possible interference arrival at directions and Ω_I the interference bandwidth. Define

$$\mathbf{R}_{\text{av}} = \frac{1}{2\pi} \int_{\theta \in \Theta_I} \int_{\omega \in \Omega_I} \mathbf{d}(\theta, \omega)\mathbf{d}^H(\theta, \omega)\, d\omega,\, d\theta \tag{4-146}$$

Any valid $\mathbf{R}_I$ is given by a sum (integral) of outer products of $\mathbf{d}(\theta, \omega)$, $\theta \in \Theta_I$, $\omega \in \Omega_I$ (see (4-69)). Thus, the columns of all $\mathbf{R}_I$ must lie in the space spanned by the columns of $\mathbf{R}_{\text{av}}$. If we choose the columns of $\mathbf{T}$ as a basis for the space spanned by the columns of $\mathbf{C}_n^H \mathbf{R}_{\text{av}} \mathbf{C}_n$, then the columns of $\mathbf{V}_p$ associated with any likely $\mathbf{R}_I$ lie in the space spanned by the columns of $\mathbf{T}$. This design approach, termed the eigenstructure method, is presented in [43].

The rank of $\mathbf{C}_n^H \mathbf{R}_{\text{av}} \mathbf{C}_n$ is very important since it determines the number of adaptive degrees of freedom required to guarantee fully adaptive interference cancellation. In general, the rank depends on the array geometry, $\mathbf{C}_n$, Ω_I, and Θ_I and must be evaluated numerically for the particular case of interest. Several examples are given in [43]. This method typically results in a larger number of adaptive weights than are necessary. For example, if $\mathbf{R}_I$ is known, then fully adaptive performance is obtained using one adaptive degree of freedom by choosing $\mathbf{T} = \mathbf{w}_n$ in (4-115). In this case $\mathbf{w}_a = 1$. However, the eigenstructure method requires at least p adaptive degrees of freedom.

It is tempting to choose the columns of $\mathbf{T}$ as the eigenvectors corresponding only to the largest eigenvalues to minimize the squared error between the fully adaptive weight space and the space spanned by $\mathbf{T}$. However, this can result in dramatic performance breakdown as illustrated in [45]. If $\mathbf{R}_I$ is known, (4-90) indicates that the cancellation associated with the ith mode depends on the product of $\lambda_i/(\lambda_i + \sigma_n^2)$ and $\cos^2\phi_i$. Note that if σ_n^2 is very small, then the cancellation associated with the ith mode is almost independent of eigenvalue size.

4-11-2 Power Minimization Based Design

The second method to be discussed is the power minimization approach suggested in [42]. In this approach the goal is to choose the $\mathbf{T}$ which minimizes the interference and noise output power over a set of likely interference scenarios. Let the likely interference environments be parameterized by a vector $\mathbf{q}$ and assume that the parameters in $\mathbf{q}$ are drawn from a set of likely environments denoted by Q. In general, the parameters in $\mathbf{q}$ could include the number of interferers, interferer directions, interferer spectral densities, white noise level, etcetera. For example, suppose the likely interference environment is assumed to consist of two interferers of unknown direction with known spectral density shape in unit power white noise. Let q_1 and q_2 denote the power levels and q_3 and q_4 the directions of the two interferers with $\mathbf{q} = [q_1 \quad q_2 \quad q_3 \quad q_4]$. We have

$$\mathbf{R}_n(\mathbf{q}) = q_1\mathbf{R}_o(q_3) + q_2\mathbf{R}_o(q_4) + \mathbf{I} \tag{4-147}$$

where $\mathbf{R}_o(\theta)$ is the covariance matrix of a unit interferer from direction θ. It appears as if $\mathbf{q}$ must contain a large number of parameters to adequately describe realistic interference environments. However, after presenting the design method it will be argued that there are actually relatively few parameters that significantly affect the design of $\mathbf{T}$.

The output power due to noise and interference is expressed as a function of $\mathbf{T}$ as

$$P_n(\mathbf{q}, \mathbf{T}) = \mathbf{w}_q^H\mathbf{R}_n(\mathbf{q})\mathbf{w}_q - \mathbf{w}_q^H\mathbf{R}_n(\mathbf{q})\mathbf{C}_n\mathbf{T}(\mathbf{T}^H\mathbf{C}_n^H\mathbf{R}_n(\mathbf{q})\mathbf{C}_n\mathbf{T})^{-1}\mathbf{T}^H\mathbf{C}_n^H\mathbf{R}_n(\mathbf{q})\mathbf{w}_q \tag{4-148}$$

The goal is to minimize the average value of $P_n(\mathbf{q}, \mathbf{T})$ for $\mathbf{q} \in Q$

$$\min_{\mathbf{T}}\left\{\int_{\mathbf{q}\in Q} P_n(\mathbf{q}, \mathbf{T})\, d\mathbf{q}\right\} \tag{4-149}$$

A nonnegative weighting function can be inserted in the integrand to emphasize some interference scenarios over others. Note that there are other functions of $P_n(\mathbf{q}, \mathrm{T})$ that are logical candidates for minimization, for example, minimize the maximum value of $P_n(\mathbf{q}, \mathbf{T})$ for $\mathbf{q} \in Q$. However, solutions to other forms of optimization problems have not been reported. In fact the nonlinear manner in which $\mathbf{T}$ enters into the expression for $P_n(\mathbf{q}, \mathbf{T})$ has prevented a direct solution to (4-149). Direct numerical optimization of (4-149) is probably not feasible due to the number of unknown parameters to be optimized. For example, if $\mathbf{T}$ maps from 100 adaptive degrees of freedom to 20, then the minimization in (4-149) must occur over 2000 free parameters.

These difficulties have led to the development of a suboptimal sequential approach in which the columns of $\mathbf{T}$ are designed one at a time, with each new column dependent on the previously chosen columns. Let $\mathbf{T}_{i+1} = [\mathbf{T}_i \,|\, \mathbf{t}_i]$, $i = 2, \ldots, M$ where $\mathbf{T}_i$ represents the i columns that have already been chosen and $\mathbf{t}_i$ is the $(i+1)$st column to be designed. Here $\mathbf{T}_1 = \mathbf{t}_1$ and $\mathbf{T} = \mathbf{T}_M$. The design problem is now restated: given the matrix $\mathbf{T}_i$, find the vector $\mathbf{t}_i$ that satisfies

$$\min_{\mathbf{t}_i}\left\{\int_{\mathbf{q}\in Q_i} P_n(\mathbf{q}, \mathbf{t}_i)\, d\mathbf{q}\right\}, \qquad i = 1, 2, \ldots, M \tag{4-150}$$

The set of interference scenarios over which $\mathbf{t}_i$ is designed, Q_i, is selected as a subset of Q. It will be argued later in this section that the best performance is obtained by splitting Q into M nonintersecting subsets.

It is straightforward to show that given $\mathbf{T}_i$, the only components of $\mathbf{t}_i$ that contribute to further minimizing $P_n(\mathbf{q}, \mathbf{t}_i)$ are linearly independent of the columns of $\mathbf{T}_i$. Thus, the number of free parameters in $\mathbf{t}_i$ are reduced by defining $\mathbf{t}_i = \bar{\mathbf{T}}_i\mathbf{b}$. Here $\mathbf{b}$ is an $N - K - i$ dimensioned vector representing the free parameters in $\mathbf{t}_i$. The columns of the $N - K$ by $N - K - i$ matrix $\bar{\mathbf{T}}_i$ form a basis for the complement of the space spanned by the columns of $\mathbf{T}_i$. Denote $P_n(\mathbf{q}, \mathbf{t}_i)$ as $P_n(\mathbf{q}, \mathbf{b})$.

The modular decomposition derived in Section 4-10 is now used to express $P_n(\mathbf{q}, \mathbf{b})$ as a function of $\mathbf{T}_i$ and $\mathbf{b}$. Partition $\mathbf{C}_n\mathbf{T}_{i+1}$ as $[\mathbf{H} \,|\, \mathbf{S}\mathbf{b}]$ where $\mathbf{H} = \mathbf{C}_n\mathbf{T}_i$ and $\mathbf{S} = \mathbf{C}_n\bar{\mathbf{T}}_i$ to obtain the modular structure depicted in Fig. 4-17. The matrices $\mathbf{G}_1(\mathbf{q})$ and $\mathbf{G}_2(\mathbf{q})$ represent the adaptive weights associated with the first and second modules, respectively. This decomposition naturally separates the effect of the columns of $\mathbf{T}_a$ that are already designed, represented by $\mathbf{H}$, from the column of interest, $\mathbf{S}\mathbf{b}$. $P_n(\mathbf{q}, \mathbf{b})$ is expressed as

$$P_n(\mathbf{q}, \mathbf{b}) = \mathbf{w}_q^H \mathbf{R}_{z_2}(\mathbf{q}, \mathbf{b}) \mathbf{w}_q \tag{4-151}$$

where (4-144) implies

$$\mathbf{R}_{z_2}(\mathbf{q}, \mathbf{b}) = \mathbf{R}_{z_1}(\mathbf{q}) - \frac{\mathbf{R}_{z_1}(\mathbf{q})\mathbf{S}\mathbf{b}\mathbf{b}^H\mathbf{S}^H\mathbf{R}_{z_1}(\mathbf{q})}{\mathbf{b}^H\mathbf{S}^H\mathbf{R}_{z_1}(\mathbf{q})\mathbf{S}\mathbf{b}} \tag{4-152}$$

and

$$\mathbf{R}_{z_1}(\mathbf{q}) = \mathbf{R}_n(\mathbf{q}) - \mathbf{R}_n(\mathbf{q})\mathbf{H}(\mathbf{H}^H\mathbf{R}_n(\mathbf{q})\mathbf{H})^{-1}\mathbf{H}^H\mathbf{R}_n(\mathbf{q}) \tag{4-153}$$

Substitution of (4-152) into (4-151) gives

$$P_n(\mathbf{q}, \mathbf{b}) = \mathbf{w}_q^H\mathbf{R}_{z_1}(\mathbf{q})\mathbf{w}_q - \frac{\mathbf{w}_q^H\mathbf{R}_{z_1}(\mathbf{q})\mathbf{S}\mathbf{b}\mathbf{b}^H\mathbf{S}^H\mathbf{R}_{z_1}(\mathbf{q})\mathbf{w}_q}{\mathbf{b}^H\mathbf{S}^H\mathbf{R}_{z_1}(\mathbf{q})\mathbf{S}\mathbf{b}} \tag{4-154}$$

The first term, $\mathbf{w}_q^H\mathbf{R}_{z1}(\mathbf{q})\mathbf{w}_q$, corresponds to the noise and interference power at the output of a beamformer based on $\mathbf{H}$ and is independent of $\mathbf{b}$. Thus, the criteria for choosing $\mathbf{t}_i$,

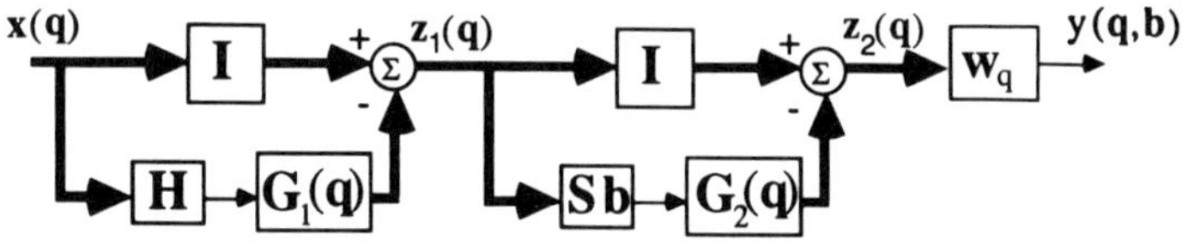

Figure 4-17 Modular GSC structure indicating separation of columns of $\mathbf{T}_a$ that have already been designed, $\mathbf{H}$, from the column being designed, $\mathbf{S}\mathbf{b}$.

$$\min_{\mathbf{b}}\left\{\int_{\mathbf{q}\in Q_i} P_n(\mathbf{q},\mathbf{b})\,d\mathbf{q}\right\} \tag{4-155a}$$

is equivalent to

$$\min_{\mathbf{b}}\left\{\int_{\mathbf{q}\in Q_i} \frac{\mathbf{w}_q^H\mathbf{R}_{z_1}(\mathbf{q})\mathbf{S}\mathbf{b}\mathbf{b}^H\mathbf{S}^H\mathbf{R}_{z_1}(\mathbf{q})\mathbf{w}_q}{\mathbf{b}^H\mathbf{S}^H\mathbf{R}_{z_1}(\mathbf{q})\mathbf{S}\mathbf{b}}\,d\mathbf{q}\right\} \tag{4-155b}$$

The presence of $\mathbf{b}$ in the denominator of the integrand prevents a direct solution to (4-155b), unless Q_i represents a single interference scenario in which case the integral disappears. Designing only a single column of $\mathbf{T}$ at a time has significantly reduced the number of free parameters so that a numerical solution to (4-155a) may now be practical. Examples of designs based on numerical optimization are given in [42]. However, an approximate analytic solution to (4-155a) remains of interest. An approximate solution provides an initial guess for numerical optimization and may perform sufficiently well that numerical optimization is unnecessary.

Additional insight to the design process is obtained by combining $\mathbf{w}_q$ with the second section of the modular structure given in Fig. 4-17 to generate the two stage structure depicted in Fig. 4-18. The first section in Fig. 4-18, $\mathbf{I}-\mathbf{H}\mathbf{G}_1(q)$, is unchanged and removes the components of the input data $\mathbf{x}(\mathbf{q})$ associated with $\mathbf{H}_1$. The output of the first section is $\mathbf{z}_1(\mathbf{q})=[\mathbf{I}-\mathbf{H}\mathbf{G}_1(\mathbf{q})]^H\mathbf{x}(\mathbf{q})$. The second section, $\mathbf{w}_q-\mathbf{S}\mathbf{b}w(\mathbf{q})$, is an adaptive beamformer with a single adaptive weight, $w(\mathbf{q})=\mathbf{G}_2(\mathbf{q})\mathbf{w}_q$. $\mathbf{b}$ determines the space in which adaptation occurs. This new structure indicates that $P_n(\mathbf{q},\mathbf{b})$ in (4-154) is also expressed as

$$\begin{aligned}P_n(\mathbf{q},\mathbf{b}) &= E\{|(\mathbf{w}_q-\mathbf{S}\mathbf{b}w(\mathbf{q}))^H\mathbf{z}_1(\mathbf{q})|^2\}\\ &= (\mathbf{w}_q-\mathbf{S}\mathbf{b}w(\mathbf{q}))^H\mathbf{R}_{z_1}(\mathbf{q})(\mathbf{w}_q-\mathbf{S}\mathbf{b}w(\mathbf{q}))\end{aligned} \tag{4-156}$$

The weight $w(\mathbf{q})$ can also be obtained by minimizing $P_n(\mathbf{q},\mathbf{b})$ with respect to $w(q)$

$$w(\mathbf{q}) = \frac{\mathbf{b}^H\mathbf{S}^H\mathbf{R}_{z_1}(\mathbf{q})\mathbf{w}_q}{\mathbf{b}^H\mathbf{S}^H\mathbf{R}_{z_1}(\mathbf{q})\mathbf{S}\mathbf{b}} \tag{4-157}$$

Substitution of (4-157) into (4-156) leads to (4-154).

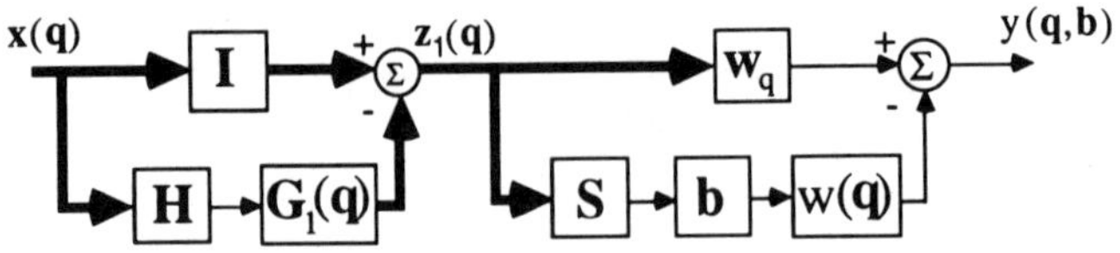

Figure 4-18 Alternate modular GSC structure indicating separation of columns of $\mathbf{T}_a$ that have already been designed, $\mathbf{H}$, from the column being designed, $\mathbf{Sb}$.

The fact that $w(\mathbf{q})$ is a function of $\mathbf{q}$ and involves $\mathbf{b}$ in the denominator is the reason that an analytic solution to (4-155b) is not evident. If $w(\mathbf{q})$ is a constant independent of $\mathbf{q}$, then $P_n(\mathbf{q}, \mathbf{b})$ is given by

$$P_n(\mathbf{q}, \mathbf{b}) = (\mathbf{w}_q - \mathbf{S}\mathbf{b})^H \mathbf{R}_{z_1}(\mathbf{q})(\mathbf{w}_q - \mathbf{S}\mathbf{b}) \tag{4-158}$$

where $w(\mathbf{q}) = 1$ is assumed for convenience. Using (4-158) as $P_n(\mathbf{q}, \mathbf{b})$ in (4-155a) gives a quadratic problem in $\mathbf{b}$. The solution satisfies

$$(\mathbf{S}^H \mathbf{R}_{z_1}^{\text{av}} \mathbf{S})\mathbf{b} = \mathbf{S}^H \mathbf{R}_{z_1}^{\text{av}} \mathbf{w}_q \tag{4-159}$$

where

$$\mathbf{R}_{z_1}^{\text{av}} = \int_{\mathbf{q} \in Q_i} \mathbf{R}_{z_1}(\mathbf{q}) \, d\mathbf{q} \tag{4-160}$$

Note that if the set of interference environments represented by Q_i is sufficiently small, then $w(\mathbf{q})$ should be approximately constant on $\mathbf{q} \in Q_i$ and the solution to (4-159) leads to a nearly optimal analytic solution of (4-155b). This is the motivation for the recommendation to select the Q_i as nonintersecting subsets of Q. It can be shown that if Q_i represents a single interference scenario, then the solution for $\mathbf{b}$ in (4-159) results in a partially adaptive beamformer that obtains fully adaptive interference cancellation in Q_i, even if only one adaptive degree of freedom is used.

The concept of interference cancellation via response matching developed in Section 4-7 provides important insight into the sensitivity of this partially adaptive beamformer design philosophy to deviations of the actual interference environment from the set Q. Response matching concepts also indicate which of the potential interference environment parameters are important to represent in $\mathbf{q}$. Note that if $\mathbf{q}$ represents a large number of parameters, then it is computationally impractical to evaluate (4-160). Equation (4-72) represents the response matching phenomena and is repeated here for convenience

$$P_n = \frac{1}{2\pi} \int_{\theta \in \Theta} \int_{\omega \in \Omega} S(\theta, \omega) |q(\theta, \omega) - \mathbf{w}_a^H \mathbf{a}(\theta, \omega)|^2 \, d\omega \, d\theta + \sigma_n^2 \mathbf{w}_q^H \mathbf{w}_q + \sigma_n^2 \mathbf{w}_a^H \mathbf{w}_a \tag{4-161}$$

Recall that $q(\theta, \omega) = \mathbf{w}_q^H \mathbf{d}(\theta, \omega)$ and $\mathbf{a}(\theta, \omega) = \mathbf{T}^H \mathbf{C}_n^H \mathbf{d}(\theta, \omega)$. Equation (4-161) indicates that the design problem is to choose $\mathbf{T}$ so that the response of the adaptive branch, $\mathbf{w}_a^H \mathbf{a}(\theta, \omega)$, is best able to match the nonadaptive branch response, $q(\theta, \omega)$, with a $\mathbf{w}_a$ of small norm over the likely set of $S(\theta, \omega)$ and σ_n^2 represented by Q. It is often reasonable to assume that the interference consists of a sum of independent interferers located at points in space, that is, $S(\theta, \omega) = \Sigma_{i=1}^d S(\theta_i, \omega)\delta(\theta - \theta_i)$. This assumption is utilized in the discussion below, although many of the comments are also applicable to more general scenarios.

The size of σ_n^2 varies the importance of response matching and of the norm of $\mathbf{w}_a$. If a partially adaptive beamformer capable of good interference cancellation is desired, then the response matching component of P_n must be emphasized while designing $\mathbf{T}$. This implies that σ_n^2 should be fixed to a relatively small value and need not be represented in $\mathbf{q}$. Simulations support this reasoning. Performance is observed to be very good if the actual σ_n^2 is larger than assumed during design of $\mathbf{T}$, but is very poor if the actual σ_n^2 is smaller than assumed during design of $\mathbf{T}$. The power of the interferers also affects the tradeoff between response matching and the norm of $\mathbf{w}_a$. It is therefore reasonable to fix the powers to a large (relative to σ_n^2) level to emphasize response matching when designing $\mathbf{T}$.

Good response matching requires that $\mathbf{T}$ be designed over the frequencies and directions from which interference is likely to arrive. If interference energy appears at directions or frequencies other than those assumed during design, performance is at best uncertain. The system bandwidth is usually known and may determine the range of frequencies of interest. On the other hand, the interferer directions θ_i, $i = 1, 2, \ldots, d$ are usually unknown and must be presented in Q. If the physical configuration of the environment in which the beamformer operates limits the set of possible interferer directions, this information may also be incorporated in Q.

The shape of $S(\theta_i, \omega)$ emphasizes the response matching at some frequencies more heavily than others. If no prior knowledge of interferer spectral shape is available, a conservative strategy is to assume $S(\theta_i, \omega)$ is constant over ω so that response matching is equally important at all frequencies. The number of interferers, d, determines the number of directions in which good response matching must simultaneously occur. If the actual number of interferers is less than the number assumed during design of $\mathbf{T}$, then reasonable performance is expected. However, performance is at best uncertain if the actual number of interferers is greater than the number assumed when $\mathbf{T}$ is designed. Limited simulations suggest that the partially adaptive beamformer's performance degrades gracefully as the number of interferers is increased. A conservative strategy is to fix the value for d used to design $\mathbf{T}$ at the maximum number of interferers expected.

The preceding discussion suggests that Q should represent variation in the interferer directions and possibly the number of interferers. Response matching arguments indicate that $\mathbf{T}$ is robust to variation in the other interference environment parameters, provided the values assumed during design are intelligently chosen.

4-11-3 Examples

The array and beamformer used in the simulations at the end of Section 4-4 is employed here to illustrate partially adaptive beamformer design and performance. Recall that the desired signal direction sine is 0.1; eleven constraints are used to ensure unit gain and linear phase over the frequency range $3\pi/5 \leqslant |\omega| \leqslant 4\pi/5$ resulting in a fully adaptive beamformer with seventy-nine adaptive weights. The eigenstructure and power minimization methods are now employed to design partially adaptive beamformers. In both cases the interferers are assumed to have direction sines between -0.2 and 1.

In the eigenstructure examples, the matrix $\mathbf{R}_{\text{av}}$ is computed according to (4-146) with Θ_I and Ω_I defined as $-0.2 \leqslant \sin\theta \leqslant 1$ and $3\pi/5 \leqslant |\omega| \leqslant 4\pi/5$, respectively. A plot of the eigenvalues of $\mathbf{C}_n^H \mathbf{R}_{\text{av}} \mathbf{C}_n$ is given in Fig. 4-19. The behavior of the eigenvalues does not provide a clear indication of the rank of $\mathbf{C}_n^H \mathbf{R}_{\text{av}} \mathbf{C}_n$, so the number of adaptive weights is determined based on the fraction of the total energy represented by the largest M eigenvalues

$$f_\lambda = \frac{\sum_{i=1}^{M} \lambda_i}{\sum_{i=1}^{79} \lambda_i} \tag{4-162}$$

The columns of $\mathbf{T}$ are the eigenvectors of $\mathbf{C}_n^H \mathbf{R}_{\text{av}} \mathbf{C}_n$ corresponding to the largest M eigenvalues. The criterion $f_\lambda \geqslant 0.9999$ gives $M = 45$. A partially adaptive beamformer based on $M = 13$ is also designed to facilitate comparison of eigenstructure and power minimization design strategies. If $M = 13$, then $f_\lambda = 0.866$.

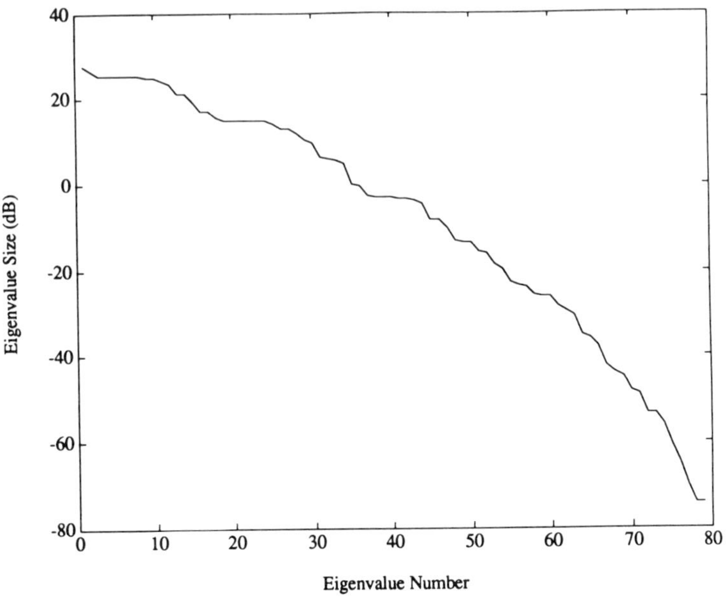

Figure 4-19 Eigenvalues of $\mathbf{C}_n^H \mathbf{R}_{\text{av}} \mathbf{C}_n$ for eigenstructure design example.

In the power minimization example, a partially adaptive beamformer with thirteen adaptive weights is designed using the approximate solution (4-159) to the optimization criterion of (4-150). For the design procedure, the interference environment is assumed to consist of two broadband interferers with constant power spectral density on $3\pi/5 \leq |\omega| \leq 4\pi/5$ in uncorrelated noise. The interferers have direction sines between -0.2 and 1. The ratio of interferer power to white noise power is assumed to be 40 dB. Table 4-1 lists the range of interferer directions over which each column of **T** is designed and the design order.

Partially adaptive beamformer interference cancellation performance is evaluated for three different types of interference scenarios. In the first case the actual environment is assumed to consist of two broadband interferers with constant power spectral density on $3\pi/5 \leq |\omega| \leq 4\pi/5$ in uncorrelated noise with interference to noise ratios of 30 dB and 40 dB. The difference between fully and partially adaptive array gain in dB is evaluated for 256 different combinations of interferer locations with direction sines in the intervals -0.2 to 0.025 and 0.175 to 1. The fully adaptive beamformer array gain ranges from 17 dB to 34 dB with 175 of the 256 scenarios having array gain greater than 30 dB. Table 4-2 summarizes the array gain loss associated with reducing the number of adaptive weights for three different partially adaptive beamformers.

In the second case partially adaptive beamformer performance is evaluated assuming the actual environment consists of a single broadband interferer with constant power spectral density on $3\pi/5 \leq |\omega| \leq 4\pi/5$ in uncorrelated noise with an interferer to noise ratio of 40 dB. The difference between fully and partially adaptive array gain in dB is evaluated for 16 different interferer locations on the interval -0.2 to 0.025 and 0.175 to 1.

TABLE 4-1 Regions Used to Design Columns of T. Interferer Locations Given as Direction Sines

Column Number	Interferer 1 Start	Interferer 1 Stop	Interferer 2 Start	Interferer 2 Stop
1	0.50	1.00	0.50	1.00
2	0.26	0.49	0.50	1.00
3	0.26	0.49	0.26	0.49
4	0.17	0.25	0.66	1.00
5	0.17	0.25	0.44	0.65
6	0.17	0.25	0.26	0.42
7	0.17	0.25	0.17	0.25
8	−0.20	−0.08	0.50	1.00
9	−0.07	0.02	0.50	1.00
10	−0.20	0.02	0.26	0.49
11	−0.20	−0.08	0.17	0.25
12	−0.07	0.02	0.17	0.25
13	−0.20	0.02	−0.20	0.02

TABLE 4-2 Difference Between Fully and Partially Adaptive Array Gain in dB for 256 Different Combinations of Interferer Locations

	Number of cases		
Loss in dB	Eigenstructure (45 weights)	Eigenstructure (13 weights)	Power Min (13 weights)
0–1	249	8	150
1–2	6	14	59
2–3	0	5	31
3–4	1	12	15
4–5	0	20	1
>5	0	197	0
Maximum loss (dB)	3.21	34.5	4.01
Average loss (dB)	0.08	11.8	1.19

Here the fully adaptive beamformer array gain varies from 22 dB to 34 dB with 13 scenarios having array gain greater than 30 dB. Table 4-3 summarizes the array gain loss associated with reducing the number of adaptive weights for three different partially adaptive beamformers.

For the third case the actual interference environment is again assumed to consist of two broadband interferers in uncorrelated noise with relative power levels of 30 dB and 40 dB. The 30 dB interferer has a constant power spectral density on $3\pi/5 \leq |\omega| \leq 4\pi/5$, but the 40 dB interferer has raised a cosine power spectral density, $1 + \cos(10\omega - 7\pi/10)$, for $3\pi/5 \leq |\omega| \leq 4\pi/5$. The difference between fully and partially adaptive array gain in dB is evaluated for 256 different combinations of interferer locations in the interval -0.2 to 0.025 and 0.175 to 1. Here the fully adaptive beamformer array gain varies from 20 dB to 34 dB with 187 scenarios having array gain greater than 30 dB. Table 4-4 summarizes the array gain loss associated with reducing the number of adaptive weights for three different partially adaptive beamformers.

TABLE 4-3 Difference Between Fully and Partially Adaptive Array Gain in dB for 16 Different Interferer Locations

	Number of cases		
Loss in dB	Eigenstructure (45 weights)	Eigenstructure (13 weights)	Power Min (13 weights)
0–1	16	6	15
1–2	0	3	1
>2	0	7	0
Maximum loss (dB)	0.10	27.9	1.80
Average loss (dB)	0.01	5.30	0.49

TABLE 4-4 Difference Between Fully and Partially Adaptive Array Gain in dB for 256 Different Combinations of Interferer Locations

	Number of cases		
Loss in dB	Eigenstructure (45 weights)	Eigenstructure (13 weights)	Power Min (13 weights)
0–1	253	15	164
1–2	3	13	53
2–3	0	17	26
3–4	0	21	13
4–5	0	19	0
>5	0	171	0
Maximum loss (dB)	1.51	31.0	3.74
Average loss (dB)	0.05	10.5	1.08

As a final example the performance of all three partially adaptive beamformers is evaluated for the five interferer example introduced in Section 4-4. Recall that the interference environment consists of a 40 dB interferer with raised cosine power spectral density, $1 + \cos(10\omega - 7\pi/10)$ on $3\pi/5 \leq |\omega| \leq 4\pi/5$ at direction sine 0.4, a 30 dB interferer with constant power spectral density on $3\pi/5 \leq |\omega|4\pi/5$ at direction sine 0.85, a 40 dB interferer with constant power spectral density on $7\pi/10 \leq |\omega| \leq 4\pi/5$ at direction sine −0.2, a 50 dB interferer with constant power spectral density on $79\pi/100 \leq |\omega| \leq 4\pi/5$ at direction sine 0.7, and a 40 dB interferer with constant power spectral density on $7\pi/10 \leq |\omega| \leq 71\pi/100$ at direction sine −0.15 in uncorrelated noise. The interferer power levels are given relative to the uncorrelated noise level. The array gain of the fully adaptive beamformer is 38.34 dB. The 13 weight power minimization designed partially adaptive beamformer gives an array gain of 32.84 dB for a loss of 5.5 dB. The 13 and 45 weight eigenstructure designed partially adaptive beamformers give array gains of 27.39 dB and 38.15 dB for losses of 10.95 dB and 0.19 dB, respectively.

These examples illustrate that significant reductions in the number of adaptive degrees of freedom can be achieved without dramatic losses in interference cancellation capability for broadband interference environments. Furthermore, they indicate that the power minimization design procedure is not highly sensitive to certain differences between the actual interference environment and that assumed during design of **T**. The response matching analysis given previously in this section suggests that the power minimization design is robust with respect to the three variations between actual and assumed environments reported in Tables 4-2–4-4 (different power levels, spectral shapes, and a reduced number of interferers). The relative performance degrades somewhat in the five interferer example. The eigenstructure design procedure results in excellent performance under all conditions provided enough adaptive weights are used. Thirteen weights are not sufficient to obtain reasonable performance using the eigenstructure design criterion.

Thus far in this section the emphasis has been on designing **T** to achieve good interference cancellation. The weight vector $\mathbf{w}_q$ also affects interference cancellation. Griffiths and Buckley [22] have shown that $\mathbf{w}_q$ is uniquely defined in fully adaptive beamformers; the components of $\mathbf{w}_q$ in the space spanned by the columns of **C** cannot be changed without violating the constraints. Furthermore, adding components to $\mathbf{w}_q$ other than those in the space spanned by the columns **C** does not change steady state performance because the adaptive branch effectively subtracts them out. However, in a partially adaptive beamformer $\mathbf{w}_q$ is not uniquely defined and can be changed by adding components in the space orthogonal to that spanned by the columns of **C** and $\mathbf{T}_a$ This freedom is exploited by Van Veen [46] to choose $\mathbf{w}_q$ such that the quiescent beamformer response approximates a desired quiescent response or to minimize the average interference and noise output power over a likely set of interference scenarios. The latter is similar in spirit to the method discussed in this section for designing $\mathbf{T}_a$, although an exact solution to the minimization is obtained because the interference and noise output power is a quadratic function of $\mathbf{w}_q$.

As a final comment for this section, note that the framework used to parameterize unknown interference environments in both the eigenstructure and power minimization methods can be extended in principle to represent other unknown characteristics in the beamforming system. For example, with very large arrays the antennas may be mounted on a semi-rigid structure, in which case the location of each antenna is unknown. Also, the response of each antenna as a function of direction and frequency may not be exactly known. It is straightforward to design **T** to optimize performance over likely ranges of these parameters, although there does not appear to be any results of this type published to date. One suspects that these variations are very significant since they change $\mathbf{d}(\theta, \omega)$ and thus influence the response matching capabilities of the beamformer.

4-12 CORRELATED INTERFERENCE AND SIGNAL CANCELLATION

The performance of an adaptive beamformer is sensitive to correlation or coherence between the desired signal and one or more of the interfering signals. An interferer is said to be correlated with the desired signal when the cross correlation between the desired signal and interferer is nonzero. Correlated interference can occur in practice as a result of multipath propagation or intelligent jamming. Recall that the adaptive weights are chosen to minimize output power. When the interference and desired signals are uncorrelated the beamformer minimizes output power by attenuating the interferers. However, when the interference and desired signals are correlated the beamformer achieves its goal of minimum output power by processing the interference in such a way as to cancel the portion of the signal that is correlated with the interference.

The problem of signal cancellation was discussed in the context of the MSC in Section 4-3. If the signal is present in the auxiliary antennas, then the MSC minimizes output power by choosing the adaptive weights to cancel the signal. Signal cancellation due to correlated interference is a similar phenomenon. The outputs of the signal blocking matrix $\mathbf{T}_a$ in the GSC correspond to the auxiliary channels in the MSC. Although $\mathbf{T}_a$ blocks the desired signal, the outputs of $\mathbf{T}_a$ are correlated with the desired signal. The adaptive weights exploit this correlation to cancel the signal.

Let the data vector $\mathbf{x}$ consist of a desired signal $\mathbf{s}$ and an interference/noise component $\mathbf{n}$, that is, $\mathbf{x} = \mathbf{s} + \mathbf{n}$. If the desired signal and interference are correlated, then $E\{\mathbf{sn}^H\} \neq \mathbf{0}$ and

$$\mathbf{R}_x = \mathbf{R}_s + \mathbf{R}_{sn} + \mathbf{R}_{ns} + \mathbf{R}_n \tag{4-163}$$

where $\mathbf{R}_s = E\{\mathbf{ss}^H\}$, $\mathbf{R}_{sn} = E\{\mathbf{sn}^H\}$, $\mathbf{R}_{ns} = E\{\mathbf{ns}^H\}$, and $\mathbf{R}_n = E\{\mathbf{nn}^H\}$. The orthogonality of the signal blocking matrix $\mathbf{T}_a$ and the desired signal $\mathbf{s}$ imply that $\mathbf{T}_a^H\mathbf{R}_s = \mathbf{0}$, $\mathbf{T}_a^H\mathbf{R}_{sn} = \mathbf{0}$, and $\mathbf{R}_{ns}\mathbf{T}_a = \mathbf{0}$. Substituting (4-163) into the expression for the partially adaptive weight vector $\mathbf{w}_a$ (4-65) and utilizing the orthogonality of $\mathbf{T}_a$ and $\mathbf{s}$ gives the expression for the adaptive weights

$$\mathbf{w}_a = (\mathbf{T}_a^H\mathbf{R}_n\mathbf{T}_a)^{-1}\mathbf{T}_a^H(\mathbf{R}_n + \mathbf{R}_{ns})\mathbf{w}_q \tag{4-164}$$

The presence of correlation between $\mathbf{s}$ and $\mathbf{n}$ changes $\mathbf{w}_a$. Note that (4-164) is analogous to (4-115), the expression for $\mathbf{w}_a$ obtained using the sample covariance matrix.

In order to illustrate the effect of nonzero $\mathbf{R}_{ns}$ on the beamformer output, we consider the MSE at the beamformer output. Recall from Section 4-4 that the MSE, $e = E\{|y_d - y|^2\}$, is given in (4-52) as

$$e = \mathbf{w}^H\mathbf{R}_n\mathbf{w} \tag{4-165}$$

Substitute the GSC representation for $\mathbf{w}$, $\mathbf{w}_q - \mathbf{T}_a\mathbf{w}_a$, with $\mathbf{w}_a$ given by (4-164), and simplify to obtain

$$\begin{aligned} e = {} & \mathbf{w}_q^H\mathbf{R}_n\mathbf{w}_q - \mathbf{w}_q^H\mathbf{R}_n\mathbf{T}_a(\mathbf{T}_a^H\mathbf{R}_n\mathbf{T}_a)^{-1}\mathbf{T}_a^H\mathbf{R}_n\mathbf{w}_q \\ & + \mathbf{w}_q^H\mathbf{R}_{sn}\mathbf{T}_a(\mathbf{T}_a^H\mathbf{R}_n\mathbf{T}_a)^{-1}\mathbf{T}_a^H\mathbf{R}_{ns}\mathbf{w}_q \end{aligned} \tag{4-166}$$

The first two terms on the right hand side of (4-166) represent P_n (4-67b), the interference and noise power at the beamformer output. Define

$$e_s = \mathbf{w}_q^H\mathbf{R}_{sn}\mathbf{T}_a(\mathbf{T}_a^H\mathbf{R}_n\mathbf{T}_a)^{-1}\mathbf{T}_a^H\mathbf{R}_{ns}\mathbf{w}_q \tag{4-167}$$

so that $e = P_n + e_s$. The term e_s is the MSE corresponding to signal cancellation and is a result of the correlation between signal and interference.

The MSE in (4-166) bears a striking resemblance to the sample MSE (4-119) that results when the sample covariance matrix is used to estimate $\mathbf{R}_x$. The signal cancellation MSE e_s (4-167) directly corresponds with the sample MSE due to the signal presence $\hat{e}_s$ (4-121), which also represents a signal cancellation phenomenon. Note that $\mathbf{X}_n\mathbf{X}_s^H$ is the sample cross covariance matrix corresponding to $\mathbf{R}_{ns}$. Even though the data in $\mathbf{X}_n$ and $\mathbf{X}_s$ are uncorrelated, the sample cross variance matrix $\mathbf{X}_n\mathbf{X}_s^H$ will be nonzero unless L, the number of data vectors, is sufficiently large. Recall that the mean of $\hat{e}_s$ decreases as L increases (see 4-137)). As L increases the correlation between signal and interference as measured by the sample cross covariance matrix $\mathbf{X}_n\mathbf{X}_s^H$ decreases resulting in reduced signal cancellation.

Methods for eliminating signal cancellation associated with correlated interference generally focus on reducing the correlation between signal and interference. We only mention several of these here. Shan and Kailath [41] utilize spatial smoothing to reduce the correlation between the signal and interferer. The array is divided into partially overlapping subarrays and the beamformer weights are based on a covariance matrix formed as an average of the covariance matrices of each subarray. Correlation between the signal and an interferer results in a field that is nonstationary in space. Averaging in space has the effect of reducing the correlation. This technique is limited to arrays with geometrically uniform configurations. Furthermore, the effective aperture of the array is reduced to the aperture of the subarray.

Frequency domain averaging is used to reduce correlation between signal and interference in Zhu and Wang [50] and Yang and Kaveh [49]. The output of each antenna is passed through a bank of narrowband filters that decompose the data into a set of subbands. A transformation is applied to data in each subband to translate all the subband center frequencies to a common frequency. The weights for each subband are determined based on the average of the transformed covariance matrices for each subband. The averaging operation reduces the correlation between signals and interferences by a factor proportional to the operating bandwidth. This method is clearly limited to broadband beamforming. Knowledge of the source and interferer directions is required in order to determine the transformation applied to the subband data. In practice these are unknown and must be estimated from the received data.

4-13 SUMMARY

A beamformer forms a weighted combination of the outputs of spatially separate antennas in order to separate propagating signals that originate from different locations. The weights determine the spatial and temporal filtering characteristics of the beamformer. In an adaptive beamformer the weights are adjusted in response to the data received at the antennas for the purpose of optimizing the beamformer's response in an environment with changing or unknown characteristics. A common criterion for optimizing the weights is minimization of beamformer output power or variance. The goal of minimizing output power is to minimize the contributions of interference and noise to the output. In the MSC, the desired signal must be weak or absent when the weights are determined to prevent the weights from cancelling the signal along with the interference. The LCMV approach minimizes output power subject to a set of linear constraints on the weight vector. These constraints are often used to control the response of the beamformer in the direction of the desired signal to prevent the weights from cancelling the desired signal.

There are three different types of linear constraints: point, derivative, and eigenvector constraints. Point constraints restrict the response at specified points of location and frequency. Derivative constraints are typically utilized in conjunction with point constraints and force the derivatives of the response to be zero at specified points of location and frequency. Eigenvector constraints are usually used to control the response over a range of location or frequency and are obtained by minimizing the mean squared error between the desired and actual beamformer response. Quadratic constraints are used to control beamformer response and/or ensure robustness to uncertainties such as antenna location or source direction. A drawback of quadratic constraints is the increased complexity necessary to determine the beamformer's weights.

The GSC separates the weight vector into constrained and unconstrained components. The unconstrained components represent the beamformer's adaptive degrees of freedom and are adjusted using standard adaptive algorithms. The GSC structure is useful for implementation and analysis of LCMV beamformers. In a partially adaptive beamformer the number of adaptive degrees of freedom are reduced by restricting the adaptive weights in the GSC to lie within a subspace. A disadvantage of partially adaptive beamforming is decreased interference cancellation capability. The modular GSC implementation is a decomposition of the GSC into a product of adaptive modules, each representing separate sets of adaptive degrees of freedom. This modular structure facilitates simultaneous implementation of beamformers based on differing numbers of adaptive degrees of freedom or, alternatively, facilitates dynamically varying the number of adaptive degrees of freedom.

Insight into the interference cancellation process is obtained using the GSC decomposition. The response of the adaptive component of the GSC must match the response of the constrained component over the interference frequency range and directions in order to cancel the interference. The presence of white noise limits the acceptable norm for the adaptive weight vector used to achieve response matching. If the white noise is very weak relative to the interferers, then near complete cancellation of narrowband interferers located at discrete points in space is accomplished using one adaptive degree of freedom per interferer. With broadband or angularly extended interference, the level of cancellation and number of adaptive degrees of freedom required for good cancellation are strongly dependent on the subspace spanned by the adaptive component.

The rates at which the output power and mean squared error converge to steady state values are dependent on the number of adaptive degrees of freedom. Convergence rates increase as the number of adaptive degrees of freedom decrease. Analysis indicates the presence of the signal during estimation of the covariance matrix results in an excess mean squared error whose mean is given by the product of the signal power and number of degrees of freedom divided by the number of data vectors used to estimate the covariance matrix.

The motivation for partially adaptive beamforming is to reduce the computational burden associated with adaptive algorithms and to improve the adaptive convergence rate. However, partially adaptive beamformers are generally not capable of the same level of interference cancellation as fully adaptive beamformers. Thus, the partially adaptive beamformer design problem is to select the adaptive degrees of freedom that minimize the interference cancellation performance loss. One approach to selecting the degrees of freedom is based on the eigenstructure of the interference covariance matrix; another minimizes the average interference output power over a range of likely interference scenarios. Optimal degree of freedom selection is not as critical in narrowband beamforming as in broadband beamforming because at high interference to noise ratios narrowband interference cancellation depends only on the number of adaptive weights.

Interference that is correlated with the desired signal arises due to multipath propagation or intelligent jamming and creates a special problem for conventional adaptive beamformers. The beamformer exploits the correlation to not only cancel the interference but at the same time cancel the desired signal. A number of techniques have been proposed to alleviate this problem. Several are based on averaging to destroy the correlation between the signal and interference. However, elimination of signal cancellation associated with correlated interference still remains an important research topic.

ACKNOWLEDGMENT

I have benefitted immensely from numerous discussions with students and colleagues on various aspects of adaptive beamforming. Keith Burgess and Bruce Williams thoroughly reviewed an earlier version of this chapter and provided numerous suggestions to improve the presentation. I also gratefully acknowledge the Army Research Office (grant DAAL03-89-K-0141) and the National Science Foundation (award MIP-8958559) for supporting much of the work presented here.

REFERENCES

[1] S. P. Applebaum (1966) "Adaptive arrays", Report SURC TR 66-001, Syracuse University Research Corporation.

[2] S. P. Applebaum and D. J. Chapman (1976) "Adaptive arrays with main beam constraints", *IEEE Trans. Antennas Propag.*, **24** 650–662.

[3] A. B. Baggeroer (1976) "Confidence intervals for regression (MEM) spectral estimates", *IEEE Trans. Inform. Theory*, **22**, 534–545.

[4] E. Brookner and J. M. Howell (1986) "Adaptive-adaptive array processing", *Proc. IEEE*, **74** 602–604.

[5] K. M. Buckley (1987) "Spatial/spectral filtering with linearly constrained minimum variance beamformers", *IEEE Trans. Acoust., Speech, Signal Process.*, **35**, 249–266.

[6] K. M. Buckley and L. J. Griffiths (1986) "An adaptive generalized sidelobe canceller with derivative constraints", *IEEE Trans. Antennas Propag.*, **34**, 311–319.

[7] J. Capon and N. R. Goodman (1970) "Probability distributions for estimators of the frequency-wavenumber spectrum", *Proc. IEEE*, **58**, 1785–1786.

[8] D. J. Chapman (1976) "Partial adaptivity for the large array", *IEEE Trans. Antennas Propag.*, **24**, 685–696.

[9] R. T. Compton (1988) *Adaptive Antennas: Concepts and Performance*, Prentice Hall, Englewood Cliffs, NJ.

[10] H. Cox (1973) "Resolving power and sensitivity to mismatch of optimum array processors", *J. Acoust. Soc. Am.*, **53**, 771–785.

[11] H. Cox, R. M. Zeskind, and M. M. Owen (1987) "Robust adaptive beamforming", *IEEE Trans. Acoust., Speech, Signal Process.*, **35**, 1365–1376.

[12] L. Elden (1977) "Algorithms for the regularization of ill-conditioned least squares problems", *Bit*, 17, 134–145.

[13] L. Elden and R. Schrieber (1986) "An application of systolic arrays to linear discrete ill-posed problems", *SIAM J. Sci. Statist. Comput.*, **7**, 892–903.

[14] M. H. Er and A. Cantoni (1983) "Derivative constraints for broadband element space antenna array processors", *IEEE Trans. Acoust., Speech, Signal Process.*, **31**, 1378–1393.

[15] M. H. Er and A. Cantoni (1985) "An alternative formulation for an optimum beamformer with robustness capability", *Proc. IEE* Part F, **132**, 447–460.

[16] O. L. Frost (1972) "An algorithm for linearly constrained adaptive array processing", *Proc. IEEE*, **60**, 926–935.

[17] W. F. Gabriel (1976) "Adaptive arrays – an introduction", *Proc. IEEE*, **64**, 239–272.

[18] W. F. Gabriel (1986) "Using spectral estimation techniques in adaptive processing antenna systems", *IEEE Trans. Antennas Propag.*, **34**, 291–300.

[19] M. W. Ganz, R. L. Moses, and S. L. Wilson (1990) "Convergence of the SMI and the diagonally loaded SMI algorithms with weak interference", *IEEE Trans. Antennas Propag.*, **38**, 394–399.

[20] G. H. Golub and C. F. Van Loan (1989) *Matrix Computations*, 2nd edition, John Hopkins, Baltimore, MD.

[21] N. R. Goodman (1963) "Statistical analysis based on a certain multivariate Gaussian distribution", *Ann. Math. Statist.*, **34**, 152–177.

[22] L. J. Griffiths and K. M. Buckley (1987) "Quiescent pattern control in linearly constrained adaptive arrays", *IEEE Trans. Acoust., Speech, Signal Process.*, **35**, 917–926.

[23] L. J. Griffiths and C. W. Jim (1982) "An alternate approach to linearly constrained adaptive beamforming", *IEEE Trans. Antennas Propag.*, **30**, 27–34.

[24] R. J. Hanson and C. L. Lawson (1969) "Extensions and applications of the Householder algorithm for solving linear least squares problems", *Math. Comput.*, **23**, 787–812.

[25] S. Haykin (1991) *Adaptive Filter Theory*, Second Edition, Prentice Hall, Englewood Cliffs, NJ.

[26] P. W. Howels (1959) "Intermediate frequency sidelobe canceller", U.S. Patent 3202990.

[27] J. E. Hudson (1981) *Adaptive Array Principles*, Peter Peregrinus, New York.

[28] T. Kailath (1980) *Linear Systems*, Prentice Hall, Englewood Cliffs, NJ, Appendix.

[29] E. J. Kelly (1986) "An adaptive detection algorithm", *IEEE Trans. Aerosp. Electon. Systems*, **22**, 115–127.

[30] E. J. Kelly and K. M. Forsythe (1989) "Adaptive detection and parameter estimation for multidimensional signal models", MIT Lincoln Laboratory Technical Report 848.

[31] C. L. Lawson and R. J. Hanson (1974) *Solving Least Squares Problems*, Prentice Hall, Englewood Cliffs, NJ.

[32] T. C. Liu and B. D. Van Veen (1991) "A modular structure for implementation of linearly constrained minimum variance beamformers", *IEEE Trans. Signal Process.*, **39**, 2343–2346.

[33] J. D. Marr (1986) "A selected bibliography on adaptive antenna arrays", *IEEE Trans. Aerosp. Electron Systems*, **22**, 781–798.

[34] R. A. Monzingo and T. W. Miller (1980) *Introduction to Adaptive Arrays*, Wiley, New York.

[35] D. R. Morgan (1978) "Partially adaptive array techniques", *IEEE Trans. Antennas Propag.*, **26**, 823–833.

[36] R. J. Muirhead (1982) *Aspects of Multivariate Statistical Theory*, Wiley, New York, Chap. 3.

[37] N. L. Owsley (1973) "A recent trend in adaptive spatial processing for sensor arrays: constrained adaptation", in *Signal Processing*, J. W. R. Griffiths et al. (Eds.), pp. 591–604.

[38] N. L. Owsley (1985) "Sonar array processing", in *Array Signal Processing*, S. Haykin (Ed.), Prentice Hall, Englewood Cliffs, NJ.

[39] C. M. Rader and A. O. Steinhardt (1986) "Hyperbolic householder transformations", *IEEE Trans. Acoust., Speech, Signal Process.*, **34**, 1589–1602.

[40] I. S. Reed, J. D. Mallet, and L. E. Brennan (1974) "Rapid convergence rate in adaptive arrays", *IEEE Trans. Aerosp. Electron. Systems*, **10**, 853–863.

[41] T. J. Shan and T. Kailath (1985) "Adaptive beamforming for coherent signals and interference", *IEEE Trans. Acoust., Speech, Signal Process.*, **33**, 527–536.

[42] B. D. Van Veen and R. A. Roberts (1987) "Partially adaptive beamformer design via output power minimization", *IEEE Trans. Acoust., Speech, Signal Process.*, **35**, 1524–1532.

[43] B. D. Van Veen (1988) "Eigenstructure based partially adaptive array design", *IEEE Trans. Antennas Propag.*, **36**, 357–362.

[44] B. D. Van Veen and K. M. Buckley (1988) "Beamforming: a versatile approach to spatial filtering", *IEEE Acoust., Speech, Signal Process. Mag.* **5**, 4–24.

[45] B. D. Van Veen (1989) "An analysis of several partially beamformer designs", *IEEE Trans. Acoust., Speech, Signal Process.*, **37**, 192–203.

[46] B. D. Van Veen (1990) "Optimization of quiescent response in partially adaptive beamformers", *IEEE Trans. Acoust., Speech, Signal Process.*, **38**, 471–477.

[47] B. D. Van Veen (1991) "Minimum Variance Beamforming With Soft Response Constraints", *IEEE Trans. Signal Process.*, **39**, pp. 1964–1972.

[48] B. D. Van Veen (1991) "Adaptive convergence of linearly constrained beamformers based on the sample covariance matrix", *IEEE Trans. Signal Process.*, **39**, pp. 1470–1473.

[49] J. Yang and M. Kaveh (1990) "Coherent signal-subspace transformation beamformer", *Proc. IEE*, Part F, **137**, 267–275.

[50] J. X. Zhu and H. Wang (1990) "Adaptive beamforming for correlated signal and interference: a frequency domain smoothing approach", *IEEE Trans. Acoust., Speech, Signal Process.*, **38**, 193–195.

[51] G. Zorpette (1986) "Monitoring the tests", *IEEE Spectrum*, **23**, 57–66.

5

BEAMSPACE ML BEARING ESTIMATION FOR ADAPTIVE PHASED ARRAY RADAR

MICHAEL D. ZOLTOWSKI

School of Electrical Engineering,
Purdue University,
West Lafayette, USA

5-1 INTRODUCTION

In the search mode, a radar system probes the search zone by sequentially transmitting pulses in stepped angular increments. Once an echo is detected the system converts to track mode. In track mode, the pulse repetition rate is increased so that the range and bearing of a target may be updated in a near continuous fashion. Typically, a number of tracks are monitored simultaneously. In this chapter, we consider the problem of estimating the target bearing while in track mode wherein a new estimate is generated with each new pulse.

Consider that a target is detected by forming a beam in the receive mode having a pointing angle equal to that of the transmit beam. Due to the finite dimensions of the radar array, the receive beam exhibits a mainlobe and a number of sidelobes. As measures are taken to achieve low sidelobes, we can say with some degree of confidence that the target is angularly located within the mainlobe of the receive beam. The width of the mainlobe is

Adaptive Radar Detection and Estimation, Edited by Simon Haykin and Allan Steinhardt.
ISBN 0-471-54468-X

inversely proportional to the length of the array aperture. The arc length of the angular sector encompassed by the mainlobe increases as the square of the range of the target. Thus, depending on the array aperture and, more importantly, the target range, this angular sector may correspond to a fairly large swath of space. Another measurement is needed to determine the bearing of the target more precisely. The typical mode of operation is to form a sum beam and a difference beam in the receive mode. In this chapter, we develop a number of maximum likelihood (ML) based bearing estimation schemes for use in various operational environments.

By way of introduction, Section 5-2 reviews the narrowband signal model described previously in Chapters 3 and 4. The beam pattern generated by weighting and summing the antenna element outputs is formulated as the Discrete Spatial Fourier Transform of the weighting sequence. Properties of phase-only steered beams formed in the receive mode are examined.

In laying the groundwork for later sections, Section 5-3 develops the ML estimator of the bearing of a single target given the outputs of two classical, phase-only steered beams formed in the receive mode. The respective pointing angles of these two beams are selected so that the two beams have all but one null in common. It is shown that through judicious exploitation of this property, the ML estimate of the target bearing may be formulated in terms of a simple trigonometric expression which is exact in the case of no noise. Computational simplicity is of utmost importance under monopulse operating conditions in which the target coordinates are updated at or near the pulse repetition rate. The common nulls property is the cornerstone of the remaining developments in the chapter.

The bearing estimator developed in Section 5-3 is of limited utility in actual radar systems due to the high sidelobes associated with classical, phase-only steered beams. Low sidelobe levels are important for filtering out unwanted returns such as clutter. Accordingly, Section 5-4 develops the ML estimator of the bearing of a single target given the respective outputs of sum and difference beams formed as linear combinations of classical, phase-only steered beams formed in the receive mode. It is shown that sum and difference beams having sidelobe levels comparable to those achieved with aperture tapering according to either the Hanning or Hamming windows may be synthesized via an appropriate linear combination of three classical, phase-only steered beams having all but two nulls in common. Again, judicious exploitation of the fact that the resulting sum and difference beams have all but two nulls in common facilitates formulation of the ML estimate of the target bearing in terms of a simple trigonometric expression.

In general, a sum beam is formed employing a symmetric (even) aperture taper while a difference beam is formed employing an anti-symmetric (odd) aperture taper. For the sake of completeness, Section 5.5 examines some approximations made in practice in order to achieve computationally simple bearing estimation schemes in the case of sum and difference beams that do not possess common nulls. A classical result involving the ratio of the difference beam to the sum beam is derived.

The low sidelobe levels achieved with sum and difference beams formed as linear combinations of classical, phase-only steered beams in accordance with the development in Section 5-3 may not be adequate in strong interference environments. For example, in some cases it may be necessary to form a broad null in both the sum and difference beams over an angular region in which clutter returns are strong. Alternatively, or, in addition, it may be necessary to modify the quiescent nulls of the sum and difference beams so that each exhibits a null in the direction of each of a number of strong interfering point sources. Section 5-6 develops simple procedures for constructing sum and difference beams having all but two nulls in common with a subset of the common nulls aligned with the bearings of strong interferences. Measures are taken in order to maintain low sidelobe levels as best as possible. As in the nonadaptive case, the commonality of all but two of the nulls amongst the two beams facilitates estimation of the bearing of the target of interest via a simple trigonometric expression. The various procedures examined differ in accordance with the information available regarding the interference environment. It is shown that if estimates of the interference directions are available, by applying spectral estimation during a period in which the radar is passively listening, for example, adaptive–adaptive [11] sum and difference beams exhibiting the desired properties may be synthesized via linear combinations of phase-only steered beams formed in the receive mode.

The chapter concludes with the development and analysis of a ML based bearing estimation scheme applicable in the case of mainbeam interference. Low-angle radar tracking is cited as a classical example of a mainbeam interference problem. In this case, echoes return from a target flying at a low altitude over a fairly smooth surface via a specular path as well as by a direct path. Due to the small angular separation between the direct and specular path signals, the placement of a null in the direction of the specular path signal causes the pointing angle of the sum beam, and hence, the direction of maximum gain, to squint away from the angle of the direct path signal. As the angular separation between the two signals becomes a smaller and smaller fraction of a beamwidth, the direct path signal becomes more and more buried in the noise leading to track breaking. Section 5-7 develops the ML estimator of the respective bearings of the direct and specular path signals given the outputs of three classical, phase-only steered beams having all but two nulls in common. Again, the commonality of nulls leads to a simple estimation scheme applicable under monopulse operating conditions. The performance of this scheme is analyzed in terms of the dependence on the relative phase difference between the direct and specular path signals. Frequency diversity is explored as a means for overcoming this phase dependence. The efficacy of incorporating the multipath geometry as a priori information is also examined.

5-2 ARRAY SIGNAL MODEL AND CLASSICAL BEAMFORMING

5-2-1 Narrowband Signal Model for Linear Array

Consider a linear array of N antenna elements equispaced by the amount d as depicted in Fig. 5-1. The target under surveillance at a particular instant in time is assumed to be in the far field of the array giving rise to a planewave model. If the target is not in the far field, knowledge of the array geometry and the coordinates of the range bin under scrutiny allows one to compute a phase correction to be applied to each element for the purpose of focusing into the far field [47].

A coordinate system is defined with the origin located at the center of the array aperture. The N elements of the array are located along the x-axis symmetrically positioned about the origin. If N is odd, the origin is located at the center element. If N is even, the origin is not located at an element but located between the two middle elements. Let the unit vector in the direction of the x-axis be denoted $\hat{\mathbf{i}}$. Further, let $\hat{\mathbf{r}}_0$ denote the unit position vector of the target. The target bearing, denoted θ_0, is the angle between $\hat{\mathbf{r}}_0$ and a line normal to the array situated in the plane spanned by $\hat{\mathbf{r}}_0$ and $\hat{\mathbf{i}}$. Let f be a frequency which lies in the passband of the bandpass filter at the front end of each element. Based on the discussion in Chapter 3, the $N \times 1$ array response vector for this uniformly spaced linear array scenario is

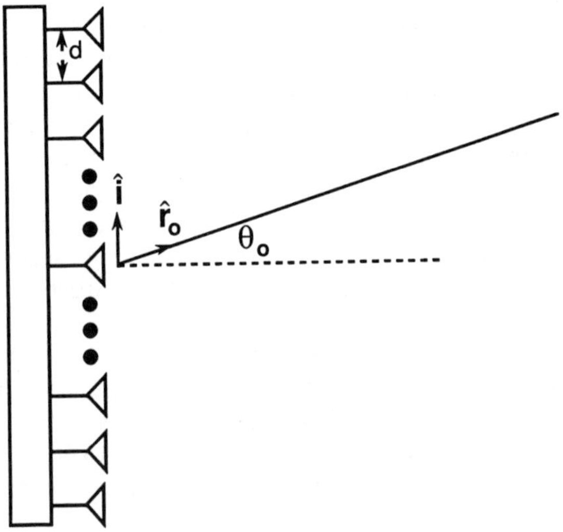

Figure 5-1 Array geometry and bearing coordinate system.

$$\mathbf{d}(\theta, f) = \begin{bmatrix} \exp\left[-j \dfrac{N-1}{2} 2\pi \dfrac{f}{c} d \sin\theta\right] \\ \exp\left[-j \dfrac{N-3}{2} 2\pi \dfrac{f}{c} d \sin\theta\right] \\ \vdots \\ \exp\left[j \dfrac{N-3}{2} 2\pi \dfrac{f}{c} d \sin\theta\right] \\ \exp\left[j \dfrac{N-1}{2} 2\pi \dfrac{f}{c} d \sin\theta\right] \end{bmatrix} \tag{5-1}$$

where c is the speed of light, the speed of electromagnetic waves in free space. Note, as a consequence of the choice of origin as the array center, the array response vector exhibits conjugate centrosymmetry. This property is described mathematically as

$$\tilde{\mathbf{I}}_N \mathbf{d}(\theta, f) = \mathbf{d}^*(\theta, f) \tag{5-2}$$

where $\tilde{\mathbf{I}}_N$ is the $N \times N$ reverse permutation matrix

$$\tilde{\mathbf{I}}_N = \begin{bmatrix} 0 & 0 & & 1 \\ 0 & 0 & & 0 \\ \vdots & \vdots & \cdot^{\cdot^{\cdot}} & \vdots \\ 0 & 1 & & 0 \\ 1 & 0 & & 0 \end{bmatrix} \qquad (N \times N) \tag{5-3}$$

As we shall see shortly, the conjugate centrosymmetry of $\mathbf{d}(\theta, f)$ gives rise to a real-valued beam pattern when $\mathbf{d}(\theta, f)$ evaluated at some specific angle is used as a beamforming weight vector. Note that $\tilde{\mathbf{I}}_N$ described by (5-3) is symmetric and equal to its own inverse, that is, $\tilde{\mathbf{I}}_N^T = \tilde{\mathbf{I}}_N$ and $\tilde{\mathbf{I}}_N \tilde{\mathbf{I}}_N = \mathbf{I}_N$. The latter property will prove useful at various points in our development.

In the receive mode, radar engineers typically work in the $u = \sin\theta$ domain [46, 47]. The mapping $u = \sin\theta$ is one-to-one if θ is restricted to the angular interval $-90° < \theta < 90°$, corresponding to the so called visible region. We may thus equivalently denote the array response vector as $\mathbf{d}(u, f)$. The bearing estimation schemes presented in this chapter will be developed primarily in the u domain. However, beam patterns and simulation results will be presented in the θ domain.

We initially consider narrowband operation at the frequency for which the elements are spaced by a half-wavelength. This frequency is denoted f_0. Multifrequency or wideband operation is discussed in Section 5-4-3. Half-wavelength spacing between elements is the spatial analogue of the Nyquist rate and avoids ambiguities with regard to the bearing angle over the visible region. Under these conditions the array response vector is simply denoted $\mathbf{d}(u)$ and may be expressed as

$$\mathbf{d}(u) = \begin{bmatrix} \exp\left[-j\,\frac{N-1}{2}\,\pi u\right] \\ \exp\left[-j\,\frac{N-3}{2}\,\pi u\right] \\ \vdots \\ \exp\left[j\,\frac{N-3}{2}\,\pi u\right] \\ \exp\left[j\,\frac{N-1}{2}\,\pi u\right] \end{bmatrix} \tag{5-4}$$

We will also invoke the narrowband signal model. Consider a single target at a bearing angle of θ_0. Further, let B denote the width of the passband of the narrowband filter at the front end of each element. It is assumed that B is commensurate with the bandwidth of the echo return. As discussed in Chapter 3, the narrowband signal model is appropriate if the following condition is met [36]

$$\frac{B}{f_0}\,Nd \sin\theta_0 \ll 1 \tag{5-5}$$

The basic assumption is that $B \ll f_0$ and that N is not too large. Also, it should be noted that the maximum swing from boresite in the case of a phased array is typically $\pm 60°$ [10].

Lwt $\mathbf{x}(n)$ denote the $N \times 1$ data snapshot vector. The kth element of $\mathbf{x}(n)$, denoted $\mathbf{x}_k(n)$, is the sample value at discrete time n of the analytic signal obtained after complex baseband demodulation of the signal received at the kth element, $k = 1, 2, \ldots, N$. Consider a single target at a bearing angle of θ_0. Assuming the narrowband signal model to be valid, $\mathbf{x}(n)$ may be expressed as

$$\mathbf{x}(n) = c_0(n)\mathbf{d}(u_0) + \mathbf{n}(n) \tag{5-6}$$

where $c_0(n)$ is the sample value of the complex envelope of the echo return measured at the nth snapshot. $\mathbf{n}(n)$ is an $N \times 1$ noise vector. The kth element of $\mathbf{n}(n)$, denoted $\mathbf{n}_k(n)$, is the sample value at discrete time n of the complex envelope of the narrowband noise present at the output of the kth receiver, $k = 1, \ldots, N$. The elements of $\mathbf{n}(n)$ constitute receiver generated noise and, as a consequence, are modeled as independent random variables. Other types of noise such as strong point interferers and clutter will be considered at a later point.

The characteristic feature of the narrowband signal model is that the complex envelope of the echo return is temporally coincident across the array; this is signified by the fact that $c_0(n)$ in (5-6) is a scalar quantity. The phase angle of $c_0(n)$ is the phase of the echo return at the array center. The role of $\mathbf{d}(u_0)$ in (5-6), therefore, is to account for the linear phase variation across the array due to the planar nature of the returning wavefront and the linear structure of the array.

5-2-2 Classical Beamforming

In the receive mode, a beam is formed as the weighted sum of each of the array element outputs. Let $\mathbf{w}$ denote an $N \times 1$ weight vector. The output of the spatial filter created by $\mathbf{w}$ is computed as

$$y(n) = \mathbf{w}^H \mathbf{x}(n) \tag{5-7}$$

Thus, in conforming to the definition of the inner product in a complex-valued Hilbert space, the complex conjugate of the kth element of $\mathbf{w}$, $\mathbf{w}_k^*$, is the weight applied to the output of the kth element. The spatial frequency response or array pattern associated with a specific set of weights is the spatial analogue of the Discrete Time Fourier Transform (DTFT) [47]:

$$f(u) = \sum_{k=0}^{N-1} \mathbf{w}_{k+1} \exp\left[-j\left(k - \frac{N-1}{2}\right)\pi u\right] = \mathbf{d}^H(u)\mathbf{w} \tag{5-8}$$

The far right hand side of (5-8) follows from the definition of $\mathbf{d}(u)$ in (5-4). Note that the Discrete Spatial Fourier Transform (DSFT) defined by (5-8) is periodic with period 2, and is typically plotted over the interval $-1 \leqslant u \leqslant 1$. This is in contrast to the DTFT of a discrete-time domain signal which is periodic with period 1.

The use of $\mathbf{d}(u)$ evaluated at a specific value of u, say $u = u_C$, as a weight vector is referred to as classical beamforming. The array pattern associated with $\mathbf{w} = \mathbf{d}(u_c)$ is simply evaluated as

$$\begin{aligned} f(u) = \mathbf{d}^H(u)\mathbf{d}(u_C) &= \sum_{k=0}^{N-1} \exp\left[-j\left(k - \frac{N-1}{2}\right)\pi(u - u_C)\right] \\ &= S_N(u - u_C) \end{aligned} \tag{5-9}$$

where $S_N(u)$ is defined as

$$S_N(u) = \frac{\sin\left(\frac{N}{2}\pi u\right)}{\sin\left(\frac{1}{2}\pi u\right)} \tag{5-10}$$

$S_N(u)$ is sometimes referred to as the periodic sinc function. Note that $f(u) = \mathbf{d}^H(u)\mathbf{d}(u_C)$ is a real-valued function of u regardless of the pointing angle u_C. Hence, $f(u) = \mathbf{d}^H(u)\mathbf{d}(u_C) = \mathbf{d}^H(u_C)\mathbf{d}(u)$. Also note that the output of the spatial filter created by $\mathbf{w} = \mathbf{d}(u_C)$, $\mathbf{y}(n) = \mathbf{d}^H(u_C)\mathbf{x}(n)$, is the DSFT of $\mathbf{x}(n)$ evaluated at $u = u_C$.

Returning to the case of a single target, consider the output of an arbitrary spatial filter when the input is the snapshot vector described by (5-6). Assume the receiver generated noise at each element to be independent and identically distributed (i.i.d.) such that $E\{\mathbf{n}(n)\mathbf{n}^H(n)\} = \sigma_n^2\mathbf{I}_N$. Since the magnitude of each component of $\mathbf{d}(u_0)$ is unity, the SNR at each element prior to beamforming is $\text{SNR}_E = |c(n)|^2/\sigma_n^2$. The SNR at the output of the spatial filter or beamformer is

$$\text{SNR}_B = \frac{|c_0(n)\mathbf{w}^H\mathbf{d}(u_0)|^2}{E\{|\mathbf{w}^H\mathbf{n}(n)|^2\}} = |c_0(n)|^2\,\frac{\mathbf{w}^H\mathbf{d}(u_0)\mathbf{d}^H(u_0)\mathbf{w}}{\sigma_n^2\mathbf{w}^H\mathbf{w}} \tag{5-11}$$

At this point, we refer to a classical theorem from linear algebra [37].

Theorem 5-1 The vector $\mathbf{x}$ which maximizes (minimizes) $\mathbf{x}^H\mathbf{B}\mathbf{x}/\mathbf{x}^H\mathbf{x}$, where $\mathbf{B}$ is Hermitian, is that eigenvector of $\mathbf{B}$ corresponding to the largest (smallest) eigenvalue.

This theorem will be invoked throughout. With regard to (5-11), the $N \times N$ matrix $\mathbf{d}(u_0)\mathbf{d}^H(u_0)$ has only one nonzero eigenvalue equal to $\mathbf{d}^H(u_0)\mathbf{d}(u_0) = N$ with corresponding eigenvector equal to $\mathbf{d}(u_0)$. Hence, maximum SNR at the beamformer output is achieved with $\mathbf{w} = \mathbf{d}(u_0)$. This result is not surprising since with this set of weights, the signal component of each element output is added up in-phase.

5-3 BEARING ESTIMATION FOR MONOPULSE RADAR

5-3-1 Two-Dimensional Beamspace Manifold

The model for the snapshot vector described by (5-6) assumes that there is a single target intercepted by the transmit pulse over a given range bin. If the array is uniformly illuminated with phase-only beam steering on transmit, the transmit beam pattern is the same as the receive pattern described by (5-8). Let u_C be the pointing angle of the transmit beam for a particular transmission period. In the receive mode, consider that a beam is formed with the weight vector $\mathbf{w} = \mathbf{d}(u_c)$ and that an echo is detected. Due to the directivity of both the transmit and receive beams, we say with some degree of confidence that a target is angularly located in the general vicinity of the mainlobe and possibly the first two sidelobes of the transmit/receive beam pattern corresponding to the interval $u_C - 4/N < u < u_C + 4/N$. Depending on N and, more importantly, the target range, this interval may correspond to a fairly large angular swath of space. Another measurement is needed to determine the bearing of the target more precisely. Hence, we consider the problem of estimating the target bearing given the outputs of two beams formed on return [46]. In phased array radar, the typical mode of operation is to form a sum beam and a difference beam on return. We first develop the maximum likelihood (ML) estimator of the target bearing given the outputs of a related pair of beams referred to as the left and right beams. A discussion of the use of sum and difference beams will follow.

In the initial phase of our study, we consider the left beam formed with $\mathbf{w}_L = \mathbf{d}(u_L)$, where $u_L = u_C - 1/N$, and the right beam formed with $\mathbf{w}_R = \mathbf{d}(u_R)$, where $u_R = u_C + 1/N$. These two vectors comprise the $N \times 2$ beamforming matrix

$$\mathbf{W}_{B2} = [\mathbf{d}(u_L) \,\vdots\, \mathbf{d}(u_R)] = [\mathbf{w}_L \,\vdots\, \mathbf{w}_R] \tag{5-12}$$

The angular positions of the left and right beams were chosen for a number of reasons. First, in conventional phased array radar with analog beamforming, the pointing angle of the transmit beam is typically stepped in increments of one-half the null-to-null beamwidth during scan mode [10, 46]. The null-to-null beamwidth of the array pattern achieved with uniform illumination described by (5-10) is $4/N$. Hence, the angular separation $2/N$ corresponds to one-half the null-to-null beamwidth. Second, with $u_R - u_L = 2/N$, $\mathbf{d}(u_L)$ and $\mathbf{d}(u_R)$ are orthogonal. This property is established by invoking (5-9):

$$\mathbf{d}^H(u_L)\mathbf{d}(u_R) = S_N(u_R - u_L) = \frac{\sin\left(\frac{N}{2}\pi\frac{2}{N}\right)}{\sin\left(\frac{1}{2}\pi\frac{2}{N}\right)} = 0 \tag{5-13}$$

Example 5-1 As an illustrative example, consider an $N = 21$ element array with $u_L = 0$ and $u_R = 2/21$. In terms of θ, the respective pointing angles are $\theta_L = 0$ and $\theta_R = \sin^{-1}(2/21) = 5.46°$. The respective normalized beam patterns as a function of θ are depicted in Fig. 5-2. The dB scale in Fig. 5-2(b) reveals the -13 dB peak sidelobes occurring with either beam due to the rectangular taper. Note that the pointing angle of the left beam is located at a null of the right beam and vice versa. More importantly, though, note that the two beams have $N - 2 = 21 - 2 = 19$ nulls in common. This property will be exploited shortly.

The $N \times 2$ beamforming matrix $\mathbf{W}_{B2}$ described by (5-12) may be interpreted as performing a transformation from N-dimensional element space to two-dimensional beamspace according to

$$\mathbf{x}_{B2}(n) = \mathbf{W}_{B2}^H \mathbf{x}(n) = \begin{bmatrix} \mathbf{d}^H(u_L)\mathbf{x}(n) \\ \mathbf{d}^H(u_R)\mathbf{x}(n) \end{bmatrix} \tag{5-14}$$

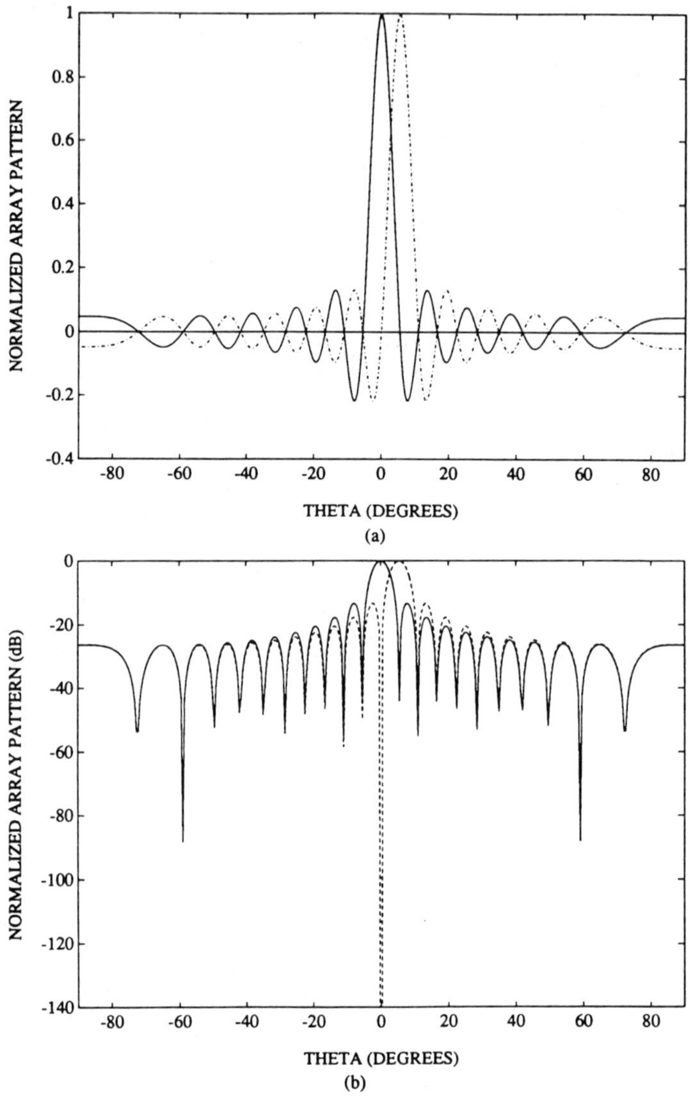

Figure 5-2 Left and right beam patterns with rectangular amplitude window and pointing angles of $u_L = 0$ ($\theta_L = 0°$) and $u_R = 2/N$ ($\theta_R = 5.46°$), respectively, where $N = 21$.

$\mathbf{x}_{B2}(n)$ is referred to as the beamspace domain snapshot vector and may be expressed in a format similar to that in (5-6) for $\mathbf{x}(n)$, now referred to as the element space snapshot vector.

$$\mathbf{x}_{B2}(n) = c_0(n)\begin{bmatrix} \mathbf{d}^H(u_L)\mathbf{d}(u_0) \\ \mathbf{d}^H(u_R)\mathbf{d}(u_0) \end{bmatrix} + \mathbf{W}_{B2}^H\mathbf{n}(n) = c_0(n)\mathbf{b}_{B2}(u_0) + \mathbf{n}_B(n) \tag{5-15}$$

where

$$\mathbf{b}_{B2}(u) = \begin{bmatrix} \mathbf{d}^H(u_L)\mathbf{d}(u) \\ \mathbf{d}^H(u_R)\mathbf{d}(u) \end{bmatrix} = \begin{bmatrix} S_N(u - u_L) \\ S_N(u - u_R) \end{bmatrix} \tag{5-16}$$

is referred to as the beamspace manifold vector. Recall that $S_N(u)$ was defined previously in (5-10). Note that $\mathbf{b}_{B2}(u)$ is a 2×1 real-valued vector for all u. $\mathbf{n}_B(n) = \mathbf{W}_{B2}^H \mathbf{n}(n)$ is referred to as the beamspace domain noise vector and, in general, is complex valued.

If N, the number of antenna elements, is relatively large, each of the two components of $\mathbf{n}_B(n)$ are weighted sums of a large number of random noise variates. In light of the assumed independence of the noise in element space, it is reasonable to invoke the Central Limit Theorem and approximate the noise in beamspace as being Gaussian distributed. Note that $E\{\mathbf{n}_B(n)\mathbf{n}_B^H(n)\} = \mathbf{W}_{B2}^H E\{\mathbf{n}(n)\mathbf{n}^H(n)\}\mathbf{W}_{B2}$. If we further assume the noise at each element is of equal power such that $E\{\mathbf{n}(n)\mathbf{n}^H(n)\} = \sigma_n^2 \mathbf{I}_N$, it follows from the orthogonality of the left and right beamforming vectors that $E\{\mathbf{n}_B(n)\mathbf{n}_B^H(n)\} = N\sigma_n^2 \mathbf{I}_2$. Note that if the noise at each element is not exactly equal but of roughly the same magnitude, it may be argued that the two components of $\mathbf{n}_B(n)$ are approximately independent and of equal power [12]. Correspondingly, the development of the ML estimator is based on the assumption that $\mathbf{n}_B(n)$ is a multivariate Gaussian random vector distributed according to $N(\mathbf{0}, N\sigma_n^2 \mathbf{I}_2)$.

5-3-2 Development of Two-Dimensional Beamspace Domain ML Bearing Estimator

To distinguish it from the element space ML estimator, the ML estimator of u_0 given the outputs of the left and right beams is referred to as the two-dimensional beamspace domain maximum likelihood (2D-BDML) estimator. In addition to the assumptions described above, the complex amplitude $c_0(n)$ is assumed to be unknown but deterministic. The appropriate likelihood function is then

$$\underset{u_0, c_0(n)}{\text{Maximize}} \; \frac{1}{\pi^2} \frac{1}{N^2\sigma_n^4} \exp\left[- \frac{1}{N\sigma_n^2} \|\mathbf{x}_{B2}(n) - \mathbf{b}_{B2}(u_0)c_0(n)\|^2 \right] \tag{5-17}$$

Maximization of the natural logarithm of the likelihood function, with additive constants discarded, leads to minimization of the negative of the exponent in (5-17).

$$\underset{u_0, c_0(n)}{\text{Minimize}} \; \|\mathbf{x}_{B2}(n) - c_0(n)\mathbf{b}_{B2}(u_0)\|^2 \tag{5-18}$$

At this point, the 2D-BDML estimator involves a search over the 2-D parameter space $\{u_0, c_0(n)\}$. One way of performing this search is to vary u_0 setting the value of the objective function for a specific value of u_0 equal to the smallest possible value over the domain of $c_0(n)$. For a specific value of u_0, the value of $c_0(n)$ which minimizes the objective function in (5-18) is the least square error solution $c_{0,\mathrm{LS}}(n) = \mathbf{b}_{B2}^T(u_0)\mathbf{x}_{B2}(n) / \mathbf{b}_{B2}^T(u_0)\mathbf{b}_{B2}(u_0)$. Substitution of $c_{0,\mathrm{LS}}(n)$ in (5-18) yields, after some algebraic manipulation, an objective function to be minimized over u_0 only

$$\underset{u_0}{\text{Minimize}} \left\| \left\{ \mathbf{I} - \frac{\mathbf{b}_{B2}(u_0)\mathbf{b}_{B2}^T(u_0)}{\mathbf{b}_{B2}^T(u_0)\mathbf{b}_{B2}(u_0)} \right\} \mathbf{x}_{B2}(n) \right\|^2 = \mathbf{x}_{B2}^H(n)\mathbf{P}_B^{\perp}(u_0)\mathbf{x}_{B2}(n) \tag{5-19}$$

where $\mathbf{P}_B^{\perp}(u_0) = \mathbf{I} - \mathbf{b}_{B2}(u_0)\mathbf{b}_{B2}^T(u_0) / \mathbf{b}_{B2}^T(u_0)\mathbf{b}_{B2}(u_0)$.

Note that for a specific value of u_0, $\mathbf{P}_B^{\perp}(u_0)$ is a real-valued, 2×2 projection operator. Two main properties of a projection operator are symmetry, $\mathbf{P}_B^{\perp T}(u_0) = \mathbf{P}_B^{\perp}(u_0)$, and idempotence, $\mathbf{P}_B^{\perp}(u_0)\mathbf{P}_B^{\perp}(u_0) = \mathbf{P}_B^{\perp}(u_0)$. Since $\mathbf{P}_B^{\perp}(u_0)\mathbf{b}_{B2}(u_0) = \mathbf{0}_2$, it follows that $\mathbf{P}_B^{\perp}(u_0)$ is a projection operator onto a 1-D space equal to the orthogonal complement of the 1-D space spanned by $\mathbf{b}_{B2}(u_0)$. For a specific value of u_0, $\mathbf{P}_B^{\perp}(u_0)$ may thus be expressed as $\mathbf{P}_B^{\perp}(u_0) = \mathbf{v}\mathbf{v}^T/\mathbf{v}^T\mathbf{v}$, where $\mathbf{v}$ is a real-valued 2×1 vector orthogonal to $\mathbf{b}_{B2}(u_0)$. Prompted by these observations, consider transforming the search over $\mathbf{u}_0$ into a search over a real-valued 2×1 vector of unit length related to u_0 as

$$\mathbf{v}^T\mathbf{b}_{B2}(u_0) = 0 \tag{5-20}$$

This transformation of search variables is justified if we can show that for every value of u_0 in the interval $-1 < u_0 < 1$, there is a unique 2×1 real-valued vector of unit length satisfying (5-20), that is, orthogonal to $\mathbf{b}_{B2}(u_0)$. We will show shortly that this is indeed the case except for a subset of the interval $-1 < u_0 < 1$ of measure zero.

The motivation for this transformation of search variables is that it allows us to convert the nonlinear search over u_0 in (5-19) into a simple eigenvalue/eigenvector problem. This is achieved via the substitution of $\mathbf{P}_B^{\perp}(u_0) = \mathbf{v}\mathbf{v}^T/\mathbf{v}^T\mathbf{v}$ in (5-19):

$$\underset{\mathbf{v}}{\text{Minimize}} \frac{\mathbf{x}_{B2}^H(n)\mathbf{v}\mathbf{v}^T\mathbf{x}_{B2}(n)}{\mathbf{v}^T\mathbf{v}} = \frac{\mathbf{v}^T \operatorname{Re}\{\mathbf{x}_{B2}(n)\mathbf{x}_{B2}^H(n)\}\mathbf{v}}{\mathbf{v}^T\mathbf{v}} \tag{5-21}$$

The expression for the objective function on the far right hand side of (5-21) is deduced by invoking the fact that $\mathbf{v}$ is real and the fact that $\mathbf{x}_{B2}(n)\mathbf{x}_{B2}^H(n)$ is a Hermitian matrix. As a consequence, the real part of $\mathbf{x}_{B2}(n)\mathbf{x}_{B2}^H(n)$ is symmetric while the imaginary part is antisymmetric. It is easily shown that if $\mathbf{A}$ is a real-valued, antisymmetric matrix, that is, $\mathbf{A}^T = -\mathbf{A}$, and $\mathbf{v}$ is real, then $\mathbf{v}^T\mathbf{A}\mathbf{v} = 0$. Returning to (5-21), Theorem 5-1 dictates that the solution for $\mathbf{v}$ is that eigenvector of $\operatorname{Re}\{\mathbf{x}_{B2}(n)\mathbf{x}_{B2}^H(n)\}$ associated with the smaller of the two eigenvalues.

We now validate the claim that the relationship in (5-20) represents a one-to-one mapping between the set of all 2×1 distinct, real-valued vectors $\mathbf{v}$ of unit length and u_0 for $-1 < u_0 < 1$. Substituting $\mathbf{v} = [v_1 \quad v_2]^T$ and the component-wise expression for $\mathbf{b}_{B2}(u_0)$ in (5-16) into (5-20) yields

$$v_1 S_N(u_0 - u_L) + v_2 S_N(u_0 - u_R) = 0 \qquad \text{(5-22)}$$

Since $u_R = u_L + 2/N$, $\sin[N\pi(u_0 - u_R)/2] = \sin[N\pi(u_0 - u_L)/2 - \pi] = -\sin[N\pi(u_0 - u_L)/2]$. Invoking this fact, (5-22) may be expressed, after some algebraic manipulation, as

$$\frac{v_2}{v_1} = \frac{\sin\left(\frac{1}{2}\pi(u_0 - u_R)\right)}{\sin\left(\frac{1}{2}\pi(u_0 - u_L)\right)} = \frac{\sin\left(\frac{1}{2}\pi(u_0 - u_C) - \frac{\pi}{2N}\right)}{\sin\left(\frac{1}{2}\pi(u_0 - u_C) + \frac{\pi}{2N}\right)} \qquad \text{(5-23)}$$

where $u_C = (u_L + u_R)/2 = u_L + 1/N = u_R - 1/N$. Simple trigonometry yields the final relationship

$$u_0 = u_C + \frac{2}{\pi}\tan^{-1}\left\{\left[\frac{v_1 + v_2}{v_1 - v_2}\right]\tan\left(\frac{\pi}{2N}\right)\right\} \qquad \text{(5-24)}$$

The 2D-BDML estimate of u_0 is determined by substituting the components of the eigenvector of $\text{Re}\{\mathbf{x}_{B2}(n)\mathbf{x}_{B2}^H(n)\}$ associated with the smaller of the two eigenvalues.

To illustrate that (5-24) represents a one-to-one transformation, recall that $\mathbf{v}$ is a unit length vector such that the point (v_1, v_2) lies on the unit circle in the v_1-v_2 plane. Consider Fig. 5-3. We need to show that as the point (v_1, v_2) varies over a 180° arc of the unit circle in a continuous fashion, the variation in u_0 according to (5-24) is continuous and over an interval of length 2. To this end, consider the following five test cases cited in Fig. 5-3:

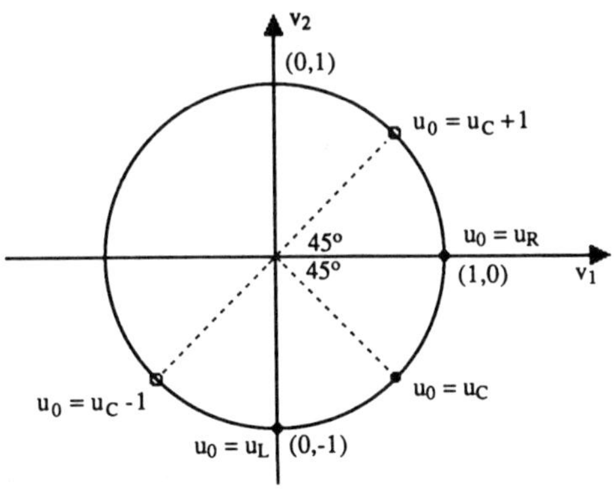

Figure 5-3 Mapping between target bearing and unit-length vector orthogonal to corresponding beamspace manifold.

Case 1. As (v_1, v_2) approaches clockwise the point $(-1\sqrt{2}, -1/\sqrt{2})$, $(v_1 + v_2)/(v_1 - v_2)$ approaches $-\infty$ and u_0 according to (5-24) approaches $u_C - 1$.

Case 2. With $(v_1, v_2) = (0, -1)$, substitution in (5-24) yields $u_0 = u_C - 1/N = u_L$.

Case 3. With $(v_1, v_2) = (1/\sqrt{2}, -1/\sqrt{2})$, substitution in (5-24) yields $u_0 = u_C$.

Case 4. With $(v_1, v_2) = (1, 0)$, substitution in (5.24) yields $u_0 = u_C + 1/N = u_R$.

Case 5. As (v_1, v_2) approaches counterclockwise the point $(1/\sqrt{2}, 1/\sqrt{2})$, $(v_1 + v_2)/(v_1 - v_2)$ approaches ∞ and u_0 according to (5-24) approaches $u_C + 1$.

The nonlinear character of the mapping is apparent. The interval $u_L < u_0 < u_R$ of length $2/N$ is mapped into a segment of the unit circle having an arc length of 90°. On the other hand, the interval $u_C - 1 < u_0 < u_L$ of length $1 - 1/N$ is mapped into a segment of the unit circle having an arc length of 45° as is the interval $u_R < u_0 < u_C + 1$, also of length $1 - 1/N$. This makes sense, of course, since the angular interval $u_L < u_0 < u_R$ is effectively emphasized in the beamforming process described by (5-14). In the practical case of noisy data, the range of interest for u_0 is the interval $u_L - 1/N < u_0 < u_R + 1/N$ corresponding to the interval between the lower 3 dB point of the left beam and the upper 3 dB point of the right beam.

To instill confidence in the 2D-BDML estimator, consider the ideal case of noiseless data. Recall the $\mathbf{v}$ is computed as that eigenvector of $\text{Re}\{\mathbf{x}_{B2}(n)\mathbf{x}_{B2}^H(n)\}$ associated with the smaller of the two eigenvalues. With $\mathbf{n}_B(n) = \mathbf{0}$ in (5-15), $\text{Re}\{\mathbf{x}_{B2}(n)\mathbf{x}_{B2}^H(n)\} = |c_0(n)|^2\mathbf{b}_{B2}(u_0)\mathbf{b}_{B2}^T(u_0)$. This 2×2 real-valued matrix has one nonzero eigenvalue equal to $|c_0(n)|^2\mathbf{b}_{B2}^T(u_0)\mathbf{b}_{B2}(u_0)$ with corresponding eigenvector equal to $\mathbf{b}_{B2}(u_0)$. In accordance with 2D-BDML, $\mathbf{v}$ is the eigenvector associated with the eigenvalue zero. Since the eigenvectors of a symmetric matrix are mutually orthogonal, it follows that $\mathbf{v}$ is orthogonal to $\mathbf{b}_{B2}(u_0)$. Five different target locations are considered in Table 5-1. The respective target locations considered in cases II, III, and IV are exactly the same as those considered in cases 2, 3, and 4, respectively, discussed above. In case II, the target is located at the pointing angle of the left beam corresponding to a null of the right beam. Hence, $\mathbf{b}_{B2}(u_0) = \mathbf{b}_{B2}(u_L) = [N \quad 0]^T$ such that $\mathbf{v} = [0 \quad -1]^T$ as indicated. It follows from the discussion of case 2 above that 2D-BDML yields the correct value of $\hat{u}_0 = u_L$ under these conditions. Similar comments hold with regard to case IV in which the target is located at the pointing angle of the right beam. In case III, the target is located at $u_0 = u_C$ where the response of the left beam is the same as that of the right beam. In this case, the two components of $\mathbf{b}_{B2}(u_0)$ are equal such that $\mathbf{v} = [(1/\sqrt{2} \quad -1/\sqrt{2}]^T$. Again, 2D-BDML provides the correct value of $u_0 = u_C$.

TABLE 5-1

Case	Target Angle	Beamspace Manifold Vector	Unit Vector Orthogonal to $\mathbf{b}_{B2}(u_0)$
I	$u_0 = u_L - \frac{2}{N}$	$\mathbf{b}_{B2}\left(u_L - \frac{2}{N}\right) = \begin{bmatrix} 0 \\ 0 \end{bmatrix}$	Ambiguity
II	$u_0 = u_L$	$\mathbf{b}_{B2}(u_L) = \begin{bmatrix} N \\ 0 \end{bmatrix}$	$\mathbf{v} = \begin{bmatrix} 0 \\ -1 \end{bmatrix}$
III	$u_0 = u_C$	$\mathbf{b}_{B2}(u_C) = \frac{1}{\sin\left(\frac{\pi}{2N}\right)} \begin{bmatrix} 1 \\ 1 \end{bmatrix}$	$\mathbf{v} = \begin{bmatrix} \frac{1}{\sqrt{2}} \\ -\frac{1}{\sqrt{2}} \end{bmatrix}$
IV	$u_0 = u_R$	$\mathbf{b}_{B2}(u_R) = \begin{bmatrix} 0 \\ N \end{bmatrix}$	$\mathbf{v} = \begin{bmatrix} 1 \\ 0 \end{bmatrix}$
V	$u_0 = u_R + \frac{2}{N}$	$\mathbf{b}_{B2}\left(u_R + \frac{2}{N}\right) = \begin{bmatrix} 0 \\ 0 \end{bmatrix}$	Ambiguity

In both case I and case V, the target is located at a null of both the left and right beams such that $\mathbf{b}_{B2}(u_0) = [0 \quad 0]^T$. Recall Example 5-1 where it was observed that if $u_R - u_L = 2/N$, the left and right beam patterns have $N - 2$ null locations in common. This phenomenon is depicted in Fig. 5-2 for the case of $N = 21$, $u_L = 0$ and $u_R = 2/N$. If $\mathbf{b}_{B2}(u_0) = [0 \quad 0]^T$, we can only stipulate that the target is located at one of these common null positions. Of course, the point is moot as a target located at one of these positions will not be detected.

5-3-3 Polynomial Formulation

Invoking the common nulls property of the left and right beams, it is possible to derive (5-24) from (5-20) in an alternative fashion based on the roots of appropriately formed left and right polynomials. This alternative derivation is included as it provides insight that will prove valuable in the development of procedures for constructing adaptive left and right beams with prescribed nulls in the directions of strong point interferers.

Recall that the beamspace manifold is defined as $\mathbf{b}_{B2}(u) = \mathbf{W}_{B2}^H \mathbf{d}(u)$. Thus, the relationship $\mathbf{v}^T \mathbf{b}_{B2}(u) = 0$ in (5-20) may be alternatively expressed as

$$\mathbf{v}^T \mathbf{W}_{B2}^H \mathbf{d}(u) = v_1 \mathbf{d}^H(u_L)\mathbf{d}(u) + v_2 \mathbf{d}^H(u_R)\mathbf{d}(u) = 0 \tag{5-25}$$

Invoking the definition of $\mathbf{d}(u)$ in (5-4) and factoring out $\exp[-j\pi \frac{1}{2}(N-1) u]$ yields

$$\exp\left[-j\pi \frac{N-1}{2} u\right]\left\{v_1 \sum_{k=0}^{N-1} \mathbf{d}_k^*(u_L) \exp[jk\pi u] + v_2 \sum_{k=0}^{N-1} \mathbf{d}_k^*(u_R) \exp[jk\pi u]\right\} = 0 \tag{5-26}$$

Defining the $(N-1)$th order polynomials $w_L^*(z)$ and $w_R^*(z)$ as

$$w_L^*(z) = \sum_{k=0}^{N-1} \mathbf{d}_k^*(u_L) z^k \tag{5-27a}$$

$$w_R^*(z) = \sum_{k=0}^{N-1} \mathbf{d}_k^*(u_R) z^k \tag{5-27b}$$

(5-25) may be interpreted as dictating that $z = \exp[j\pi\hat{u}_0]$ be determined as a root of

$$w_{LR}^*(z) = v_1 w_L^*(z) + v_2 w_R^*(z) \tag{5-28}$$

$w_L^*(z)$ is referred to as the left polynomial and $w_R^*(z)$ as the right polynomial.

Let us examine the roots of the left and right polynomials. The roots of $w_L^*(z)$ are related to the nulls of the left beam pattern according to the transformation $z = \exp[j\pi u]$. The left beam pattern is $S_N(u - u_L)$ where $S_N(u)$ is defined in (5-10). Hence, the nulls of the left beam pattern occur at $u = u_L \pm m2/N$ for integer $m \neq 0$ satisfying $-1 \leq u_L \pm m2/N \leq 1$, corresponding to those nulls that lie within the visible region. It follows that the roots of $w_L^*(z)$ occur at $z_m = \exp[j\pi(u_L + m2/N)]$, $m = 1, \ldots, N-1$. Similarly, the nulls of the right beam pattern occur at $u = u_R \pm m2/N$ for integer $m \neq 0$ satisfying $-1 \leq u_R \pm m2/N \leq 1$ such that the roots of $w_R^*(z)$ occur at $z_m = \exp[j\pi(u_R + m2/N)]$, $m = 1, \ldots, N-1$. Since $u_R = u_L + 2/N$, it follows that the left and right beams have $N-2$ nulls in common as asserted previously. This also implies that $w_L^*(z)$ and $w_R^*(z)$ have $N-2$ roots in common located at $z_m = \exp[j\pi(u_L + m2/N)]$, $m = 2, \ldots, N-1$.

Let $c_{B2}^*(z)$ denote the monic polynomial of order $N-2$ whose roots are these $N-2$ common roots, that is,

$$c_{B2}^*(z) = \prod_{m=2}^{N-1} (z - \exp[j\pi(u_L + m2/N)]) \tag{5-29}$$

Given the definition of $\mathbf{d}(u)$ in (5-4), it follows that the coefficient of the highest power in $w_L^*(z)$ defined in (5-27a) is $\exp[-j\pi\frac{1}{2}(N-1)u_L]$. Similarly, the coefficient of the highest power in $w_R^*(z)$ defined in (5-27b) is $\exp[-j\pi\frac{1}{2}(N-1)u_R]$. Therefore,

$$w_L^*(z) = \exp\left[-j\pi \frac{N-1}{2} u_L\right](z - \exp[j\pi u_R])c_{B2}^*(z) \tag{5-30a}$$

$$w_R^*(z) = \exp\left[-j\pi \frac{N-1}{2} u_R\right](z - \exp[j\pi u_L])c_{B2}^*(z) \qquad \text{(5-30b)}$$

Substituting in (5-28) and factoring yields

$$\begin{aligned} w_{LR}^*(z) &= v_1 w_L^*(z) + v_2 w_R^*(z) \\ &= \left\{v_1 \exp\left[-j\pi \frac{N-1}{2} u_L\right](z - \exp[j\pi u_R]) \right. \qquad \text{(5-31)} \\ &\quad \left. + v_2 \exp\left[-j\pi \frac{N-1}{2} u_R\right](z - \exp[j\pi u_L])\right\} c_{B2}^*(z) \end{aligned}$$

It is apparent that the root of interest, $\hat{z}_0 = \exp[j\pi \hat{u}_0]$, is the root of the first order polynomial in brackets. Denoting this polynomial as $e(z) = e_0 + e_1 z$, the coefficients may be expressed as

$$\begin{aligned} e_0 = -v_1 \exp[j\pi u_R] \exp\left[-j\pi \frac{N-1}{2} u_L\right] - v_2 \exp[j\pi u_L] \\ \times \exp\left[-j\pi \frac{N-1}{2} u_R\right] \end{aligned} \qquad \text{(5-32a)}$$

$$e_1 = v_1 \exp\left[-j\pi \frac{N-1}{2} u_L\right] + v_2 \exp\left[-j\pi \frac{N-1}{2} u_R\right] \qquad \text{(5-32b)}$$

The root of $e(z) = e_0 + e_1 z$ is simply $\hat{z}_0 = -e_0/e_1$. Substituting the expressions for e_0 and e_1 above and invoking the fact that $u_R = u_L + 2/N$ yields after some simple algebraic manipulation

$$\hat{z}_0 = \left\{\frac{v_1 \exp[j\pi/2N] - v_2 \exp[-j\pi/2N]}{v_1 \exp[-j\pi/2N] - v_2 \exp[j\pi/2N]}\right\} \exp[j\pi u_C] \qquad \text{(5-33)}$$

where $u_C = (u_L + u_R)/2$ as defined previously.

Recall that $\hat{z}_0 = \exp[j\pi \hat{u}_0]$. $\hat{u}_0$ may be determined from $\hat{z}_0$ in the obvious manner. Since v_1 and v_2 are real, the quantity in brackets on the right hand side (5-33) is of the form γ/γ^* and, hence, lies on the unit circle. As the phase angle of $\hat{z}_0$ is $\pi \hat{u}_0$, (5-33) dictates $\hat{u}_0 = u_C + 2 \arg\{\gamma\}/\pi$, where $\gamma = v_1 \exp[j\pi/2N] - v_2 \exp[-j\pi/2N] = (v_1 - v_2) \cos(\pi/2N) + j(v_1 + v_2) \sin(\pi/2N)$. Hence,

$$\arg\{\gamma\} = \tan^{-1}\left\{\left[\frac{v_1 + v_2}{v_1 - v_2}\right] \tan\left(\frac{\pi}{2N}\right)\right\} \qquad \text{(5-34)}$$

and substitution in $\hat{u}_0 = u_C + 2 \arg\{\gamma\}/\pi$ yields the previously derived relationship between u_0 and $\mathbf{v} = [v_1 \quad v_2]^T$ described by (5-24).

5-4 CLASSICAL TAPERING AND SUM/DIFFERENCE BEAMS

5-4-1 Sum/Difference Beam Properties

In conventional monopulse radar systems, the bearing of a particular target is typically estimated given the respective outputs of so called sum and difference beams [46]. Let $\mathbf{s}(u_C)$ and $\mathbf{a}(u_C)$ denote sum and difference beamforming vectors, respectively; u_C is the pointing angle of each of the two beams. The general form of $\mathbf{s}(u_C)$ and $\mathbf{a}(u_C)$, respectively, is

$$\mathbf{s}(u_C) = \mathbf{D}_s\mathbf{d}(u_C), \qquad \text{where } \mathbf{D}_s = \text{diag}\{s_0, s_1, \ldots, s_1, s_0\} \quad (N \times N) \tag{5-35a}$$

$$\mathbf{a}(u_C) = j\mathbf{D}_a\mathbf{d}(u_C), \qquad \text{where } \mathbf{D}_a = \text{diag}\{a_0, a_1, \ldots, -a_1, -a_0\} \quad (N \times N) \tag{5-35b}$$

The diagonal elements of $\mathbf{D}_s$ are real and represent a symmetric taper applied across the array aperture in forming the sum beam. The diagonal elements of $\mathbf{D}_a$ are real and represent an antisymmetric taper applied across the array aperture in forming the difference beam. Note that $\mathbf{s}(u_C)$ and $\mathbf{a}(u_C)$ are both conjugate centrosymmetric, that is, $\tilde{\mathbf{I}}_N\mathbf{s}(u_C) = \mathbf{s}^*(u_C)$ and $\tilde{\mathbf{I}}_N\mathbf{a}(u_C) = \mathbf{a}^*(u_C)$. As a consequence, the sum and difference beam patterns defined as

$$\text{sum}(u) = \mathbf{d}^H(u)\mathbf{s}(u_C) \triangleq \text{sum beam pattern} \tag{5-36a}$$

$$\text{diff}(u) = \mathbf{d}^H(u)\mathbf{a}(u_C) \triangleq \text{difference beam pattern} \tag{5-36b}$$

are each a real-valued function of u. More importantly, though sum(u) is an even function of u about $u = u_C$, that is, sum$(u_C + \delta) = $ sum$(u_C - \delta)$ while diff(u) is an odd function of u about $u = u_C$, that is, diff$(u_C + \delta) = -$diff$(u_C - \delta)$ due to the antisymmetry of the taper. The latter observation implies that diff$(u_C) = 0$.

Note, if the factor j on the right hand side of (5-35b) was not included in the definition of $\mathbf{a}(u_C)$, the difference beam pattern would be a purely imaginary function of u. That is, defining $\mathbf{a}_r(u_C) = \mathbf{D}_a\mathbf{d}(u_C)$, $d^H(u)\mathbf{a}_r(u_C)$ is a purely imaginary function of u. For our development here, we will find it useful to include the factor j. It should be pointed out, however, that this is not convention. However, with regard to implementation, the difference is minor. If in practice the difference beam is formed employing the weight vector $\mathbf{a}_r(u_c)$ instead of $\mathbf{a}(u_C)$, the result is simply an interchange of the real and imaginary parts, that is, $\text{Re}\{\mathbf{a}^H(u_C)\mathbf{x}(\mathbf{n})\} = \text{Im}\{\mathbf{a}_r^H(u_C)\mathbf{x}(\mathbf{n})\}$ and $\text{Im}\{\mathbf{a}^H(u_C)\mathbf{x}(\mathbf{n})\} = -\text{Re}\{\mathbf{a}_r^H(u_C)\mathbf{x}(\mathbf{n})\}$.

5-4-2 Linear Combinations of Two Orthogonal Classical Beams

As an example, consider sum and difference beamforming vectors defined as

$$\mathbf{s}(u_C) = \frac{1}{\sqrt{2}}\{\mathbf{d}(u_L) + \mathbf{d}(u_R)\} \tag{5-37a}$$

$$\mathbf{a}(u_C) = \frac{1}{\sqrt{2}}\{\mathbf{d}(u_L) - \mathbf{d}(u_R)\} \tag{5-37b}$$

where $u_L = u_C - 1/N$ and $u_R = u_C + 1/N$.

With regard to the structure of $\mathbf{s}(u_C)$ and $\mathbf{a}(u_C)$ in (5.35a) and (5-35b), respectively, the sum beamforming vector defined by (5-37a) corresponds to a cosine taper while the difference beamforming vector defined by (5-37b) corresponds to a sine taper. The former claim is validated by analyzing a specific component of $\mathbf{s}(u_C)$:

$$\begin{aligned}
\mathbf{s}_k(u_C) &= \frac{1}{\sqrt{2}}\{\mathbf{d}_k(u_L) + \mathbf{d}_k(u_R)\} \\
&= \frac{1}{\sqrt{2}}\exp\left[-j\left(k - \frac{N-1}{2}\right)\pi\left(u_C - \frac{1}{N}\right)\right] \\
&\quad + \frac{1}{\sqrt{2}}\exp\left[-j\left(k - \frac{N-1}{2}\right)\pi\left(u_C + \frac{1}{N}\right)\right] \\
&= \sqrt{2}\cos\left[\frac{\pi}{N}\left(k - \frac{N-1}{2}\right)\right]\exp\left[-j\left(k - \frac{N-1}{2}\right)\pi u_C\right], \\
&\quad k = 0, 1, \ldots, N-1
\end{aligned} \tag{5-38a}$$

Likewise, it is easily shown that for $k = 0, 1, \ldots, N-1$,

$$\mathbf{a}_k(u_C) = j\sqrt{2}\,\sin\left[\frac{\pi}{N}\left(k - \frac{N-1}{2}\right)\right]\exp\left[-j\left(k - \frac{N-1}{2}\right)\pi u_C\right] \tag{5-38b}$$

Example 5-1 (continued) The array patterns generated by sum and difference beamforming weight vectors defined by (5-37) with $N = 21$ and $u_C = 1/N$ ($\theta_C = 2.73°$) are shown in Fig. 5-4. Note, in this case $u_L = 0$ ($\theta_L = 0°$) and $u_R = 2/21$ ($\theta_R = 5.46°$). In effect, the sum beam pattern is the addition of the left and right beam patterns depicted in Fig. 5-2. Likewise, the difference beam pattern is the difference between the two patterns, that is, the right beam subtracted from the left beam. Note the reduced sidelobes of the sum beam relative to that associated with either the left or right beams are depicted in Fig. 5-2. This may be explained by observing that the

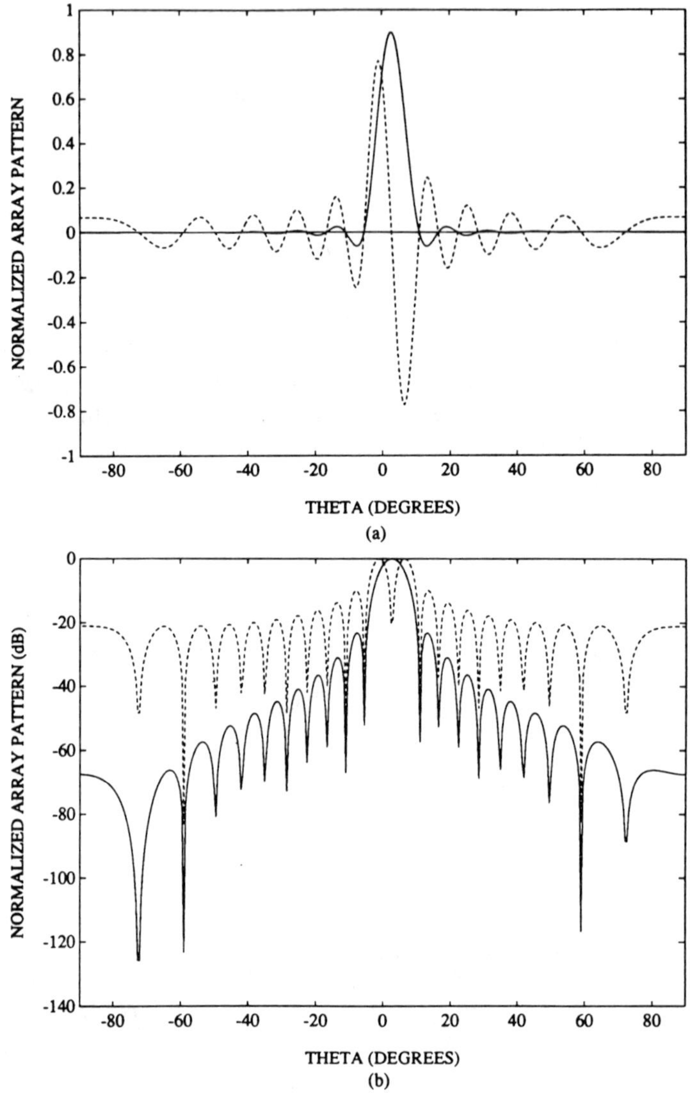

Figure 5-4 Sum and difference patterns formed from left and right beams pictured in Fig. 5-2. Sum beam, cosine window; difference beam, half-cycle sine window; pointing angle, $u_c = 1/N$ ($\theta_c = 2.73°$), where $N = 21$.

left and right beams depicted in Fig. 5-2 are lined up such that co-located sidelobes are 180° out-of-phase. This is a result of the $2/N$ spacing between the beams. Thus, the addition of the two beams yields greatly reduced sidelobes relative to that achieved with a rectangular window. The farther we move away from the respective pointing angles of the left and right beams, the more the respective sidelobes approach the condition of being equal in magnitude. The result is a fairly dramatic roll-off in sidelobe level. Unfortunately, this is not the case with the difference beam pattern. Since co-located sidelobes are opposite in sign, the difference between them yields increased sidelobe levels. Note that the difference beam is an odd function of u about $u = u_C$ as indicated previously.

We now consider the 2D-BDML estimate of the target bearing u_0, given as data the respective outputs of the sum and difference beams defined by (5-37). In so doing, we obtain the not so surprising result that the 2D-BDML bearing estimate, given the outputs of sum and difference beams defined by (5-37), is exactly equal to the 2D-BDML bearing estimate, given the outputs of the left and right beams defined by (5-12). This is an indication that an equivalence exists between the use of left and right beams and the use of sum and difference beams, if the one pair of beams is constructed from the other in accordance with (5-37).

Similar to the definition of the $N \times 2$ beamforming matrix $\mathbf{W}_{B2}$ in (5-12), define $\mathbf{W}_{T2}$ as

$$\mathbf{W}_{T2} = [\mathbf{s}(u_C) \,\vdots\, \mathbf{a}(u_C)] \tag{5-39}$$

From the definitions in (5-37) above, $\mathbf{W}_{T2}$ is related to $\mathbf{W}_{B2}$ defined in (5-12) according to

$$\mathbf{W}_{T2} = \mathbf{W}_{B2}\mathbf{T} \tag{5-40}$$

where

$$\mathbf{T} = \begin{bmatrix} \frac{1}{\sqrt{2}} & \frac{1}{\sqrt{2}} \\ \frac{1}{\sqrt{2}} & \frac{-1}{\sqrt{2}} \end{bmatrix} \tag{5-41}$$

Note that $\mathbf{T}$ is an orthogonal matrix equal to its own inverse, that is, $\mathbf{TT} = \mathbf{I}_2$. Combined with the fact that the two columns of $\mathbf{W}_{B2}$ arc orthogonal, it follows that the two columns of $\mathbf{W}_{T2}$ are orthogonal as well.

Following a development similar to that for the case of left and right beams with pointing angles $u_L = u_C - 1/N$ and $u_R = u_C + 1/N$, respectively, the 2D-BDML estimate of u_0 given the sum and difference beam outputs is obtained via a two-step procedure. In the first step, $\mathbf{v}_T$ is computed as that eigenvector of $\text{Re}\{\mathbf{x}_{T2}(n)\mathbf{x}_{T2}^H(n)\}$, where $\mathbf{x}_{T2}(n) = \mathbf{W}_{T2}^H\mathbf{x}(n)$, associated with the smaller of the two eigenvalues. $\hat{u}_0$ is then determined as the solution to $\mathbf{v}_T^T\mathbf{b}_{T2}(u) = 0$, where $\mathbf{b}_{T2}(u) = \mathbf{W}_{T2}^H\mathbf{d}(u)$, in the vicinity of $u = u_C$. This last step may be formulated as determining $\hat{u}_0$ according to a formula similar to (5-24) for the case of left and right beams. To achieve such, note that since

$\mathbf{W}_{T2} = \mathbf{W}_{B2}\mathbf{T}$, $\mathbf{b}_{T2}(u) = \mathbf{W}_{T2}^H \mathbf{d}(u)$ and $\mathbf{b}_{B2}(u) = \mathbf{W}_{B2}^H \mathbf{d}(u)$ are related as $\mathbf{b}_{T2}(u) = \mathbf{T}^T \mathbf{b}_{B2}(u)$. Hence, $\mathbf{v}_T^T \mathbf{b}_{T2}(u)$ may be alternatively expressed as

$$\mathbf{v}_T^T \mathbf{b}_{T2}(u) = \mathbf{v}_T^T \mathbf{T}^T \mathbf{b}_{B2}(u) = \{\mathbf{T}\mathbf{v}_T\}^T \mathbf{b}_{B2}(u) = 0 \qquad (5\text{-}42)$$

As (5-20) leads to (5-24), it follows from (5-42) that the 2D-BDML estimate of u_0 given the sum and difference beam outputs is given by (5-24) with $\mathbf{v} = [v_1 \quad v_2]^T$ replaced by

$$\mathbf{v}_T' = \begin{bmatrix} v_{T1}' \\ v_{T2}' \end{bmatrix} = \mathbf{T} \begin{bmatrix} v_{T1} \\ v_{T2} \end{bmatrix} \qquad (5\text{-}43)$$

The equivalency is established by observing that since $\mathbf{W}_{T2} = \mathbf{W}_{B2}\mathbf{T}$, $\mathbf{x}_{T2}(n) = \mathbf{W}_{T2}^H \mathbf{x}(n)$ and $\mathbf{x}_{B2}(n) = \mathbf{W}_{B2}^H \mathbf{x}(n)$ are related as $\mathbf{x}_{T2}(n) = \mathbf{T}^T \mathbf{x}_{B2}(n)$. Hence,

$$\mathrm{Re}\{\mathbf{x}_{T2}(n)\mathbf{x}_{T2}^H(n)\} = \mathbf{T}^T \, \mathrm{Re}\{\mathbf{x}_{B2}(n)\mathbf{x}_{B2}^H(n)\}\mathbf{T} \qquad (5\text{-}44)$$

Since $\mathbf{T}$ is an orthogonal matrix, the right hand side of (5-44) represents a similarity transformation. Thus, the eigenvectors of $\mathrm{Re}\{\mathbf{x}_{T2}(n)\mathbf{x}_{T2}^H(n)\}$ are related to those of $\mathrm{Re}\{\mathbf{x}_{B2}(n)\mathbf{x}_{B2}^H(n)\}$ through the transformation $\mathbf{T}$. Specifically, $\mathbf{v}_T = \mathbf{T}\mathbf{v}$ where $\mathbf{v}_T$ is the smallest eigenvector of $\mathrm{Re}\{\mathbf{x}_{T2}(n)\mathbf{x}_{T2}^H(n)\}$ and $\mathbf{v}$ is the smallest eigenvector of $\mathrm{Re}\{\mathbf{x}_{B2}(n)\mathbf{x}_{B2}^H(n)\}$. Here smallest eigenvector refers to that eigenvector associated with the smallest eigenvalue. As a consequence of these observations, it follows that $\mathbf{v}_T'$ defined by (5-43) is identically equal to $\mathbf{v}$, since $\mathbf{v}_T' = \mathbf{T}\mathbf{v}_T = \mathbf{T}\mathbf{T}\mathbf{v} = \mathbf{v}$. Here we have invoked the property $\mathbf{T}\mathbf{T} = \mathbf{I}_2$. This proves the assertion that the 2D-BDML bearing estimate, given the outputs of sum and difference beams defined by (5-37), is exactly equal to the 2D-BDML bearing estimate, given the outputs of the left and right beams defined by (5-12).

5-4-3 Linear Combinations of Three Orthogonal Classical Beams

The beams generated by cosine and sine illuminations represent one specific example of a sum and difference beam pair. The sidelobe levels associated with each of these two illuminations may not be low enough to combat the pejorative effects of clutter and/or jammers. Note that both of these illuminations were achieved via a linear combination of two classical beamforming vectors with $2/N$ separation between respective pointing angles. Reduced sidelobe levels may be achieved by employing linear combinations of more than two classical beamforming vectors. For example, as we shall demonstrate, both the Hanning (raised cosine) and Hamming windows may be achieved by employing an appropriate linear combination of three classical beamforming vectors with $2/N$ separation between adjacent beams.

Whereas the peak sidelobe occurring with the cosine window is at −23.5 dB relative to the peak of the mainlobe, the peak sidelobe occurring with the Hanning window relative to its respective mainlobe peak is at −32 dB. In addition, the sidelobe roll-off realized with the Hanning window is as steep as that realized with the cosine window. In contrast, the sidelobe level realized with the Hamming window is relatively flat over the visible region with the peak sidelobe at −45 dB relative to the mainlobe peak. The price paid for the reduced sidelobes achieved with either the Hanning or Hamming windows is a 50% increase in mainlobe width relative to that associated with the cosine window.

As stated above, the Hanning and Hamming windows may be achieved via a linear combination of three classical beamforming vectors with pointing angles equispaced by $2/N$. This leads us to define the $N \times 3$ beamforming matrix

$$\mathbf{W}_{B3} = [\mathbf{d}(u_L) \,\vdots\, \mathbf{d}(u_C) \,\vdots\, \mathbf{d}(u_R)] = [\mathbf{w}_{BL} \,\vdots\, \mathbf{w}_{BC} \,\vdots\, \mathbf{w}_{BR}] \tag{5-45}$$

where $u_L = u_C - 2/N$ and $u_R = u_C + 2/N$. Note that the columns of $\mathbf{W}_{B3}$ are mutually orthogonal; this may be proved by invoking (5-13). The first, second, and third columns of $\mathbf{W}_{B3}$ generate the so-called left, center, and right beams, respectively. For later purposes, the corresponding left, center, and right polynomials are respectively defined as

$$w^*_{BL}(z) = \sum_{k=0}^{N-1} \mathbf{d}^*{}_k(u_L) z^k \tag{5-46a}$$

$$w^*_{BC}(z) = \sum_{k=0}^{N-1} \mathbf{d}^*{}_k(u_C) z^k \tag{5-46b}$$

$$w^*_{BR}(z) = \sum_{k=0}^{N-1} \mathbf{d}^*{}_k(u_R) z^k \tag{5-46c}$$

Whereas left and right beams with $2/N$ separation have $N-2$ nulls in common, the left, center, and right beams generated by the columns of $\mathbf{W}_{B3}$ have $N-3$ nulls in common. Figure 5-5 displays this phenomcnon for the case of $N = 21$ and $u_C = 0$. It is thus easily established that the left, center, and right polynomials defined in (5-46) have $N-3$ roots in common on the unit circle located at $z_m = \exp[j\pi(u_C + m2/N)]$, $m = 2, \ldots, N-2$. Let $c^*_{B3}(z)$ denote the monic polynomial of order $N-3$ whose roots are these $N-3$ common roots, that is,

$$c^*_{B3}(z) = \prod_{m=2}^{N-2} (z - \exp[j\pi(u_C + m2/N)]) \tag{5-47}$$

It follows that $w^*_{BL}(z)$, $w^*_{BC}(z)$, and $w^*_{BR}(z)$ may be factored as

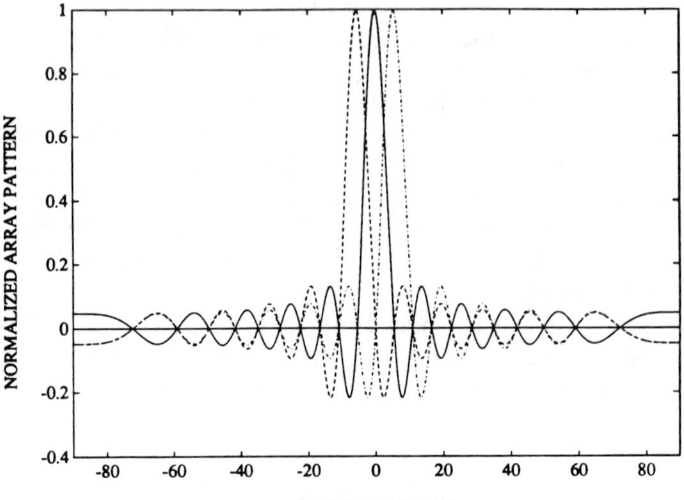

(a)

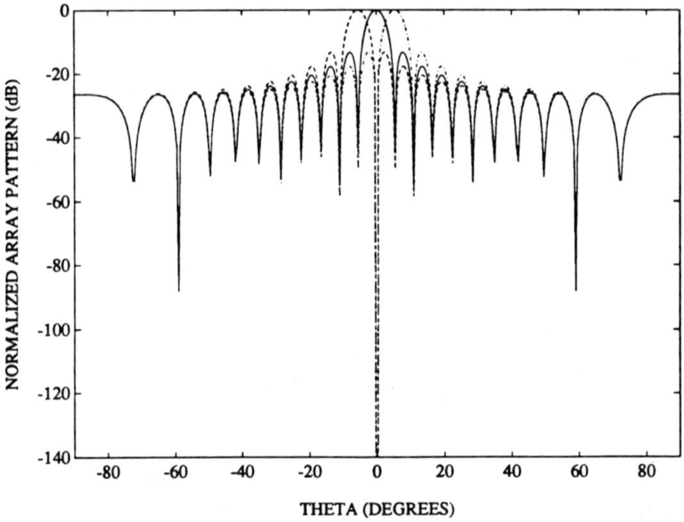

(b)

Figure 5-5 Left, center, and right beam patterns with rectangular amplitude window and pointing angles of $u_L = -2/N$ ($\theta_L = -5.46°$), $u_C = 0$ ($\theta_C = 0°$), and $u_R = 2/N$ ($\theta_R = 5.46°$), respectively, where $N = 21$.

$$w^*_{BL}(z) = c^*_{B3}(z)q^*_{BL}(z) \tag{5-48a}$$

where

$$q^*_{BL}(z) = \exp\left[-j\pi\,\frac{N-1}{2}\,u_L\right](z - \exp[j\pi u_C])(z - \exp[j\pi u_R])$$

$$w^*_{BC}(z) = c^*_{B3}(z)q^*_{BC}(z) \tag{5-48b}$$

where

$$q_{BC}^*(z) = \exp\left[-j\pi \frac{N-1}{2} u_C\right](z - \exp[j\pi u_L])(z - \exp[j\pi u_R])$$

$$w_{BR}^*(z) = c_{B3}^*(z) q_{BR}^*(z) \tag{5-48c}$$

where

$$q_{BR}^*(z) = \exp\left[-j\pi \frac{N-1}{2} u_R\right](z - \exp[j\pi u_C])(z - \exp[j\pi u_L])$$

We shall make use of these factorizations shortly.

Consider the construction of sum and difference beamforming vectors according to

$$\mathbf{W}_{H2} = [\mathbf{s}(u_C) \,\vdots\, \mathbf{a}(u_C)] = \mathbf{W}_{B3}\mathbf{H} \tag{5-49}$$

where

$$\mathbf{H} = \begin{bmatrix} 0.5 & 0.5 \\ 1 & 0 \\ 0.5 & -0.5 \end{bmatrix} \tag{5-50}$$

$\mathbf{s}(u_C)$ constructed according to (5-49) may be expressed in the form described by (5-35a) where the taper is the Hanning window. This is established by examining the kth component of $\mathbf{s}(u_C)$ invoking the componentwise definition of $\mathbf{d}(u)$ in (5-4):

$$\begin{aligned} \mathbf{s}_k(u_C) &= 0.5\mathbf{d}_k(u_L) + \mathbf{d}_k(u_C) + 0.5\mathbf{d}_k(u_R) \\ &= \left\{1 + \cos\left[\frac{2\pi}{N}\left(k - \frac{N-1}{2}\right)\right]\right\} \exp\left[-j\left(k - \frac{N-1}{2}\right)\pi u_C\right] \qquad \text{(5-51a)} \\ &= \left\{1 - \cos\left[\frac{2\pi}{N}\left(k + \tfrac{1}{2}\right)\right]\right\} \exp\left[-j\left(k - \frac{N-1}{2}\right)\pi u_C\right], \\ & k = 0, 1, \ldots, N-1 \end{aligned}$$

Similarly, it is easily established that the taper associated with $\mathbf{a}(u_C)$ defined by (5-49) is a sine illumination:

$$\begin{aligned} \mathbf{a}_k(u_C) &= 0.5\mathbf{d}_k(u_L) - 0.5\mathbf{d}_k(u_R), \\ &= j \sin\left[\frac{2\pi}{N}\left(k - \frac{N-1}{2}\right)\right] \exp\left[-j\left(k - \frac{N-1}{2}\right)\pi u_C\right] \qquad \text{(5-51b)} \\ & k = 0, 1, \ldots, N-1 \end{aligned}$$

Comparing (5-38b) with (5-51b), we see that with $u_R - u_L = 2/N$ a half-cycle of the sine appears across the array aperture, whereas with

$u_R - u_L = 4/N$ a full-cycle of the sine appears across the array aperture. The sharp discontinuity occurring at either end of the window in the former case is thus avoided in the latter case yielding reduced sidelobes. This claim is substantiated in the example below.

Example 5-1 (continued) The array patterns generated by sum and difference beamforming weight vectors defined by (5-49) with $N = 21$ and $u_C = 0$ ($\theta_C = 0°$) are shown in Fig. 5-6. Note that the two beams have $N - 3 = 18$ nulls in common. For illustrative purposes, the three beams generated by $\mathbf{d}(u_L)$, $\mathbf{d}(u_C)$, and $\mathbf{d}(u_R)$, respectively, where $u_L = -2/21$ ($\theta_L = -5.46°$) and $u_R = 2/21$ ($\theta_L = 5.46°$), are displayed in Fig. 5-5. Note the reduced sidelobes of the difference beam shown in Fig. 5-6 relative to the difference beam pattern shown in Fig. 5-4. In addition to the argument above regarding the discontinuity at the ends of the window, this may alternatively be explained by observing that with $u_R - u_L = 4/N$ the respective left and right beams depicted in Fig. 5-5 are lined up such that co-located sidelobes are *in-phase*. Thus, the difference between this particular pair of left and right beams yields reduced sidelobes relative to that achieved with a rectangular window.

We now consider the 2D-BDML estimate of the target bearing u_0 given as data the respective outputs of the sum and difference beams defined by (5-49). A development similar to that in Section 5-3 once again leads to a two-step procedure in which the first step is to compute $\mathbf{v}_H$ as the smallest eigenvector of the 2×2 matrix $\text{Re}\{\mathbf{x}_{H2}(n)\mathbf{x}_{H2}^H(n)\}$, where $\mathbf{x}_{H2}(n) = \mathbf{W}_{H2}^H\mathbf{x}(n)$ with $\mathbf{W}_{H2}$ defined by (5-49). $\hat{u}_0$ is then determined as the solution to $\mathbf{v}_H^T\mathbf{b}_{H2}(u) = 0$, where $\mathbf{b}_{H2}(u) = \mathbf{W}_{H2}^H\mathbf{d}(u)$, in the vicinity of $u = u_C$. This last step may be formulated in terms of finding $\hat{z}_0 = \exp[j\pi\hat{u}_0]$ as the root of a quadratic equation. The appropriate development is provided below.

Invoking the definition of $\mathbf{b}_{H2}(u)$ in (5-49), observe

$$\mathbf{v}_H^T\mathbf{b}_{H2}(u) = \mathbf{v}_H^T\mathbf{H}^T\mathbf{W}_{B3}^H(u)\mathbf{d}(u) = \{\mathbf{H}\mathbf{v}_H\}^T\mathbf{W}_{B3}^H(u)\mathbf{d}(u) = 0 \qquad (5\text{-}52)$$

Hence,

$$\mathbf{v}_H^T\mathbf{b}_{H2}(u) = v'_{H1}\mathbf{d}^H(u_L)\mathbf{d}(u) + v'_{H2}\mathbf{d}^H(u_C)\mathbf{d}(u) + v'_{H3}\mathbf{d}^H(u_R)\mathbf{d}(u) = 0 \qquad (5\text{-}53)$$

where v'_{Hi}, $i = 1, 2, 3$, are the three components of $\mathbf{v}'_H = \mathbf{H}\mathbf{v}_H$. Similar to the sequence of steps which lead from (5-25) to (5-28), determination of $\hat{u}_0$ as the solution to the above equation is equivalent to determining $\hat{z}_0 = \exp[j\pi\hat{u}_0]$ as a root of the polynomial

$$w(z) = v'_{H1}w_{BL}^*(z) + v'_{H2}w_{BC}^*(z) + v'_{H3}w_{BR}^*(z) \qquad (5\text{-}54)$$

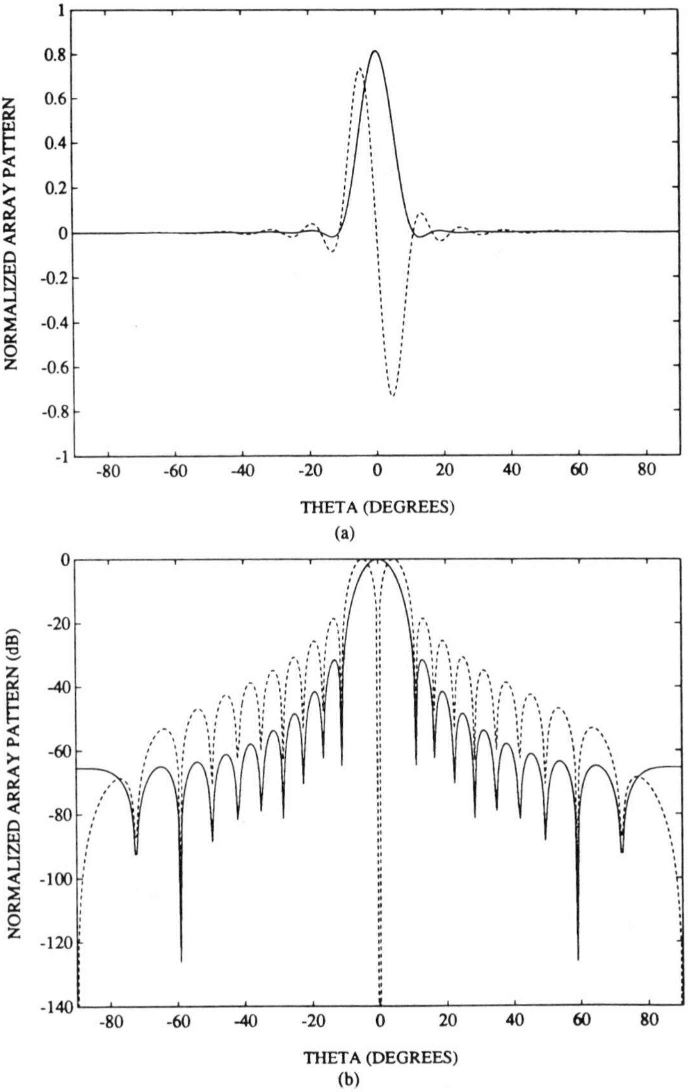

Figure 5-6 Sum and difference patterns formed from linear combination of left, center, and right beams pictured in Fig. 5-5. Sum beam, Hanning window; difference beam, full-cycle sine window; pointing angle, $u_C = 0$ ($\theta_C = 0°$).

Exploiting the factorizations of $w^*_{BL}(z)$, $w^*_{BC}(z)$, and $w^*_{BR}(z)$ in (5-48a), (5-48b), and (5-48c), respectively, it follows that $\hat{z}_0 = \exp[j\pi\hat{u}_0]$ is one of the two roots of the quadratic equation

$$\begin{aligned} e(z) &= v'_{H1} q^*_{BL}(z) + v'_{H2} q^*_{BC}(z) + v'_{H3}(z) q^*_{BR}(z) \\ &= e_0 + e_1 z + e_2 z^2 \end{aligned} \tag{5-55}$$

Based on the coefficients of the second order polynomials $q^*_{BL}(z)$, $q^*_{BC}(z)$, and $q^*_{BR}(z)$ defined in (5.48a), (5-48b), and (5-48c), respectively, each e_i, $i = 0, 1, 2$, in (5-55) may be expressed in terms of a linear combination of the three components of $\mathbf{v}'_H$. Correspondingly, defining $\mathbf{e} = [e_0 \quad e_1 \quad e_2]^T$ as the coefficient vector for $e(z)$, simple manipulation allows us to express $\mathbf{e}$ in terms of a 3×3 transformation on $\mathbf{v}'_H$ as $\mathbf{e} = \mathbf{Q}^*_{B3}\mathbf{v}'_H$, where

$$\mathbf{Q}^*_{B3} = \begin{bmatrix} -\exp[j\pi(u_C + 1/N)] & \exp[j\pi u_C] & -\exp[j\pi(u_C - 1/N)] \\ 2\cos(\pi/N) & -2\cos(2\pi/N) & 2\cos(\pi/N) \\ -\exp[-j\pi(u_C + 1/N)] & \exp[-j\pi u_C] & -\exp[-j\pi(u_C - 1/N)] \end{bmatrix} \tag{5-56}$$

Recall that $\mathbf{v}'_H = \mathbf{H}\mathbf{v}_H$, where $\mathbf{H}$ is defined in (5-50) and $\mathbf{v}_H$ is the smallest eigenvector of the 2×2 matrix $\text{Re}\{\mathbf{x}_{H2}(n)\mathbf{x}^H_{H2}(n)\}$. Hence, the coefficient vector for $e(z)$ in (5-55) may be expressed in terms of $\mathbf{v}_H$ directly as $\mathbf{e} = \mathbf{Q}^*_{B3}\mathbf{H}\mathbf{v}_H$. Direct substitution yields the following expressions for the components of $\mathbf{e}$:

$$e_0 = \exp[j\pi u_C]\left\{v_{H1}\left[1 - \cos\left(\frac{\pi}{N}\right)\right] - jv_{H2}\sin\left(\frac{\pi}{N}\right)\right\} \tag{5-57a}$$

$$e_1 = 2v_{H1}\left\{\cos\left(\frac{\pi}{N}\right) - \cos\left(\frac{2\pi}{N}\right)\right\} \tag{5-57b}$$

$$e_2 = \exp[-j\pi u_C]\left\{v_{H1}\left[1 - \cos\left(\frac{\pi}{N}\right)\right] + jv_{H2}\sin\left(\frac{\pi}{N}\right)\right\} \tag{5-57c}$$

Observe that $e_2 = e^*_0$ and that e_1 is real. As stated previously, $\hat{z}_0 = \exp[j\pi\hat{u}_0]$ is determined as one of the two roots of $e(z)$. Two questions arise: (1) are the roots of $e(z)$ on the unit circle and, if so, (2) which one should be equated to $\hat{z}_0 = \exp[j\pi\hat{u}_0]$? Let us examine the roots of $e(z)$.

To this end, let z_1 and z_2 denote the two roots of $e(z) = e_0 + e_1 z + e^*_0 z^2$, where e_1 is real. The two roots of $e(z)$ are simply $z_{1,2} = (-e_1 \pm \sqrt{e_1^2 - 4|e_0|^2})/2e^*_0$. It is easy to show that the two roots lie on the unit circle if and only if $|e_1/e_0| \leq 2$. For practical values of N, that is, for N fairly large, $\cos(\pi/N) \approx \cos(2\pi/N)$ such that $|e_1| \ll 1$, where e_1 is computed according to (5-57b). (Note that $\mathbf{v}_H$ is normalized to have unity magnitude such that $|\mathbf{v}_{H1}| \leq 1$.) We may thus deduce that for practical values of N, the roots of $e(z)$ lie on the unit circle. This answers the first question posed above. With regard to the second question, note that simple algebra dictates that the sum of the roots is $z_1 + z_2 = -e_1/e^*_0$. Again, $|e_1| \ll 1$ for practical N such that $z_1 \approx -z_2$ or $z_1 \approx \exp[j\pi]z_2$. Thus, for practical N one root lies on the opposite side of the unit circle from the other root. We now show that one of the roots lies in the vicinity of $\hat{z}_c = \exp[j\pi\hat{u}_C]$.

Since we have argued that for practical N the two roots of $e(z)$ lie on the unit circle, consider the values of u_0 which satisfy

$$e_0 + e_1 \exp[j\pi u_0] + e_0^* \exp[j2\pi u_0] = 0 \tag{5-58}$$

To facilitate determination of the solution to (5-58), express e_0 in polar form as $r_0 \exp[j\phi_0]$, where r_0 and ϕ_0 are the magnitude and phase of e_0, respectively. From (5-57a)

$$r_0 = |e_0| = \sqrt{v_{H1}^2\left[1 - \cos\left(\frac{\pi}{N}\right)\right]^2 + v_{H2}^2 \sin^2\left(\frac{\pi}{N}\right)} \tag{5-59a}$$

$$\phi_0 = \arg\{e_0\} = \pi u_C - \tan^{-1}\left\{\frac{v_{H2}\sin\left(\frac{\pi}{N}\right)}{v_{H1}\left[1 - \cos\left(\frac{\pi}{N}\right)\right]}\right\}$$

$$= \pi u_C - \tan^{-1}\left\{\frac{v_{H2}}{v_{H1}\tan\left(\frac{\pi}{2N}\right)}\right\} \tag{5-59b}$$

Substituting $e_0 = r_0 \exp[j\phi_0]$ and multiplying both sides of (5-58) by $\exp[-j\pi u_0]$ yields

$$r_0 \exp[j\phi_0]\exp[-j\pi u_0] + e_1 + r_0 \exp[-j\phi_0]\exp[j\pi u_0] = 0 \tag{5-60}$$

Simple algebraic manipulation of (5-60) yields

$$\cos(\pi u_0 - \phi_0) = -\frac{e_1}{2r_0} \tag{5-61}$$

such that the two solutions are

$$u_0 = \frac{1}{\pi}\left\{\phi_0 \pm \cos^{-1}\left(\frac{e_1}{2r_0}\right)\right\}$$

$$= u_C - \frac{1}{\pi}\tan^{-1}\left\{\frac{v_{H2}}{v_{H1}\tan\left(\frac{\pi}{2N}\right)}\right\} \pm \frac{1}{\pi}\cos^{-1}\left(\frac{e_1}{2r_0}\right) \tag{5-62}$$

where we have substituted (5-59b) and invoked the even symmetry of the cosine function. The choice of the plus or minus sign on the right hand side of (5-62) is dictated by the practical consideration that u_0 be in the vicinity of u_C. Note that as N approaches infinity, $\tan^{-1}\{v_{H2}/(v_{H1}\tan(\pi/2N))\}$ approaches $\pi/2$ and $\cos^{-1}\{e_1/(2r_0)\}$ approaches $\pm\pi/2$. Hence, for N large one of the solutions is in the vicinity of u_C while the other is in the vicinity of $u_C - 1$. This is in agreement with our previous observation concerning the two roots of $e(z)$. If we normalize $\mathbf{v}_H$ such that $v_{H1} \geqslant 0$, then $e_1 \geqslant 0$ and $\hat{u}_0$ is obtained with the + before the $\cos^{-1}\{\cdot\}$ term in (5-62).

The procedure for estimating the target bearing when the sum beam is formed using a Hanning taper and the difference beam is formed using a full-cycle sine taper is summarized. First, $\mathbf{v}_H$ is computed as the smallest eigenvector of the 2×2 matrix $\mathrm{Re}\{\mathbf{x}_{H2}(n)\mathbf{x}_{H2}^H(n)\}$, where $\mathbf{x}_{H2}(n) = \mathbf{W}_{H2}^H \mathbf{x}(n)$ with $\mathbf{W}_{H2}$ defined by (5-49). $\hat{u}_0$ is then determined according to (5-62). If one desires to employ a Hamming taper rather than a Hanning taper to form the sum beam, the only modification to this procedure is to replace $\mathbf{H}$ in (5-50) by

$$\mathbf{H}' = \begin{bmatrix} 0.46/2 & 0.5 \\ 0.54 & 0 \\ 0.46/2 & -0.5 \end{bmatrix} = \begin{bmatrix} 0.23 & 0.5 \\ 0.54 & 0 \\ 0.23 & -0.5 \end{bmatrix} \tag{5-63}$$

Similar to (5-51a), the reader may verify that weighting the left, center, and right classical beams generated by the columns of $\mathbf{W}_{B3}$ in (5-45) in accordance with the first column of $\mathbf{H}'$ above is equivalent to applying a Hamming taper across the array aperture, that is,

$$\begin{aligned} \mathbf{s}_k(u_C) &= 0.23\mathbf{d}_k(u_L) + 0.54\mathbf{d}_k(u_C) + 0.23\mathbf{d}_k(u_R) \\ &= 2\left\{0.54 - 0.46\cos\left[\frac{2\pi}{N}\left(k + \tfrac{1}{2}\right)\right]\right\}\exp\left[-j\left(k - \frac{N-1}{2}\right)\pi u_C\right], \\ &\quad k = 0, 1, \ldots, N-1 \end{aligned} \tag{5-64}$$

Example 5-1 (continued) The array patterns generated by sum and difference beamforming weight vectors defined by (5-49) with $\mathbf{H}$ replaced by $\mathbf{H}'$ above, corresponding to a Hamming taper in element space, are shown in Fig. 5-7. As in previous examples, $N = 21$ and $u_C = 0$ ($\theta_C = 0°$). The sidelobe level of the sum beam is observed to be relatively flat; each of the sidelobes peak at approximately -45 dB with respect to the mainlobe peak. This is in contrast to the sidelobe response achieved with Hanning tapering displayed in Fig. 5-6 wherein the sidelobe peaks start at -32 dB but roll off to -65 dB at the ends of the visible region. The width of the mainlobe is the same in each case. Finally, note that the sum and difference beams in Fig. 5-7 have $N - 3 = 18$ nulls in common.

Summarizing, construction of both the sum beam and the difference beam as a judiciously weighted combination of three classical beams with respective pointing angles equispaced by $2/N$ yields both reduced sidelobes and a simple, closed-form procedure for estimating the target bearing which is exact in the case of no noise. The low sidelobe levels are due to judicious exploitation of the fact that the beam patterns associated with two classical beamformers are lined up such that co-located sidelobes are 180° out-of-phase when the respective pointing angles are separated by $2/N$ and in-phase when the respective pointing angles are separated by $4/N$. The computational simplicity of the procedure is due to judicious exploitation of

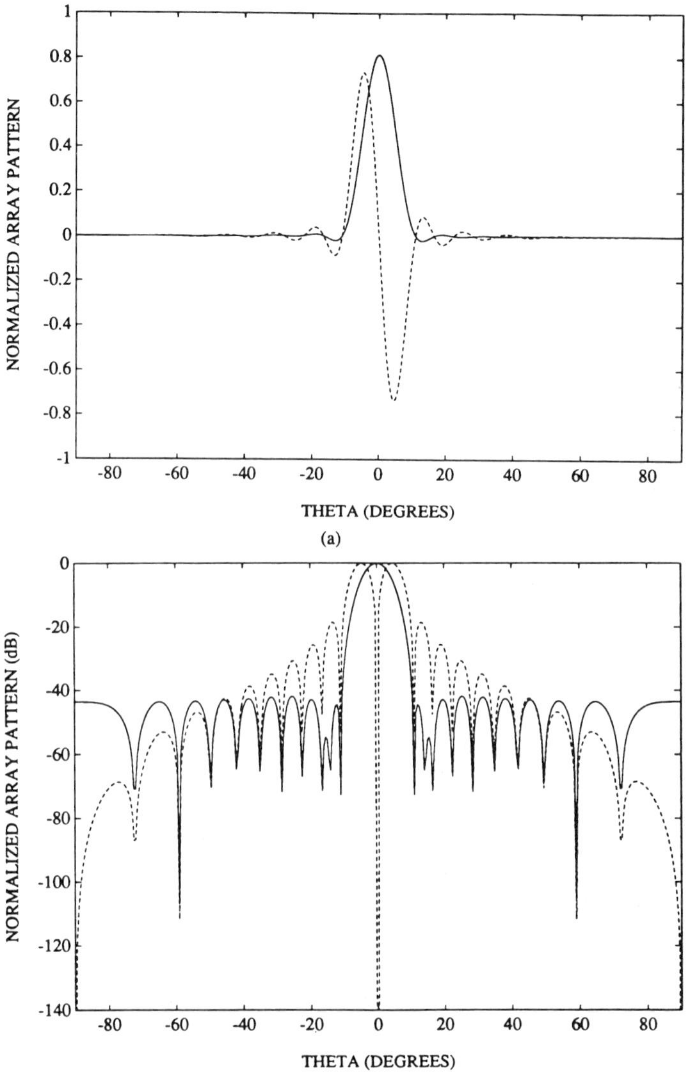

Figure 5-7 Sum and difference patterns formed from linear combination of left, center, and right beams pictured in Fig. 5-5. Sum beam, Hamming window; difference beam, full-cycle sine window; pointing angle, $u_C = 0$ ($\theta_C = 0°$).

the fact that three classical beams with pointing angles equispaced by $2/N$ have $N-3$ nulls in common. Further reductions in sidelobe levels may be achieved by employing linear combinations of a larger number of classical beams with pointing angles equispaced by $2/N$ for both the sum and difference beams. For example, the Blackman window may be emulated via appropriate weighting of five such beams to yield a sum beam whose sidelobe is at -60 dB relative to the mainlobe peak. The price paid, however, is a widening of the mainlobe. Also, in the case of five beams, the number of common nulls is $N-5$ enabling us to only simplify the final step to that of solving a fourth-order polynomial or quartic equation. Notwithstanding, this argument illustrates how the procedure developed in this section may be generalized to accommodate any aperture taper which may be equivalently implemented as a linear combination of classical beams with pointing angles equispaced by $2/N$.

However, there are other classes of aperture tapers employed in practice that do not fall into this category. Examples include the Dolph–Chebyshev window and the Taylor window [46], which are commonly used for sum beam formation, and the Bayliss window, which is commonly used for difference beam formation. The Dolph–Chebyshev window yields equal height sidelobes and provides the narrowest mainlobe for a prescribed sidelobe level [47]. However, a plot of the magnitude of the Dolph–Chebyshev (element) weights as a function of aperture location reveals that it initially decreases as we move in either direction away from the aperture center but then rises again at both ends. For analog beamforming purposes, this is difficult to implement in practice. The Taylor window is a suboptimum variation of the Dolph–Chebyshev window which allows one to specify a number of equal height sidelobes adjacent to the mainlobe at a level commensurate with a monotonic decrease in the magnitude of the (element) weights as we move away from the aperture center. The respective heights of the remaining sidelobes are lower than that of the equal height sidelobes and fall off monotonically. The Bayliss window is a similar design procedure for constructing the difference beam. Although sum and difference beams generated using appropriate Taylor and Bayliss windows, respectively, may have a number of nulls in common, the number of nulls not in common among the two beams may be relatively large negating the computational simplicity of the procedure outlined in this section. For the sake of completeness, the next section examines some simple schemes used in practice that do not rely on the common nulls property.

5-5 SIMPLE TECHNIQUES FOR GENERAL SUM/DIFFERENCE BEAM PROCESSING

The bearing estimation techniques developed in this section do not require the sum and difference beams to have all but a few nulls in common. The only requirements are that the sum and difference beamforming vectors have the structure described by (5-35a) and (5-35b), respectively, and that the respective aperture tapers yield reasonable sum and difference beams. As a starting point, denote the sum and difference beam outputs as Σ and Δ, respectively, that is,

$$\Sigma = \mathbf{s}^H(u_C)\mathbf{x}(n) \tag{5-65a}$$

$$\Delta = \mathbf{a}^H(u_C)\mathbf{x}(n) \tag{5-65b}$$

Σ and Δ are the first and second components, respectively, of $\mathbf{x}_{SD}(n) = \mathbf{W}_{SD}^H\mathbf{x}(n)$ where

$$\mathbf{W}_{SD} = [\mathbf{s}(u_C) \,\vdots\, \mathbf{a}(u_C)] \tag{5-66}$$

A development similar to that which lead from (5-17) to (5-21) once again leads to a two-step procedure in which the first step is to compute $\mathbf{v}$ as the solution to

$$\underset{\mathbf{v}}{\text{Minimize}} \frac{\mathbf{v}^T \operatorname{Re}\{\mathbf{x}_{SD}(n)\mathbf{x}_{SD}^H(n)\}\mathbf{v}}{\mathbf{v}^T\mathbf{v}} \tag{5-67}$$

$\hat{u}_0$ is then determined as the solution to $\mathbf{v}^T\mathbf{b}_{SD}(u) = 0$, where $\mathbf{b}_{SD}(u) = \mathbf{W}_{SD}^H\mathbf{d}(u)$, in the vicinity of $u = u_C$. We expand upon these two steps.

The solution to the optimization problem in (5-67) is, of course, to choose $\mathbf{v}$ as the smallest eigenvector of $\operatorname{Re}\{\mathbf{x}_{SD}(n)\mathbf{x}_{SD}^H(n)]$. Note, however, that the objective function in (5-67) is invariant to a scale change in $\mathbf{v}$. This is consistent with the fact that an eigenvector is only unique to within a scalar multiple. It is thus permissible to set one of the components of $\mathbf{v}$ equal to some fixed value and minimize (5-67) with respect to the other component. To this end, fix the second component equal to unity, that is, $v_2 = 1$. This is motivated by the observation that with $u_0 = u_C$ and no noise, $\Delta = 0$ and the smallest eigenvector of $\operatorname{Re}\{\mathbf{x}_{SD}(n)\mathbf{x}_{SD}^H(n)\}$ is $\mathbf{v} = [0 \quad 1]^T$. Expressing $\operatorname{Re}\{\mathbf{x}_{SD}(n)\mathbf{x}_{SD}^H(n)\}$ in (5-67) in terms of Σ and Δ as

$$\operatorname{Re}\{\mathbf{x}_{SD}(n)\mathbf{x}_{SD}^H(n)\} = \begin{bmatrix} \Sigma\Sigma^* & \operatorname{Re}\{\Delta\Sigma^*\} \\ \operatorname{Re}\{\Delta\Sigma^*\} & \Delta\Delta^* \end{bmatrix} \tag{5-68}$$

and substituting $v_2 = 1$ yields

$$\underset{v_1}{\text{Minimize}} \frac{\Sigma\Sigma^* v_1^2 + 2\operatorname{Re}\{\Delta\Sigma^*\}v_1 + \Delta\Delta^*}{v_1^2 + 1} \tag{5-69}$$

Taking the derivative with respect to v_1 and equating to zero yields

$$v_1 = \frac{-2\operatorname{Re}\{\Delta/\Sigma\}}{1 - |\Delta/\Sigma|^2 + \sqrt{(1 + (\Delta/\Sigma)^2)(1 + (\Delta/\Sigma)^{*2})}} \tag{5-70}$$

Note that this depends on the ratio of the difference beam to the sum beam; this is a classical result.

Given $\mathbf{v} = [v_1 \quad 1]^T$ with v_1 determined according to (5-70), the corresponding value of u_0 is determined in accordance with the relationship $\mathbf{v}^T\mathbf{b}_{SD}(u_0) = 0$. Substituting $\mathbf{b}_{SD}(u) = \mathbf{W}_{SD}^H\mathbf{d}(u)$, where $\mathbf{W}_{SD} = [\mathbf{s}(u_C) \,\vdots\, \mathbf{a}(u_C)]$ yields

$$\mathbf{v}^T\mathbf{b}_{SD}(u_0) = v_1\mathbf{s}^H(u_C)\mathbf{d}(u_0) + \mathbf{a}^H(u_C)\mathbf{d}(u_0) = 0 \tag{5-71}$$

As discussed previously, this development does not require the sum and difference beam patterns to have either $N-2$ or $N-3$ nulls in common. We thus cannot reduce the problem at hand to that of solving a simple linear or quadratic equation. One way radar engineers proceed is to construct a look-up table or calibration curve [46] by solving (5-71) for v_1, that is,

$$v_{1C} = \text{cal}(u_0 - u_C)\,, \qquad \text{where } \text{cal}(u_0) = -\frac{\mathbf{a}^H(0)\mathbf{d}(u_0)}{\mathbf{s}^H(0)\mathbf{d}(u_0)} \tag{5-72}$$

Here we have invoked the fact that the respective shapes of the sum and difference beams in the $u = \sin\theta$ domain are translationally invariant with respect to pointing angle. As a consequence, $\mathbf{s}^H(u_C)\mathbf{d}(u_0)$ and $\mathbf{a}^H(u_C)\mathbf{d}(u_0)$ depend only on the difference between u_0 and u_C. Note that $\text{cal}(u_0)$ is a (real-valued) odd function of u_0.

In an actual implementation, the calibration curve $\text{cal}(u_0)$ is discretized and stored in memory on a computer in ordered pairs of the form (u_0, v_{1C}), where $v_{1C} = \text{cal}(u_0)$. The grid of u_0 values should correspond to an angular interval of length equal to one or two beamwidths centered at $u_0 = 0$. Due to the odd symmetry of $\text{cal}(u_0)$, it is only necessary to store half of these values. Given a pair of sum and difference beam outputs, the target bearing is determined as follows. First, v_1 is determined according to (5-70). The sign of v_1 dictates the sign of the difference $u_0 - u_C$. The magnitude of the difference $u_0 - u_C$ is then estimated by locating the two nearest values of v_{1C} on the grid and interpolating between the two corresponding values of u_0. Again, the fact that the respective shapes of the sum and difference beams in the $u = \sin\theta$ domain are translationally invariant with respect to pointing angle is important. Otherwise, a different calibration curve would have to be stored for each transmission point angle. This observation will have significance in the development of adaptive sum and difference beamforming procedures in the next section.

For simplification purposes, two modifications to the above two-step procedure are sometimes made in practice. First, v_1 is sometimes approximated as

$$v_1 \approx -\text{Re}\left\{\frac{\Delta}{\Sigma}\right\} \tag{5-73}$$

This is motivated by the following observation. In the case of no noise and no other sources, $\Sigma = c_0(n)\mathbf{s}^H(u_C)\mathbf{d}(u_0)$ and $\Delta = c_0(n)\mathbf{a}^H(u_C)\mathbf{d}(u_0)$. Recall that if $\mathbf{w}$ is conjugate centrosymmetric, the quantity $\mathbf{w}^H\mathbf{d}(u)$ is a real-valued function of u. Hence, under these conditions, the ratio Δ/Σ is real-valued and the expression for v_1 in (5-70) simplifies to $-\Delta/\Sigma$. Second, the sum beam pattern $\mathbf{s}^H(u_C)\mathbf{d}(u)$ is sometimes approximated as a raised cosine in the vicinity of $u = u_C$ while the difference beam pattern $\mathbf{a}^H(u_C)\mathbf{d}(u)$ is approximated as a sine in the vicinity of $u = u_C$. This yields the following approximation:

$$-\frac{\mathbf{a}^H(u_C)\mathbf{d}(u)}{\mathbf{s}^H(u_C)\mathbf{d}(u)} \approx -\eta\left\{\frac{\sin[\gamma(u-u_C)]}{1+\cos[\gamma(u-u_C)]}\right\} = \eta \tan\left[\frac{\gamma}{2}(u-u_C)\right] \tag{5-74}$$

where we have invoked simple trigonometry. Equating the far right hand side of (5-74) with $v_1 = -\text{Re}\{\Delta/\Sigma\}$ yields, after some algebraic manipulation, a simple expression for determining the value of u_0 corresponding to a particular pair of sum and difference beam outputs

$$\hat{u}_0 = u_C + \alpha \tan^{-1}\left(\beta \,\text{Re}\left\{\frac{\Delta}{\Sigma}\right\}\right) \tag{5-75}$$

where $\alpha = 2/\gamma$ and $\beta = 1/\eta$. The constants α and β depend on the particular tapers used to form the sum and difference beams. This approximation, therefore, allows one to avoid the use of a calibration curve. It should be noted that although in general (5-75) represents an approximation, it is exact in the case where $\mathbf{s}(u_C)$ and $\mathbf{a}(u_C)$ are defined by (5-40). Drawing on the development in Section 5-4, the reader may verify that the appropriate constants are $\alpha = 2/\pi$ and $\beta = \tan(2\pi/N)$.

5-6 BEARING ESTIMATION INCORPORATING INTERFERENCE CANCELLATION

Consider the target detection/bearing estimation problem for a specific transmission angle. Although the majority of the energy is contained within the mainlobe of the transmitted beam, RF energy is actually radiated at all angles due to sidelobe leakage. At those transmission angles for which a significant portion of the radiated energy intercepts either the ground, sea, or ionosphere via either a direct route or after reflection from the target of interest, the problem of clutter arises [46]. This underlines the importance of achieving low sidelobes in the formation of the sum and difference beams on return. However, aperture tapering in accordance with the procedures prescribed in the previous sections may not be adequate to combat the pejorative effects of clutter. In some cases, it may be necessary to form a broad null in both the sum and difference beams over an angular region in which the clutter returns are strong.

Another problem of concern is that of jamming. In certain scenarios, RF energy is directed at the radar array from external sources for the purpose of deliberately disrupting the tracking system. Interfering signals may arrive at the array unintentionally as well; this is particularly true at the lower end of the RF spectrum where a number of commercial and private radio sources operate. Again, the low sidelobe levels achieved with aperture tapering may not be sufficient to counteract a strong interferer. It may be necessary to modify the quiescent nulls of the sum and difference beams so that each exhibits a null in the direction of each strong interfering source [2, 7, 9–11, 17–19, 22–24, 41, 48–51]. If the interfering sources are in motion, the locations of these nulls must be updated in an adaptive fashion.

Assume for the moment that the angular positions of the strong interferers are known and fixed for a given scan period. For a given pointing angle, sum and difference beams are constructed with each exhibiting a null in each of the interfering directions. The locations of these nulls should remain invariant as the pointing angle is stepped to new positions in the process of scanning. As a consequence, the shape of the sum beam varies with pointing angle as does the shape of the difference beam. This is in contrast to the nonadaptive case studied previously in which the respective shapes of the sum and difference beams were translationally invariant with respect to pointing angle. (Recall that we are working in the $u = \sin\theta$ domain.) Thus, the adaptive situation ostensibly calls for a different calibration curve for each pointing angle. Moreover, these calibration curves have to be updated as the directions of the interferers are updated. It is thus apparent that the calibration curve approach is not practicable in an adaptive situation. In addition, the displacement of nulls from their quiescent positions affects the shape of the sum and difference beams rendering the approximation in (5-74) suspect. Thus, the formula in (5-75) is not generally applicable in the adaptive case [17, 18].

The simple, closed-form bearing estimation procedures developed in Section 5-4 resulted from the judicious exploitation of the fact that the sum and difference beams considered there had either $N-2$ or $N-3$ nulls in common. In this section, we develop a procedure for constructing adaptive sum and difference beams that retains this common nulls property and thus facilitates determination of the target bearing via a simple trigonometric expression similar to (5-62). We commence with a matrix formulation of linear convolution critical to the development.

5-6-1 Notation and Matrix Representation of Linear Convolution

We here introduce notation which will be invoked throughout the remainder of this chapter. Let $p(z)$ be a polynomial of order $L-1$ defined as

$$p(z) = p_0 + p_1 z + p_2 z^2 + \cdots + p_{L-1} z^{L-1} \qquad \text{(5-76a)}$$

$\{\mathbf{p}\}$ then represents a causal sequence of length L expressed as

$$\{\mathbf{p}\} = \{p_0 \quad p_1 \quad p_2 \quad \cdots \quad p_{L-1}\} \tag{5-76b}$$

Finally, $\mathbf{p}$ represents the $L \times 1$ coefficient vector associated with $p(z)$ constructed as

$$\mathbf{p} = [p_0 \quad p_1 \quad p_2 \quad \cdots \quad p_{L-1}]^T \tag{5-76c}$$

Thus, as long as one of these entities is defined, the other two are defined automatically. For example, if the vector $\mathbf{p}$ is defined, the polynomial $p(z)$ and the sequence $\{\mathbf{p}\}$ are both automatically defined.

In this notational convention, let $\mathbf{q}$ be an $M \times 1$ vector defined similar to $\mathbf{p}$ in (5-76c). This automatically defines $q(z)$, a polynomial of order $M-1$, and $\{\mathbf{q}\}$, a causal sequence of length M. Let $r(z)$ be the product of $p(z)$ and $q(z)$, that is, $r(z) = p(z)q(z)$, a polynomial of order $N = M + L - 1$. The set composed of the roots of $r(z)$ is equal to the union of the set composed of the roots of $p(z)$ with the set composed of the roots of $q(z)$. Also, the sequence $\{\mathbf{r}\}$ is the linear convolution of the sequence $\{\mathbf{p}\}$ with the sequence $\{\mathbf{q}\}$, that is,

$$r_n = \sum_{k=0}^{L-1} p_k q_{n-k} = \sum_{k=0}^{M-1} q_k p_{n-k}, \qquad n = 0, 1, \ldots, M + L - 2 \tag{5-77}$$

As indicated, $\{\mathbf{r}\}$ is a sequence of length $N = M + L - 1$. We thus have the following polynomial, sequence, and vector representations associated with either the product of two polynomials or, equivalently, the linear convolution of two causal sequences:

$$r(z) = p(z)q(z) \tag{5-78a}$$

$$\{\mathbf{r}\} = \{\mathbf{p}\} * \{\mathbf{q}\} \tag{5-78b}$$

$$\begin{bmatrix} r_0 \\ r_1 \\ r_2 \\ \vdots \\ r_{M+L-2} \end{bmatrix} = \begin{bmatrix} p_0 & 0 & & 0 \\ p_1 & p_0 & & 0 \\ \vdots & \vdots & \ddots & \vdots \\ p_{L-1} & p_{L-2} & & 0 \\ 0 & p_{L-1} & & 0 \\ 0 & 0 & & p_0 \\ \vdots & \vdots & & \vdots \\ 0 & 0 & & p_{L-1} \end{bmatrix} \begin{bmatrix} q_0 \\ q_1 \\ q_2 \\ \vdots \\ q_{M-1} \end{bmatrix} \tag{5-78c}$$

Note that the matrix representation of linear convolution in (5-78c) follows from the definition of linear convolution in (5-77). For notational simplicity, we will find it convenient in the development to follow to express (5-78c) in compact form as

$$\mathbf{r} = \begin{bmatrix} \mathbf{p} & 0 & & 0 \\ 0 & \mathbf{p} & & 0 \\ 0 & 0 & & 0 \\ \vdots & \vdots & \ddots & \vdots \\ 0 & 0 & & 0 \\ 0 & 0 & & \mathbf{p} \end{bmatrix} \mathbf{q} \tag{5-79}$$

where it is understood that (5-79) is shorthand for (5-78c).

5-6-2 Properties of $N \times 3$ Classical Matrix Beamformer

To facilitate the present development, we here summarize the nonadaptive bearing estimation scheme developed in Section 5-4.

ML Bearing Estimation Based on Classical Sum/Difference Beams

1. Form $\mathbf{x}_{H2}(n) = \mathbf{H}^T \mathbf{W}_{B3}^H \mathbf{x}(n)$, where

$$\mathbf{H} = \begin{bmatrix} & \vdots & \frac{1}{\sqrt{2}} \\ \frac{\mathbf{h}_s}{\|\mathbf{h}_s\|_2} & \vdots & 0 \\ & \vdots & -\frac{1}{\sqrt{2}} \end{bmatrix}, \qquad \text{Hanning:} \quad \mathbf{h}_s = \begin{bmatrix} 0.25 \\ 0.5 \\ 0.25 \end{bmatrix}$$

$$\text{or Hamming:} \quad \mathbf{h}_s = \begin{bmatrix} 0.23 \\ 0.54 \\ 0.23 \end{bmatrix}$$

2. Compute $\mathbf{v}_H = [v_{H1} \quad v_{H2}]^T$ as smallest EVEC of $\mathrm{Re}\{\mathbf{x}_{H2}(n)\mathbf{x}_{H2}^H(n)\}$. Normalize so that $v_{H1} \geq 0$.
3. Compute $\mathbf{e} = [e_0 \quad e_1 \quad e_0^*]^T = \mathbf{Q}_{B3}^* \mathbf{H} \mathbf{v}_H$.
4. With $e_0 = r_0 \exp[j\phi_0]$, compute

$$\hat{u}_0 = \frac{1}{\pi}\left\{\phi_0 + \cos^{-1}\left(\frac{e_1}{2r_0}\right)\right\}$$

Recall that $\mathbf{W}_{B3}$ and $\mathbf{Q}_{B3}^*$ were defined in (5-45) and (5-56), respectively. Note that the two columns of $\mathbf{H}$ are normalized to have the same length, equal to unity, so that the expected power of the noise at the sum and difference beam outputs is equal. Also, note that the two columns of $\mathbf{H}$ are orthogonal such that the noise at each of the two beam outputs is uncorrelated. These two properties of the beamspace noise were assumed in the development of the ML bearing estimation procedure above. They may be verified as follows. Recall that the receiver generated noise is modeled as being independent from element to element and of equal power such that $E\{\mathbf{n}(n)\mathbf{n}^H(n)\} = \sigma_n^2\mathbf{I}_N$. Hence, $E\{\mathbf{n}_{H2}(n)\mathbf{n}_{H2}^H(n)\} = \sigma_n^2\mathbf{W}_{H2}^H\mathbf{W}_{H2}$, where $\mathbf{n}_{H2}(n) = \mathbf{W}_{H2}^H\mathbf{n}(n)$ and $\mathbf{W}_{H2} = \mathbf{W}_{B3}\mathbf{H}$. Invoking the mutual orthogonality of the columns of $\mathbf{W}_{B3}$, that is,

$$\mathbf{W}_{B3}^H\mathbf{W}_{B3} = N\mathbf{I}_3 \tag{5-80}$$

yields $E\{\mathbf{n}_{H2}(n)\mathbf{n}_{H2}^H(n)\} = N\sigma_n^2\mathbf{I}_2$ which substantiates the former claims regarding the beamspace noise.

Our approach to the interference problem is to construct an $N \times 3$ interference cancellation matrix beamformer (ICMBF), denoted $\mathbf{W}_{I3}$, sufficiently similar to $\mathbf{W}_{B3}$ so that the only modifications to the procedure above are (i) replace $\mathbf{W}_{B3}$ by $\mathbf{W}_{I3}$ and (ii) replace $\mathbf{Q}_{B3}^*$ by a matrix $\mathbf{Q}_{I3}^*$ related to $\mathbf{W}_{I3}$ the same way $\mathbf{Q}_{B3}^*$ is related to $\mathbf{W}_{B3}$. The major difference between $\mathbf{W}_{B3}$ and $\mathbf{W}_{I3}$ is that the beam patterns associated with the first, second and third columns of $\mathbf{W}_{I3}$, referred to as the adaptive left, center and right beams, respectively, will be constructed so that each exhibits a null in the direction of each interfering source. Adaptive sum and difference beams are then constructed as linear combinations of the beams generated by $\mathbf{W}_{I3}$ in exactly the same way that classical sum and difference beams are constructed as linear combinations of the beams generated by $\mathbf{W}_{B3}$. In order that $\mathbf{W}_{I3}$ emulates $\mathbf{W}_{B3}$, we examine the properties of $\mathbf{W}_{B3}$ critical to the efficacy of the ML bearing estimation scheme summarized above.

One important property of $\mathbf{W}_{B3}$ is that described by (5-80). Equation (5-80) states that the columns of $\mathbf{W}_{B3}$ are mutually orthogonal and have equal 2-norm. As discussed previously, this has important implications with regard to the beamspace noise. Another important property is the fact that $w_L^*(z)$, $w_C^*(z)$, and $w_R^*(z)$, the polynomials associated with the first, second, and third columns, respectively, of $\mathbf{W}_{B3}^*$, have $N-3$ roots in common. This property enabled us to convert the problem of determining u_0 as a solution $\mathbf{v}_H^T\mathbf{b}_{H2}(u) = 0$ to an equivalent problem of finding $z_0 = \exp[j\pi u_0]$ as a root of the second order polynomial $e(z)$ described by (5-55). Ultimately, the problem simplified to that of evaluating the bearing estimation formula in (5-62). To facilitate the construction of an ICMBF $\mathbf{W}_{I3}$ possessing this $N-3$ common roots/nulls property, we provide here an alternative derivation of the expression for the coefficient vector for $e(z)$, $\mathbf{e} = \mathbf{Q}_{B3}^*\mathbf{H}\mathbf{v}_H$.

Since $w_{BL}^*(z)$, $w_{BC}^*(z)$, and $w_{BR}^*(z)$ have $N-3$ roots in common, it follows that each is the product of a common roots polynomial of order $N-3$ with a polynomial of order 2. The corresponding polynomial products are delineated in (5-48) with the common roots polynomial $c_{B3}^*(z)$ defined in (5-47). Thus, each of the columns of $\mathbf{W}_{B3}^*$ is the linear convolution of the coefficient sequence $\{\mathbf{c}_{B3}^*\}$ with a sequence of length 3. In accordance with (5-48),

$$\{\mathbf{w}_{BL}^*\} = \{\mathbf{c}_{B3}^*\} * \{\mathbf{q}_{BL}^*\} \tag{5-81a}$$

$$\{\mathbf{w}_{BC}^*\} = \{\mathbf{c}_{B3}^*\} * \{\mathbf{q}_{BC}^*\} \tag{5-81b}$$

$$\{\mathbf{w}_{BR}^*\} = \{\mathbf{c}_{B3}^*\} * \{\mathbf{q}_{BR}^*\} \tag{5-81c}$$

Invoking the matrix representation of linear convolution, these relationships allow us to factor $\mathbf{W}_{B3}^*$ as

$$\mathbf{W}_{B3}^* = \mathbf{C}_{B3}^* \mathbf{Q}_{B3}^* \tag{5-82}$$

where $\mathbf{Q}_{B3}^*$ is defined in (5-56) and $\mathbf{C}_{B3}^*$ is the $N \times 3$ banded, Toeplitz matrix

$$\mathbf{C}_{B3}^* = \begin{bmatrix} \mathbf{c}_{B3}^* & 0 & 0 \\ 0 & \mathbf{c}_{B3}^* & 0 \\ 0 & 0 & \mathbf{c}_{B3}^* \end{bmatrix} \tag{5-83}$$

Note that the expression for $\mathbf{C}_{B3}^*$ above is in the shorthand notation introduced in (5-79).

Use of the factorization of $\mathbf{W}_{B3}^*$ in (5-82) allows us to express the inner product $\mathbf{v}_H^T \mathbf{b}_{H2}(u_0)$ as

$$\mathbf{v}_H^T \mathbf{b}_{H2}(u_0) = \mathbf{v}_H^T \mathbf{H}^T \mathbf{W}_{B3}^H \mathbf{d}(u_0) = \{\mathbf{Q}_{B3} \mathbf{H} \mathbf{v}_H\}^H \mathbf{C}_{B3}^H \mathbf{d}(u_0) \tag{5-84}$$

Invoking the banded, Toeplitz structure of $\mathbf{C}_{B3}$, the mth component of the 3×1 vector $\mathbf{C}_{B3}^H \mathbf{d}(u_0)$ is the inner product between $\mathbf{c}_{B3}$ and a subvector of $\mathbf{d}(u_0)$ of length $N-2$ beginning with the mth component, $m = 1, 2, 3$. The latter quantity may be obtained by operating on $\mathbf{d}(u_0)$ with the selection matrix $\tilde{\mathbf{J}}_L^{(m)}$ defined as

$$\tilde{\mathbf{J}}_L^{(m)} = [\mathbf{O}_{L \times (m-1)} \,\vdots\, \mathbf{I}_{L \times L} \,\vdots\, \mathbf{O}_{L \times (N-L-m+1)}], \qquad m = 1, 2, \ldots, N-L+1 \tag{5-85}$$

with $L = N-2$. Invoking the definition of $\mathbf{d}(u)$ in (5-4), it is easily shown that premultiplication of $\mathbf{d}(u)$ by $\tilde{\mathbf{J}}_L^{(m)}$ yields the interesting result

$$\tilde{\mathbf{J}}_L^{(m)} \mathbf{d}(u; N) = \exp\left[-j \frac{N-L}{2} \pi u\right] \exp[j(m-1)\pi u] \mathbf{d}(u; L), \quad m = 1, 2, \ldots, N-L+1 \tag{5-86}$$

where $\mathbf{d}(u; J)$ for positive integer J is simply a generalization of $\mathbf{d}(u)$, defined in (5-4), for length J rather than N, that is,

$$\mathbf{d}(u; J) = \begin{bmatrix} \exp\left[j \dfrac{J-1}{2} \pi u \right] \\ \exp\left[-j \dfrac{J-3}{2} \pi u \right] \\ \vdots \\ \exp\left[j \dfrac{J-3}{2} \pi u \right] \\ \exp\left[j \dfrac{J-1}{2} \pi u \right] \end{bmatrix} \tag{5-87}$$

The result in (5-86) allows us to simplify the far right hand side of (5-84) as

$$\begin{aligned} \mathbf{v}_H^T \mathbf{b}_{H2}(u_0) &= \{\mathbf{Q}_{B3} \mathbf{H} \mathbf{v}_H\}^H \mathbf{C}_{B3}^H \mathbf{d}(u_0) \\ &= \mathbf{c}_{B3}^H \mathbf{d}(u_0; N-2) \{\mathbf{Q}_{B3}^* \mathbf{H} \mathbf{v}_H\}^T \mathbf{d}(u_0; 3) = 0 \end{aligned} \tag{5-88}$$

The quantity $\mathbf{c}_{B3}^H \mathbf{d}(u; N-2)$ is a scalar function of u which is only equal to zero for those values of u corresponding to one of the $N-3$ common nulls. Thus, (5-88) implies $\mathbf{e}^T \mathbf{d}(u_0; 3) = 0$, where $\mathbf{e} = [e_0 \quad e_1 \quad e_0^*]^T = \mathbf{Q}_{B3}^* \mathbf{H} \mathbf{v}_H$. Invoking the definition of $\mathbf{d}(u_0; J)$ in (5-87) with $J = 3$, $\mathbf{e}^T \mathbf{d}(u_0; 3) = e_0 \exp[-j\pi u_0] + e_1 + e_0^* \exp[j\pi u_0]$ and algebraic manipulation yields the simple bearing estimation formula in (5-62).

It is apparent from this development that the applicability of the bearing estimation formula in (5-62) in the case of interference cancellation is contingent upon the ability to factor $\mathbf{W}_{I3}^*$ as $\mathbf{C}_{I3}^* \mathbf{Q}_{I3}^*$, where $\mathbf{C}_{I3}^*$ possesses the banded-Toeplitz structure described by (5-83). If $\mathbf{W}_{I3}^*$ may be factored as such, $\mathbf{Q}_{I3}^*$ takes on the role of $\mathbf{Q}_{B3}^*$. The procedure for constructing $\mathbf{W}_{I3}$, to be developed shortly, will simultaneously provide $\mathbf{Q}_{I3}^*$ as well.

Another important property of $\mathbf{W}_{B3}$ is the conjugate centrosymmetry exhibited by each of its columns. This property is described by the relationship

$$\tilde{\mathbf{I}}_N \mathbf{W}_{B3} = \mathbf{W}_{B3}^* \tag{5-89}$$

where $\tilde{\mathbf{I}}_N$ is the reverse permutation matrix defined in (5-3). The fact that the beamspace manifold vector $\mathbf{b}_{H2}(u) = \mathbf{H}^T \mathbf{W}_{B3}^H \mathbf{d}(u)$ is a real-valued, 2×1 vector function of u is a direct consequence of this property. This claim is substantiated by the following sequence of manipulations:

$$\mathbf{b}_{H2}(u) = \mathbf{H}^T \mathbf{W}_{B3}^H \mathbf{d}(u) = \mathbf{H}^T \mathbf{W}_{B3}^H \tilde{\mathbf{I}}_N \tilde{\mathbf{I}}_N \mathbf{d}(u) = \mathbf{H}^T \mathbf{W}_{B3}^T \mathbf{d}^*(u) = \mathbf{b}_{H2}^*(u) \tag{5-90}$$

where we have used the trick $\tilde{\mathbf{I}}_N \tilde{\mathbf{I}}_N = \mathbf{I}_N$ and the fact that $\mathbf{H}$ is a real-valued

3×2 matrix. Incorporation of the fact that $\mathbf{b}_{H2}(u)$ is real-valued in the transformation of search variables from u_0 to $\mathbf{v}_H$ satisfying $\mathbf{v}_H^T \mathbf{b}_{H2}(u_0) = 0$ dictated that $\mathbf{v}_H$ be computed as the smallest eigenvector of $\mathrm{Re}\{\mathbf{x}_{H2}(n)\mathbf{x}_{H2}^H(n)\}$. As a consequence, $\mathbf{v}_H$ is real-valued and the imaginary component of the cross-correlation between the noise at the sum and difference beam outputs is suppressed.

Another consequence of the conjugate centrosymmetric property of $\mathbf{W}_{B3}$ described by (5-89) is that in the factorization described by (5-82), $\mathbf{C}_{B3}^*$ and $\mathbf{Q}_{B3}^*$ may be normalized so that each of the three columns of $\mathbf{Q}_{B3}^*$ exhibits conjugate centrosymmetry. This property is exhibited in the expression for $\mathbf{Q}_{B3}$ in (5-56). Combined with the fact that $\mathbf{v}_H$ is real-valued as is $\mathbf{H}$, the property $\tilde{\mathbf{I}}_3 \mathbf{Q}_{B3}^* = \mathbf{Q}_{B3}$ forces $\mathbf{e} = \mathbf{Q}_{B3}^* \mathbf{H} \mathbf{v}_H$ to be conjugate centrosymmetric as well. This property of $\mathbf{e}$ was important in the development of the ML bearing estimator since, combined with some practical considerations, it ensured that the two roots of $e(z)$ lay on the unit circle. In addition, the conjugate centrosymmetry of $\mathbf{e}$ was exploited in arriving at the simple bearing estimation formula in (5-62).

Another feature of $\mathbf{W}_{B3}$ exploited in the development of the ML bearing estimation scheme has to deal with the sidelobe phase relationships among the corresponding left, center, and right beams. Specifically, the left and center beams are lined up such that co-located sidelobes are 180° out-of-phase. The same is true with respect to the center and right beams. On the other hand, the left and right beams are lined up such that co-located sidelobes are in-phase. As discussed in Section 5-4, the low sidelobe levels achieved with the Hanning and Hamming windows may be attributed to judicious exploitation of these sidelobe relationships. We will attempt to construct the ICMBF $\mathbf{W}_{I3}$ so that the adaptively formed left, center, and right beams exhibit these same sidelobe phase relationships.

5-6-3 Interference Cancellation

In order to avert track breaking, it is necessary that both the sum beam and the difference beam exhibit nulls in the directions of strong interference. Let u_{Ik}, $k = 1, \ldots, K$, denote the directions in which nulls are to be formed. Some of these angles may correspond to the bearings of strong point interferers. Alternatively, if an interference is spatially distributed, as with clutter, each beam should exhibit a suitable number of point nulls over the angular extent of the continuous source. Note that, as in the adaptive beamforming procedures discussed in Chapters 3 and 4, it is typically not necessary to estimate these directions explicitly. We will return to this point shortly. As discussed previously, our approach is to construct an $N \times 3$ ICMBF $\mathbf{W}_{I3}$ exhibiting those properties of $\mathbf{W}_{B3}$ cited above. Let the columns of $\mathbf{W}_{I3}$ be denoted as

$$\mathbf{W}_{I3} = [\mathbf{w}_{IL} \,\vdots\, \mathbf{w}_{IC} \,\vdots\, \mathbf{w}_{IR}] \tag{5-91}$$

Adaptive sum and difference beams are then constructed as linear combinations of $\mathbf{w}_{IL}$, $\mathbf{w}_{IC}$, and $\mathbf{w}_{IR}$, in the same way that classical sum and difference beams are constructed as linear combinations of $\mathbf{w}_{BL}$, $\mathbf{w}_{BC}$, and $\mathbf{w}_{BR}$, the columns of $\mathbf{W}_{B3}$.

To this end, consider the following optimization problem:

$$\underset{\mathbf{W}_{I3}}{\text{Minimize}} \ \|\mathbf{W}_{I3} - \mathbf{W}_{B3}\|_F \tag{5-92}$$

subject to

$$\mathbf{W}_{I3}^H \mathbf{d}(u_{Ik}) = \begin{bmatrix} 0 \\ 0 \\ 0 \end{bmatrix}, \qquad k = 1, 2, \ldots, K$$

$$\mathbf{W}_{I3}^H \mathbf{W}_{I3} = N\mathbf{I}_3$$

$$\tilde{\mathbf{I}}_N \mathbf{W}_{I3} = \mathbf{W}_{I3}^*$$

$$\mathbf{W}_{I3} = \mathbf{C}_{I3}\mathbf{Q}_{I3}$$

where

$$\mathbf{C}_{I3} = \begin{bmatrix} \mathbf{c}_{I3} & 0 & 0 \\ 0 & \mathbf{c}_{I3} & 0 \\ 0 & 0 & \mathbf{c}_{I3} \end{bmatrix} \quad \text{and} \quad \tilde{\mathbf{I}}_3 \mathbf{Q}_{I3} = \mathbf{Q}_{I3}^*$$

Again, the expression for $\mathbf{C}_{I3}$ is in the shorthand notation of (5-79) such that $\mathbf{C}_{I3}$ has the same banded, Toeplitz structure exhibited by $\mathbf{C}_{B3}$. Given a solution to this optimization problem, the only modifications to the bearing estimation procedure outlined at the beginning of this section are (i) replace $\mathbf{W}_{B3}$ by $\mathbf{W}_{I3}$ and (ii) replace $\mathbf{Q}_{B3}^*$ by $\mathbf{Q}_{I3}^*$.

The objective function in (5-92) was chosen as a means of achieving the sidelobe phase relationships exhibited by the beams generated by $\mathbf{W}_{B3}$. The basis for this claim is as follows. First, note that the Frobinius norm [20] of a matrix is a nonnegative number. Since for nonnegative x, x^2 increases monotonically as x increases, we may replace the objective function in (5-92) by its square without altering the solution to the constrained optimization problem. Invoking the definition of the Frobinius norm [20], the square of the objective function in (5-92) may be expressed as

$$\|\mathbf{W}_{I3} - \mathbf{W}_{B3}\|_F^2 = \|\mathbf{w}_{IL} - \mathbf{w}_{BL}\|_2^2 + \|\mathbf{w}_{IC} - \mathbf{w}_{BC}\|_2^2 + \|\mathbf{w}_{IR} - \mathbf{w}_{BR}\|_2^2 \tag{5-93}$$

Now, recall, in accordance with (5-8), the beam pattern generated by an arbitrary weight vector $\mathbf{w}$, $\mathbf{d}^H(u)\mathbf{w}$, is the Discrete Space Fourier Transform (DSFT) of $\{\mathbf{w}\}$. The following variation of Parseval's relationship thus holds:

$$\int_{-1}^{1} |\mathbf{d}^H(u)\mathbf{w}|^2 \, du = \|\mathbf{w}\|_2^2 \tag{5-94}$$

Invoking this relationship, the square of the objective function in (5-92) may be alternatively expressed as

$$\begin{aligned}\|\mathbf{W}_{I3} - \mathbf{W}_{\mathrm{B3}}\|_{\mathrm{F}}^2 = &\int_{-1}^{1} |\mathbf{d}^H(u)\mathbf{w}_{IL} - \mathbf{d}^H(u)\mathbf{w}_{BL}|^2 \, du \\ &+ \int_{-1}^{1} |\mathbf{d}^H(u)\mathbf{w}_{IC} - \mathbf{d}^H(u)\mathbf{w}_{BC}|^2 \, du \\ &+ \int_{-1}^{1} |\mathbf{d}^H(u)\mathbf{w}_{IR} - \mathbf{d}^H(u)\mathbf{w}_{BR}|^2 \, du\end{aligned} \tag{5-95}$$

Thus, the square of the objective function in (5-92) is equal to the sum of the square of the difference between each corresponding pair of beam patterns integrated over the visible region. Minimization of this quantity under the common nulls constraint yields adaptively formed left, center, and right beams having the same sidelobe phase relationships exhibited by classical left, center, and right beams with pointing angles equispaced by $2/N$. This will be demonstrated in examples presented at the end of this section.

To facilitate a simple procedure for satisfying the common nulls constraint in (5-92), the following approach is taken. The weight vector associated with the center beam, $\mathbf{w}_{IC}$, is constructed first. $\mathbf{w}_{IL}$ and $\mathbf{w}_{IR}$ are then constructed based on $\mathbf{w}_{IC}$. $\mathbf{w}_{IC}$ is determined as that weight vector which minimizes the expected power of the receiver generated noise present at the beamformer output subject to a number of null constraints and a unity gain constraint in the direction u_C. In addition to the null constraints associated with interfering sources, two additional null constraints are imposed. Ultimately, these produce the two nulls of the center beam which are not members of the set of $N-3$ nulls common to the three beams formed by the columns of $\mathbf{W}_{I3}$. In accordance with the objective function in (5-92), the positions of these two nulls are chosen to be same as the locations of the two uncommon nulls associated with the center beam in the case of classical beamforming: $u = u_C - 2/N$ and $u = u_C + 2/N$. Under the assumption that none of the interferences is located within a beamwidth of the center beam, this serves to maintain the integrity of the mainlobe of the center beam relative to that achieved with $\mathbf{w}_{BC}$.

Several methods for constructing $\mathbf{w}_{IC}$ will be presented shortly. Assume for the moment, however, that $\mathbf{w}_{IC}$ has been generated satisfying the above specifications. Given that the two corresponding uncommon nulls are located at $u = u_C - 2/N$ and $u = u_C + 2/N$, the $(N-3)$th order common roots polynomial $c_{I3}(z)$ may be obtained via the polynomial ratio

$$c_{I3}(z) = \frac{w_{IC}(z)}{q_{IC}(z)}, \qquad \text{where } q_{IC}(z) = q_{BC}(z) \tag{5-96}$$

Recall that $q_{BC}(z)$ was defined previously in (5-48b). Accordingly, the length $N-2$ sequence $\{\mathbf{c}_{I3}\}$ may be computed by simply deconvolving the length 3 sequence $\{\mathbf{q}_{IC}\} = \{\mathbf{q}_{BC}\}$ out of the length N sequence $\{\mathbf{w}_{IC}\}$. The common nulls constraint then dictates that each of the two sequences $\{\mathbf{w}_{IL}\}$ and $\{\mathbf{w}_{IR}\}$ must be the linear convolution of $\{\mathbf{c}_{I3}\}$ with a sequence of length 3, that is,

$$\{\mathbf{w}_{IL}\} = \{\mathbf{c}_{I3}\} * \{\mathbf{q}_{IL}\} \tag{5-97a}$$

$$\{\mathbf{w}_{IR}\} = \{\mathbf{c}_{I3}\} * \{\mathbf{q}_{IR}\} \tag{5-97b}$$

In this approach, the remaining step is to determine the length three sequences $\{\mathbf{q}_{IL}\}$ and $\{\mathbf{q}_{IR}\}$ such that $\mathbf{W}_{I3} = \mathbf{C}_{I3}\mathbf{Q}_{I3}$, where

$$\mathbf{Q}_{I3} = [\mathbf{q}_{IL} \mid \mathbf{q}_{IC} \mid \mathbf{q}_{IR}] \tag{5-98}$$

minimizes the objective function in (5-92) under the constraint that the columns of $\mathbf{W}_{I3}$ be mutually orthogonal and conjugate centrosymmetric. A simple procedure for accomplishing this will be developed shortly.

5-6-4 Construction of the Center Beam

We briefly describe four procedures for constructing $\mathbf{w}_{IC}$. Each of the procedures is a variation on the Minimum Variance beamforming principle discussed in Chapter 3. The five procedures differ in accordance with the information available regarding the interference environment.

1. Adaptive-Adaptive Processing. This method assumes that the user has explicitly selected the K directions u_{Ik}, $k = 1, \ldots, K$, in which nulls are to be formed. These may be determined by applying spectral estimation during a period in which the radar is passively listening [11, 17–19]. Recently developed parametric spectral estimators may be employed to accurately locate the directions of strong point jammers [8, 26, 33, 35, 39, 42–44, 54, 56]. In the case of clutter, a priori information such as the geometry of the environment may dictate the locations of extended nulls. Given the null directions, we may construct the corresponding array response vectors $\mathbf{d}(u_{IK})$, $k = 1, \ldots, K$. Recall that nulls are to be formed in the directions $u = u_C - 2/N$ and $u = u_C + 2/N$ as well for reasons discussed above. Application of the Minimum Variance principle espoused in Chapter 4 under these constraints leads to the following constrained optimization problem:

$$\underset{\mathbf{w}_{IC}}{\text{Minimize}}\ E\{|\mathbf{w}_{IC}^H\mathbf{n}(n)|^2\} = \sigma_n^2\|\mathbf{w}_{IC}\|^2 \tag{5-99}$$

$$\text{subject to}\quad \mathbf{A}_{IC}^H\mathbf{w}_{IC} = \boldsymbol{\delta}$$

where σ_n^2 is the expected power of the receiver generated noise at each element, $\boldsymbol{\delta} = [0\ \ 0\ \ \cdots\ \ 0\ \ 1]^T$, a $(K+3)\times 1$ vector, and $\mathbf{A}_{IC}$ is an $N \times (K+3)$ matrix defined as

$$\mathbf{A}_{IC} = [\mathbf{d}(u_{I1}) \mid \cdots \mid \mathbf{d}(u_{IK}) \mid \mathbf{d}(u_C - 2/N) \mid \mathbf{d}(u_C + 2/N) \mid \mathbf{d}(u_C)] \tag{5-100}$$

Translated, (5-99) dictates that $\mathbf{w}_{IC}$ be computed as that weight vector which minimizes the expected power of the receiver generated noise at the output of the beamformer subject to a unity gain constraint in the direction u_C and a null constraint in each of the directions $u_L = u_C - 2/N$, $u_R = u_C + 2/N$, and u_{Ik}, $k = 1, \ldots, K$.

The constraint equation in (5-99), $\mathbf{A}_{IC}^H\mathbf{w}_{IC} = \boldsymbol{\delta}$, is an underdetermined system of equations. Hence, the solution is not unique. The objective function in (5-99) dictates that $\mathbf{w}_{IC}$ be the minimum norm solution. This may be expressed as [37]

$$\mathbf{w}_{IC} = \mathbf{A}_{IC}(\mathbf{A}_{IC}^H\mathbf{A}_{IC})^{-1}\boldsymbol{\delta} \tag{5-101}$$

Note that this solution exhibits conjugate centrosymmetry as desired. This may be proved by invoking the conjugate centrosymmetry of each of the columns of $\mathbf{A}_{IC}$.

$$\tilde{\mathbf{I}}_N\mathbf{w}_{IC} = \tilde{\mathbf{I}}_N\mathbf{A}_{IC}(\mathbf{A}_{IC}^H\tilde{\mathbf{I}}_N\tilde{\mathbf{I}}_N\mathbf{A}_{IC})^{-1}\boldsymbol{\delta} = \mathbf{A}_{IC}^*(\mathbf{A}_{IC}^T\mathbf{A}_{IC}^*)^{-1}\boldsymbol{\delta} = \mathbf{w}_{IC}^* \tag{5-102}$$

Again, the relationship $\tilde{\mathbf{I}}_N\tilde{\mathbf{I}}_N = \mathbf{I}_N$ has proved useful.

The solution in (5-101) has an interesting interpretation. Denote $\mathbf{g} = (\mathbf{A}_{IC}^H\mathbf{A}_{IC})^{-1}\boldsymbol{\delta}$, a $(K+3)\times 1$ vector. Note that $\mathbf{g}$ may be computed as either the last column of the inverse of the $(K+3)\times(K+3)$ real-valued matrix $\mathbf{A}_{IC}^H\mathbf{A}_{IC}$ or the solution to the system of equations $(\mathbf{A}_{IC}^H\mathbf{A}_{IC})\mathbf{g} = \boldsymbol{\delta}$. With $\mathbf{g}$ defined as such, (5-101) may be alternatively expressed as

$$\mathbf{w}_{IC} = \mathbf{A}_{IC}\mathbf{g} = \sum_{k=1}^{K} g_k\mathbf{d}(u_{Ik}) + g_{K+1}\mathbf{d}(u_C - 2/N) + g_{K+2}(u_C + 2/N) + g_{K+3}\mathbf{d}(u_C) \tag{5-103}$$

Now, in accordance with the constraint in (5-99), $\mathbf{w}_{IC}$ is orthogonal to each of the vectors $\mathbf{d}(u_C - 2/N)$, $\mathbf{d}(u_C + 2/N)$ and $\mathbf{d}(u_{Ik})$, $k = 1, \ldots, K$. It follows, therefore, that (5-103) may be interpreted as the last step of a Gram–Schmidt procedure [20, 37] in which the components of $\mathbf{d}(u_C)$ along each of these vectors is subtracted off. Accordingly, $\mathbf{w}_{IC}$ may be alternatively computed as follows. Denote $\mathbf{A}_{ILR} = [\mathbf{d}(u_{I1}) \vdots \cdots \vdots \mathbf{d}(u_{IK}) \vdots \mathbf{d}(u_C - 2/N) \,|\, \mathbf{d}(u_C + 2/N)]$, an $N \times (K+2)$ matrix. Construct an orthonormal basis for the $(K+2)$-dimensional subspace equal to the span of the columns of $\mathbf{A}_{ILR}$. In light of the poor numerical properties of the classical Gram–Schmidt procedure [20], it is recommended that this be accomplished by performing a QR decomposition on $\mathbf{A}_{ILR}$ utilizing Householder transformations [20]. Let $\{\mathbf{e}_1, \mathbf{e}_2, \ldots, \mathbf{e}_{K+2}\}$ represent the orthonormal basis constructed accordingly. $\mathbf{w}_{IC}$ is determined, to within a scalar multiple, by subtracting from $\mathbf{d}(u_C)$ its projection onto this $(K+2)$-dimensional space via

$$\mathbf{w}_{IC} = \alpha \left\{ \mathbf{d}(u_C) - \sum_{i=1}^{K+2} \{\mathbf{e}_i^H \mathbf{d}(u_C)\} \mathbf{e}_i \right\} \tag{5-104}$$

α may be determined by scaling $\mathbf{w}_{IC}$ so that $\mathbf{a}^H(u_C)\mathbf{w}_{IC} = 1$.

Note that $\mathbf{w}_{IC}$ may also be determined by performing a QR decomposition on $\mathbf{A}_{IC}$ rather than $\mathbf{A}_{ILR}$, where $\mathbf{A}_{IC} = [\mathbf{A}_{ILR} \,|\, \mathbf{d}(u_C)]$ in accordance with (5-100). To this end, the QR factorization of $\mathbf{A}_{IC}$ is expressed as $\mathbf{QR}$ where $\mathbf{Q}$ is an $N \times (K+3)$ matrix whose columns form an orthonormal basis for range$\{\mathbf{A}_{IC}\}$ and $\mathbf{R}$ is a $(K+3) \times (K+3)$ upper triangular matrix. As $\mathbf{A}_{IC} = \mathbf{QR}$, it follows that $\mathbf{A}_{IC}\mathbf{R}^{-1} = \mathbf{Q}$. Note that the inverse of a square upper triangular matrix is upper triangular as well. Combined with the orthonormality of the columns of $\mathbf{Q}$, it follows that the last column of $\mathbf{A}_{IC}\mathbf{R}^{-1}$ is orthogonal to the first $K+2$ columns of $\mathbf{A}_{IC}$. In turn, it follows from the discussion concerning (5-103) that $\mathbf{w}_{IC}$ is proportional to the last column of $\mathbf{A}_{IC}\mathbf{R}^{-1}$ equal to the last column of $\mathbf{Q}$. That is, given the QR factorization of $\mathbf{A}_{IC}$ as described above $\mathbf{w}_{IC}$ is proportional to the last column of $\mathbf{Q}$.

2. Eigenstructure Beamforming. As indicated, Method 1 explicitly requires estimates of the interference directions u_{Ik}, $k = 1, \ldots, K$. Alternatively, the MUltiple SIgnal Classification (MUSIC) [43, 44] algorithm may be applied during a period in which the radar is passively listening to estimate an orthonormal basis for the K-dimensional subspace range $\{\mathbf{d}(u_{I1}), \ldots, \mathbf{d}(u_{IK})\}$. In MUSIC, this is estimated as the span of the K principal eigenvectors of an $N \times N$ sample correlation matrix formed during the passive listening period [17–19]. Let $\mathbf{x}(n)$, $n = 1, \ldots, M$, denote M snapshots collected during this period. The $N \times N$ sample correlation matrix, denoted $\hat{\mathbf{R}}_{xx}$, may be formed as

$$\hat{\mathbf{R}}_{xx} = \frac{1}{M} \sum_{n=1}^{M} \mathbf{x}(n)\mathbf{x}^H(n) \tag{5-105}$$

Since $\hat{\mathbf{R}}_{xx}$ is Hermitian, its K principal eigenvectors are orthonormal. A simple two-step Gram–Schmidt procedure may then be implemented to expand to an orthonormal basis of $K+2$ vectors encompassing the space spanned by $\mathbf{d}(u_C+2/N)$ and $\mathbf{d}(u_C-2/N)$ as well. The orthonormal basis of $K+2$ vectors thus obtained is then used in (5-104) to construct $\mathbf{w}_{IC}$.

Note that the inner product between two conjugate centrosymmetric vectors is real-valued. Also, the addition of two conjugate centrosymmetric vectors yields a conjugate centrosymmetric vector. It follows, therefore, that the eigenvector-based Gram–Schmidt procedure described above will yield a $\mathbf{w}_{IC}$ exhibiting conjugate centrosymmetry if the $K+2$ vectors comprising the orthonormal basis used in (5-104) are each conjugate centrosymmetric. This will be the case if we use the principal eigenvectors of the forward-backward averaged (FBAVG'd) sample correlation matrix [17, 18, 55] defined as

$$\hat{\mathbf{R}}_{FB} = \tfrac{1}{2}\{\hat{\mathbf{R}}_{XX} + \tilde{\mathbf{I}}_N \hat{\mathbf{R}}_{XX}^* \tilde{\mathbf{I}}_N\} \tag{5-106}$$

as opposed to those of $\hat{\mathbf{R}}_{XX}$ itself. The conjugate centrosymmetry of each of the eigenvectors of $\text{Re}\{\hat{\mathbf{R}}_{FB}\}$ is easily established as follows. First, note $\hat{\mathbf{R}}_{FB}$ in (5-106) satisfies $\tilde{\mathbf{I}}_N \hat{\mathbf{R}}_{FB} \tilde{\mathbf{I}}_N = \hat{\mathbf{R}}_{FB}^*$. With this in mind, let λ_i and $\mathbf{e}_i$ be an eigenvalue-eigenvector pair of $\hat{\mathbf{R}}_{FB}$, that is, $\hat{\mathbf{R}}_{FB}\mathbf{e}_i = \lambda_i \mathbf{e}_i$. This implies $\tilde{\mathbf{I}}_N \hat{\mathbf{R}}_{FB}^* \tilde{\mathbf{I}}_N \mathbf{e}_i = \lambda_i \mathbf{e}_i$. Conjugating and multiplying both sides of this equality by $\tilde{\mathbf{I}}_N$ yields $\hat{\mathbf{R}}_{FB}\tilde{\mathbf{I}}_N \mathbf{e}_i^* = \lambda_i \tilde{\mathbf{I}}_N \mathbf{e}_i^*$. Of course, λ_i is real since $\hat{\mathbf{R}}_{FB}$ is Hermitian. Under the practical assumption that the eigenvalues of $\hat{\mathbf{R}}_{FB}$ are unique, it follows that the corresponding eigenvectors are unique to within a scalar multiple. Hence, $\tilde{\mathbf{I}}_N \mathbf{e}_i = \gamma_i \mathbf{e}_i^*$. Since $\tilde{\mathbf{I}}_N$ is unitary, and thus preserves two norm, $|\gamma_i| = 1$. It follows, therefore, that we can always normalize the eigenvectors of $\hat{\mathbf{R}}_{FB}$ so that they exhibit conjugate centrosymmetry, that is, satisfy $\tilde{\mathbf{I}}_N \mathbf{e}_i = \mathbf{e}_i^*$.

Inherent in the eigenvector based procedure described above is the assumption that the number of interferers, K, is known. A number of statistically based procedures have been proposed for estimating this quantity via an examination of the eigenvalues of $\hat{\mathbf{R}}_{XX}$ or $\hat{\mathbf{R}}_{FB}$ [38]. These include the Akaike Information Criterion (AIC) and the Minimum Description Length (MDL) principle. A discussion of these procedures may be found in [26, 38, 39].

3. Minimum Variance Beamforming. The eigenvector based beamforming procedure requires determination of the number of interferers and an eigenanalysis of the array data. Alternatively, $\mathbf{w}_{IC}$ may be computed according to a slight modification of the Minimum Variance (MV) beamforming procedure [16] described in Chapter 4. In contrast to MUSIC based beamforming, MV does not require determination of the number of sources. In addition, MV requires the inverse of the sample correlation matrix as opposed to an EVD of such. As a consequence, the task of updating MV beamforming weights as a function time, for the purpose of adapting to a changing interference environment, is much less computationally complex than that of updating the eigenvector-based beamforming weights.

The application of the MV principle in the computation of $\mathbf{w}_{IC}$ leads to the following constrained optimization problem:

$$\underset{\mathbf{w}_{IC}}{\text{Minimize}}\ E\{|\mathbf{w}_{IC}^H\mathbf{x}(n)|^2\} = \mathbf{w}_{IC}^H\hat{\mathbf{R}}_{XX}\mathbf{w}_{IC}$$
$$\text{subject to}\quad \mathbf{W}_{B3}^H\mathbf{w}_{IC} = \mathrm{f} \quad \text{and} \quad \tilde{\mathbf{I}}_N\mathbf{w}_{IC} = \mathbf{w}_{IC}^* \tag{5-107}$$

Here $\mathbf{f} = [0 \quad 1 \quad 0]^T$ such that $\mathbf{w}_{IC}$ is chosen to minimize the expected power at the beamformer output subject to a unity gain constraint in the direction $u = u_C$, and a null constraint in each of the directions $u = u_C - 2/N$ and $u = u_C + 2/N$. As discussed previously, the latter two null constraints facilitate simple construction of the left and right beams.

Since $\mathbf{w}_{IC}$ is constrained to be conjugate centrosymmetric, $\mathbf{w}_{IC}^H\hat{\mathbf{R}}_{XX}\mathbf{w}_{IC} = \mathbf{w}_{IC}^T\tilde{\mathbf{I}}_N\hat{\mathbf{R}}_{XX}\tilde{\mathbf{I}}_N\mathbf{w}_{IC}^* = \mathbf{w}_{IC}^H\tilde{\mathbf{I}}_N\hat{\mathbf{R}}_{XX}^*\tilde{\mathbf{I}}_N\mathbf{w}_{IC}$. Thus, the objective function in (5-107) may be alternatively expressed as

$$\underset{\mathbf{w}_{IC}}{\text{Minimize}}\ \tfrac{1}{2}\{\mathbf{w}_{IC}^H\hat{\mathbf{R}}_{XX}\mathbf{w}_{IC} + \mathbf{w}_{IC}^H\tilde{\mathbf{I}}_N\hat{\mathbf{R}}_{XX}^*\tilde{\mathbf{I}}_N\mathbf{w}_{IC}\} = \mathbf{w}_{IC}^H\hat{\mathbf{R}}_{FB}\mathbf{w}_{IC}$$
$$\text{subject to}\quad \mathbf{W}_{B3}^H\mathbf{w}_{IC} = \mathbf{f} \quad \text{and} \quad \tilde{\mathbf{I}}_N\mathbf{w}_{IC} = \mathbf{w}_{IC}^* \tag{5-108}$$

where $\hat{\mathbf{R}}_{FB}$ is the FBAVG'd sample correlation matrix defined in (5-106). Now, the constraint equation $\mathbf{W}_{B3}^H\mathbf{w}_{IC} = \mathbf{f}$ is an underdetermined system of equations. Hence, the solution is not unique. The objective function in (5-108) dictates that $\mathbf{w}_{IC}$ be the minimum norm solution in a Hilbert space in which the norm of a vector $\mathbf{x}$ is defined as $\langle \mathbf{x}, \mathbf{x}\rangle = \mathbf{x}^H\hat{\mathbf{R}}_{FB}\mathbf{x}$. This may be expressed as [37]

$$\mathbf{w}_{IC} = \hat{\mathbf{R}}_{FB}^{-1}\mathbf{W}_{B3}(\mathbf{W}_{B3}^H\hat{\mathbf{R}}_{FB}^{-1}\mathbf{W}_{B3})^{-1}\mathbf{f} \tag{5-109}$$

Fortuitously, this solution exhibits conjugate centrosymmetry. This is easily established as follows. Recall that $\tilde{\mathbf{I}}_N\hat{\mathbf{R}}_{FB}\tilde{\mathbf{I}}_N = \hat{\mathbf{R}}_{FB}^*$. Inverting both sides of this equality yields $\tilde{\mathbf{I}}_N\hat{\mathbf{R}}_{FB}^{-1}\tilde{\mathbf{I}}_N = \hat{\mathbf{R}}_{FB}^{-1*}$. Hence,

$$\begin{aligned}\tilde{\mathbf{I}}_N\mathbf{w}_{IC} &= \tilde{\mathbf{I}}_N\hat{\mathbf{R}}_{FB}^{-1}\tilde{\mathbf{I}}_N\tilde{\mathbf{I}}_N\mathbf{W}_{B3}(\mathbf{W}_{B3}^H\tilde{\mathbf{I}}_N\tilde{\mathbf{I}}_N\hat{\mathbf{R}}_{FB}^{-1}\tilde{\mathbf{I}}_N\tilde{\mathbf{I}}_N\mathbf{W}_{B3})^{-1}\mathbf{f}\\ &= \hat{\mathbf{R}}_{FB}^{-1*}\mathbf{W}_{B3}^*(\mathbf{W}_{B3}^T\hat{\mathbf{R}}_{FB}^{-1*}\mathbf{W}_{B3}^*)^{-1}\mathbf{f} = \mathbf{w}_{IC}^*\end{aligned} \tag{5-110}$$

Thus, $\mathbf{w}_{IC}$ given by (5-109) is indeed the solution to the optimization problem in (5-107).

Although computationally expedient schemes exist for updating the MV beamforming weights in accordance with (5-109) [39], it should be noted that MV beamforming does not yield hard nulls in the directions of the interferers. Fortunately, as discussed in Chapter 4, the depth of a given null is proportional to the strength or SNR of the corresponding interferer.

4. Interference Polynomial Beamforming. An alternative procedure for computing $\mathbf{w}_{IC}$ requires only the coefficient vector of the *interference polynomial* [9], $i(z)$, defined as

$$\begin{aligned} i(z) &= \alpha_I(z - \exp[j\pi u_{I1}])(z - \exp[j\pi u_{I2}]) \cdots (z - \exp[j\pi u_{IK}]) \\ &= i_0 + i_1 z + i_2 z^2 + \cdots + i_K z^K \end{aligned} \tag{5-111}$$

where α_I is that scalar which makes the coefficient vector $\mathbf{i} = [i_0 \quad i_1 \quad \cdots \quad i_{IK}]^T$ conjugate centrosymmetric. That is, in the procedure to be described it is not necessary to root $i(z)$. Examples of spectral estimation algorithms which provide a direct estimate of the coefficient vector $\mathbf{i}$ include the Iterative Quadratic Maximum Likelihood (IQML) algorithm [8, 33] and the Estimation of Signal Parameters by Rotational Invariance Techniques (ESPRIT) algorithm [42, 56]. In the latter case, $\mathbf{i}$ may be estimated as the coefficient vector of a characteristic equation associated with a generalized eigenvalue problem.

Given the coefficient vector for the interference polynomial, the next step in this procedure is to convolve the length $K+1$ sequence $\{\mathbf{i}\}$ with the length 3 sequence $\{\mathbf{q}_{BC}\}$, the coefficient vector for $\mathbf{q}_{BC}(z)$ defined in (5-48b). This yields a length $K+3$ sequence denoted $\{\mathbf{i}_c\}$, that is, $\{\mathbf{i}_c\} = \{\mathbf{i}\} * \{\mathbf{q}_{BC}\}$. Note that the convolution of two conjugate centrosymmetric sequences yields a conjugate centrosymmetric sequence. $\{\mathbf{w}_{IC}\}$ is then computed as the convolution of $\{\mathbf{i}_c\}$ with some sequence of length $N-K-2$, denoted arbitrarily as $\{\mathbf{g}_c\}$. In matrix form,

$$\mathbf{w}_{IC} = \mathbf{I}_c \mathbf{g}_c \tag{5-112}$$

where $\mathbf{I}_c$ is the $N \times (N-K-2)$ banded, Toeplitz matrix

$$\mathbf{I}_c = \begin{bmatrix} \mathbf{i}_c & 0 & & 0 \\ 0 & \mathbf{i}_c & & 0 \\ 0 & 0 & \cdots & 0 \\ \vdots & \vdots & & \vdots \\ 0 & 0 & & \mathbf{i}_c \end{bmatrix} \tag{5-113}$$

Again, the expression for $\mathbf{I}_c$ above is in the shorthand notation introduced in (5-79).

Construction of $\mathbf{w}_{IC}$ according to (5-112) guarantees that the $(N-1)$th order polynomial $w_{IC}(z)$ has roots at $z_{Ik} = \exp[j\pi u_{Ik}]$, $k = 1, \ldots, K$, $z_L = \exp[j\pi(u_C - 2/N)]$, and $z_R = \exp[j\pi(u_C + 2/N)]$. As a consequence, the beam produced by $\mathbf{w}_{IC}$ is guaranteed to exhibit a null in each of the interference directions u_{Ik}, $k = 1, \ldots, K$, and in the directions $u_L = u_C - 2/$

N and $u_R = u_C + 2/N$ as required. In conformance with the MV principle, the remaining degrees of freedom inherent in the selection of $\mathbf{g}_c$ should be used to minimize the expected power of the receiver generated noise at the beamformer output subject to a unity gain constraint in the direction u_C. As observed in the objective function in (5-99), the former is equivalent to minimizing the norm of $\mathbf{w}_{IC} = \mathbf{I}_c\mathbf{g}_c$. Accordingly, the $(N - K - 2) \times 1$ vector $\mathbf{g}_c$ is determined as the solution to

$$\underset{\mathbf{g}_c}{\text{Minimize}} \; \|\mathbf{w}_{IC}\|^2 = \mathbf{g}_c^H \mathbf{I}_c^H \mathbf{I}_c \mathbf{g}_c \tag{5-114}$$

$$\text{subject to} \quad \mathbf{w}_{IC}^H \mathbf{d}(u_C) = \mathbf{g}_c^H \mathbf{I}_c^H \mathbf{d}(u_C) = 1$$

(5-114) may be solved via the method of Lagrange multipliers. This leads to the result that $\mathbf{g}_c$ be determined as the solution to the $(N - K - 2) \times (N - K - 2)$ system of equations

$$\mathbf{I}_c^H \mathbf{I}_c \mathbf{g}_c = \lambda_{IC} \mathbf{I}_c^H \mathbf{d}(u_C) \tag{5-115}$$

where λ_{IC} is a Lagrange multiplier. The value of λ_{IC} is dictated by compliance with the unity gain constraint in (5-114). Expressing the solution as $\mathbf{g}_c = \lambda_{IC}(\mathbf{I}_c^H\mathbf{I}_c)^{-1}\mathbf{I}_c^H\mathbf{d}(u_C)$ yields $\lambda_{IC} = 1/(\mathbf{d}^H(u_C)\mathbf{P}_{I_c}\mathbf{d}(u_C))$, where $\mathbf{P}_{I_c} = \mathbf{I}_c(\mathbf{I}_c^H\mathbf{I}_c)^{-1}\mathbf{I}_c^H$. Note that $\mathbf{P}_{I_c}$ is Hermitian such that λ_{IC} is real-valued.

As a consequence of the banded, Toeplitz structure of $\mathbf{I}_c$, $\mathbf{I}_c^H\mathbf{I}_c$ is an $(N - K - 2) \times (N - K - 2)$ Toeplitz–Hermitian matrix. Hence, (5-115) may be alternatively solved in a computationally efficient manner via the Levinson–Durbin recursion [39]. In this mode of operation, λ_{IC} in (5-115) is set equal to unity and the solution obtained via the Levinson–Durbin recursion simply scaled to comply with the unity gain constraint in the direction u_C. Note that the inputs to the Levinson–Durbin recursion are the first column of $\mathbf{I}_c^H\mathbf{I}_c$ and the vector $\mathbf{I}_c^H\mathbf{d}(u_C)$. The components of the first column of $\mathbf{I}_c^H\mathbf{I}_c$ are simply the first $N - K - 2$ autocorrelation values associated with the sequence $\{\mathbf{i}_c\}$. Likewise, the components of $\mathbf{I}_c^H\mathbf{d}(u_C)$ are the first $N - K - 2$ values of the cross correlation between the sequence $\{\mathbf{i}_c\}$ and the sequence $\{\mathbf{d}(u_C)\}$. The computation may be further reduced by realizing that the solution $\mathbf{g}_c$ is conjugate centrosymmetric. As a consequence, execution of the Levinson–Durbin algorithm may be terminated at iteration number $(N - K - 2)/2$, in the case of $N - K - 2$ even, or iteration number $(N - K - 1)/2$, in the case of $N - K - 2$ odd. Proof of the conjugate centrosymmetry of $\mathbf{g}_c$ is achieved by invoking the Toeplitz–Hermitian structure of $\mathbf{I}_c^H\mathbf{I}_c$, the conjugate centrosymmetry of $\mathbf{d}(u_C)$, and the fact that λ_{IC} is real. The proof is left to the reader. Thus, $\{\mathbf{w}_{IC}\} = \{\mathbf{i}_c\} * \{\mathbf{g}_c\}$ is a conjugate centrosymmetric sequence and $\mathbf{w}_{IC}$ is a conjugate centrosymmetric vector.

An insight into interference polynomial beamforming is provided by identifying $\mathbf{w}_{IC} = \mathbf{I}_c\mathbf{g}_c$ in (5-115). With this substitution, (5-115) may be expressed as $\mathbf{I}_c^H \mathbf{w}_{IC} = \lambda_{IC} \mathbf{I}_c^H \mathbf{d}(u_C)$. From this it may be deduced that the solution for $\mathbf{w}_{IC}$ obtained through this procedure is attempting to emulate the classical beamformer $\mathbf{w}_{BC} = \mathbf{d}(u_C)$ as best as possible under the given null constraints. This is in accordance with the objective function in (5-92).

5-6-5 Construction of Interference Cancellation Matrix Beamformer

Given $\mathbf{w}_{IC}$ computed according to one of these four procedures, $\mathbf{c}_{I3}$ is readily determined in accordance with (5-96). As discussed previously, the length $N-2$ sequence $\{\mathbf{c}_{I3}\}$ is computed by simply deconvolving the length 3 sequence $\{\mathbf{q}_{BC}\}$, defined by (5-48b), out of the length N sequence $\{\mathbf{w}_{IC}\}$. Note that $\{\mathbf{w}_{IC}\}$ and $\{\mathbf{q}_{BC}\}$ are each conjugate centrosymmetric such that $\{\mathbf{c}_{I3}\}$ is conjugate centrosymmetric as well. The sequence $\{\mathbf{c}_{I3}\}$ may thus be computed according to the following simple recursion:

$$\begin{aligned} c_{I3}(k) &= \exp[-j\pi u_C]\{w_{IC}(k) + 2\cos(2\pi/N)c_{I3}(k-1) \\ &\qquad - \exp[j\pi u_C]c_{I3}(k-2)\} \\ c_{I3}(N-3-k) &= c_{I3}^*(k)\,, \qquad k = 0, 1, \ldots, I,\ N \text{ even: } I = (N-2)/2; \\ &\qquad\qquad\qquad N \text{ odd: } I = (N-1)/2 \end{aligned}$$

where

$$c_{I3}(-1) = c_{I3}(-2) = c_{I3}(-3) = 0 \tag{5-116}$$

The sequence $\{\mathbf{c}_{I3}\}$ obtained from this recursion is that used to construct the $N \times 3$ banded, Toeplitz matrix $\mathbf{C}_{I3}$ to be employed in solving (5-92).

At this point, $\mathbf{W}_{I3} = \mathbf{C}_{I3}\mathbf{Q}_{I3}$ with $\mathbf{C}_{I3}$ and the middle column of $\mathbf{Q}_{I3}$, $\mathbf{q}_{IC}$, determined. In addition, the constraint $\mathbf{W}_{I3}^H \mathbf{d}(u_{Ik}) = [0 \quad 0 \quad 0]^T$, $k = 1, 2, \ldots, K$, is satisfied. Thus, the remaining quantities, the first and third columns of $\mathbf{Q}_{I3}$, are determined in accordance with the following constrained optimization problem derived from (5-92):

$$\begin{aligned} \underset{\mathbf{Q}_{I3}}{\text{Minimize}}\ &\|\mathbf{W}_{B3} - \mathbf{W}_{I3}\|_F^2 = \|\mathbf{W}_{B3} - \mathbf{C}_{I3}\mathbf{Q}_{I3}\|_F^2 \\ \text{subject to}\ &\mathbf{W}_{I3}^H \mathbf{W}_{I3} = \mathbf{Q}_{I3}^H \{\mathbf{C}_{I3}^H \mathbf{C}_{I3}\} \mathbf{Q}_{I3} = N\mathbf{I}_3 \\ &\tilde{\mathbf{I}}_3 \mathbf{Q}_{I3} = \mathbf{Q}_{I3}^* \end{aligned} \tag{5-117}$$

To facilitate a simple procedure for solving (5-117), $\mathbf{Q}_{I3}$ is factored as $\mathbf{Q}_{I3} = \mathbf{Q}\mathbf{T}_Q$ where $\mathbf{Q}$ and $\mathbf{T}_Q$ are 3×3 matrices. $\mathbf{Q}$ is utilized to comply with the orthogonality constraint independently of the objective function. The objective function is then minimized with respect to $\mathbf{T}_Q$ under the constraint that $\mathbf{T}_Q$ be unitary so as to maintain compliance with the orthogonality constraint. This two-step solution procedure allows us to reformulate the constrained optimization problem in (5-92) in terms of a simple subspace rotation problem referred to as the *orthogonal Procrustes problem* [20]. A simple, closed-form solution exists for this problem. The conjugate centrosymmetry constraint is satisfied surreptitiously. The appropriate development is provided below.

Let $\mathbf{q}_L$, $\mathbf{q}_C$, and $\mathbf{q}_R$ denote the first, second, and third columns of $\mathbf{Q}$, respectively, that is, $\mathbf{Q} = [\mathbf{q}_L \mid \mathbf{q}_C \mid \mathbf{q}_R]$. The orthogonality constraint is satisfied if the set $\{\mathbf{q}_L, \mathbf{q}_C, \mathbf{q}_R\}$ forms an orthogonal basis for the 3-D Hilbert space with inner product defined as $\langle \mathbf{x}, \mathbf{y} \rangle = \mathbf{x}^H \mathbf{C}_{I3}^H \mathbf{C}_{I3} \mathbf{y}$. As the objective function in (5-117) dictates that $\mathbf{W}_{I3} = \mathbf{C}_{I3}\mathbf{Q}\mathbf{T}_Q$ be close to $\mathbf{W}_{B3}$ in a Frobinius norm sense, it seems logical to employ the Gram–Schmidt procedure to construct such a basis using $\{\mathbf{q}_{BL}, \mathbf{q}_{BC}, \mathbf{q}_{BR}\}$ as the initial basis set. Recall that $\mathbf{q}_{BL}$, $\mathbf{q}_{BC}$, and $\mathbf{q}_{BR}$ are the first, second, and third columns, respectively, of $\mathbf{Q}_{B3}$ defined in (5-56). In light of the fact that $\mathbf{q}_{IC}$ is predetermined to be $\mathbf{q}_{BC}$, $\mathbf{q}_{BC}$ is used as the starting point for the Gram–Schmidt procedure yielding

$$\mathbf{q}_C' = \mathbf{q}_{BC} \tag{5-118a}$$

$$\mathbf{q}_L' = \mathbf{q}_{BL} - \frac{\mathbf{q}_C'^H \mathbf{C}_{I3}^H \mathbf{C}_{I3} \mathbf{q}_{BL}}{\mathbf{q}_C'^H \mathbf{C}_{I3}^H \mathbf{C}_{I3} \mathbf{q}_C'} \mathbf{q}_C' \tag{5-118b}$$

$$\mathbf{q}_R' = \mathbf{q}_{BR} - \frac{\mathbf{q}_C'^H \mathbf{C}_{I3}^H \mathbf{C}_{I3} \mathbf{q}_{BR}}{\mathbf{q}_C'^H \mathbf{C}_{I3}^H \mathbf{C}_{I3} \mathbf{q}_C'} \mathbf{q}_C' - \frac{\mathbf{q}_L'^H \mathbf{C}_{I3}^H \mathbf{C}_{I3} \mathbf{q}_{BR}}{\mathbf{q}_L'^H \mathbf{C}_{I3}^H \mathbf{C}_{I3} \mathbf{q}_L'} \mathbf{q}_L' \tag{5-118c}$$

$$\mathbf{q}_L = N \frac{\mathbf{q}_L'}{\|\mathbf{q}_L'\|_2}, \qquad \mathbf{q}_C = N \frac{\mathbf{q}_C'}{\|\mathbf{q}_C'\|_2}, \qquad \mathbf{q}_R = N \frac{\mathbf{q}_R'}{\|\mathbf{q}_R'\|_2} \tag{5-118d}$$

With $\mathbf{Q} = [\mathbf{q}_L \,\vdots\, \mathbf{q}_C \,\vdots\, \mathbf{q}_R]$ the columns of $\mathbf{W}_Q = \mathbf{C}_{I3}\mathbf{Q}$ are mutually orthogonal with the 2-norm of each equal to N.

Invoking the fact that $\mathbf{C}_{I3}^H \mathbf{C}_{I3}$ is Toeplitz–Hermitian, such that $\tilde{\mathbf{I}}_3 \mathbf{C}_{I3}^H \mathbf{C}_{I3} \tilde{\mathbf{I}}_3 = \{\mathbf{C}_{I3}^H \mathbf{C}_{I3}\}^*$, and the fact that $\mathbf{q}_{BL}$, $\mathbf{q}_{BC}$, and $\mathbf{q}_{BR}$ defined by (5-48) are each conjugate centrosymmetric, it is easy to show that $\mathbf{q}_L$, $\mathbf{q}_C$, and $\mathbf{q}_R$ constructed according to (5-118) are each conjugate centrosymmetric as well. As a consequence, the columns of $\mathbf{W}_Q = \mathbf{C}_{I3}\mathbf{Q}$ are each conjugate centrosymmetric. Therefore, $\mathbf{W}_Q = \mathbf{C}_{I3}\mathbf{Q}$ exhibits the following two properties:

$$\mathbf{W}_Q^H \mathbf{W}_Q = N\mathbf{I}_3 \tag{5-119a}$$

$$\tilde{\mathbf{I}}_N \mathbf{W}_Q = \mathbf{W}_Q^* \tag{5-119b}$$

The remaining step then is to compute $\mathbf{T}_Q$ as the solution to

$$\underset{\mathbf{T}_Q}{\text{Minimize}}\ \|\mathbf{W}_{B3} - \mathbf{W}_{I3}\|_F^2 = \|\mathbf{W}_{B3} - \mathbf{W}_Q\mathbf{T}_Q\|_F^2 \qquad \text{subject to} \quad \mathbf{T}_Q^H\mathbf{T}_Q = \mathbf{T}_Q\mathbf{T}_Q^H = \mathbf{I}_3 \tag{5-120}$$

As discussed previously, the unitary constraint on $\mathbf{T}_Q$ serves to maintain orthogonality and equality of 2-norm amongst the columns of $\mathbf{W}_{I3} = \mathbf{W}_Q\mathbf{T}_Q$. Equation (5-120) is in the form of a classical subspace rotation problem referred to as the orthogonal Procrustes problem. Golub and Van Loan [20] describe a solution to this problem based on the Singular Value Decomposition (SVD) [20, 37] of the matrix $\mathbf{W}_{B3}^H\mathbf{W}_Q$, which in this case is 3×3. Since each of the columns of $\mathbf{W}_{B3}$ and $\mathbf{W}_Q$ is conjugate centrosymmetric, it turns out that $\mathbf{W}_{B3}^H\mathbf{W}_Q$ is real-valued as evidenced by

$$\mathbf{W}_{B3}^H\mathbf{W}_Q = \mathbf{W}_{B3}^H\tilde{\mathbf{I}}_N\tilde{\mathbf{I}}_N\mathbf{W}_Q = \mathbf{W}_{B3}^T\mathbf{W}_Q^* = (\mathbf{W}_{B3}^H\mathbf{W}_Q)^* \tag{5-121}$$

Thus, the left and right singular vectors of $\mathbf{W}_{B3}^H\mathbf{W}_Q$ are real-valued quantities. With this in mind, let $\mathbf{W}_{B3}^H\mathbf{W}_Q = \mathbf{U}\mathbf{\Sigma}\mathbf{V}^T$ be the SVD. The solution to (5-120) is then $\mathbf{T}_Q = \mathbf{U}\mathbf{V}^T$, a real-valued, orthogonal matrix. Note, since $\mathbf{T}_Q = \mathbf{U}\mathbf{V}^T$ is real-valued, the columns of $\mathbf{W}_{I3} = \mathbf{W}_Q\mathbf{T}_Q$ are each conjugate centrosymmetric as evidenced by

$$\tilde{\mathbf{I}}_N\mathbf{W}_{I3} = \tilde{\mathbf{I}}_N\mathbf{W}_Q\mathbf{T}_Q = \mathbf{W}_Q^*\mathbf{T}_Q = \mathbf{W}_{I3}^* \tag{5-122}$$

As a consequence of this and previous observations, it follows that $\mathbf{W}_{I3} = \mathbf{C}_{I3}\mathbf{Q}\mathbf{T}_Q$ satisfies all of the constraints imposed in (5-92).

A summary of the procedure for constructing the ICMBF $\mathbf{W}_{I3}$ is outlined below.

Summary of Method for Constructing ICMBF $\mathbf{W}_{I3}$

1a. Construct $\mathbf{w}_{IC}$ according to one of four methods proposed:
   (i) Adaptive-adaptive processing
   (ii) Eigenstructure beamforming
   (iii) Minimum variance beamforming
   (iv) Interference polynomial beamforming

2a. Form length $N-2$ sequence $\{\mathbf{c}_{I3}\}$ by deconvolving length 3 sequence $\{\mathbf{q}_{BC}\}$ out of length N sequence $\{\mathbf{w}_{IC}\}$ according to (5-116)

2b. Construct $N \times 3$ banded, Toeplitz matrix $\mathbf{C}_{I3}$ according to (5-92)

3a. Construct orthonormal basis $\{\mathbf{q}_L, \mathbf{q}_C, \mathbf{q}_R\}$ for 3-D Hilbert space with inner product $\langle \mathbf{x}, \mathbf{y}\rangle = \mathbf{x}^H\mathbf{C}_{I3}^H\mathbf{C}_{I3}\mathbf{y}$ according to Gram–Schmidt procedure outlined in (5-118)

3b. With $\mathbf{Q} = [\mathbf{q}_L \,\vdots\, \mathbf{q}_C \,\vdots\, \mathbf{q}_R]$, construct $\mathbf{W}_Q = \mathbf{C}_{I3}\mathbf{Q}$

4a. Compute 3×3 real-valued SVD: $\mathbf{W}_{B3}^H\mathbf{W}_Q = \mathbf{U}\mathbf{\Sigma}\mathbf{V}^T$

4b. Form $\mathbf{T}_Q = \mathbf{U}\mathbf{V}^T$

5. ICMBF: $\mathbf{W}_{I3} = \mathbf{W}_Q\mathbf{T}_Q = \mathbf{C}_{I3}\mathbf{Q}\mathbf{T}_Q$. Also: $\mathbf{Q}_{I3} = \mathbf{Q}\mathbf{T}_Q$

With $\mathbf{W}_{B3}$ replaced by $\mathbf{W}_{I3}$ and $\mathbf{Q}_{B3}^*$ replaced by $\mathbf{Q}_{I3}^*$, the ML bearing estimation procedure incorporating interference cancellation is the same as that employing classical beamforming summarized at the beginning of this section.

One of the claims made in the previous development is that this procedure yields adaptively formed left, center, and right beams exhibiting the same sidelobe phase characteristics exhibited by left, center, and right beams created by classical beamforming. This claim is substantiated in the following examples.

Example 5-2 Consider again the $N = 21$ element array used in Example 5-1. Figure 5-8(a) displays the three beams generated by the 21×3 classical matrix beamformer, $\mathbf{W}_{B3}$, with a center pointing angle of $\theta_c = 0°$ ($u_C = 0$). In this case, the pointing angles of the left and right beams are $\theta_L = -5.46°$ ($u_L = -2/21$) and $\theta_R = 5.46°$ ($u_R = 2/21$), respectively. In the first example, the interference scenario consists of $K = 2$ interferers angularly located at $\theta_{I1} = 13.5°$ and $\theta_{I2} = 25°$ as indicated in Fig. 5-8(a). Note that the first interferer is located at the peak of the largest sidelobe of the right beam. This interferer is thus located at the peak of the first sidelobe of a classical sum beam formed according to either the Hanning or Hamming prescription as a linear combination of the three beams plotted in Fig. 5-5(a). The second interferer is located at the peak of the third sidelobe of a classical sum beam constructed in this manner.

Figure 5-8(b) displays the three beams generated by the ICMBF $\mathbf{W}_I$ constructed according to the method outlined previously with the interference parameters cited above. Since the null directions were explicitly specified, the adaptive-adaptive processing option in step 1a was selected. Note that each beam exhibits a null in each of the $K = 2$ interference directions as desired. Also, the three beams have $N - 3 = 18$ nulls in common and exhibit the same sidelobe phase relationships exhibited by the three classical beams pictured in Fig. 5.5(a). Specifically, co-located sidelobes associated with either the left and center beams or the center and right beams are 180° out-of-phase, while co-located sidelobes associated with the left and right beams are in-phase. Adaptive sum and difference beams formed as linear combinations of these three beams are plotted in Fig. 5-9. The sum beam was formed as the sum of the left and right beams with twice the center beam in accordance with the Hanning prescription. The difference beam was formed as the difference between the left and right beams. Note that despite the fact that the two nulls were induced at peak sidelobes of the respective nonadaptive beams, the majority of the sidelobes of the sum beam are at approximately −55 dB with respect to the mainlobe peak; the peak sidelobe is at −35 dB. Also, the sidelobe roll-off exhibited by the difference beam is similar to that obtained in the nonadaptive case as exhibited in Fig. 5-6.

The next example illustrates the efficacy of the adaptive sum/difference beamforming scheme in combating a continuous interference such as a

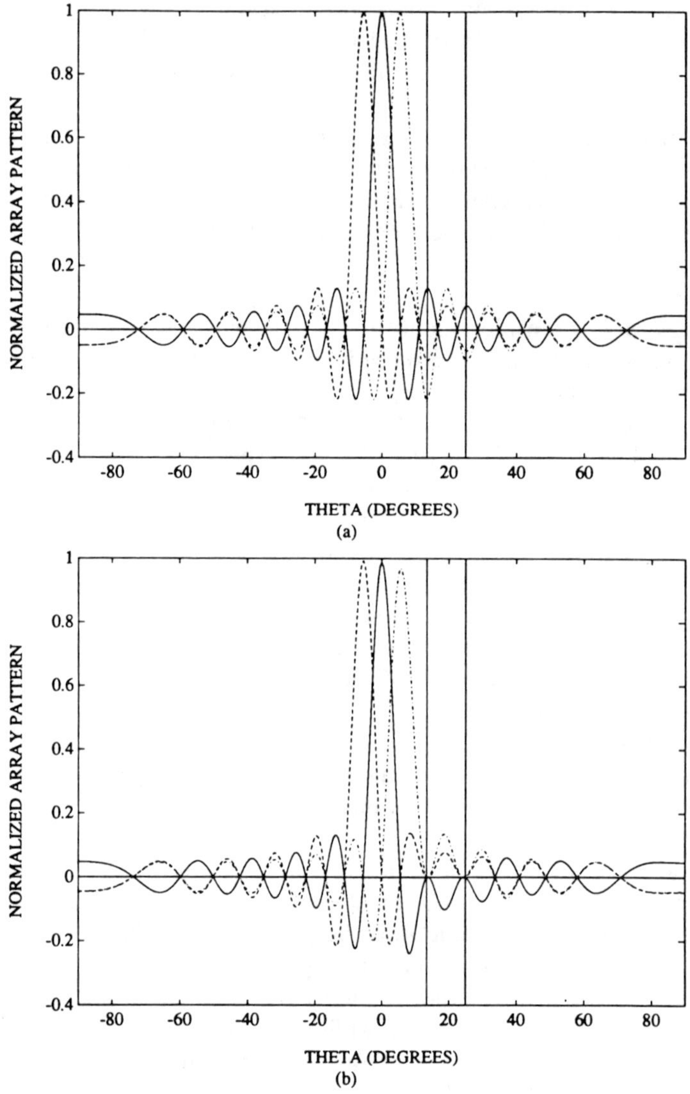

Figure 5-8 Adaptively formed left, center, and right beam patterns with interferers located at $\theta_{I1} = 13.5°$ and $\theta_{I2} = 25°$ corresponding to peak sidelobes.

diffuse multipath arrival. The assumed interference scenario consisted of an extended source occupying an interval encompassing the largest sidelobe of the nonadaptive sum beam obtained with Hamming weighting. To generate a broad null, five point nulls equispaced in the u-domain were induced over this interval. The location of the nulls are indicated in Fig. 5-10 along with the adaptive left, center, and right beams constructed according to the method outlined previously. Again, the three beams have $N - 3 = 18$ nulls in

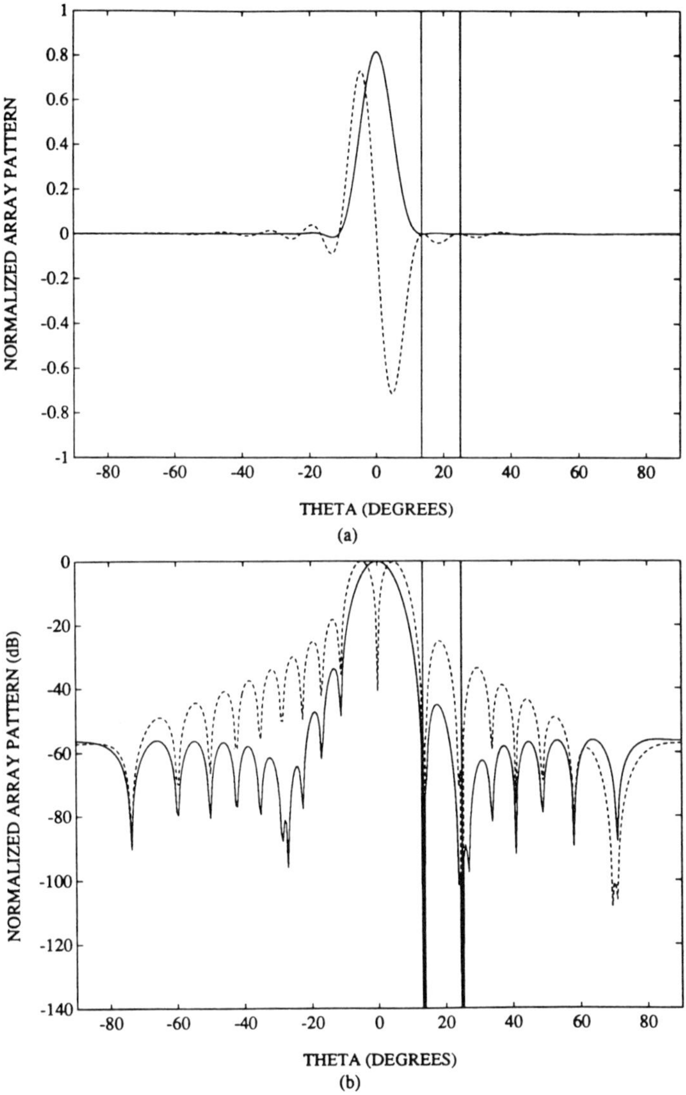

Figure 5-9 Sum and difference patterns formed from linear combination of adaptive left, center, and right beams pictured in Fig. 5-8. Sum beam, Hanning based weighting; difference beam, full-cycle sine based weighting; pointing angle, $u_C = 0$ ($\theta_C = 0°$).

common and exhibit the sidelobe phase relationships exhibited by the three classical beams pictured in Fig. 5-8(a). A sum beam formed via Hamming based weighting of these three beams is plotted in Fig. 5-11 along with a difference beam formed as the difference between the left and right beams. Despite the effective blanking of the largest sidelobe of the classical sum beam formed via Hamming tapering, the sidelobe level of the adaptive sum

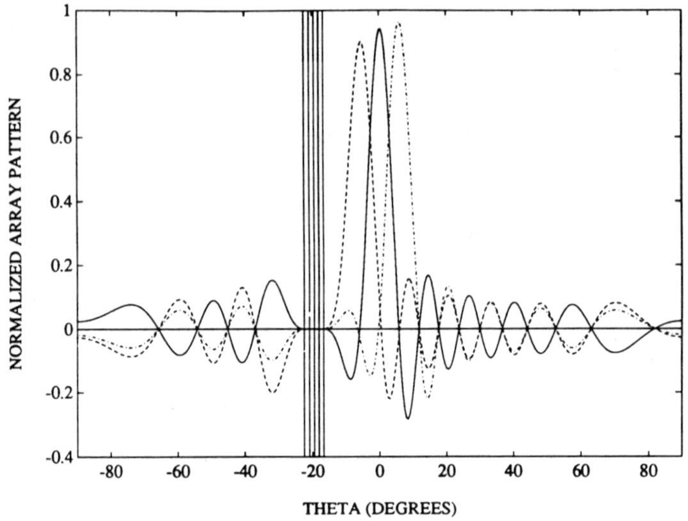

Figure 5-10 Adaptively formed left, center, and right beam patterns with nulls induced at $\theta_{I1} = -22.4°$, $\theta_{I2} = -20.9°$, $\theta_{I3} = -19.5°$, $\theta_{I4} = -18°$, and $\theta_{I5} = -16.6°$ to yield a broad null.

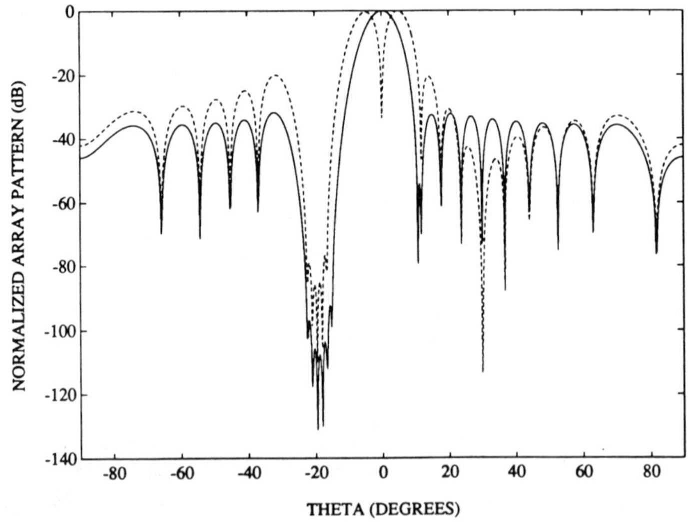

Figure 5-11 Sum and difference patterns formed from linear combination of adaptive left, center, and right beams pictured in Fig. 5-10. Sum beam, Hamming based weighting; difference beam, full-cycle sine based weighting; pointing angle, $u_C = 0$ ($\theta_C = 0°$).

beam is observed to be relatively flat, hovering around −37 dB with respect to the mainlobe peak. The sidelobe level of the adaptive difference beam is also relatively equalized with respect to the nonadaptive case.

The above examples demonstrate the efficacy of the objective function in (5-92) in preserving the low sidelobe level achieved with classical sum and difference beams in the process of incorporating interference null constraints. It should be kept in mind, however, that the most important property is the $N-3$ common nulls as this facilitates bearing estimation via a simple trigonometric formula without approximation.

5-7 THREE-DIMENSIONAL BEAMSPACE ML BEARING ESTIMATION WITH MAINBEAM INTERFERENCE

5-7-1 Low-Angle Radar Tracking Problem

As we have seen, the bearing of a single target within the mainlobe of the transmitted beam may be estimated via a ML procedure based on a 2-D beamspace defined by sum and difference beams. In the presence of strong interference, the sum and difference beams are formed adaptively such that ideally each exhibits a null in the direction of each interferer. Judicious construction of adaptive sum and difference beams according to the method outlined in Section 5-6 enables one to nullify the effect of interferers and nevertheless estimate the target bearing via the simple trigonometric expression in (5-62). However, if an interfering source is angularly located within the mainlobe of the sum beam, the performance of the adaptive sum and difference beam method degrades severely. In attempting to place a null in the angular location of the interferer, the pointing angle of the sum beam, and hence, the direction of maximum gain, squints away from the angular position of the target under surveillance. As the angular separation between the target and the interferer becomes a smaller and smaller fraction of a beamwidth, the drop in target SNR at the adaptive sum beam output becomes more and more severe. The target SNR at the adaptive difference beam output is affected similarly. Ultimately this phenomenon causes track breaking [6, 14, 17, 18, 53].

The low-angle radar tracking problem [3–6, 13, 15, 21, 26–28, 30–31, 53–54, 57–59] is a classic example of a mainbeam interference problem. In low-angle radar tracking, echoes return from a target angularly located near broadside to a vertical array via a specular path as well as by a direct path. The specular path signal arises due to the relative proximity of both the target and the array to a smooth reflecting surface, a body of water, for example. The scenario is depicted in Fig. 5-12. Due to the low elevation angle of the target, the direct and specular path signals arrive within a fraction of a beamwidth near broadside. Beamwidth is here defined as the angular distance between 3 dB points of the mainlobe of a beam formed

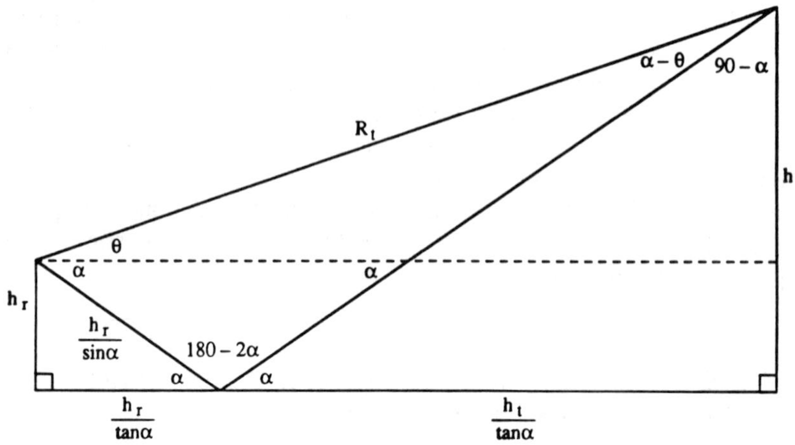

Figure 5-12 Multipath geometry for low-angle radar tracking scenario for the case of a flat earth model.

with a rectangular window, that is, no taper. In the $u = \sin\theta$ domain, this distance is $2/N$ and is referred to as the Rayleigh resolution limit. The specular path signal thus represents an interfering signal angularly located within a beamwidth of the signal of interest, the direct path signal. It should be noted that the differential between the direct and specular path lengths is small enough such that the two respective signals arrive overlapping in time. As a consequence, the two cannot be distinguished according to range bin.

Rather than attempting to place a null in the direction of the specular path signal, it is apparent that it must be treated as arising from a second source located at the image of the target. The bearing estimation problem in low-angle radar tracking then represents a classical example of a real world problem in which it is necessary to resolve two sources angularly separated by less than the Rayleigh resolution limit. As single target tracking is accomplished very simply and effectively in a 2-D beamspace, it is natural to explore the efficacy of operating in a 3-D beamspace to contend with the low-angle radar tracking problem. We develop here a ML estimator of the target bearing based in a 3-D beamspace as a natural extension of the 2D-BDML procedure developed in Section 5-3.

5-7-2 Bisector Angle Determination from Multipath Geometry

Consider the geometry of the low-angle radar tracking scenario in the case of a flat earth model as depicted in Fig. 5-12. The target of interest is flying at a relatively low altitude over the sea surface. R_t denotes the range of the target, h_r denotes the height of the center of the receiving array above sea level, and h_t denotes the height of the target above sea level, that is, the target altitude. As discussed previously, echoes return to the radar site via a specular path as well as by a direct path. θ is the angle of the direct path ray, that is, the target bearing, while α is the angle of the specular path ray measured downward from broadside to the array. The goal is to estimate θ.

The geometry of the low-angle radar tracking scenario represents a priori information that should be incorporated into any scheme for estimating the direct path angle. We show here that the bisector angle between the direct and specular path rays at the array, $\theta_B = (\theta - \alpha)/2$, may be determined from knowledge of h_r and R_t. Note that determination of the target range, R_t, is inherently a time delay estimation problem and may be accomplished independently of bearing determination.

The other angles indicated in Fig. 5-12 may be determined from α and θ in accordance with Snell's law of reflection and the fact that the sum of the angles in a triangle is 180°. Given this geometric information, the law of sines dictates

$$\frac{h_r/\sin\alpha}{\sin(\alpha-\theta)} = \frac{R_t}{\sin(180-2\alpha)} \tag{5-123}$$

Substitution of the trigonometric identity $\sin(180-2\alpha) = \sin(2\alpha) = 2\sin\alpha\cos\alpha$ coupled with simple algebraic manipulation yields

$$\sin(\alpha-\theta) = 2\,\frac{h_r}{R_t}\cos\alpha \tag{5-124}$$

In the low-angle radar scenario, α is quite small such that $\cos\alpha \approx 1$. Incorporating this approximation yields

$$\theta_B = \frac{\theta-\alpha}{2} \approx -\frac{1}{2}\sin^{-1}\left\{2\,\frac{h_r}{R_t}\right\} \tag{5-125}$$

Finally, invoking the fact that $h_r/R_t \ll 1$, the result in (5-125) may be expressed in the $u = \sin\theta$ domain as

$$u_B \approx -\frac{h_r}{R_t} \tag{5-126}$$

where $u_B = \frac{1}{2}\{\sin\theta - \sin\alpha\}$.

Now, R_t is only known to within a certain tolerance based on the dimensions of range bin in which the target is located. Let ΔR_t be the error in the range estimate. For the practical case where $\Delta R_t \ll R_t$

$$\frac{h_r}{R_t+\Delta R_t} = \frac{h_r}{R_t}\left\{\frac{1}{1+\Delta R_t/R_t}\right\} \approx \frac{h_r}{R_t}\left\{1-\frac{\Delta R_t}{R_t}\right\} = \frac{h_r}{R_t} - \frac{\Delta R_t}{R_t^2} \tag{5-127}$$

Thus, the additive correction to the bisector angle estimate due to inaccurate knowledge of the target range is inversely proportional to the square of the target range!! Notwithstanding, the effect of an error in the bisector angle estimate on the performance of the 3D-BDML bearing estimator developed in Section 5-7-3 will be examined in Section 5-7-4.

5-7-3 Development of Three-Dimensional Beamspace ML Bearing Estimator

Here we develop a 3-D Beamspace Domain Maximum Likelihood (3D-BDML) bearing estimator for low-angle radar tracking as a natural extension of the 2D-BDML bearing estimation procedure developed in Section 5-3. Two versions of 3D-BDML are presented. The version of the algorithm which incorporates no a priori information with regard to the multipath geometry is referred to as Nonsymmetric 3D-BDML. In contrast, the version which incorporates a priori information in the form of the bisector angle is referred to as Symmetric 3D-BDML. Nonsymmetric 3D-BDML is developed first.

Array Data Model. Let θ_1 and θ_2 denote the bearings of the direct and specular path signals, respectively. In terms of the notation in Fig. 5-12, $\theta_1 = \theta$ and $\theta_2 = -\alpha$. The $N \times 1$ snapshot vector, $\mathbf{x}(n)$, may be expressed as

$$\mathbf{x}(n) = c_1(n)\mathbf{d}(u_1) + c_2(n)\mathbf{d}(u_2) + \mathbf{n}(n) \tag{5-128}$$

where $c_1(n)$ is the sample value of the complex envelope of the direct path echo at the nth snapshot and $u_1 = \sin\theta_1$. Recall that the array response vector $\mathbf{d}(u)$ in (5-4) is defined such that the phase angle of $c_1(n)$ is that measured at the center of the array. $c_2(n)$ and u_2 are defined similarly with respect to the specular path signal. As in (5-6) corresponding to the single target case, the elements of $\mathbf{n}(n)$ constitute the complex noise present at each antenna output at the nth snapshot.

As discussed previously, the presence of two signals in the vicinity of the pointing angle of the transmit beam necessitates the formation of three beams on return, that is, operation in a 3-D beamspace. Since the two signals are very closely spaced in angle, resolution is an issue. Although tapering at the element level reduces sidelobes, it also yields a wider mainlobe relative to that achieved with classical, phase-only steered beamforming. Thus, for the low-angle radar tracking problem, we consider the ML estimate of the target bearing given the outputs of three classical, phase-only steered beams with respective pointing angles equispaced by $2/N$. In mathematical terms, a 3×1 beamspace snapshot vector, denoted $\mathbf{x}_{B3}(n)$, is formed by operating on the snapshot vector $\mathbf{x}(n)$ with the $N \times 3$ classical matrix beamformer $\mathbf{W}_{B3}$ defined in (5-45):

$$\mathbf{x}_{B3}(n) = \mathbf{W}_{B3}^H\mathbf{x}(n) \tag{5-129}$$

Defining the 3×1 beamspace manifold vector $\mathbf{b}_{B3}(u)$ as

$$\mathbf{b}_{B3}(u) = \mathbf{W}_{B3}^H \mathbf{d}(u) = \begin{bmatrix} \mathbf{d}^H(u_L)\mathbf{d}(u) \\ \mathbf{d}^H(u_C)\mathbf{d}(u) \\ \mathbf{d}^H(u_R)\mathbf{d}(u) \end{bmatrix} = \begin{bmatrix} S_N(u - u_L) \\ S_N(u - u_C) \\ S_N(u - u_R) \end{bmatrix} \tag{5-130}$$

where $u_L = u_C - 2/N$ and $u_R = u_C + 2/N$, it follows from (5-128) that $\mathbf{x}_{B3}(n)$ may be expressed as

$$\mathbf{x}_{B3}(n) = [\mathbf{b}_{B3}(u_1) \,\vdots\, \mathbf{b}_{B3}(u_2)] \begin{bmatrix} c_1(n) \\ c_2(n) \end{bmatrix} + \mathbf{W}_{B3}^H \mathbf{n}(n)$$

$$= \mathbf{B}\mathbf{c}(n) + \mathbf{n}_{B3}(n) \tag{5-131}$$

where, as implied, $\mathbf{B} = [\mathbf{b}_{B3}(u_1) \,\vdots\, \mathbf{b}_{B3}(u_2)]$, $\mathbf{c}(n) = [c_1(n) \quad c_2(n)]^T$, and $\mathbf{n}_{B3}(n) = \mathbf{W}_{B3}^H \mathbf{n}(n)$.

As a consequence of the symmetry of the left and right pointing angles about the center pointing angle and the fact that $\sin(N\pi u/2)/\sin(\pi u/2)$ is an even function of u, it is easily verified that $\mathbf{b}_{B3}(u)$ exhibits the following properties.

$$\tilde{\mathbf{I}}_3 \mathbf{b}_{B3}(u) = \mathbf{b}_{B3}(2u_C - u) \tag{5-132a}$$

$$\tilde{\mathbf{I}}_3 \mathbf{b}_{B3}(u_C + \Delta) = \mathbf{b}_{B3}(u_C - \Delta) \tag{5-132b}$$

These properties will be exploited in the development and analysis of the Symmetric 3D-BDML estimator.

Development of Nonsymmetric 3D-BDML. In the development of Nonsymmetric 3D-BDML, we do not assume the pointing angle of the center beam to be set equal to the bisector angle estimated according to (5-126). As a consequence, we have to contend with a 2-D parameter estimation problem. That is, although we are typically only interested in the target bearing equal to the direct path angle, we must simultaneously estimate the specular path angle as well. This is in contrast to the situation when the interferences are not within the mainlobe of the transmission beam and are nulled out via adaptive sum/difference beamforming. Again, we do not attempt to place a null in the direction of the specular path angle due to the corresponding loss in SNR at the beamformer outputs associated with the direct path. The development of Nonsymmetric 3D-BDML parallels the development of 2D-BDML in Section 5-3.

Based on the assumptions discussed previously regarding the receiver generated noise at the element level, the development of 3D-BDML is based on modeling $\mathbf{n}_{B3}(n)$ as a multivariate Gaussian random vector distributed according to $N(\mathbf{0}, N\sigma_n^2 \mathbf{I}_3)$. As a consequence, the 3D-BDML estimates of u_1 and u_2 may be formulated as the solution to the following nonlinear least squares problem,

$$\underset{u_1, u_2, \mathbf{c}(n)}{\text{Minimize}} \; \|\mathbf{x}_{B3}(n) - \mathbf{B}(u_1, u_2)\mathbf{c}(n)\|^2 \tag{5-133}$$

where $\mathbf{B}(u_1, u_2) = [\mathbf{b}_{B3}(u_1) \mid \mathbf{b}_{B3}(u_2)]$. Similar to the simplifying step which lead from (5-18) to (5-19) in the development of 2D-BDML, we here invoke separability and substitute in (5-133) the respective least square error solution $\mathbf{c}_{LS}(n) = [\mathbf{B}^T\mathbf{B}]^{-1}\mathbf{B}^T\mathbf{x}_{B3}(n)$ to obtain an objective function to be minimized over u_1 and u_2 only:

$$\underset{u_1, u_2}{\text{Minimize}}\ \mathbf{x}_{B3}^H(n)\mathbf{P}_{\mathrm{B}}^{\perp}(u_1, u_2)\mathbf{x}_{B3}(n) \tag{5-134}$$

where $\mathbf{P}_B^{\perp}(u_1, u_2) = \mathbf{I}_3 - \mathbf{B}[\mathbf{B}^T\mathbf{B}]^{-1}\mathbf{B}^T$.

$\boldsymbol{P}_B^{\perp}(u_1, u_2)$ is the projection operator onto the 1-D space orthogonal to the span of $\mathbf{b}_{B3}(u_1)$ and $\mathbf{b}_{B3}(u_2)$. Hence, $\mathbf{P}_{\mathrm{B}}^{\perp}(u_1, u_2) = \mathbf{v}\mathbf{v}^T/\mathbf{v}^T\mathbf{v}$ where $\mathbf{v}$ is a real-valued 3×1 vector related to u_1 and u_2 according to

$$\mathbf{v}^T\mathbf{b}_{B3}(u_i) = 0\,, \qquad i = 1, 2 \tag{5-135}$$

Similar to the development of 2D-BDML, substitution of $\mathbf{P}_B^{\perp}(u_1, u_2) = \mathbf{v}\mathbf{v}^T/\mathbf{v}^T\mathbf{v}$ in (5-134) converts the search over u_1 and u_2 into a search over $\mathbf{v}$.

$$\underset{\mathbf{v}}{\text{Minimize}}\ \frac{\mathbf{x}_{B3}^H(n)\mathbf{v}\mathbf{v}^T\mathbf{x}_{B3}(n)}{\mathbf{v}^T\mathbf{v}} = \frac{\mathbf{v}^T\,\mathrm{Re}\{\mathbf{x}_{B3}(n)\mathbf{x}_{B3}^H(n)\}\mathbf{v}}{\mathbf{v}^T\mathbf{v}} \tag{5-136}$$

The solution for $\mathbf{v}$ in (5-136) is the smallest eigenvector of the 3×3 matrix $\mathrm{Re}\{\mathbf{x}_{B3}(n)\mathbf{x}_{B3}^H(n)\}$.

Given this $\mathbf{v}$, the Nonsymmetric 3D-BDML estimates of u_1 and u_2 are determined in accordance with the relationship in (5-135). Substituting $\mathbf{v} = [v_1 \quad v_2 \quad v_3]^T$ and $\mathbf{b}_{B3}(u_i) = \mathbf{W}_{B3}^H\mathbf{d}(u_i)$ into (5-135) and expanding yields

$$\begin{aligned}\mathbf{v}^T\mathbf{b}_{B3}(u_i) = \mathbf{v}^T\mathbf{W}_{B3}^H\mathbf{d}(u_i) &= v_1\mathbf{d}^H(u_L)\mathbf{d}(u_i) + v_2\mathbf{d}^H(u_C)\mathbf{d}(u_i) + v_3\mathbf{d}^H(u_R)\mathbf{d}(u_i)\\ &= 0\,, \qquad i = 1, 2\end{aligned} \tag{5-137}$$

Comparing (5-137) with (5-53), it follows that $\hat{z}_i = \exp[j\pi\hat{u}_i]$, $i = 1, 2$, may be determined as the two roots of the second-order polynomial $e(z)$ with coefficient vector $\mathbf{e} = \mathbf{Q}_{B3}^*\mathbf{v}$, that is, as the two roots of $e(z) = e_0 + e_1 z + e_0^* z^2$ where

$$e_0 = \exp[j\pi u_C]\{-v_1\exp[j\pi/N] + v_2 - v_3\exp[-j\pi/N]\} \tag{5-138a}$$

$$e_1 = 2(v_1 + v_3)\cos(\pi/N) - 2v_2\cos(2\pi/N) \tag{5-138b}$$

Recall that in the case of single target tracking with sum and difference beams defined by (5-49) the two roots of $e(z)$ formed with the coefficient vector $\mathbf{e} = \mathbf{Q}_{B3}^*\mathbf{H}\mathbf{v}_H$, where $\mathbf{v}_H$ is the smallest eigenvector of the 2×2 matrix $\mathrm{Re}\{\mathbf{x}_{H2}(n)\mathbf{x}_{H2}^H(n)\}$, are on opposite sides of the unit circle and that we are only interested in one of these roots. In contrast, here we are interested in both roots of $e(z)$ formed with the coefficient vector $\mathbf{e} = \mathbf{Q}_{B3}^{*}\mathbf{v}$, where $\mathbf{v}$ is the smallest eigenvector of the 3×3 matrix $\mathrm{Re}\{\mathbf{x}_{B3}(n)\mathbf{x}_{B3}^{H}(n)\}$. Due to the small angular separation between the direct and specular path signal arrivals, the two roots are in close proximity to each other and ideally lie on the unit circle. As before, the two roots lie on the unit circle if and only if $|e_1/e_0| \leq 2$. If the two roots do not lie on the unit circle, they are related as $z_2 = 1/z_1^*$ due to the conjugate centrosymmetry of the coefficients of $e(z)$. In this case, the phase angle of each root is the same indicating that the direct and specular path signals are not resolved.

A summary of Nonsymmetric 3D-BDML is then as follows. Compute $\mathbf{v} = [v_1 \quad v_2 \quad v_3]^T$ as that eigenvector of $\mathrm{Re}\{\mathbf{x}_{B3}(n)\mathbf{x}_{B3}^{H}(n)\}$ associated with the smallest eigenvalue. Form the coefficients of the quadratic polynomial $e(z)$ according to the expressions in (5-138). If $|e_1/e_0| \geq 2$, the direct and specular path signals are not resolved. Otherwise, with $e_0 = r_0 \exp[j\phi_0]$, $\hat{u}_1 = \{\phi_0 + \cos^{-1}[e_1/(2r_0)]\}/\pi$, and $\hat{u}_2 = \{\phi_0 - \cos^{-1}[e_1/(2r_0)]\}/\pi$.

Symmetric 3D-BDML Bearing Estimator. The development of Nonsymmetric 3D-BDML is easily modified to incorporate an a priori estimate of the bisector angle $u_B = (u_1 + u_2)/2$. With $u_C = u_B = (u_1 + u_2)/2$ in (5-130), invocation of the property of $\mathbf{b}_{B3}(u)$ described by (5-132) yields the following relationship between $\mathbf{b}_{B3}(u_1)$ and $\mathbf{b}_{B3}(u_2)$

$$\boldsymbol{b}_{B2}(u_2) = \tilde{\mathbf{I}}_3\mathbf{b}_{B3}(u_1) \tag{5-139}$$

The translation of (5-135) for the case of $u_C = u_B = (u_1 + u_2)/2$ yields the following pair of relationships between $\mathbf{v}$ and u_1:

$$\mathbf{v}^T\mathbf{b}_{B3}(u_1) = 0 \quad \text{and} \quad \mathbf{v}^T\tilde{\mathbf{I}}_3\mathbf{b}_{B3}(u_1) = 0 \tag{5-140}$$

These relationships imply that $\mathbf{v}$ must satisfy $\tilde{\mathbf{I}}_3\mathbf{v} = \mathbf{v}$, that is, $\mathbf{v}$ must exhibit centrosymmetry. Hence, the condition $u_C = u_B$ dictates the incorporation of the constraint $\tilde{\mathbf{I}}_3\mathbf{v} = \mathbf{v}$ into the optimization problem described by (5-136).

$$\underset{\mathbf{v}}{\text{Minimize}} \; \frac{\mathbf{v}^T \mathrm{Re}\{\mathbf{x}_{B3}(n)\mathbf{x}_{B3}^{H}(n)\}\mathbf{v}}{\mathbf{v}^T\mathbf{v}} \qquad \text{subject to} \quad \tilde{\mathbf{I}}_3\mathbf{v} = \mathbf{v} \tag{5-141}$$

Let $\hat{\mathbf{R}}_{bb}$ denote the 3×3 single snapshot correlation matrix

$$\hat{\mathbf{R}}_{bb} = \mathbf{x}_{B3}(n)\mathbf{x}_{B3}^{H}(n) \tag{5-142}$$

Since $\mathbf{v}$ is constrained to be centrosymmetric, the objective function in (5-141) may be alternatively expressed as

$$\underset{\mathbf{v}}{\text{Minimize}} \; \frac{\frac{1}{2}\mathbf{v}^T[\text{Re}\{\hat{\mathbf{R}}_{bb}\} + \tilde{\mathbf{I}}_3 \,\text{Re}\{\hat{\mathbf{R}}_{bb}\}\tilde{\mathbf{I}}_3]\mathbf{v}}{\frac{1}{2}\mathbf{v}^T[\mathbf{I} + \tilde{\mathbf{I}}_3\tilde{\mathbf{I}}_3]\mathbf{v}} = \frac{\mathbf{v}^T \,\text{Re}\{\hat{\mathbf{R}}_{bb}^{fb}\}\mathbf{v}}{\mathbf{v}^T\mathbf{v}} \qquad \text{(5-143)}$$
$$\text{subject to} \quad \tilde{\mathbf{I}}_3\mathbf{v} = \mathbf{v}$$

where $\hat{\mathbf{R}}_{bb}^{fb}$ is defined in terms of $\hat{\mathbf{R}}_{bb}$ in (5-142) as

$$\hat{\mathbf{R}}_{bb}^{fb} = \tfrac{1}{2}\{\hat{\mathbf{R}}_{bb} + \tilde{\mathbf{I}}_3\hat{\mathbf{R}}_{bb}\tilde{\mathbf{I}}_3\} \qquad \text{(5-144)}$$

$\hat{\mathbf{R}}_{bb}^{fb}$ will be referred to as the forward-backward averaged beamspace sample correlation matrix formed from a single snapshot.

It is easily shown that two of the eigenvectors of $\text{Re}\{\hat{\mathbf{R}}_{bb}^{fb}\}$ exhibit centrosymmetry while the third exhibits centroantisymmetry. This assertion is easily established as follows. First, note that $\tilde{\mathbf{I}}_3 \,\text{Re}\{\hat{\mathbf{R}}_{bb}^{fb}\}\tilde{\mathbf{I}}_3 = \text{Re}\{\hat{\mathbf{R}}_{bb}^{fb}\}$. With this in mind, let λ_i and $\mathbf{v}_i$ be an eigenvalue-eigenvector pair of $\hat{\mathbf{R}}_{bb}^{fb}$, that is, $\text{Re}\{\hat{\mathbf{R}}_{bb}^{fb}\}\mathbf{v}_i = \lambda_i\mathbf{v}_i$. This implies $\tilde{\mathbf{I}}_3 \,\text{Re}\{\hat{\mathbf{R}}_{bb}^{fb}\}\tilde{\mathbf{I}}_3\mathbf{v}_i = \lambda_i\mathbf{v}_i$. Multiplying both sides of this equality by $\tilde{\mathbf{I}}_3$ yields $\text{Re}\{\hat{\mathbf{R}}_{bb}^{fb}\}\tilde{\mathbf{I}}_3\mathbf{v}_i = \lambda_i\tilde{\mathbf{I}}_3\mathbf{v}_i$. Under the practical assumption that the eigenvalues of $\text{Re}\{\hat{\mathbf{R}}_{bb}^{fb}\}$ are unique, it follows that the corresponding eigenvectors are unique to within a scalar multiple. Hence, $\tilde{\mathbf{I}}_3\mathbf{v}_i = \gamma_i\mathbf{v}_i$ which is an eigenvalue-eigenvector relationship. It is easily determined that $\tilde{\mathbf{I}}_3$ only has two distinct eigenvalues equal to $+1$ or -1. A simple dimensionality argument then yields the end result that two of the eigenvectors of $\text{Re}\{\hat{\mathbf{R}}_{bb}^{fb}\}$ satisfy $\tilde{\mathbf{I}}_3\mathbf{v}_i = \mathbf{v}_i$ while the third satisfies $\tilde{\mathbf{I}}_3\mathbf{v}_i = -\mathbf{v}_i$.

Thus, the minimizing $\mathbf{v}$ in (5-143) is that centrosymmetric eigenvector of $\text{Re}\{\hat{\mathbf{R}}_{bb}^{fb}\}$ associated with the smaller eigenvalue. Given $\mathbf{v} = [v_1 \;\; v_2 \;\; v_1]^T$ computed as such, we may proceed to find the Symmetric 3D-BDML estimate of u_1 using the same approach as in Nonsymmetric 3D-BDML. $z_1 = \exp[j\pi u_1]$ and $z_2 = \exp[j\pi u_2]$ are found as the two roots of $e(z) = e_0 + e_1 z + e_0^* z^2$ having coefficients prescribed by (5-138) with $v_3 = v_1$:

$$e_0 = \exp[j\pi u_C]\{v_2 - 2v_1\cos(\pi/N)\} \qquad \text{(5-145a)}$$

$$e_1 = 4v_1\cos(\pi/N) - 2v_2\cos(2\pi/N) \qquad \text{(5-145b)}$$

If $|e_1/e_0| \geq 2$ the two roots of $e(z)$ have the same phase angle equal to $\pi u_C = \pi u_B$ with the respective magnitudes forming a reciprocal pair. As in the nonsymmetric case, this is an indication that the direct and specular path signals have not been resolved. If $|e_1/e_0| < 2$, the two roots lie on the unit circle symmetrically positioned about $z = \exp[j\pi u_B]$. Note that the expression for e_0 in (5-145a) is in the form $e_0 = r_0 \exp[j\phi_0]$ where $r_0 = v_2 - 2v_1 \cos(\pi/N)$ and $\phi_0 = \pi u_C$. Hence,

$$\hat{u}_1 = u_C + \frac{1}{\pi}\cos^{-1}\left\{\frac{4v_1\cos(\pi/N) - 2v_2\cos(2\pi/N)}{v_2 - 2v_1\cos(\pi/N)}\right\} \qquad \text{(5-146)}$$

A summary of Symmetric 3D-BDML is then as follows. Compute $\mathbf{v} = [v_1 \quad v_2 \quad v_1]^T$ as that centrosymmetric eigenvector of $\mathrm{Re}\{\hat{\mathbf{R}}_{bb}^{fb}\}$, where $\hat{\mathbf{R}}_{bb}^{fb}$ is defined in (5-144), associated with the smaller eigenvalue. Form the coefficients of the quadratic polynomial $e(z)$ according to the expressions in (5-145). If $|e_1/e_0| \geq 2$, the direct and specular path signals are not resolved. Otherwise, the bearing of the direct path angle is estimated according to (5-146).

5-7-4 Effects of Forward-Backward Averaging

Let the complex amplitude associated with the direct path signal be expressed as $c_1(n) = A_1(n)\exp[j\phi_1(n)]$. In accordance with the definition of $\mathbf{d}(u)$, $\phi_1(n)$ is the phase of the direct path signal measured at the center of the array at the nth snapshot. $\phi_2(n)$ is defined similarly with respect to the direct path signal, that is, $c_2(n) = A_2(n)\exp[j\phi_2(n)]$. Further, let ρ_c denote the complex reflection coefficient defined as

$$\rho_c = \frac{c_2(n)}{c_1(n)} = \frac{A_2(n)}{A_1(n)} \exp[j(\phi_2(n) - \phi_1(n))] = \rho \exp[j\Delta\Psi] \qquad (5\text{-}147)$$

where $\Delta\Psi = \phi_2(n) - \phi_1(n)$ is the relative phase difference between the direct and specular path signals at the center of the array; $\rho = A_2(n)/A_1(n)$. We show here that Nonsymmetric 3D-BDML breaks down when $\Delta\Psi$ is either 0° or 180° while Symmetric 3D-BDML exhibits no such breakdown phenomenon.

To this end, consider the execution of Nonsymmetric 3D-BDML with noiseless data. Invoking the expression for $\mathbf{x}_{B3}(n)$ in (5-131) with $\mathbf{n}_{B3}(n) = \mathbf{0}_3$, $\hat{\mathbf{R}}_{bb}$ in (5-142) may be expressed as

$$\mathbf{R}_{bb} = \mathbf{x}_{B3}(n)\mathbf{x}_{B3}^H(n) = \mathbf{B}\mathbf{R}_{ss}\mathbf{B}^T \qquad (5\text{-}148)$$

where $\mathbf{B} = [\mathbf{b}_{B3}(u_1) \,\vdots\, \mathbf{b}_{B3}(u_2)]$ and

$$\mathbf{R}_{ss} = \mathbf{c}(n)\mathbf{c}^H(n) = A_1^2(n)\begin{bmatrix} 1 & \rho\exp[-j\Delta\Psi] \\ \rho\exp[j\Delta\Psi] & \rho^2 \end{bmatrix} \qquad (5\text{-}149)$$

In Nonsymmetric 3D-BDML, $\mathbf{v}$ is computed as the eigenvector of $\mathrm{Re}\{\hat{\mathbf{R}}_{bb}\}$ associated with the smallest eigenvalue. Since $\mathbf{B}$ is real, it follows that

$$\mathrm{Re}\{\mathbf{R}_{bb}\} = \mathbf{B}\,\mathrm{Re}\{\mathbf{R}_{ss}\}\mathbf{B}^T \qquad (5\text{-}150)$$

where

$$\mathrm{Re}\{\mathbf{R}_{ss}\} = A_1^2(n)\begin{bmatrix} 1 & \rho\cos(\Delta\Psi) \\ \rho\cos(\Delta\Psi) & \rho^2 \end{bmatrix} \qquad (5\text{-}151)$$

As long as $u_1 \neq u_2$ and neither angle is aligned with one of the common nulls, the 3×2 matrix $\mathbf{B}$ is of rank 2. As a consequence, rank $(\text{Re}\{\mathbf{R}_{bb}\}) =$ rank$(\text{Re}\{\mathbf{R}_{ss}\})$. As $\det(\text{Re}\{\mathbf{R}_{ss}\}) = \rho^2\{1 - \cos(\Delta\Psi)\}$, $\text{Re}\{\mathbf{R}_{ss}\}$ is of full rank equal to 2 so long as $\Delta\Psi$ is not equal to either 0° or 180°. Thus, as long as $\Delta\Psi \neq 0°$ and $\Delta\Psi \neq 180°$, $\text{rank}(\text{Re}\{\mathbf{R}_{bb}\}) = 2$ such that zero is the smallest eigenvalue of $\text{Re}\{\mathbf{R}_{bb}\}$ of multiplicity one. Further, $\text{range}\{\text{Re}\{\mathbf{R}_{bb}\}\} = \text{range}\{\mathbf{B}\} = \text{span}\{\mathbf{b}_{B3}(u_1), \mathbf{b}_{B3}(u_2)\}$ such that the eigenvector associated with the eigenvalue zero is orthogonal to both $\mathbf{b}_{B3}(u_1)$ and $\mathbf{b}_{B3}(u_2)$. Thus, as long as $\Delta\Psi$ is not equal to either 0° or 180°, execution of Nonsymmetric 3D-BDML with noiseless data provides the true values of u_1 and u_2.

When $\Delta\Psi$ is equal to either 0° or 180°, $\text{Re}\{\mathbf{R}_{ss}\}$ is of rank 1 such that $\text{range}\{\text{Re}\{\mathbf{R}_{bb}\}\} = \text{range}\{\mathbf{B}\mathbf{c}(n)\} = \text{span}\{\mathbf{b}_{B3}(u_1) \pm \rho\mathbf{b}_{B3}(u_2)\}$ where + is for the case $\Delta\Psi = 0°$ and $-$ is for the case $\Delta\Psi = 180°$. Under these conditions, zero is an eigenvalue of $\text{Re}\{\mathbf{R}_{bb}\}$ of multiplicity two and the corresponding 2-D eigenspace is orthogonal to the 1-D space spanned by $\mathbf{b}_{B3}(u_1) \pm \rho\mathbf{b}_{B3}(u_2)$. As a consequence, neither of the two vectors chosen to span the 2-D eigenspace associated with $\lambda_{\min} = 0$ is orthogonal to $\mathbf{b}_{B3}(u_1)$ and $\mathbf{b}_{B3}(u_2)$ individually. Thus, the method breaks down when either $\Delta\Psi = 0°$ or $\Delta\Psi = 180°$. This phenomenon is demonstrated in the simulations presented in Section 5-7-6.

We next consider the execution of Symmetric 3D-BDML with noiseless data. In the first analysis, we consider u_C to be exactly equal to the bisector angle u_B. Symmetric 3D-BDML dictates that $\mathbf{v}$ be computed as that centrosymmetric eigenvector of $\text{Re}\{\hat{\mathbf{R}}_{bb}^{fb}\}$ associated with the smaller eigenvalue, where $\hat{\mathbf{R}}_{bb}^{fb}$ is defined in (5-144). Recall that with $u_C = u_B$, the two columns of $\mathbf{B}$ are related as $\mathbf{b}_{B3}(u_2) = \tilde{\mathbf{I}}_3\mathbf{b}_{B3}(u_1)$ such that $\tilde{\mathbf{I}}_3\mathbf{B}\tilde{\mathbf{I}}_2 = \mathbf{B}$. Exploitation of this property allows us to express the noiseless version of $\text{Re}\{\hat{\mathbf{R}}_{bb}^{fb}\}$ as

$$\begin{aligned}\text{Re}\{\mathbf{R}_{bb}^{fb}\} &= \frac{1}{2}\,\text{Re}\{\mathbf{R}_{bb} + \tilde{\mathbf{I}}_3\mathbf{R}_{bb}\tilde{\mathbf{I}}_3\} \\ &= \frac{1}{2}\,\{\mathbf{B}\,\text{Re}\{\mathbf{R}_{ss}\}\mathbf{B}^T + \tilde{\mathbf{I}}_3\mathbf{B}\tilde{\mathbf{I}}_2\tilde{\mathbf{I}}_2\,\text{Re}\{\mathbf{R}_{ss}\}\tilde{\mathbf{I}}_2\tilde{\mathbf{I}}_2\mathbf{B}^T\tilde{\mathbf{I}}_3\} \qquad (5\text{-}152) \\ &= \mathbf{B}\,\frac{1}{2}\,\text{Re}\{\mathbf{R}_{ss} + \tilde{\mathbf{I}}_2\mathbf{R}_{ss}\tilde{\mathbf{I}}_2\}\mathbf{B}^T\end{aligned}$$

where we have invoked the property $\tilde{\mathbf{I}}_2\tilde{\mathbf{I}}_2 = \mathbf{I}_2$. Thus, $\text{Re}\{\mathbf{R}_{bb}^{fb}\}$ may be expressed in the form $\mathbf{B}\,\text{Re}\{\mathbf{R}_{ss}^{fb}\}\mathbf{B}^T$, where

$$\mathbf{R}_{ss}^{fb} = \frac{1}{2}\,\{\mathbf{R}_{ss} + \tilde{\mathbf{I}}_2\mathbf{R}_{ss}\tilde{\mathbf{I}}_2\} = A_1^2(n)\begin{bmatrix} \dfrac{1+\rho^2}{2} & \rho\cos\Delta\Psi \\ \rho\cos\Delta\Psi & \dfrac{1+\rho^2}{2} \end{bmatrix} \qquad (5\text{-}153)$$

where we have substituted (5-149) for $\mathbf{R}_{ss}$. In contrast to the situation with $\text{Re}\{\mathbf{R}_{ss}\}$ in (5-151), which is of rank one for all values of ρ when $\Delta\Psi$ is either 0° or 180°, $\mathbf{R}_{ss}^{fb}$ is of rank two except when either $\Delta\Psi = 0°$ and, at the same time, $\rho = 1$, or when $\Delta\Psi = 180°$ and, at the same time, $\rho = 1$. That is, in terms of the complex reflection coefficient, $\rho_c = \rho \exp[j\Delta\Psi]$, the only values of ρ_c for which $\text{Re}\{\mathbf{R}_{ss}^{fb}\}$ in (5-153) is of rank one are $\rho_c = 1$ and $\rho_c = -1$. Thus, as long as ρ_c is neither 1 or -1, execution of Symmetric 3D-BDML with the noiseless data provides the true value of u_1.

As a practical matter, the magnitude of ρ_c, ρ, is always less than one due to losses incurred at the surface of reflection [4, 46] and the differential in path lengths. Notwithstanding, Symmetric 3D-BDML does not break down when $\Delta\Psi = 0°$ and $\rho = 1$, that is, when $\rho_c = 1$, despite the rank deficiency problem. In this case, $\text{Re}\{\mathbf{R}_{bb}^{fb}\} = \mathbf{B}\,\text{Re}\{\mathbf{R}_{ss}^{fb}\}\mathbf{B}^T$ is of rank one such that zero is the smallest eigenvalue of multiplicity 2 with the corresponding 2-D eigenspace orthogonal to the 1-D space spanned by $\mathbf{b}_{B3}(u_1) + \mathbf{b}_{B3}(u_2) = \mathbf{b}_{B3}(u_1) + \tilde{\mathbf{I}}_3\mathbf{b}_{B3}(u_1)$. Since $\mathbf{b}_{B3}(u_1) + \tilde{\mathbf{I}}_3\mathbf{b}_{B3}(u_1)$ is centrosymmetric, one of the eigenvectors associated with $\lambda_{\min} = 0$ is centrosymmetric while the other is centroantisymmetric. In accordance with Symmetric 3D-BDML, $\mathbf{v}$ is chosen as the centrosymmetric eigenvector associated with $\lambda_{\min} = 0$. Invoking the centrosymmetry of $\mathbf{v}$ yields

$$\begin{aligned}\mathbf{v}^T\{\mathbf{b}_{B3}(u_1) + \mathbf{b}_{B3}(u_2)\} &= \mathbf{v}^T\{\mathbf{b}_{B3}(u_1) + \tilde{\mathbf{I}}_3\mathbf{b}_{B3}(u_1)\}\\ &= 2\mathbf{v}^T\mathbf{b}_{B3}(u_1) = 0\end{aligned} \tag{5-154}$$

which indicates that $\mathbf{v}$ is orthogonal to $\mathbf{b}_{B3}(u_1)$. This implies that in the no noise case, Symmetric 3D-BDML provides the true value of u_1 even in the extreme case of $\Delta\Psi = 0°$ and $\rho = 1$.

Note that the argument above signified by (5-154) does not work for the case of $\Delta\Psi = 180°$ and $\rho = 1$, that is, $\rho_c = -1$. Symmetric 3D-BDML breaks down under these extreme conditions. In general, poor performance is obtained in the case of $\Delta\Psi = 180°$ when $\rho \approx 1$ due to the severe signal cancellation occurring across a large portion of the array. The use of frequency diversity to overcome this problem is explored in Section 5-7-5.

Effect of Error in Bisector Angle Estimate. The formula for the bisector angle given by (5-126) was based on a flat-earth model and the approximation that $\cos(\alpha) = \cos(\theta_2) \approx 1$. Although the bisector angle estimate provided by (5-126) is quite accurate, an assessment of the sensitivity of Symmetric 3D-BDML to an error in the bisector angle estimate is in order. Although Symmetric 3D-BDML is a nonlinear estimator, a simple analysis leads to the intuitively satisfying conjecture that an error in the bisector angle estimate translates into a bias in the Symmetric 3D-BDML estimator of the same magnitude. The argument is as follows.

Consider $u_C \neq u_B$ such that $u_C = u_B + \delta$. It is assumed that the deviation

δ is a very small fraction of a beamwidth, that is, $\delta \ll 2/N$. Now, since u_B is the true bisector angle, $u_1 = u_B + \Delta$ and $u_2 = u_B - \Delta$. Invoking the property of $\mathbf{b}_{B3}(u)$ in (5-132) yields

$$\begin{aligned}\tilde{\mathbf{I}}_3\mathbf{b}_{B3}(u_1) &= \mathbf{b}_{B3}(2u_C - u_1) \\ &= \mathbf{b}_{B3}(2u_B + 2\delta - u_B - \Delta) \\ &= \mathbf{b}_{B3}(u_B - \Delta + 2\delta) \\ &= \mathbf{b}_{B3}(u_2 + 2\delta)\end{aligned} \tag{5-155}$$

Similarly, $\tilde{\mathbf{I}}_3\mathbf{b}_{B3}(u_2) = \mathbf{b}_{B3}(u_1 + 2\delta)$. Hence, in the case of $u_C \neq u_B$, the noiseless form of $\text{Re}\{\mathbf{R}_{bb}^{fb}\}$ may be expressed as

$$\begin{aligned}\text{Re}\{\mathbf{R}_{bb}^{fb}\} = {} & \tfrac{1}{2}[\mathbf{b}_{B3}(u_1) \,\vdots\, \mathbf{b}_{B3}(u_2)]\,\text{Re}\{\mathbf{R}_{ss}\}[\mathbf{b}_{B3}(u_1) \,\vdots\, \mathbf{b}_{B3}(u_2)]^T \\ & + \tfrac{1}{2}[\mathbf{b}_{B3}(u_2 + 2\delta) \,\vdots\, \mathbf{b}_{B3}(u_1 + 2\delta)]\,\text{Re}\{\mathbf{R}_{ss}\} \\ & \times [\mathbf{b}_{B3}(u_2 + 2\delta) \,|\, \mathbf{b}_{B3}(u_1 + 2\delta)]^T\end{aligned} \tag{5-156}$$

The forward-backward averaging process has effectively created two artificial sources, one at $u = u_1 + 2\delta$ and another at $u = u_2 + 2\delta$. In the practical case where $\rho \approx 1$, the four sources, the two actual sources and the two artificial ones, are of nearly equal strength. In the practical case where δ is a very small fraction of a beamwidth, it is conjectured that Symmetric 3D-BDML cannot resolve the actual source at $u = u_1$ and the artificial source at $u = u_1 + 2\delta$ and, in light of their nearly equal power, yields an estimate equal to the center value of $u = u_1 + \delta$. This, in turn, leads to the conjecture that an error in the bisector angle estimate translates into a bias in the Symmetric 3D-BDML estimator of the same magnitude. This conjecture is validated by simulations presented in Section 5-7-6.

Operation With No Specular Multipath. It is of interest to examine the efficacy of Symmetric 3D-BDML where there is no measurable specular multipath component at the receiving array as would be the case in a rough sea surface. We here show that Symmetric 3D-BDML performs properly under these conditions. The previous analysis provides a simple means for examining how Symmetric 3D-BDML performs when no specular multipath component is present. The absence of a specular multipath component is signified by setting ρ equal to zero. With $\rho = 0$ in (5-152) and (5-153), $\text{Re}\{\mathbf{R}_{bb}^{fb}\}$ may be expressed as $\mathbf{B}\,\text{Re}\{\mathbf{R}_{ss}^{fb}\}\mathbf{B}^T$, where $\mathbf{R}_{ss}^{fb} = A_1^2(n)\mathbf{I}_2$ and

$$\mathbf{B} = [\mathbf{b}_{B3}(u_1) \,\vdots\, \tilde{\mathbf{I}}_3\mathbf{b}_{B3}(u_1)] = [\mathbf{b}_{B3}(u_1) \,\vdots\, \tilde{\mathbf{I}}_3\mathbf{b}_{B3}(2u_C - u_1)] \tag{5-157}$$

where we have invoked the property of $\mathbf{b}_{B3}(u)$ described by (5-132). This implies that despite the absence of specular multipath, the process of forward-backward averaging according to (5-144) effectively creates an artificial source at $u = 2u_C - u_1$ of equal power. Thus, even without a specular multipath component present, Symmetric 3D-BDML must nevertheless resolve two sources angularly separated by $|u_1 - (2u_C - u_1)| = 2|u_C - u_1|$. Hence, the closer u_C is to u_1, the harder Symmetric 3D-BDML must work to resolve the actual source and the artificial source. This phenomenon is illustrated in simulations presented in Section 5-7-6. Nevertheless, since $\mathbf{R}_{ss}^{fb} = A_1^2(n)\mathbf{I}_2$ is of rank 2, it follows that in the noiseless case Symmetric 3D-BDML provides the true value of u_1.

5-7-5 Frequency Diversity

Advances in radar technology have progressed to the point where the use of frequency diversity in tracking systems has become increasingly more commonplace [1]. Depending on the system hardware, the pulses at the various frequencies may be transmitted simultaneously and/or in rapid succession corresponding to frequency hopping. An example of an operational radar system where frequency diversity is employed is the Multi-parameter Adaptive Radar System (MARS) described by Kezys and Haykin [31]. This experimental bistatic radar array consists of a 32-element, horizontally polarized linear array operating coherently over the band 8.05–12.36 GHz. Each antenna element is followed by two receiver channels allowing for simultaneous reception on two separate frequencies: one fixed at 10.2 GHz and the other agile over the band 8.05–12.36 GHz in 30 MHz steps. The MARS radar system is cited as an example of what is feasible in the way of multifrequency transmission with current technology.

There are a number of advantages to employing frequency diversity for tracking purposes. For our purposes here, frequency diversity translates into phase diversity, that is, diversity in the phase difference occurring at the center of the array. This is due to the difference in the number of wavelengths traveled between the direct and specular path signals at the different frequencies. Accordingly, multifrequency operation diminishes the pejorative effect of a 180° phase difference at any one transmission frequency. With judicious processing, the use of multiple frequencies also allows us to achieve a large effective SNR. This is accomplished by coherently combining the additive component of the 3×3 beamspace correlation matrix at each frequency due solely to the direct and specular path signals; the additive components of the beamspace correlation matrix at each frequency due to receiver noise and cross products between signal and noise are incoherently combined. The coherent combination of the signal-only (no noise) component of the beamspace correlation matrix at each frequency is accomplished through the use of focusing transformations in accordance with the coherent signal subspace processing method of Wang and Kaveh [29, 52]. A multifrequency version of 3D-BDML incorporating coherent signal subspace processing (CSS) is developed below.

A discussion of multifrequency operation requires the introduction of some notation. The transmission frequencies are denoted f_j, $j = 1, \ldots, J$, where J is the total number of such frequencies. f_0 denotes the frequency for which the N elements of the array are spaced by a half wavelength; f_0 may or may not be one of the transmission frequencies. At frequency f_j, we allow the option of operating with a subarray of N_j contiguous elements extracted from the overall array of N elements. The formulation is general; N_j may be equal to N or it may be less than N. There are two advantages to operating with an effective subaperture equal to that associated with a subarray of N_j elements at frequency f_j. First, as we shall show, it leads to a criterion for the selection of transmission frequencies which makes the job of coherently combining the signal-only component of $\mathbf{R}_{bb}(f_j)$, the 3×3 beamspace correlation matrix formed at f_j, a very simple procedure. This criterion will be discussed shortly. Second, with $N_j < N$, there are $N - N_j + 1$ identical subarrays of $N_j < M$ contiguous elements over which $\mathbf{R}_{bb}(f_j)$ may be averaged to further diminish the sensitivity of 3D-BDML to the phase difference at any one frequency. This process, referred to as spatial smoothing [40, 45, 55], exploits the variation in the relative phase difference between the direct and specular path signals at each subarray center.

The element space manifold vector/classical beamforming vector associated with frequency f_j and a subarray of N_j contiguous elements is denoted $\mathbf{d}(u; f_j, N_j)$ and described by (5-1) with $f = f_j$, $N = N_j$, and $u = \sin\theta$. Let $\mathbf{W}_{B3}(f_j; N_j)$ denote the $N_j \times 3$ beamformer to be applied to each of $N - N_j + 1$ identical subarrays of $N_j < N$ contiguous elements at frequency f_j.

$$\mathbf{W}_{B3}(f_j; N_j) = \frac{1}{\sqrt{N_j}} [\mathbf{d}(u_C - u_L(f_j); f_j, N_j) \,|\, \mathbf{d}(u_C; f_j, N_j) \,|\, \mathbf{d}(u_C + u_R(f_j); f_j, N_j)]$$

where

$$u_L(f_j) = u_C - \frac{f_0}{f_j}\frac{2}{N_j} \quad \text{and} \quad u_R(f_j) = u_C + \frac{f_0}{f_j}\frac{2}{N_j}, \qquad j = 1, \ldots, J \tag{5-158}$$

Note that the respective pointing angles of the left and right beams depend on the frequency of transmission. For a specific frequency, the respective pointing angles of the left and right beams were chosen for two reasons. First, they serve to make the columns of $\mathbf{W}_{B3}(f_j; N_j)$ mutually orthogonal. Second, they serve to make the three polynomials formed with each of the three columns of $\mathbf{W}_{B3}(f_j; N_j)$ as a coefficient vector have $N_j - 3$ roots in common.

The first step in multifrequency 3D-BDML incorporating CSS is to form a spatially smoothed, beamspace correlation matrix, denoted $\tilde{\mathbf{R}}_{bb}(f_j)$, at each frequency in the following manner. The overall array is decomposed into $N - N_j + 1$ overlapping subarrays of $N_j \leqslant N$ contiguous elements. $\mathbf{W}_{B3}(f_j; N_j)$ is applied to each subarray, and $\tilde{\mathbf{R}}_{bb}(f_j)$ is formed as the arithmetic mean of the outer products of the $N - N_j + 1$ 3×1 beamspace snapshot vectors thus created. The second step is to apply a focusing transformation on both the left and right of each $\tilde{\mathbf{R}}_{bb}(f_j)$, $j = 1, \ldots, J$, so that the signal-only component of each may be coherently combined at some common frequency. Let the common frequency be f_k, where $f_k \in \{f_1, f_2, \ldots, f_J\}$. The focusing transformation to be applied to $\tilde{\mathbf{R}}_{bb}(f_j)$ is denoted $\mathbf{F}_j$, $j = 1, \ldots, J$. As $f_k \in \{f_1, f_2, \ldots, f_J\}$, the notation implies $\mathbf{F}_k = \mathbf{I}_3$. We will quantify the role of the focusing transformations more precisely as well as discuss methods for constructing such shortly. Given the appropriate set of focusing transformations, the CSS averaged correlation matrix, denoted $\bar{\mathbf{R}}_{bb}$, is computed via the simple sum

$$\bar{\mathbf{R}}_{bb} = \frac{1}{J} \sum_{j=1}^{J} \mathbf{F}_j \tilde{\mathbf{R}}_{bb}(f_j) \mathbf{F}_j^T \tag{5-159}$$

A summary of the multifrequency version of Nonsymmetric 3D-BDML is then as follows. Compute $\mathbf{v} = [v_1 \quad v_2 \quad v_3]^T$ as that eigenvector of $\mathrm{Re}\{\bar{\mathbf{R}}_{bb}\}$ associated with the smallest eigenvalue. Form the coefficients of $e(z) = e_0 + e_1 z + e_0^* z^2$ according to

$$e_0 = \exp\left[j \frac{f_0}{f_k} \pi u_C \right] \left\{ -v_1 \exp\left[-j \frac{f_0}{f_k} \frac{\pi}{N_k} \right] + v_2 - v_3 \exp\left[j \frac{f_0}{f_k} \frac{\pi}{N_k} \right] \right\} \tag{5-160a}$$

$$e_1 = 2(v_1 + v_3) \cos\left(\frac{f_0}{f_k} \frac{\pi}{N_k} \right) - 2v_2 \cos\left(\frac{f_0}{f_k} \frac{2\pi}{N_k} \right) \tag{5-160b}$$

where the scaling factor f_0/f_k accounts for operation at a frequency other than that for which the elements are spaced by a half wavelength. Next, compute z_1 and z_2 as the two roots of $e(z)$. The estimates of u_1 and u_2 are then $\hat{u}_1 = (f_0/f_k) \arg\{z_1\}/\pi$ and $\hat{u}_2 = (f_0/f_k) \arg\{z_2\}/\pi$.

The multifrequency version of Symmetric 3D-BDML is similar except that $\bar{\mathbf{R}}_{bb}$ is replaced by $\bar{\mathbf{R}}_{bb}^{fb} = \frac{1}{2}\{\bar{\mathbf{R}}_{bb} + \tilde{\mathbf{I}}_3 \bar{\mathbf{R}}_{bb} \tilde{\mathbf{I}}_3\}$. $\mathbf{v} = [v_1 \quad v_2 \quad v_1]^T$ is computed as that centrosymmetric eigenvector of $\mathrm{Re}\{\bar{\mathbf{R}}_{bb}^{fb}\}$ associated with the smaller eigenvalue. With $v_3 = v_1$ in (5-160), the remaining steps are the same. Of course, use of Symmetric 3D-BDML implies that u_C in (5-158) is equal to an estimate of bisector angle.

With the phase diversity achieved with frequency diversity, one might question the need for Symmetric 3D-BDML in the case of multifrequency operation. Of course, the differential in performance between the symmetric and nonsymmetric cases will depend on the specific values and total number of frequencies employed. For most practical applications in which the number of transmission frequencies is rather small (2, 3, or 4) and the interfrequency spacings not so great, Symmetric 3D-BDML can be expected to significantly outperform Nonsymmetric 3D-BDML. Simulations are presented in Section 5-7-6 backing this claim.

To quantify the role of $\mathbf{F}_j$ more precisely, let $\mathbf{b}_{B3}(u; f_j)$ denote the 3×1 beamspace manifold vector associated with frequency f_j, that is, $\mathbf{b}_{B3}(u; f_j) = \mathbf{W}_{B3}^H(f_j, N_j)\mathbf{d}(u; f_j, N_j)$, $j = 1, \ldots, J$. Given the definitions of $\mathbf{d}(u; f_j, N_j)$ and $\mathbf{W}_{B3}(f_j; N_j)$ in (5-1) and (5-158), respectively, it is easy to verify the following component-wise expression for $\mathbf{b}_{B3}(u; f_j)$:

$$\mathbf{b}_{B3}(u; f_j) = \mathbf{W}_{B3}^H(f_j; N_j)\mathbf{d}(u; f_j, N_j) = \frac{1}{N_j}\begin{bmatrix} S(u - u_L(f_j); f_j, N_j) \\ S(u - u_c; f_j, N_j) \\ S(u - u_R(f_j); f_j, N_j) \end{bmatrix} \tag{5-161}$$

where $u_L(f_j)$ and $u_R(f_j)$ are defined in (5-158) and $S(u; f_j, N_j)$ is defined as

$$S(u; f_j, N_j) = \frac{\sin\left(\frac{N_j}{2}\frac{f_j}{f_0}\pi u\right)}{\sin\left(\frac{1}{2}\frac{f_j}{f_0}\pi u\right)} \tag{5-162}$$

In accordance with the CSS methodology of Wang and Kaveh [29, 52], the focusing matrices must satisfy

$$\mathbf{F}_j\mathbf{b}(u_1; f_j) = \mathbf{b}(u_1; f_k), \qquad j = 1, \ldots, J \tag{5-163a}$$

$$\mathbf{F}_j\mathbf{b}(u_2; f_j) = \mathbf{b}(u_2; f_k), \qquad j = 1, \ldots, J \tag{5-163b}$$

Again, $\mathbf{F}_k = \mathbf{I}_3$.

A number of methods for constructing the focusing matrices have been proposed [29, 52]. We consider here the use of the Dummy Direction-Vector Constrained (DDVC) class of focusing matrices [52]. For frequency f_j, a DDVC focusing matrix [52] may be constructed as

$$\mathbf{F}_j = [\mathbf{b}(u_1; f_k) \mid \mathbf{b}(u_C; f_k) \mid \mathbf{b}(u_2; f_k)][\mathbf{b}(u_1; f_j) \mid \mathbf{b}(u_C; f_j) \mid \mathbf{b}(u_2; f_j)]^{-1}$$
$$j = 1, \ldots, J \tag{5-164}$$

Since $[\mathbf{b}(u_1; f_j) \mid \mathbf{b}(u_C; f_j) \mid \mathbf{b}(u_2; f_j)]^{-1}[\mathbf{b}(u_1; f_j) \mid \mathbf{b}(u_B; f_j) \mid \mathbf{b}(u_2; f_j)] = \mathbf{I}_3$, it follows that

$$[\mathbf{b}(u_1; f_j) \mid \mathbf{b}(u_C; f_j) \mid \mathbf{b}(u_2; f_j)]^{-1}\mathbf{b}(u_1; f_j) = \begin{bmatrix} 1 \\ 0 \\ 0 \end{bmatrix},$$

$$[\mathbf{b}(u_1; f_j) \mid \mathbf{b}(u_C; f_j) \mid \mathbf{b}(u_2; f_j)]^{-1}\mathbf{b}(u_2; f_j) = \begin{bmatrix} 0 \\ 0 \\ 1 \end{bmatrix}, \tag{5-165}$$

from which it is easily verified that $\mathbf{F}_j$ constructed according to (5-164) satisfies the conditions in (5-163a) and (5-163b). As indicated, the sole dummy pointing angle is selected to be the pointing angle of the center beam, u_C. Recall that in the symmetric form of 3D-BDML, $u_C = u_B$ where u_B is the bisector angle determined according to (5-126). Note that with $u_C = u_B = (u_1 + u_2)/2$ in (5-164) it is easily verified that $\mathbf{F}_j$ satisfies $\tilde{\mathbf{I}}_3 \mathbf{F}_j \tilde{\mathbf{I}}_3 = \mathbf{F}_j$, $j = 1, \ldots, J$. This is necessary for proper operation of the multifrequency version of Symmetric 3D-BDML.

Note that construction of the appropriate set of focusing matrices according to (5-164) requires knowledge of u_1 and u_2, that is, the angles we are trying to estimate. Accordingly, Wang and Kaveh [29, 52] propose an iterative procedure which commences with an initial set of focusing matrices based on some coarse estimates of the angles. One possibility for initialization is to choose the pointing angle of the center beam u_C as the initial estimate for both angles. Alternatively, in a tracking situation it makes sense to use the most recent bearing estimates as the initial estimates. Proceeding with the initial set of focusing matrices yields updated estimates of the angles corresponding to the first iteration. The new pair of angles are used to construct an updated set of focusing matrices which, in turn, yield the estimates of the angles at the second iteration. This procedure is iterated until the absolute value of the difference between respective angle estimates obtained at the ith and $(i+1)$th iterations is less than some threshold.

The need for focusing matrices in multifrequency 3D-BDML may be eliminated if the transmission frequencies, f_j, $j = 1, \ldots, J$, and corresponding subarray lengths, N_j, $j = 1, \ldots, J$, are selected such that the product $f_j N_j$ is the same for each frequency, that is, $f_j N_j = \text{constant}$, $j = 1, \ldots, J$. This assertion is justified by approximating each of the array patterns comprising the components of $\mathbf{b}_{B3}(u; f_j)$ in (5-161) as a sinc function in the vicinity of the respective mainlobe and first few sidelobes:

$$\mathbf{b}_{B3}(u; f_j) = \mathbf{W}_{B3}^H(f_j; N_j)\mathbf{d}(u; f_j, N_j) \approx \frac{f_0}{N_j f_j} \begin{bmatrix} \dfrac{\sin\left(\frac{\pi}{2}\frac{N_j f_j}{f_0}\left(u - u_C + \frac{f_0}{f_j}\frac{2}{N_j}\right)\right)}{\left(\frac{\pi}{2}\left(u - u_C + \frac{f_0}{f_j}\frac{2}{N_j}\right)\right)} \\[2ex] \dfrac{\sin\left(\frac{\pi}{2}\frac{N_j f_j}{f_0}(u - u_C)\right)}{\left(\frac{\pi}{2}(u - u_C)\right)} \\[2ex] \dfrac{\sin\left(\frac{\pi}{2}\frac{N_j f_j}{f_0}\left(u - u_C - \frac{f_0}{f_j}\frac{2}{N_j}\right)\right)}{\left(\frac{\pi}{2}\left(u - u_C - \frac{f_0}{f_j}\frac{2}{N_j}\right)\right)} \end{bmatrix} \tag{5-166}$$

It is apparent that if $f_jN_j = \text{constant}$, $j = 1, \ldots, J$, the beamspace manifold vector $\mathbf{b}_{B3}(u; f_j)$ is approximately identical, to a high degree of precision, for each transmission frequency!! With regard to (5-163), under these conditions $\mathbf{b}(u_i; f_j) \approx \boldsymbol{b}(u_i; f_k)$, $j = 1, \ldots, J$, $i = 1, 2$, eliminating the need for focusing matrices. In effect, with frequencies satisfying this criterion, the appropriate focusing matrices are identity matrices, that is, $\mathbf{F}_j = \mathbf{I}_3$, $j = 1, \ldots, J$. Also, observing (5-160), it is apparent that with $f_jN_j = \text{constant}$, $j = 1, \ldots, J$, any of the transmission frequencies may serve as the reference frequency f_k. Hence, CSS averaging is simply accomplished by summing the spatially smoothed beamspace correlation matrices formed at each frequency. This represents a dramatic simplification.

Some practical issues with regard to the selection of transmission frequencies are discussed in Section 5-7-6. The sensitivity of this multifrequency method to deviations in the product f_jN_j from frequency to frequency is also examined.

5-7-6 Performance in a Simulated Low-Angle Radar Environment

Computer simulations were conducted to assess the performance of the various BDML bearing estimation schemes in a simulated low-angle radar tracking environment. The following parameters were common to all test cases. First, the array employed was linear consisting of $N = 21$ elements uniformly spaced by a half wavelength at f_0. The nominal 3 dB beamwidth for this array is $2/21$ rad $= 5.46°$. The direct and specular path signals were angularly located at $\theta_1 = 1°$ and $\theta_2 = -1.5°$, respectively, corresponding to an angular separation of 0.46 beamwidths and a bisector angle of $\theta_B = -0.25°$. In the $u = \sin\theta$ domain, the relevant bisector quantity is $u_B = \frac{1}{2}\{\sin\theta_1 + \sin\theta_2\} = -0.00436$. The noise was additive, spatially white, and uncorrelated with the direct and specular path signals. The SNR of the direct path signal was 20 dB (per element). The ratio of the amplitude of the specular path signal to that of the direct path signal, ρ, was 0.9. In the case of single frequency operation at f_0, corresponding to the simulations presented in Figs. 5-13–5-15, each independent trial involved a single snapshot of data. For each simulation example, the respective performance of the particular 3D-BDML algorithm employed was examined at nine equispaced values of $\Delta\Psi$, the phase difference between the direct and specular path signals at the center element at f_0, over the interval from 0° to 180°. Multifrequency operation with three frequencies satisfying $f_jN_j = \text{constant}$, $j = 1, 2, 3$, is examined in Figs. 5-16 and 5-17. In this case, each independent trial involved the execution of the simplified multifrequency version of either Symmetric 3D-BDML or Nonsymmetric 3D-BDML given a single snapshot of data at each of the three frequencies. Finally, in all cases, sample means (SMEANs) and sample standard deviations (SSTDs) were computed from the results of 250 independent trials.

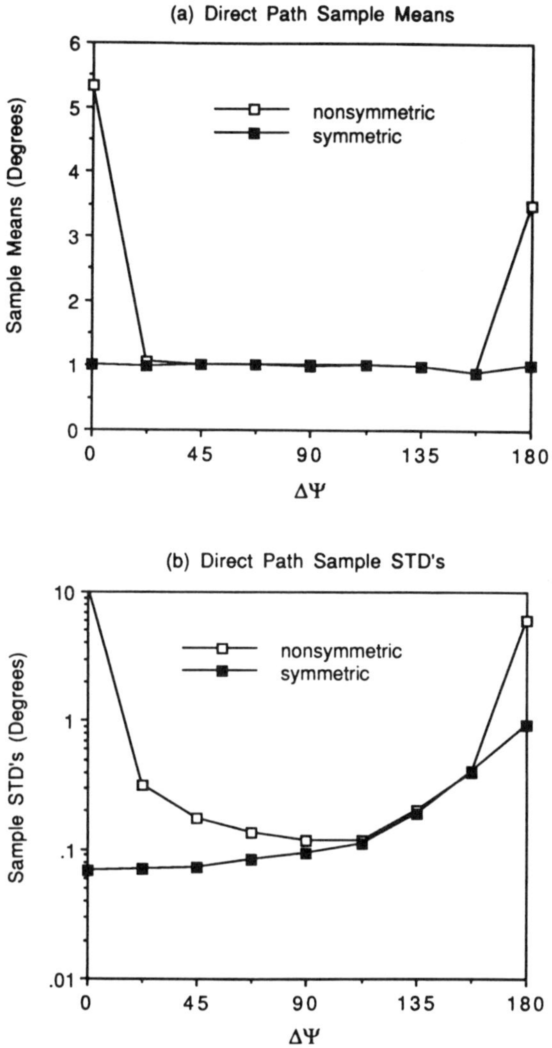

Figure 5-13 Performance comparison: Nonsymmetric 3D-BDML versus Symmetric 3D-BDML using the true bisector angle. $N = 21$ elements and single snapshot. Direct path, $\theta_1 = 1°$ and SNR = 20 dB; specular path, $\theta_2 = -1.5°$ and $\rho = 0.9$. Single frequency operation. SMEANs/SSTDs computed from 150 independent trials.

The first simulation results presented in Fig. 5-13 compare the performance of Nonsymmetric 3D-BDML with that of Symmetric 3D-BDML employing the actual bisector angle. The breakdown of Nonsymmetric 3D-BDML in the respective cases of $\Delta\Psi = 0°$ and $\Delta\Psi = 180°$ is apparent. Interestingly enough, the performance achieved with $\Delta\Psi = 0°$ is worse than that achieved with $\Delta\Psi = 180°$. No such breakdown phenomenon is observed with the Symmetric 3D-BDML estimator. However, the sample STD of the

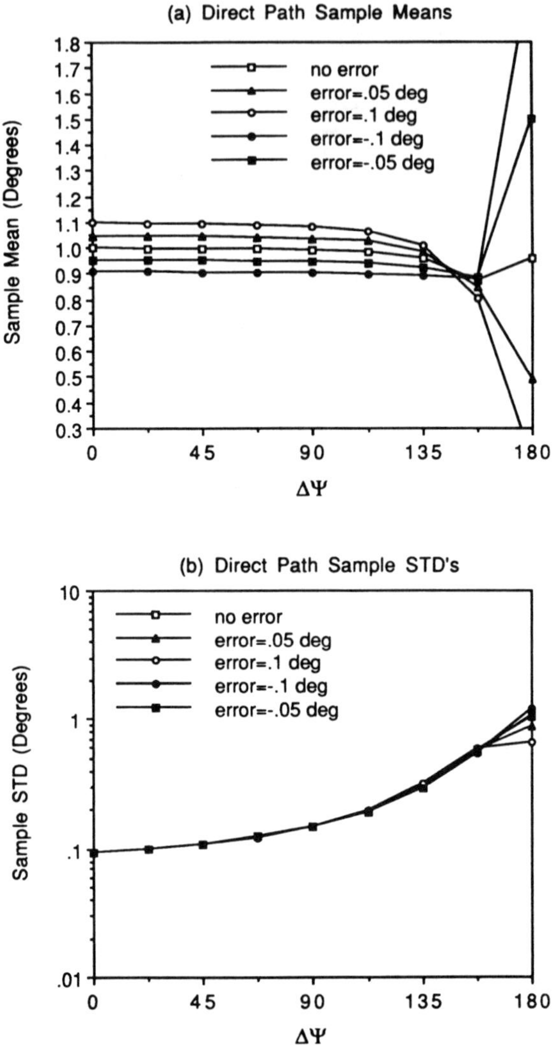

Figure 5-14 Sensitivity of Symmetric 3D-BDML to errors in the bisector angle estimate. Simulation parameters are the same as those described in the caption to Fig. 5-13.

Symmetric 3D-BDML estimator steadily increases from a value of 0.07° at $\Delta\Psi = 0°$ to a value of 0.94° at $\Delta\Psi = 180°$. The sample STD obtained with Symmetric 3D-BDML is observed to be substantially smaller than that obtained with Nonsymmetric 3D-BDML for $\Delta\Psi$ equal to 0°, 22.5°, 45°, 67.5°, and 180°.

The simulation results presented in Fig. 5-13 indicate that Symmetric 3D-BDML significantly outperforms Nonsymmetric 3D-BDML for most values of $\Delta\Psi$. As discussed previously, this may be attributed to Symmetric

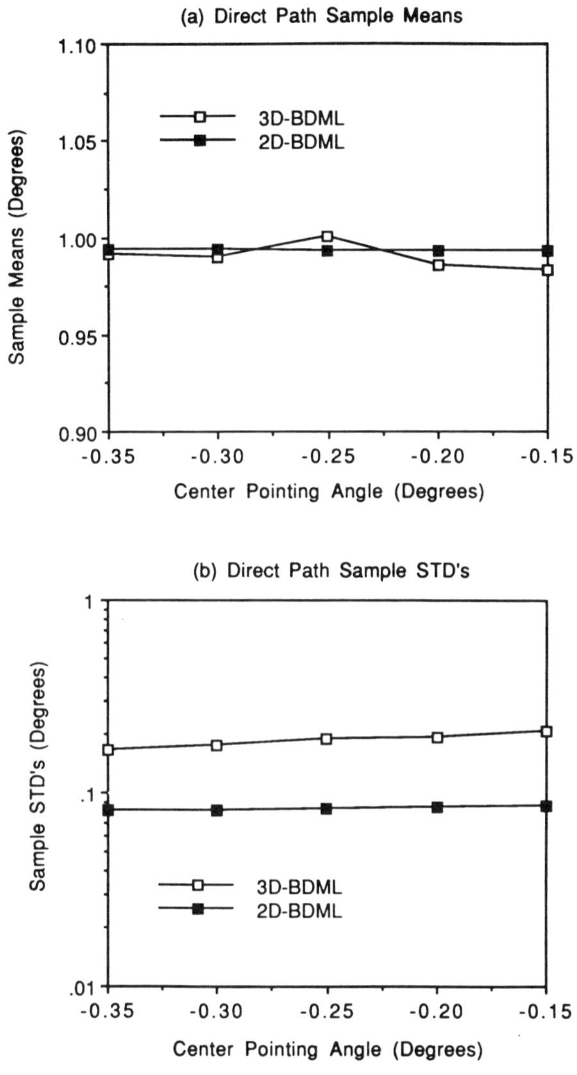

Figure 5-15 Performance comparison: Symmetric 3D-BDML versus 2D-BDML as a function of the pointing angle of the center beam in the case of no specular multipath. Simulation parameters are the same as those described in the caption to Fig. 5-13 except that $\rho = 0$.

3D-BDML's use of a priori information with regard to the bisector angle. An indication of the sensitivity of Symmetric 3D-BDML to error in the estimated bisector angle may be gleaned from observing the simulation results plotted in Fig. 5-14. For a given test case, the error cited in the legend is defined as $\theta_B - \theta_c$, where θ_c is the pointing angle of the center beam and θ_B is the actual bisector angle equal to $-0.25°$ in this example. The performance statistics plotted for the no error case are exactly the same

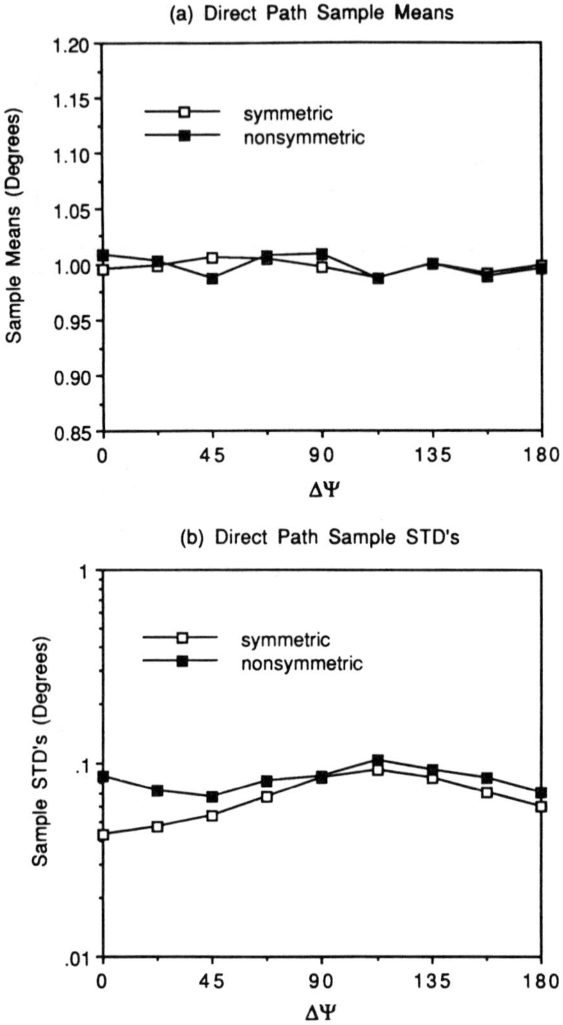

Figure 5-16 Performance comparison: Nonsymmetric 3D-BDML versus Symmetric 3D-BDML using the true bisector angle in the case of multifrequency operation. Simulation parameters are the same as those described in the caption to Fig. 5-13. Transmission frequencies: $f_1 = f_0$, $f_2 = (21/19)f_0$, and $f_3 = (21/17)f_0$, where f_0 denotes frequency for which elements are spaced by $\lambda/2$. $\Delta\Psi$ denotes phase difference at aperture center at f_0.

as those associated with the Symmetric 3D-BDML estimator plotted in Fig. 5-13. (Note the change in scale, however, between the vertical axes in Figs. 5-13(a) and 5-14(a), respectively.) For the error levels tested, which in view of the result in (5-127) correspond to fairly gross errors, very little difference is observed among the respective sample STD curves. On the other hand, an examination of the respective sample mean curves substantiates the conjecture made in Section 5-7-4 that the error in the bisector angle estimate

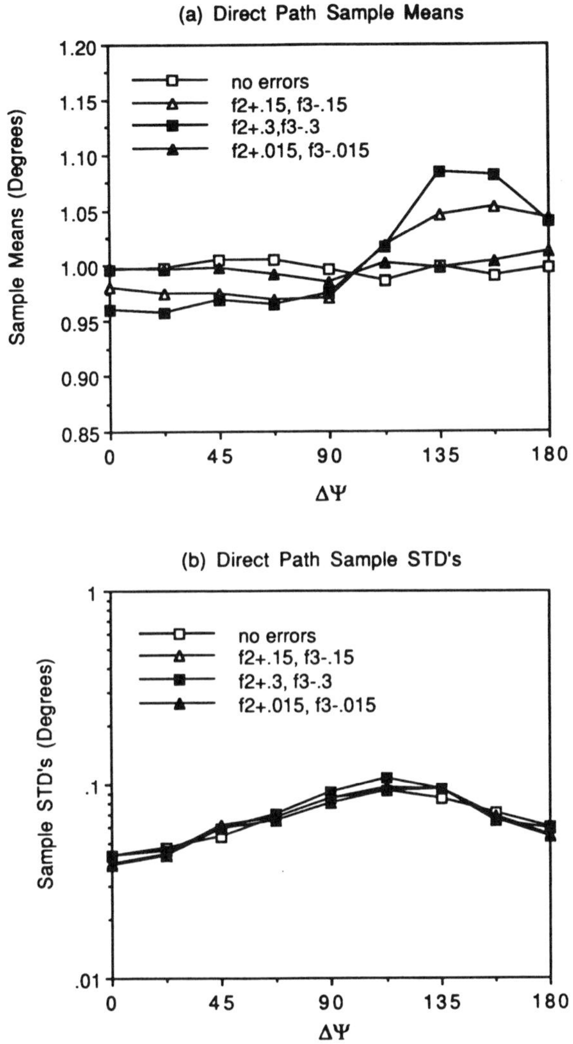

Figure 5-17 Sensitivity of Symmetric 3D-BDML in the case of multifrequency operation to deviations (in GHz) from the prescription $f_i = (N/N_i)f_0$, $i = 2, 3$, where $f_0 = 10$ GHz, $N_2 = 19$, and $N_3 = 17$, for the two auxiliary frequencies employed in the simulations presented in Fig. 5-16. Simulation parameters are the same as those described in the caption to Fig. 5-16.

translates into a bias in the Symmetric 3D-BDML estimator of the same magnitude.

As an illustrative example, consider the case $\theta_c = -0.15°$, which corresponds to an error of $\theta_B - \theta_c = 0.1°$. In accordance with the discussion in Section 5-7-4, the process of forward-backward averaging creates an artificial source at $2\theta_c - \theta_1 = 2(-0.15°) - 1° = -1.3°$ and another at $2\theta_c - \theta_2 = 2(-0.15°) - (-1.5°) = 1.2°$. Symmetric 3D-BDML cannot resolve the actual source at 1° and the artificial source at 1.2°, as they are angularly separated by 0.036 beamwidths, and in light of the nearly equal strength among the two sources, sees a single source at $\frac{1}{2}\{1° + 1.2°\} = 1.1°$. Correspondingly, we observe that with an error in the bisector angle of 0.1°, the sample mean obtained with Symmetric 3D-BDML estimator of the direct path angle is 1.1° for all of the nine phase values tested except $\Delta\Psi = 180°$.

The sample mean curves in Fig. 5-14(a) also suggest that an error in the bisector angle estimate pejoratively affects performance rather substantially in the case of $\Delta\Psi = 180°$. However, it should again be noted that the error levels tested represented gross deviations. In addition, the use of frequency diversity should remedy this problem.

The next set of simulation results examine the effect of removing the specular multipath component from the data, corresponding to $\rho = 0$. As discussed in Section 5-7-4, the process of forward-backward averaging in the execution of Symmetric 3D-BDML nevertheless creates an artificial source at $2\theta_c - \theta_1$, where θ_c is the pointing angle of the center beam. Thus, despite the absence of specular multipath, Symmetric 3D-BDML must nevertheless resolve two sources separated by $2(\theta_c - \theta_1)$. Hence, the closer θ_c is to θ_1, the harder Symmetric 3D-BDML must work to resolve the actual source and the artificial source. This translates into an increase in the sample STD of the bearing estimates as the separation between θ_1 and θ_c decreases, as illustrated in Fig. 5-15(b). The absence of a specular multipath component renders the 2D-BDML estimator developed in Section 5-3 a viable bearing estimation procedure. For purposes of comparison, sample means and sample STDs achieved with the same set of parameters but employing 2D-BDML with left and right beams pointed at $u_L = u_C - 1/N$ and $u_R = u_C + 1/N$, respectively, are also plotted in Figs. 5-15(a) and 5-15(b). With 2D-BDML, resolution is not an issue and, as a consequence, the sample STD does not vary much with respect to the location of the pointing angles of the two beams employed. Also, as expected, the sample STDs achieved with 2D-BDML are significantly lower than the respective sample STDs achieved with Symmetric 3D-BDML for each of the different beam locations tested.

The simulation results presented in Fig. 5-16 compare the performance of Nonsymmetric 3D-BDML with that of Symmetric 3D-BDML employing the true bisector angle in the case of multifrequency operation. $J = 3$ frequencies satisfying $f_j N_j = \text{constant}$ were employed with $f_1 = f_0$ and $N_1 = N$. This corresponds to no spatial smoothing and, hence, use of the full aperture at f_0. In turn, this automatically dictates that the other two frequencies satisfy $f_j N_j = f_1 N_1 = 21 f_0$, or $f_j = (21/N_j) f_0$, $j = 2, 3$, where N_j is an integer strictly less than 21. The specific selections were $N_2 = 19$ and $N_3 = 17$ yielding the frequencies $f_2 = (21/19) f_0 = 1.105 f_0$ and $f_3 = (21/17) f_0 = 1.235 f_0$. Let $\Delta\Psi_i$, $i = 1, 2, 3$, denote the phase difference occurring at the center element, modulo 360°, at the respective frequency f_i, $i = 1, 2, 3$. Further, let $\Delta\Psi_{0,T}$ denote the total phase difference between the direct and specular path signals at the center element at $f_1 = f_0$ counting integer number of wavelengths delays, that is, without the modulo by 360° operation. The values of $\Delta\Psi_i$, $i = 1, 2, 3$, were determined from $\Delta\Psi_{0,T}$ according to

$$\Delta\Psi_i = \left\{ \frac{f_i}{f_0} \{\Delta\Psi_{0,T} - 180°\} + 180°, \text{mod}(360°) \right\}, \qquad i = 1, 2, 3 \tag{5-167}$$

in accordance with the low-angle radar model described by Skolnik [46]. Note that this formula accounts for a 180° phase shift occurring at the surface of reflection. The phase shift occurring at the point of reflection is a phenomenon discussed by Skolnik [46] and Barton [4]. Note that in Fig. 5-16 (and also Fig. 5-17), $\Delta\Psi_1$ is simply denoted $\Delta\Psi$.

The performance statistics plotted in Figs. 5-16(a) and 5-16(b) demonstrate the robustness to the phase difference at any one frequency achieved with multifrequency operation. Although this is true for both Nonsymmetric 3D-BDML and Symmetric 3D-BDML, it is observed that the sample STDs achieved with Symmetric 3D-BDML are smaller than the respective sample STDs achieved with Nonsymmetric 3D-BDML for all nine values of $\Delta\Psi_1 = \Delta\Psi$ tested, with the differential between the two quite substantial for values of $\Delta\Psi$ near zero. It is noted that in all cases the sample bias achieved with Symmetric 3D-BDML is less than 0.015° and the corresponding SSTD is less than 0.07°.

To emphasize the computational simplicity of Symmetric 3D-BDML with multiple frequencies satisfying $f_jN_j = \text{constant}$, we summarize the specific form of the algorithm employed in the simulations presented in Fig. 5-16. Recall that $f_1 = f_0$ and $N_1 = N$. At $f_2 = (21/19)f_0$ $\mathbf{W}_{B3}(f_2; N_2)$ is applied to each of two overlapping subarrays of $N_2 = 19$ contiguous elements and $\tilde{\mathbf{R}}_{bb}(f_2)$ is formed as the arithmetic mean of the outer products of the two 3×1 beamspace snapshot vectors thus created. Similar processing occurs at $f_3 = (21/17)f_0$ with $N_3 = 17$ to create $\tilde{\mathbf{R}}_{bb}(f_3)$. With $\bar{\mathbf{R}}_{bb} = \{\tilde{\mathbf{R}}_{bb}(f_1) + \tilde{\mathbf{R}}_{bb}(f_2) + \tilde{\mathbf{R}}_{bb}(f_3)\}/3$, $\bar{\mathbf{R}}_{bb}^{fb}$ is formed as $\bar{\mathbf{R}}_{bb}^{fb} = \frac{1}{2}\{\bar{\mathbf{R}}_{bb} + \tilde{\mathbf{I}}_3\bar{\mathbf{R}}_{bb}\tilde{\mathbf{I}}_3\}$. $\mathbf{v} = [v_1 \;\; v_2 \;\; v_1]^T$ is then computed as that centrosymmetric eigenvector of $\text{Re}\{\bar{\mathbf{R}}_{bb}\}$ associated with the smaller eigenvalue. Finally, $\hat{u}_1$ is computed according to (5-146) since $f_1 = f_0$ and $N_1 = N$.

In a real-world application, the actual value of a transmission frequency will only match the desired value to within a certain tolerance. For example, in the MARS system described previously, the smallest increment the frequency of the agile transmitter may be stepped is 30 MHz. Suppose that specific values of f_1 and N_1 are selected, say $f_1 = f_0 = 10$ GHz and $N_1 = N = 21$, for example. Further, in determining two auxiliary frequencies to satisfy $f_jN_j = \text{constant}$, we select $N_2 = 19$ and $N_3 = 17$ as above, such that $f_2 = (21/19)f_0 = 11.05$ GHz and $f_3 = (21/17)f_0 = 12.35$ GHz. These frequencies are in the range of the agile transmitter in the MARS system. The best we can do is synthesize these frequencies to within a tolerance of ± 15 MHz =

±0.015 GHz. For the sake of illustration, consider that instead of $f_2 = 11.05$ GHz and $f_3 = 12.35$ GHz, the actual transmission frequencies are $f_2' = f_2 + 0.015 = 11.065$ GHz and $f_3' = f_3 - 0.015 = 12.335$ GHz. And, despite the fact that $f_j' N_j \neq 21 f_0$, $j = 2, 3$, we nevertheless employ the simplified multifrequency version of Symmetric 3D-BDML outlined above. The resulting performance is plotted in Fig. 5-17 and compared with the performance achieved with f_2 and f_3 equal to the desired values of 11.05 GHz and 12.35 GHz, respectively, signified by the no errors label in the legend. The difference between the respective sample mean curves is practically negligible, as is the difference between the respective sample STD curves.

To really demonstrate the robustness of the method, the magnitude of the respective error in the two auxiliary frequencies was increased by a factor of ten, to 150 MHz, yielding the actual transmission frequencies $f_2' = f_2 + 0.15 = 11.2$ GHz and $f_3' = f_3 - 0.15 = 12.2$ GHz. Again, very little difference between the respective sample STD curves is observed. However, the corresponding sample mean curve does reveal a phase dependent bias: the sample bias is negative for values of $\Delta\Psi$ less than or equal to 90° and positive for values of $\Delta\Psi$ greater than 90°. A phase dependent sample bias curve of similar shape is obtained if the magnitude of the respective error in the two auxiliary frequencies is further increased by a factor of two, to 300 MHz, yielding the actual transmission frequencies $f_2' = f_2 + 0.3 = 11.35$ GHz and $f_3' = f_3 - 0.3 = 12.05$ GHz. As might be expected, the maximum sample bias obtained with these frequencies is about twice that obtained with $f_2' = f_2 + 0.15 = 11.2$ GHz and $f_3' = f_3 - 0.15 = 12.2$ GHz.

5-7-7 Performance in an Experimental Low-Angle Radar Environment

The 3D-BDML algorithm was applied to data obtained from an experimental array deployed at the west coast of the Bruce Peninsula, Ontario, Canada, overlooking Lake Huron [31]. The array system, referred to as the Multi-parameter Adaptive Radar System (MARS), was developed at the Communications Research Laboratory at McMaster University and is described in detail in Chapter 7. A brief description is provided here for the sake of completeness. MARS consists of a vertical linear array of $N = 32$ horizontally polarized 10 dB horn antennas equispaced by 5.715 cm. The array system operates coherently over the band 8.05–12.34 GHz (X-band). The center of the array is at a height of 8.212 m above the mean water level, that is, $h_r = 8.812$ m. Each antenna element is followed by two receiver channels allowing for simultaneous reception on two separate frequencies: one fixed at 10.2 GHz and one agile over the band in 30 MHz steps. Calibration of the array was achieved via a common local test signal injection method. This involved a 6-term error model correction applied to each channel. In addition, a far field technique was employed to align the relative amplitude and phase of all the channels.

In experiments performed in October 1987, a 100 mW CW beacon served as a signal source at a distance of 4610 m from the receiving array across Lake Huron. The height of the CW transmitter, h_t, was variable between 3.5 m and 18.5 m. The scenario emulated that of a low-angle radar tracking environment in that the source signal arrived at the array via a specular path involving a bounce off of Lake Huron as well as by a direct path with the angular separation between the two a fraction of beamwidth for any transmitter height and transmission frequency. This precludes the use of conventional monopulse bearing estimation techniques.

In the specific experiment that yielded the experimental data to which 3D-BDML was applied, the transmitter height was such that the direct path angle (including a tilt angle) was 0.084° while the specular path angle was −0.303°. Measurements were performed at five different frequencies with the CW transmitter fixed at this height. The beamwidth separation between the direct and specular path signals at each of the five frequencies, 8.62 GHz, 9.76 GHz, 9.79 GHz, 11.32 GHz, and 12.34 GHz, was 0.354 BW, 0.401 BW, 0.402 BW, 0.465 BW, and 0.507 BW, respectively. At each frequency, 127 snapshots were collected over a 2 s interval. (The data was recorded with 12-bit precision at a sampling rate of 62.5 samples/s.) At each frequency, we computed the DSFT of each of five different snapshots chosen randomly from the set of 127 snapshots and plotted the corresponding square magnitude. The results are displayed in Fig. 5-18 wherein at each

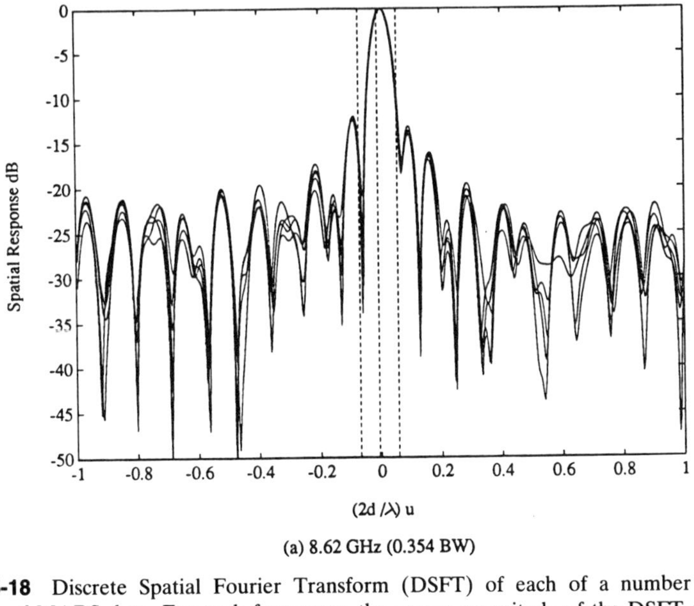

(a) 8.62 GHz (0.354 BW)

Figure 5-18 Discrete Spatial Fourier Transform (DSFT) of each of a number of single snapshots of MARS data. For each frequency, the square magnitude of the DSFT of each of five different snapshots are overlaid. The vertical dotted lines mark the pointing angles of the three beams employed in 3D-BDML.

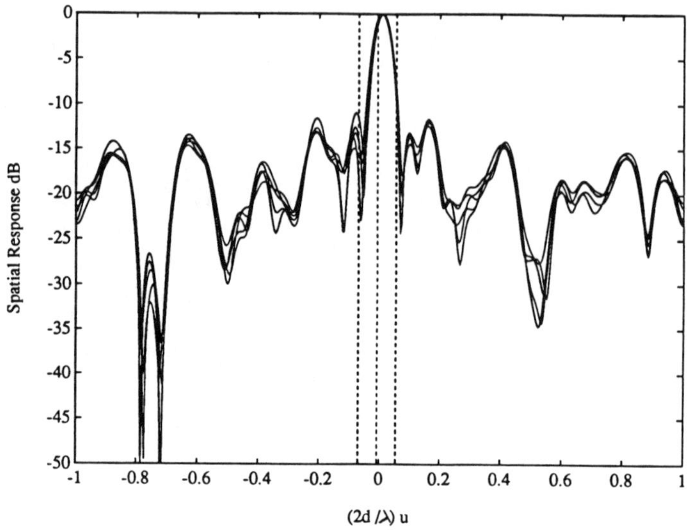

(b) 9.76 GHz (0.401 BW)

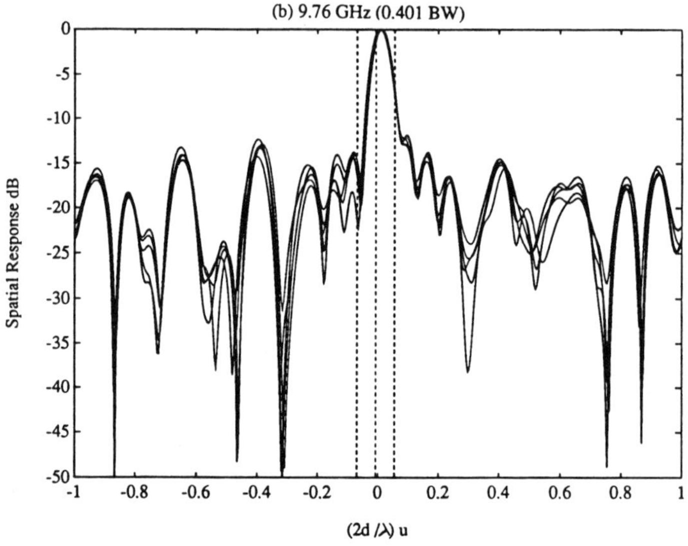

(c) 9.79 GHz (0.402 BW)

Figure 5-18 (*Continued*)

frequency the five corresponding plots are overlaid. Note for each frequency the vertical dotted lines mark the pointing angles of the three beams employed in 3D-BDML.

Note that the square magnitude of the DSFT corresponds to the conventional Bartlett spatial spectral estimator given a single snapshot in our case here. Due to the fractional beamwidth separation, the Bartlett estimator is not able to resolve the direct and specular path signals at any frequency.

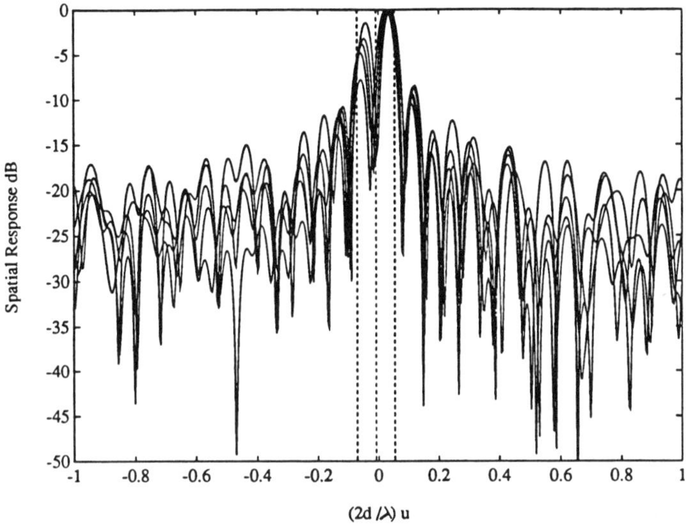

(d) 11.32 GHz (0.465 BW)

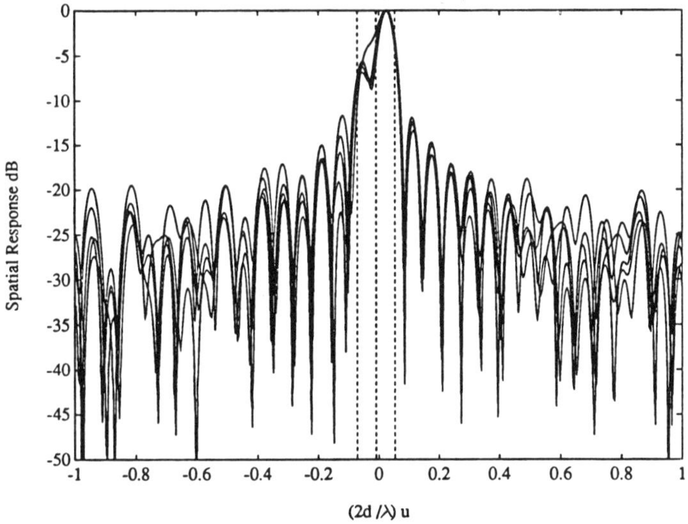

(e) 12.34 GHz (0.507 BW)

Figure 5-18 (*Continued*)

However, at both 11.32 GHz and 12.34 GHz we observe two spectral peaks in the vicinity of the actual signal arrivals near broadside. This is not an indication that the signals are resolved but is instead due to the fact that the relative phase difference between the two signals at the center of the array is near 180° at each of these two frequencies. This claim is substantiated by the following development.

The DSFT of the nth snapshot at frequency f_j is $\mathbf{d}^H(u; f_j)\,\mathbf{x}(n)$ where $\mathbf{d}(u; f_j)$ is defined by (5-1) with $f = f_j$ and $u = \sin\theta$. Consider the noiseless version of $\mathbf{x}(n)$ defined by (5-128) with $\mathbf{n}(n) = \mathbf{0}$. In this case, the DSFT of a single snapshot may be expressed as

$$\mathbf{d}^H(u; f_j)\mathbf{x}(n) = c_1(n)\mathbf{d}^H(u; f_j)\mathbf{d}(u_1; f_j) + c_2(n)\mathbf{d}^H(u; f_j)\mathbf{d}(u_2; f_j)$$

$$= c_1(n)\left\{ \frac{\sin\left(\frac{N_j}{2}\frac{f_j}{f_0}\pi(u-u_1)\right)}{\sin\left(\frac{1}{2}\frac{f_j}{f_0}\pi(u-u_1)\right)} - \rho\exp[j\Delta\Psi]\frac{\sin\left(\frac{N_j}{2}\frac{f_j}{f_0}\pi(u-u_2)\right)}{\sin\left(\frac{1}{2}\frac{f_j}{f_0}\pi(u-u_2)\right)} \right\} \tag{5-168}$$

where ρ and $\Delta\Psi$ are the amplitude ratio and relative phase difference, respectively, as defined in (5-147). Note that if $\rho = 1$ and $\Delta\Psi = 180°$, the DSFT is zero at the bisector angle $u = u_B = (u_1 + u_2)$ regardless of the angular separation $|u_1 - u_2|$! It follows in the case where ρ is near unity and $\Delta\Psi$ is near 180°, the DSFT will exhibit a soft null somewhere in between the two actual arrival angles: the closer ρ is to unity and $\Delta\Psi$ is to 180° the deeper the null. In the case where the angular separation is a fraction of a beamwidth, this null or dip in the DSFT serves to produce two spectral peaks in the general vicinity of the source angles despite the fact that the sources are not resolved.

Since the angular separation is 0.465 BW at 11.32 GHz, we know that the Bartlett spectral estimator is not able to resolve the direct and specular path signals. Thus, the dip in the DSFT near broadside in each of the five plots overlaid in Fig. 5-18(d) is a clear indication that the relative phase difference is near 180° at this frequency for this particular transmitter height. This results in signal cancellation and a low effective SNR. This manifests itself in terms of poor performance of 3D-BDML at this frequency as will be discussed shortly.

Figure 5-19 displays the results of applying Symmetric 3D-BDML to a single snapshot of data at each of the five frequencies. In all cases, the three beams were centered at the estimate of the bisector angle given by (5-127). For each frequency, the respective pointing angles of the 3 beams employed in Symmetric 3D-BDML are indicated by vertical dotted lines in the corresponding plot in Fig. 5-18. For each frequency, we have plotted in Fig. 5-19 the roots of $e(z) = e_0 + e_1 z + e_0^* z^2$ where e_0 and e_1 are computed according to (5-145a) and (5-145b), respectively, for each of the 127 different snapshots. In each plot, the angular sector demarcated by the two radial lines corresponds to the angular sector encompassed by the respective mainlobes of each of the three beams; the curve shown is a segment of the unit circle.

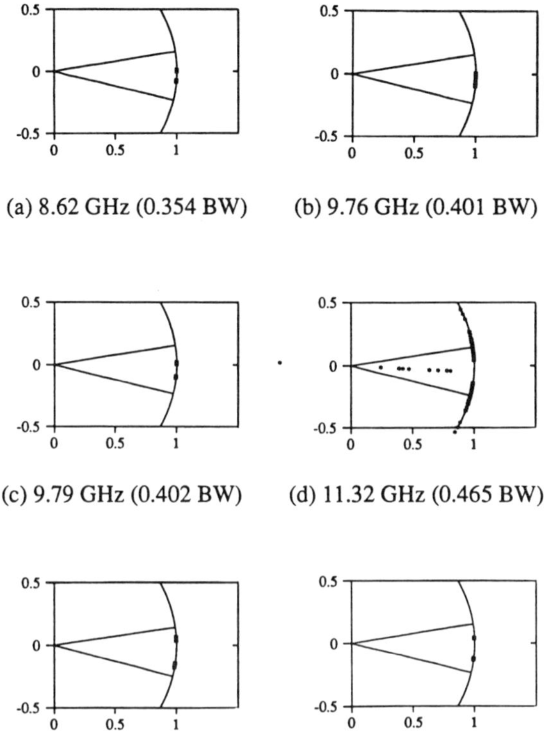

(a) 8.62 GHz (0.354 BW) (b) 9.76 GHz (0.401 BW)

(c) 9.79 GHz (0.402 BW) (d) 11.32 GHz (0.465 BW)

(e) 12.34 GHz (0.507 BW) (f) mult. freq. (f_c=9.79 GHz)

Figure 5-19 Root scatter plots obtained by applying Symmetric 3D-BDML to experimental data obtained from MARS array. For each frequency, the two signal root estimates obtained by processing each of a 127 different snapshots are overlaid. Multifrequency Symmetric 3D-BDML was employed in generating the root scatter plot in (f). In each plot, the curve shown is a segment of the unit circle.

Note that for each frequency except 11.32 GHz, Symmetric 3D-BDML yields two distinct signal roots for each of the 127 snapshots indicating that the direct and specular path signals have been resolved. At 11.32 GHz the relative phase difference was close to 180°, as discussed previously, resulting in a low effective SNR and, as a consequence, poor performance of Symmetric 3D-BDML. Note that the ideal signal roots are of the form $z_k(j) = \exp[j2\pi(f_j/c)d \sin \theta_k]$, $k = 1, 2$, $j = 1, \ldots, 5$, where c is the speed of light and d is the interelement spacing. Hence, the angular separation between the two signal roots, $2\pi(f_j/c)d|\sin \theta_1 - \sin \theta_2|$, increases with increasing frequency. This trend is indeed manifested in the various plots comprising Fig. 5-19.

Table 5.2 lists the sample means (SMEANs) and sample standard deviations (SSTDs) for both the direct and specular path angle estimates obtained at each frequency averaged over 127 independent trials. The angle estimates

TABLE 5-2 Sample Statistics Obtained with MARS Data

Frequency (3 beams)	Mean (deg) (direct: 0.0835°)	Mean (deg) (specular: −0.303°)	SSTD (deg)
8.62 GHz (0.354 BW)	0.06730	−0.27143	0.03334
9.76 GHz (0.401 BW)	−0.01300	−0.19120	0.07746
9.79 GHz (0.402 BW)	0.09375	−0.29787	0.03111
11.32 GHz (0.465 BW)	0.29730	−0.50143	0.28673
12.34 GHz (0.507 BW)	0.21942	−0.42354	0.02270
Mult. freq. ($f_k = 9.79$ GHz)	0.05323	−0.25736	0.02024

were determined from the signals roots according to the relationship $\theta_k = \sin^{-1}\{c \arg\{z_k(j)\}/(2\pi d f_j)\}$, $k = 1, 2$, $j = 1, \ldots, 5$. Note that for all frequencies except 11.32 GHz the SMEAN is within roughly a tenth of a degree of the true arrival angle for both signals and the SSTD is less than a tenth of a degree. Note that the SSTD is the same for both the direct and specular path angle estimates due to the symmetry of the estimates about the bisector angle imposed by Symmetric 3D-BDML.

Figure 5-18(f) displays the results of applying the multifrequency version of Symmetric 3D-BDML developed in Section 5-7-5 based on the coherent signal subspace method of Wang and Kaveh [52]. Focusing transformations were used to translate the signal subspace information at each frequency to a common reference frequency of $f_k = 9.79$ GHz corresponding to the median of the five transmission frequencies. For each trial run, good focusing was achieved in less than five iterations. Observe that for either signal arrival the spread of the corresponding roots in Figs. 5-18(f) is noticeably smaller than that in Figs. 5-18(a)–(e). Accordingly, we observe that the corresponding SSTD listed in Table 5.2 for the multifrequency case is smaller than that obtained at any of the individual frequencies. Note that better performance may be achieved in the multifrequency case by weighting the single frequency correlation matrices in forming the focused sum according to (5-159). In this mode of operation, the 11.32 GHz data would be weighted significantly less than the data associated with the other frequencies. The Frobinius norm of each single frequency correlation matrix may be used as a basis for a weighting scheme since this will be small at frequencies for which there is substantial signal cancellation.

5-8 SUMMARY AND CONCLUSIONS

Various beamspace ML bearing estimation procedures were developed for use in diverse operational environments. Section 5-3 dealt with the nonadaptive case corresponding to a relatively low-level interference environment. It was shown that sum and difference beams constructed as judiciously weighted combinations of three classical (phase-only steered) beams exhibited low sidelobe levels and facilitated a simple, closed-form procedure for determining the ML estimate of the target bearing. The low sidelobe levels are due to judicious exploitation of the sidelobe phase relationships amongst the three beams referred to as the left, center, and right beams. Specifically, the left and center beams are lined up such that co-located sidelobes are 180° out-of-phase. The same is true with respect to the center and right beams. On the other hand, the left and right beams are lined up such that co-located sidelobes are in-phase. The computational simplicity of 2D-BDML, a requirement under monopulse operating conditions, is due to judicious exploitation of the fact that left, center, and right beams have $N-3$ nulls in common. Ultimately, the target bearing is estimated via a simple trigonometric expression based on the ratio of the difference beam to the sum beam.

Section 5-6 developed simple procedures for constructing adaptive left, center, and right beams possessing the same sidelobe phase relationships and common nulls property exhibited by left, center, and right phase-only steered beams used in the nonadaptive case, but with a subset of the common nulls aligned with the bearings of strong interferences. Adaptive sum and difference beams are then constructed as linear combinations of these beams in the same way as classical sum and difference beams are constructed as linear combinations of classical left, center, and right beams. This facilitates a simple, closed-form procedure for estimating the target bearing identical to that for the nonadaptive case. Various procedures were proposed for constructing the adaptive center beam, from which the adaptive left and right beams are derived, based on the information available regarding the interference environment. In the adaptive-adaptive [4] processing mode, estimates of the interference directions are first obtained by applying spectral estimation during a period in which the radar is passively listening, for example. It is shown that adaptive-adaptive [4] sum and difference beams exhibiting the desired properties may be synthesized via linear combinations of the left, center, and right phase-only steered beams described above and phase-only steered beams formed in each of the interference directions. Several examples were presented demonstrating the efficacy of these procedures.

Section 5-7 developed and analyzed an extension of 2D-BDML, that is, the sum and difference beam method, referred to as 3D-BDML for bearing estimation in the presence of mainbeam interference. Low-angle radar tracking was cited as a classical example of a real-world, mainbeam interference problem. In 3D-BDML the target bearing is estimated from the respective outputs of the left, center, and right phase-only steered beams described above. Again, judicious exploitation of the common nulls property facilitates a simple estimation procedure amenable to monopulse operating conditions. In Symmetric 3D-BDML the pointing angle of the center beam is equal to the bisector angle between the direct path ray and the image ray, yielding a 1-D parameter estimation problem. The bisector angle may be accurately estimated a priori given the height of the receiving array and an estimate of the range of the target. In contrast, Nonsymmetric 3D-BDML does not account for the multipath geometry and effectively solves a 2-D parameter estimation problem. As a consequence, we find that that Nonsymmetric 3D-BDML breaks down when $\Delta\Psi$, the phase difference between the direct and specular path signals at the aperture center, is either 0° or 180°. In contrast, Symmetric 3D-BDML can theoretically handle any value of $\Delta\Psi$ with $\Delta\Psi = 0°$ yielding best performance.

The only difference between Symmetric 3D-BDML and Nonsymmetric 3D-BDML is the forward-backward averaging of the 3×3 beamspace correlation matrix performed in Symmetric 3D-BDML. If the pointing angle of the center beam is not equal to the bisector angle, the process of forward-backward averaging effectively creates artificial sources at the mirror images of the actual sources about the pointing angle of the center beam. A simple analysis of this phenomenon reveals that an error in the bisector angle estimate translates into a bias in the Symmetric 3D-BDML estimator of the same magnitude. In the case of no specular multipath, this phenomenon makes the Symmetric 3D-BDML estimator nevertheless have to resolve two sources separated by less than a beamwidth. The 2D-BDML estimator, that is, the sum and difference beam method, may be used under these conditions to achieve better performance.

Although Symmetric 3D-BDML is theoretically capable of handling any value of $\Delta\Psi$, performance in the case of $\Delta\Psi = 180°$ is rather poor due to partial signal cancellation. An obvious means for overcoming this problem is to employ frequency diversity. It was shown that if the transmission frequencies satisfy $f_j N_j = \text{constant}$, $j = 1, \ldots, J$, where $N_j \leq N$ is the number of elements comprising the subarray employed at f_j, coherent signal subspace averaging is achieved by simply summing the (spatially smoothed) beamspace correlation matrices formed at each frequency. The multifrequency version of Symmetric 3D-BDML is then simply the single frequency version executed with the coherently averaged beamspace correlation matrix thus obtained. Simulations demonstrating the method to be rather robust to relatively large deviations in the product $f_j N_j$ from frequency to frequency suggest that the procedure may indeed be practicable.

The performance achieved in applying 3D-BDML to the experimental data obtained from the MARS array demonstrates the following capabilities of 3D-BDML. First, it demonstrates the subbeamwidth resolution capability of 3D-BDML, as small as three-tenths of the Rayleigh resolution limit. Second, it showed that 3D-BDML is able to perform well with a single data snapshot demonstrating the real-time tracking capability of 3D-BDML. This may be attributed to working with the real part of beamspace sample correlation matrix and exploitation of the multipath geometry. Finally, it demonstrates the ability of 3D-BDML to integrate multifrequency data in a computationally efficient manner to achieve enhanced SNR.

ACKNOWLEDGMENTS

It should be pointed out that much of the work presented here is the result of collaboration with a former Ph.D. student, Dr. Ta-Sung Lee. It was a blessing to have had such an excellent researcher as my first Ph.D. student. In addition, the author wishes to express sincere gratitude to Dr. Vitas Kezys and Dr. Simon Haykin of the Communications Research Laboratory at McMaster University for supplying the experimental data obtained from the MARS array. Sincere thanks are also extended to Gregory M. Kautz for his work in applying the 3D-BDML algorithm to the MARS data and generating the accompanying performance statistics/plots. Finally, the financial support of the Corporate Research and Development Center of the General Electric Company is gratefully acknowledgd.

REFERENCES

[1] J. A. Adam (1988) "Pinning defense hopes on AEGIS", *IEEE Spectrum*, **25**, 24–27.

[2] C. Baird and G. Rasswiler (1976) "Adaptive sidelobe nulling using digitally controlled phase shifters", *IEEE Trans. Antennas Propag.*, **24**, 638–649.

[3] W. P. Ballance and A. G. Jaffer (1987) "Low angle direction finding based on maximum likelihood: a unification", *Conference Record of the 21st Asilomar Conference on Signals, Systems, and Computers*, pp. 119–124.

[4] D. K. Barton (1974) "Low angle radar tracking", *Proc. IEEE*, **62**, 687–704.

[5] D. K. Barton (1969) "Comment on 'Angular tracking of two closely spaced radar targets' ", *IEEE Trans. Aerosp. Electron, Systems*, **5**, 350–351.

[6] D. K. Barton (1979) "Multipath fluctuation effects in track-while-scan radar", *IEEE Trans. Aerosp. Electron. System*, **15**, 754–764.

[7] A. J. Berni (1975) "Angle of arrival estimation using an adaptive antenna array", *IEEE Trans. Aerosp. Electron. Systems*, **11**, 278–284.

[8] Y. Bresler and A. Macovski (1986) "Exact maximum likelihood parameter estimation of superimposed exponential signals in noise", *IEEE Trans. Acoust., Speech, Signal Process.*, **34**, 1081–1089.

[9] Y. Bresler, V. U. Reddy, and T. Kailath (1988) "Optimum beamforming for coherent signals and interference", *IEEE Trans. Acoust., Speech, Signal Process.*, **36**, 833–843.

[10] E. Brookner (1985) "Phased array radars", *Sci. Am.*, **252,** 94–102.

[11] E. Brookner and J. Howells (1986) "Adaptive-adaptive array processing", *Proc. IEEE*, **74**, 602–604.

[12] C. L. Byrne and A. K. Steele (1987) "Sector-focused stability for high-resolution array processing", in *Proc. 1987 IEEE Int. Conf. Acoust., Speech, and Signal Process.*, 2340–2343.

[13] B. H. Cantrell, W. B. Gordon, and G. V. Trunk (1981) "Maximum likelihood

elevation angle estimation of radar targets using subapertures", *IEEE Trans. Aerosp. Electron, Systems*, **17**, 213–221.

[14] R. C. Davis, L. E. Brennan, and L. S. Reed (1976) "Angle estimation with adaptive arrays in external noise fields", *IEEE Trans. Aerosp. Electron. Systems*, **12**, 179–186.

[15] P. R. Dax (1973) "Accurate tracking of low elevation targets over the sea with a monopulse radar", *IEE Conf. Radar—Present and Future*, London; *IEE Publ.*, **105,** 160–165.

[16] O. L. Frost (1972) "An algorithm for linearly constrained adaptive array processing", *Proc. IEEE*, **60**, 926–935.

[17] W. F. Gabriel (1982) "Tracking closely spaced multiple sources via spectral-estimation techniques", *NRL Report 8603.*

[18] W. F. Gabriel (1982) "A high-resolution target-tracking concept using spectral estimation techniques", *NRL Report 8797.*

[19] W. F. Gabriel (1985) "Using spectral estimation techniques in adaptive processing antenna systems", *NRL Report 8920.*

[20] G. H. Golub and C. F. Van Loan (1983) *Matrix Computations*, Johns Hopkins University Press, Baltimore, MD.

[21] W. B. Gordon (1983) "Improved three subaperture method for elevation angle estimation", *IEEE Trans. Aerosp. Electron. Systems*, **19**, 114–122.

[22] L. J. Griffiths (1976) "Adaptive monopulse beamforming", *Proc. IEEE*, **64**, 1260–1261.

[23] R. L. Haupt (1984) "Simultaneous nulling in the sum and difference of a monopulse array", *IEEE Trans. Antennas Propag.*, **34**, 486–493.

[24] R. L. Haupt (1984) "Adaptive nulling in monopulse antennas", *IEEE 1984 Antennas and Propagation Int. Symp. Digest*, pp. 819–822.

[25] R. L. Haupt (1985) "Null synthesis with phase and amplitude controls at the subarray outputs", *IEEE Trans. Antennas Propag.*, **33**, 505–509.

[26] S. Haykin (1985) "Radar array processing for angle of arrival estimation", *Array Signal Processing*, Prentice Hall, Egnlewood Cliffs, NJ, Chap. 4.

[27] D. D. Howard, J. Nessmith, and S. M. Sherman (1971) "Monopulse tracking errors due to multipath: cause and remedies", *IEEE EASCON '71 Rec.*, pp. 175–182.

[28] J. E. Howard (1975) "A low angle tracking system for fire control radars", in *IEEE 1975 Int. Radar Conf. Rec.*, pp. 412–417.

[29] H. Hung and M. Kaveh (1988) "Focusing matrices for coherent single-subspace processing", *IEEE Trans. Acoust., Speech, Signal Process.*, **36**, 1272–1281.

[30] J. Kesler and S. Haykin (1980) "A new adaptive antenna for elevation angle estimation in the presence of multipath", *IEEE 1980 Antenna and Propagation Int. Symp. Digest*, pp. 130–133.

[31] V. Kezys and S. Haykin (1968) "Multifrequency angle-of-arrival estimation: an experimental evaluation", in *SPIE Int. Symp. Advanced Algorithms and Architectures for Signal Process. III*, SPIE Vol. 975, pp. 98–106.

[32] A. A. Ksienski and R. B. McGhee, Jr. (1968) "A decision theoretic approach

to the angular resolution and parameter estimation problem for multiple target", *IEEE Trans. Aerosp. Electron. Systems*, **4**, 443–445.

[33] R. Kumaresan, L. L. Scharf, and A. K. Shaw (1986) "An algorithm for pole-zero modelling and spectral analysis", *IEEE Trans. Acoust., Speech, Signal Process.*, **34**, 637–640.

[34] T. S. Lee (1989) "Beamspace domain ML based low-angle radar tracking with an array of antennas", Ph.D. Dissertation, Purdue University.

[35] J. T. Mayhan and L. Niro (1987) "Spatial spectrum estimation using multiple beam antennas", *IEEE Trans. Antennas Propag.*, **35**, 897–906.

[36] R. Monzingo and T. Miller (1980) *Introduction to Adaptive Arrays*, Wiley, New York.

[37] B. Noble and J. W. Daniel (1977) *Applied Linear Algebra*, Prentice Hall, Englewood Cliffs, NJ.

[38] S. U. K. Pillai (1989) *Array Signal Processing*, Springer-Verlag, New York.

[39] J. G. Proakis and D. G. Manolakis (1988) *Introduction to Digital Signal Processing*, Macmillan, New York.

[40] V. U. Reddy, A. Paulraj, and T. Kailath (1987) "Performance analysis of the optimum beamformer in the presence of correlated sources and its behavior under spatial smoothing", *IEEE Trans. Acoust., Speech, Signal Process.*, **35**, 527–536.

[41] I. S. Reed, J. D. Mallet, and L. E. Brennan (1974) "Rapid convergence rate in adaptive arrays", *IEEE Trans. Aerosp. Electron. Systems*, **10**, 853–863.

[42] R. Roy and T. Kailath (1989) "ESPRIT—Estimation of signal parameters via rotational invariance techniques", *IEEE Trans. Acoust., Speech, Signal Process.*, **37**, 984–995.

[43] R. O. Schmidt (1979) "Multiple emitter location and signal parameter estimation", in: *Proc. RADC Spectral Estimation Workshop*, Rome, pp. 243–256.

[44] R. O. Schmidt (1986) "Multiple emitter location and signal parameter estimation", *IEEE Trans. Antennas Propag.*, **34**, 276–280.

[45] T. J. Shan, M. Wax, and T. Kailath (1985) "On spatial smoothing for direction-of-arrival estimation of coherent signals", *IEEE Trans. Acoust., Speech, Signal Process.*, **33**, 806–811.

[46] M. I. Skolnik (1980) *Introduction to Radar Systems*, McGraw-Hill, New York, 442–446.

[47] B. D. Steinberg (1976) *Principle of Aperture and Array System Design*, Wiley, New York.

[48] H. Steyskal (1983) "Simple method for pattern nulling by phase perturbation", *IEEE Trans. Antennas Propag.*, **31**, 163–166.

[49] H. Steyskal, R. A. Shore, and R. L. Haupt (1986) "Methods for null control and their effects on the radiation pattern", *IEEE Trans. Antennas Propag.*, **34**, 404–409.

[50] G. V. Trunk, B. H. Cantrell, and W. B. Gordon (1979) "Bounds on elevation error estimates of a target in multipath", *IEEE Trans. Aerosp. Electron. Systems*, **15**, 883–887.

[51] T. B. Vu (1986) "Simultaneous nulling in sum and difference patterns by amplitude control", *IEEE Trans. Antennas Propag.*, **34**, 214–218.

[52] H. Wang and M. Kaveh (1985) "Coherent single-subspace averaging for the detection and estimation of angles of arrival of multiple wide-band sources", *IEEE Trans. Acoust., Speech, Signal Process.*, **33**, 823–831.

[53] W. D. White (1974) "Low-angle radar tracking in the presence of multipath", *IEEE Trans. Aerosp. Electron. Systems*, **10**, 835–853.

[54] W. D. White (1979) "Angular spectra in radar applications", *IEEE Trans. Aerosp. Electron. Systems*, **15**, 895–899.

[55] R. T. Williams, S. Prasad, A. K. Mahalanabis, and L. H. Sibul (1988) "An improved spatial smoothing technique for bearing estimation in a multipath environment", *IEEE Trans. Acoust., Speech, Signal Process.*, **36**, 425–432.

[56] M. Zoltowski and D. Stavrinides (1989) "Sensor array signal processing via a Procrustes rotations based eigenanalysis of the ESPRIT data pencil", *IEEE Trans. Acoust., Speech, Signal Process.*, **37**, 832–861.

[57] M. Zoltowski and T. Lee (1991) "Maximum likelihood based sensor array signal processing in the beamspace domain for low-angle radar tracking", *IEEE Trans. Acoust., Speech, Signal Process.*, **39**, 656–671.

[58] M. Zoltowski and T. Lee (1991) "Beamspace ML Bearing Estimation Incorporating Low-Angle Geometry", *IEEE Trans. Aerosp. Electron. Systems*, **27**, 441–458.

[59] M. Zoltowski and T. Lee (1991) "Interference cancellation matrix beamforming for 3-D beamspace ML/MUSIC bearing estimation", *IEEE Trans. Acoust., Speech, Signal Process.*, **39**, 1858–1876.

6

THE MAP ALGORITHM FOR DIRECTION OF ARRIVAL ESTIMATION IN UNKNOWN COLORED NOISE

JAMES P. REILLY AND K. M. WONG
Communications Research Laboratory,
McMaster University,
Hamilton, Ontario, Canada

6-1 INTRODUCTION

Array signal processing has found important applications in diverse areas such as radar, sonar, communications, and seismic explorations. A problem central to sensor array processing is the estimation of the directions of arrival (DOA) of the signals. Recently, high resolution methods for this type of estimation have captured the attention of many researchers [1–10]. Most of these methods require knowledge of the number of incident signals. Since this quantity is usually unknown, a detection procedure [11–17] generally precedes the DOA estimation. Furthermore, high resolution methods for DOA estimation usually assume that the background noise is isotropic (spatially uncorrelated) or that the covariance matrix of the noise is known. In practice, however, these assumptions are often not valid. This has led to the consideration of methods which account for the effects of correlated

---

Adaptive Radar Detection and Estimation, Edited by Simon Haykin and Allan Steinhardt.
ISBN 0-471-54468-X

noise fields. One method in this respect is the covariance differencing technique due to Paulraj and Kailath [18]. In this method, two measurements of the array covariance are required; It is assumed that the unknown noise field remains spatially invariant, while the signal field undergoes some change between the two measurements. Attempts are then made to subtract out the unknown noise covariance. However, in many practical situations, only one measurement of the covariance is available. Moreover, in many applications such as sonar, spatial stationarity is often violated in a gross sense by geographic location, and in a finer sense by the proximity and state of the surface and bottom boundaries.

In this chapter, we propose an alternative approach to DOA estimation of signals in noise with unknown spatial covariance structure. The signal model we consider here consists of an N-dimensional complex data vector $\mathbf{x}(m) \in \mathscr{C}^N$ which represents the data received by a linear array of N sensors at the mth snapshot. The data vector is composed of plane-wave incident narrowband signals each of angular frequency ω_0 from K distinct sources embedded in Gaussian noise. The received data vector at the mth snapshot can be written as

$$\mathbf{x}(m) = \mathbf{D}(\boldsymbol{\phi})\mathbf{a}(m) + \boldsymbol{\nu}(m)\,, \quad m = 1, \ldots, M \tag{6-1}$$

where M is the total number of snapshots observed. In particular, when the array elements are uniformly spaced (Fig. 6-1), $L_0 \leq \pi c/\omega_0$ is the distance between two sensors, where c is the velocity of propagation. Each of the

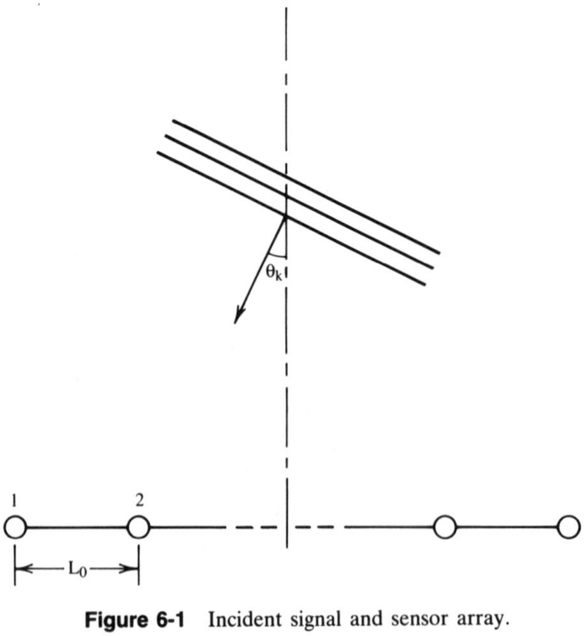

Figure 6-1 Incident signal and sensor array.

t that the noise $\boldsymbol{\nu}(m)$ is Gaussian and is independent from nother, the projected data will also be Gaussian such that the true values of signal directions, its conditional PDF is

$$^{1}) = \pi^{-MN}\{\det(\mathbf{S}^{-1})\}^{M} \exp\left[-\sum_{m=1}^{M} (\mathbf{y}^{H}(m)\mathbf{S}^{-1}\mathbf{y}(m))\right] \quad (6\text{-}18a)$$

$$= \pi^{-MN}\{\det(\mathbf{S}^{-1})\}^{M} \exp\left[-\mathrm{tr}\{\mathbf{S}^{-1}(M\hat{\mathbf{S}})\}\right] \quad (6\text{-}18b)$$

$$\hat{\mathbf{S}} = \frac{1}{M} \sum_{m=1}^{M} \mathbf{y}(m)\mathbf{y}^{H}(m) \quad (6\text{-}19)$$

Bayes' rule so that

$$(\boldsymbol{\phi}, \mathbf{S}^{-1} \mid \mathbf{Y}) = p(\mathbf{Y} \mid \boldsymbol{\phi}, \mathbf{S}^{-1})p(\boldsymbol{\phi}, \mathbf{S}^{-1})/p(\mathbf{Y}) \quad (6\text{-}20)$$

tice that $p(\mathbf{Y})$ is completely independent of the signal and $\boldsymbol{\phi}$ and $\mathbf{S}^{-1}$, thus its contribution to the *a posteriori* PDF is normalizing constant. Therefore, to arrive at the *a posteriori* the signal direction parameter only, we can use the marginal obtained by integrating (6-20) with respect to $\mathbf{S}^{-1}$ so that

$$\boldsymbol{\phi} \mid \mathbf{Y}) \propto p(\boldsymbol{\phi}) \int_{-\infty}^{\infty} p(\mathbf{Y} \mid \boldsymbol{\phi}, \mathbf{S}^{-1})p(\mathbf{S}^{-1} \mid \boldsymbol{\phi})\, d\mathbf{S}^{-1} \quad (6\text{-}21a)$$

$$\propto \int_{-\infty}^{\infty} p(\mathbf{Y} \mid \boldsymbol{\phi}, \mathbf{S}^{-1})p(\mathbf{S}^{-1} \mid \boldsymbol{\phi})\, d\mathbf{S}^{-1} \quad (6\text{-}21b)$$

at (6-21b), we have assumed that all signal directions are To evaluate the integral in (6-21b) we must obtain an e *a priori* distribution $p(\mathbf{S}^{-1} \mid \boldsymbol{\phi})$. Now, $\mathbf{S}$ is the covariance ojected noise and since we assume that we know nothing we choose a noninformative *a priori* PDF [21] $p(\mathbf{S}^{-1} \mid \boldsymbol{\phi})$ so r ignorance of the noise environment. (A brief explanation ve *a priori* PDF is given in Appendix A). Jeffreys [19] l rule for obtaining the noninformative *a priori* PDF, which s follows:

ribution for a set of parameters is taken to be proportional root of the determinant of the information matrix.

ly this rule to obtain the *a priori* PDF $p(\mathbf{S}^{-1} \mid \boldsymbol{\phi})$. The

incident sources is embedded in Gaussian noise. If each of these incident plane-wave signals impinges on the array of sensors at an angle θ_k, $k = 1, \ldots, K$, to the normal of the array, then $\mathbf{D}(\boldsymbol{\phi})$ is the $N \times K$ matrix the kth column vector of which is

$$\mathbf{d}(\phi_k) = [1 \quad e^{j\phi_k} \quad e^{j2\phi_k} \quad \cdots \quad e^{j(N-1)\phi_k}]^T \quad (6\text{-}2)$$

with

$$\phi_k = \omega_0 L_0 \sin \theta_k / c \quad (6\text{-}3)$$

and where the superscript T denotes transposition of a vector or matrix. The quantity $\mathbf{a}(m)$ is the K-dimensional complex vector whose elements represent the phasors of the signals at the mth snapshot, that is,

$$\mathbf{a}(m) = [|a_1(m)|e^{j\alpha_1(m)} \quad |a_2(m)|e^{j\alpha_2(m)} \quad \cdots \quad |a_K(m)|e^{j\alpha_K(m)}]^T \quad (6\text{-}4)$$

The noise vector $\boldsymbol{\nu}(m)$ in (6-1) is a zero-mean Gaussian vector such that

$$E[\boldsymbol{\nu}(m)] = \mathbf{0} \quad (6\text{-}5a)$$

and

$$E[\boldsymbol{\nu}(m_1)\boldsymbol{\nu}(m_2)^H] = \delta_{m_1 m_2}\boldsymbol{\Sigma} \quad (6\text{-}5b)$$

where $\delta_{m_1 m_2}$ is the Kronecker delta and the superscript H denotes conjugate transposition of a vector or matrix, and $\boldsymbol{\Sigma}$ is an unknown positive definite Hermitian matrix.

We assume that K, the total number of signals arriving at the array is known, and that all signals come from distinct directions. Then the subspace spanned by the K distinct vectors $\mathbf{d}(\phi_1), \ldots, \mathbf{d}(\phi_K)$ is designated the signal subspace $\mathcal{S}$ whereas the $(N-K)$-dimensional orthogonal complement [22] of $\mathcal{S}$ is designated the noise space $\mathcal{N}$. We first examine the projection of the received data onto the noise subspace. From the assumption of Gaussian noise, the conditional probability density function (PDF) of the projected data given the unknown signal directions ϕ and the unknown covariance matrix of the noise is first formed. The *a posteriori* PDF of the signal directions is then obtained by assigning a noninformative *a priori* PDF to the unknown noise covariance matrix using Jeffreys's invariance theory [19]. By choosing the maximum *a posteriori* (MAP) estimate of the signal directions, we arrive at a simple and novel optimization criterion in the noise subspace. In Section 2, we present the mathematical development leading to this new MAP criterion. Section 3 discusses some properties of the criterion. Section 4 examines the numerical implementation of this new MAP DOA estimation method. Sections 5 and 6 examine the asymptotic properties and the theoretical variances of the MAP estimate. Various results comparing simulation and theoretical performances are given in Section 6. Conclusions are presented in Section 7.

6-2 MATHEMATICAL DEVELOPMENT OF THE NEW METHOD

Using the signal model in Section 1, we define the data matrix $\mathbf{X} \in \mathscr{C}^{N \times M}$ as

$$\mathbf{X} = [\mathbf{x}(1) \quad \mathbf{x}(2) \quad \cdots \quad \mathbf{x}(M)] \tag{6-6}$$

Since DOA estimation of signals is performed under the condition of unknown spatially correlated noise, it seems advisable for us to examine the components of the data vector in the noise subspace so that its structure may be revealed. Accordingly, we form an orthogonal projector onto the noise subspace such that

$$\mathbf{P}(\boldsymbol{\phi}) = \mathbf{I} - \mathbf{D}(\boldsymbol{\phi})[\mathbf{D}^H(\phi)\mathbf{D}(\boldsymbol{\phi})]^{-1}\mathbf{D}^H(\boldsymbol{\phi}) \tag{6-7}$$

where $\boldsymbol{\phi} \in \mathscr{R}^K$ is the signal direction vector given by

$$\boldsymbol{\phi} = [\phi_1 \quad \phi_2 \quad \cdots \quad \phi_K]^T \tag{6-8}$$

with K known. Since $\mathbf{P}(\boldsymbol{\phi})$ is an orthogonal projector which projects a vector in the N-dimensional space onto an $(N-K)$-dimensional subspace, we can then find a set of vectors $\{\mathbf{u}_{K+1}(\boldsymbol{\phi}), \ldots, \mathbf{u}_N(\boldsymbol{\phi})\}$ which forms an $N \times (N-K)$ matrix $\mathbf{U}_\nu(\boldsymbol{\phi})$, and is an orthonormal basis of the range subspace $\mathscr{N}$ of $\mathbf{P}(\boldsymbol{\phi})$. Thus, we can write

$$\mathbf{P}(\boldsymbol{\phi}) = \mathbf{U}_\nu(\boldsymbol{\phi})\mathbf{U}_\nu^H(\boldsymbol{\phi}) \tag{6-9}$$

Consider the projection of the data vector $\mathbf{x}(m)$ onto the $(N-K)$-dimensional range subspace of $\mathbf{P}(\boldsymbol{\phi})$ spanned by $\{\mathbf{u}_{K+1}(\boldsymbol{\phi}), \ldots, \mathbf{u}_N(\boldsymbol{\phi})\}$ such that

$$\mathbf{y}(m) = \mathbf{U}_\nu^H(\boldsymbol{\phi})\mathbf{x}(m) \tag{6-10}$$

Let the projected data vector matrix be given by

$$\mathbf{Y} = [\mathbf{y}(1) \quad \cdots \quad \mathbf{y}(M)] = \mathbf{U}_\nu^H(\phi)\mathbf{X} \tag{6-11}$$

The purpose of the above transformation is ultimately to remove the dependence of the data on the amplitude components $\mathbf{a}(m)$. We now proceed to examine the problem from a Bayesian point of view [21, 26]. The Bayes estimate is the estimate that minimizes the expected cost. It is reasonable, in general, to assume that the cost depends only on the error of the estimate. Thus, if we define this error as

where $\hat{\boldsymbol{\phi}}$ is the estimated sign
average of the cost function (
arbitrary since it is a meas
subjective quality. The most
mean-square error, (ii) the
function. In this paper, we e

$C(\Delta$

where ε is a positive quantity

$E[C(\Delta\boldsymbol{\phi})] \propto \int_{-\infty}^{\infty}\int_{-\infty}^{\infty} C(\Delta\boldsymbol{\phi})p($

$= \int_{-\infty}^{\infty}\Big[1 - \int_{|\Delta\boldsymbol{\phi}|\le\varepsilon/2}$

Since $p(\mathbf{Y})$ is everywhere p
ing the inner integral which
a posteriori PDF $p(\boldsymbol{\phi}\,|\,\mathbf{Y})$ is m

where the set Ω is defined as

$\Omega = \{\phi : |\phi$

since we have assumed that th
constraint given by (6-16) ensu
Equation (6-15) is called the
is to be employed in this cha
posteriori PDF $p(\boldsymbol{\phi}\,|\,\mathbf{Y})$, we f
projected data.
We note that when $\boldsymbol{\phi}$ assum
projected data in (6-10) becom

$\mathbf{y}(m$

We therefore let

$\mathbf{S} =$

where $\boldsymbol{\Sigma}$ is defined in (6-5b). E
notation, $\mathbf{S}$ is a function of $\boldsymbol{\phi}$.

Due to the fa
one snapshot to
when $\boldsymbol{\phi}$ assumes
given by

$p(\mathbf{Y}\,|\,\boldsymbol{\phi}, \mathbf{S}$

where

We now appl

In (6-20) we no
noise paramete
no more than a
PDF containing
density function

p

where, to arriv
equally likely.
expression for t
matrix of the p
about the noise
that it reflects c
of noninformat
derived a gener
may be stated
The prior dis
to the square
We now ap

log-likelihood function is the log of the conditioned PDF of the projected data **Y** such that

$$\begin{aligned} L &= \log p(\mathbf{Y} \mid \boldsymbol{\phi}, \mathbf{S}^{-1}) \\ &= -MN \log \pi + M \log[\det(\mathbf{S}^{-1})] - \sum_{m=1}^{M} (\mathbf{y}^H(m)\mathbf{S}^{-1}\mathbf{y}(m)) \end{aligned} \tag{6-22}$$

The Fisher information matrix is then given by

$$\mathbf{J}_{S^{-1}} = -E[\nabla_S^{-1} \cdot \nabla_S^{-1} L] \tag{6-23}$$

where the matrix operator ∇_S^{-1} is defined such that the (l, m)th element is given by

$$\nabla_{s^{lm}} = \frac{1}{2}\left(\frac{\partial}{\partial s_R^{lm}} - j\frac{\partial}{\partial s_I^{lm}}\right) \tag{6-24}$$

with s^{lm} being the (l, m)th element of $\mathbf{S}^{-1}$ having s_R^{lm} and s_I^{lm} as the real and imaginary parts. We note that $\mathbf{J}_{S^{-1}}$ is an $(N-K)^2 \times (N-K)^2$ matrix. It is shown in Appendix A that

$$\det(\mathbf{J}_{S^{-1}}) = M\{\det(\mathbf{S}^{-1})\}^{-2(N-K)} \tag{6-25}$$

so that when Jeffreys's rule is applied, the noninformative *a priori* PDF of the projected noise covariance matrix can be written as

$$p(\mathbf{S}^{-1} \mid \boldsymbol{\phi}) \propto \{\det(\mathbf{S}^{-1})\}^{-(N-K)} \tag{6-26}$$

Substituting (6-18b) and (6-26) into (6-21b), the *a posteriori* PDF of the signal direction parameters becomes

$$p(\boldsymbol{\phi} \mid \mathbf{Y}) \propto \int_{-\infty}^{\infty} \{\det(\mathbf{S}^{-1})\}^{M} \exp[-\mathrm{tr}\{\mathbf{S}^{-1}(M\hat{\mathbf{S}})\}]\{\det(\mathbf{S}^{-1})\}^{-(N-K)}\, d\mathbf{S}^{-1} \tag{6-27a}$$

$$= \{\det(M\hat{\mathbf{S}})\}^{-M} \int_{-\infty}^{\infty} \{\det(M\hat{\mathbf{S}})\}^{M}\{\det(\mathbf{S}^{-1})\}^{M-N+K}\, \mathrm{etr}[-\mathbf{S}^{-1}(M\hat{\mathbf{S}})]\, d\mathbf{S}^{-1} \tag{6-27b}$$

where etr(·) denotes exp[tr(·)]. The integrand in (6-27b) can be recognized to be the complex Wishart distribution [27] with the role of $\mathbf{S}^{-1}$ and $M\hat{\mathbf{S}}$ reversed, and hence the integral is a constant. Therefore,

$$p(\boldsymbol{\phi} \mid \mathbf{Y}) \propto \{\det(\hat{\mathbf{S}})\}^{-M} = \{\det[\mathbf{U}_\nu^H(\boldsymbol{\phi})\hat{\mathbf{R}}_M\mathbf{U}_\nu(\boldsymbol{\phi})]\}^{-M} \tag{6-28}$$

where

$$\hat{\mathbf{R}}_M = \frac{1}{M}\sum_{m=1}^{M} \mathbf{x}(m)\mathbf{x}^H(m) \tag{6-29}$$

The MAP estimate seeks the maximum of (6-28) with respect to $\boldsymbol{\phi}$. Equivalently, we can write the MAP criterion as

$$\min_{\boldsymbol{\phi}\in\Omega} \det[\mathbf{U}_\nu^H(\boldsymbol{\phi})\hat{\mathbf{R}}_M\mathbf{U}_\nu(\boldsymbol{\phi})] \equiv \min_{\boldsymbol{\phi}\in\Omega} \prod_{n=1}^{N-K} \hat{\lambda}_n(M) \tag{6-30}$$

where $\hat{\lambda}_1(M) \geq \hat{\lambda}_2(M) \geq \cdots \geq \hat{\lambda}_{N-K}(M)$ are the eigenvalues of $\mathbf{U}_\nu^H(\boldsymbol{\phi})\hat{\mathbf{R}}_M\mathbf{U}_\nu(\boldsymbol{\phi})$ and the set Ω is defined in (6-16). Because of the relationship between $\mathbf{P}(\boldsymbol{\phi})$ and $\mathbf{U}_\nu(\boldsymbol{\phi})$ as expressed in (6-9), $\{\hat{\lambda}_n(M)\}$, $n = 1, \ldots, N-K$ are also the nonzero eigenvalues of $\mathbf{P}(\boldsymbol{\phi})\hat{\mathbf{R}}_M\mathbf{P}(\boldsymbol{\phi})$.

It is interesting to examine the Bayesian approach in the case when the noise is known to be spatially white, that is, when

$$\boldsymbol{\Sigma} = \sigma_\nu^2\mathbf{I} \tag{6-31}$$

In this case, we can write

$$p(\mathbf{S}^{-1} \mid \boldsymbol{\phi}) = \delta(\mathbf{S}^{-1} - \sigma_\nu^{-2}\mathbf{U}_\nu^H(\boldsymbol{\phi})\mathbf{I}^{-1}\mathbf{U}_\nu(\boldsymbol{\phi})) = \delta(\mathbf{S}^{-1} - \sigma_\nu^{-2}\mathbf{I}) \tag{6-32}$$

where $\delta(\cdot)$ is an $(N-K)^2$-dimensional impulse function, and σ_ν^2 is the noise power.

Using (6-18b) and (6-32) in (6-27a), the *a posteriori* PDF becomes

$$\begin{aligned} p(\boldsymbol{\phi} \mid \mathbf{Y}) &\propto \int_{-\infty}^{\infty} \{\det(\mathbf{S}^{-1})\}^M \operatorname{etr}[-\mathbf{S}^{-1}(M\hat{\mathbf{S}})]\delta(\mathbf{S}^{-1} - \sigma_\nu^{-2}\mathbf{I})\, d\mathbf{S}^{-1} \\ &\propto \operatorname{etr}[-\sigma_\nu^{-2}(M\hat{\mathbf{S}})] \end{aligned} \tag{6-33}$$

where, in the last step of (6-33), we have used the result of the convolution with an impulse. Applying the MAP estimation, the criterion can be written as

$$\min_{\boldsymbol{\phi}\in\Omega} \operatorname{etr}[\mathbf{U}_\nu^H(\boldsymbol{\phi})\hat{\mathbf{R}}_M\mathbf{U}_\nu(\boldsymbol{\phi})] \equiv \min_{\boldsymbol{\phi}\in\Omega} \operatorname{tr}[\mathbf{U}_\nu^H(\boldsymbol{\phi})\hat{\mathbf{R}}_M\mathbf{U}_\nu(\boldsymbol{\phi})] \equiv \min_{\boldsymbol{\phi}\in\Omega}[\operatorname{tr}(\mathbf{P}(\boldsymbol{\phi})\hat{\mathbf{R}}_M)] \tag{6-34}$$

Equation (6-34) can be recognized to be the deterministic maximum likelihood (ML) criterion for DOA estimation in spatially white noise [3, 9, 10, 36, 37]. Thus, the MAP criterion of (6-30) can be viewed as a generalized approach which includes ML estimation in spatially white noise as a particular case.

Now we can summarize rather simply the procedure of the MAP method as follows:

1. Form the data sample covariance matrix $\hat{\mathbf{R}}_M$.
2. Form the matrix $\mathbf{U}_\nu(\boldsymbol{\phi})$ which contains K unknown angles of arrival as indicated by (6-9).
3. Perform a K-dimensional optimization such that the determinant of $\mathbf{U}_\nu^H(\boldsymbol{\phi})\hat{\mathbf{R}}_M\mathbf{U}_\nu(\boldsymbol{\phi})$ is minimized as expressed in (6-30).

Efficient ways of performing the MAP estimation are discussed in Section 6-4.

6-3 PROPERTIES OF THE MAP OBJECTIVE FUNCTION

We define the objective function det $[[\mathbf{U}_\nu^H(\phi)\hat{\mathbf{R}}_M\mathbf{U}_\nu(\phi)]$ to be minimized according to (6-30) as $g(\phi, \hat{\mathbf{R}}_M)$. We have seen in the previous section that minimizing $g(\phi, \hat{\mathbf{R}}_M)$ is equivalent to minimizing the product of the nonzero eigenvalues of the $N \times N$ matrix $\mathbf{P}(\phi)\hat{\mathbf{R}}_M\mathbf{P}(\phi)$. We now define [29] the $(N-K)$th trace of an $N \times N$ matrix $\mathbf{B}$ as the sum of the $(N-K)$th order principal minors of $\mathbf{B}$, that is,

$$\mathrm{tr}_{N-K}(\mathbf{B}) = \sum_{\vartheta} \det(\mathbf{B}_{i_1, \ldots, i_K}) \tag{6-35}$$

where $\mathbf{B}_{i_1, \ldots, i_K}$ denotes a principal submatrix by deleting the i_1th, i_2th, . . . , i_Kth rows and columns of $\mathbf{B}$, and ϑ denotes the combination set of $i_1, \ldots, i_K$. In particular, $\mathrm{tr}_1(\mathbf{B}) = \mathrm{tr}(\mathbf{B})$ and $\mathrm{tr}_N(\mathbf{B}) = \det(\mathbf{B})$.

Let the characteristic polynomial of an $N \times N$ matrix $\mathbf{B}$ be written as

$$h(\lambda) = \det \mathbf{B}(\lambda) \overset{\Delta}{=} \det(\mathbf{B} - \lambda\mathbf{I}) = (-1)^N(\lambda^N + c_1\lambda^{N-1} + \cdots + c_{N-1}\lambda + c_N) \tag{6-36}$$

If we put $\lambda = 0$ in (6-36), we have

$$h(0) = (-1)^N c_N = \det(\mathbf{B}) = \mathrm{tr}_N(\mathbf{B}) \tag{6-37}$$

Differentiating (6-36) with respect to λ, and putting $\lambda = 0$, we have

$$\begin{aligned} \frac{d}{d\lambda} \det \mathbf{B}(\lambda)\big|_{\lambda=0} &= \sum_i \sum_k \frac{\partial \det \mathbf{B}(\lambda)}{\partial b_{ik}(\lambda)} \frac{db_{ik}(\lambda)}{d\lambda}\bigg|_{\lambda=0} \\ &= \sum_i \frac{\partial \det \mathbf{B}(\lambda)}{\partial b_{ii}(\lambda)} \frac{db_{ii}(\lambda)}{d\lambda}\bigg|_{\lambda=0} \\ &= -\sum_i \det \mathbf{B}_i(\lambda)\big|_{\lambda=0} \\ &= -\mathrm{tr}_{N-1}(\mathbf{B}) \end{aligned} \tag{6-38}$$

where b_{ik} denotes the (ik)th element of $\mathbf{B}$.

Hence we have

$$(-1)^{N-1} c_{N-1} = \mathrm{tr}_{N-1}(\mathbf{B}) \tag{6-39}$$

Further differentiation with respect to λ and evaluating the result at $\lambda = 0$ results in

$$(-1)^{N-K} c_{N-K} = \mathrm{tr}_{N-K}(\mathbf{B}) \tag{6-40}$$

If $\lambda_1, \ldots, \lambda_N$ are the roots of $h(\lambda)$, then from our knowledge of the theory of polynomials, the $(N-K)$th coefficient c_{N-K} of $h(\lambda)$ and the roots are related by

$$(-1)^{N-K} c_{N-K} = \sum_{N-K} \left(\prod \lambda_i \right) \tag{6-41}$$

where $\Sigma_{N-K} (\Pi\, \lambda_i)$ denotes the sum of the products of the roots taken $N-K$ at a time. If we replace $\mathbf{B}$ by $\mathbf{P}\hat{\mathbf{R}}_M\mathbf{P}$ in (6-36), and apply (6-41), we see that since there are only $N-K$ nonzero eigenvalues of $\mathbf{P}\hat{\mathbf{R}}_M\mathbf{P}$, then there is only one nonzero product of roots taken $N-K$ at a time. Thus, we can write, for $\mathbf{B} = \mathbf{P}\hat{\mathbf{R}}_M\mathbf{P}$,

$$\prod_{n=1}^{N-K} \hat{\lambda}_n(M) = (-1)^{N-K} c_{N-K} \tag{6-42}$$

where c_{N-K} is the $(N-K)$th coefficient of the characteristic polynomial of $\mathbf{P}\hat{\mathbf{R}}_M\mathbf{P}$ expanded in the same way as in (6-36).

Writing $\mathbf{B} = \mathbf{P}\hat{\mathbf{R}}_M\mathbf{P}$ in (6-40) and comparing with (6-42), we obtain

$$g(\phi, \hat{\mathbf{R}}_M) = \prod_{n=1}^{N-K} \hat{\lambda}_n(M) = \mathrm{tr}_{N-K}(\mathbf{P}\hat{\mathbf{R}}_M\mathbf{P}) \tag{6-43}$$

The property shown in (6-43) is convenient for performance analysis and will be used later. Furthermore, it also provides us with a geometrical interpretation of the criterion used by the MAP estimate.

Let the projected data be represented by

$$\frac{1}{\sqrt{M}} \mathbf{P}(\boldsymbol{\phi})\mathbf{X} = [\boldsymbol{\eta}_1 \quad \boldsymbol{\eta}_2 \quad \cdots \quad \boldsymbol{\eta}_N]^H \tag{6-44}$$

where $\boldsymbol{\eta}_n^H$ is an M-dimensional operated data row vector being the operated output, at all M snapshots, of the nth sensor. Since $\mathbf{P}(\boldsymbol{\phi})$ projects the N-dimensional vector $\mathbf{x}(m)$ onto an $(N-K)$-dimensional hyperplane, the vectors $\boldsymbol{\eta}_1, \ldots, \boldsymbol{\eta}_N$ are linearly dependent, spanning an $(N-K)$-dimensional hyperplane.

Now if we form the matrix

$$\frac{1}{M}\mathbf{P}\hat{\mathbf{R}}_M\mathbf{P} = [\boldsymbol{\eta}_1 \quad \boldsymbol{\eta}_2 \quad \cdots \quad \boldsymbol{\eta}_N]^H[\boldsymbol{\eta}_1 \quad \boldsymbol{\eta}_2 \quad \cdots \quad \boldsymbol{\eta}_N] \tag{6-45}$$

then it can easily be seen [34] that each of its $(N-K)$-dimensional principal minors (formed by deleting K of the corresponding rows and columns) is equal to the square of the volume of the $(N-K)$-dimensional parallelepipeds whose edges are the $(N-K)$ vectors $\{\boldsymbol{\eta}_i\}$ involved in the principal minor. Therefore, the minimization of the objective function $g(\boldsymbol{\phi}, \hat{\mathbf{R}}_M)$ can be interpreted as the minimization of the sum of the square of the volumes of all the parallelepipeds whose edges are formed by taking all possible combinations of $N-K$ of the vectors $\{\boldsymbol{\eta}_i; i=1, \ldots, N\}$. Since the volume of a parallelepiped not only depends on the length (strength) of the vectors, but also on the angles between them, we can see that the MAP criterion seeks to locate a noise subspace onto which the projected data vectors have minimum volume. This implies the projected data vectors have not only minimum strength but also minimum angles between them; that is, maximum correlation. It is the location of this subspace which caters for the correlation of the projected noise vectors that enables the MAP approach to function even under unknown correlated noise.

6-4 NUMERICAL IMPLEMENTATION OF THE MAP METHOD

The MAP criterion, as discussed in Section 6-2, seeks to minimize the objective function

$$g(\phi, \hat{\mathbf{R}}_M) = \det[\mathbf{U}_\nu^H(\boldsymbol{\phi})\hat{\mathbf{R}}_M\mathbf{U}_\nu(\boldsymbol{\phi})] = \prod_{n=1}^{N-K} \hat{\lambda}_n(M) \tag{6-46}$$

which can be equivalently written as

$$\min_{\phi\in\Omega} \log g(\boldsymbol{\phi}, \hat{\mathbf{R}}_M) \equiv \min_{\phi\in\Omega} \log\det[\mathbf{U}_\nu^H(\boldsymbol{\phi})\mathbf{X}\mathbf{X}^H\mathbf{U}_\nu(\boldsymbol{\phi})] \tag{6-47}$$

where $\mathbf{X}$ is the data matrix defined in (6-6). The matrix $\mathbf{U}_\nu(\boldsymbol{\phi})$ contains orthonormal vectors spanning the range subspace of $\mathbf{P}(\boldsymbol{\phi})$. To obtain the matrix $\mathbf{U}_\nu(\boldsymbol{\phi})$, we perform a QR decomposition [32] on the direction matrix $\mathbf{D}(\phi)$ defined in (6-2) such that

$$\mathbf{D}(\boldsymbol{\phi}) = \mathbf{U}(\boldsymbol{\phi})\begin{bmatrix}\boldsymbol{\Gamma}\\ \mathbf{0}_\Gamma\end{bmatrix} \tag{6-48}$$

where $\boldsymbol{\Gamma} \in \mathscr{C}^{K\times K}$ is upper triangular, $\mathbf{0}_\Gamma$ is an $(N-K)\times K$ null matrix, and $\mathbf{U}(\boldsymbol{\phi}) \in \mathscr{C}^{N\times N}$ is unitary with columns $\mathbf{u}_1(\boldsymbol{\phi}), \ldots, \mathbf{u}_N(\boldsymbol{\phi})$ and can be partitioned such that

$$\mathbf{U}(\boldsymbol{\phi}) = [\mathbf{U}_s(\boldsymbol{\phi}) \mid \mathbf{U}_\nu(\boldsymbol{\phi})] = [\mathbf{u}_1(\boldsymbol{\phi}) \cdots \mathbf{u}_K(\boldsymbol{\phi}) \mid \mathbf{u}_{K+1}(\boldsymbol{\phi}) \cdots \mathbf{u}_N(\boldsymbol{\phi})] . \tag{6-49}$$

After obtaining $\mathbf{U}_\nu(\boldsymbol{\phi})$ we can proceed on to evaluate the objective function. Suppose we now perform an LQ decomposition [32] on the matrix $\mathbf{U}_\nu^H(\boldsymbol{\phi})\mathbf{X}$ so that

$$\mathbf{U}_\nu^H(\boldsymbol{\phi})\mathbf{X} = [\mathbf{L} \quad \mathbf{0}_L]\mathbf{W} \tag{6-50}$$

where $\mathbf{L} \in \mathscr{C}^{(N-K)\times(N-K)}$ is lower triangular, $\mathbf{0}_L$ is an $(N-K)\times(M-N+K)$ null matrix, and $\mathbf{W} \in \mathscr{C}^{M\times M}$ is unitary. Then,

$$\mathbf{U}_\nu^H(\boldsymbol{\phi})\mathbf{X}\mathbf{X}^H\mathbf{U}_\nu(\boldsymbol{\phi}) = \mathbf{L}\mathbf{L}^H \tag{6-51}$$

From (6-51), it is clear that

$$\det\,[\mathbf{U}_\nu^H(\boldsymbol{\phi})\mathbf{X}\mathbf{X}^H\mathbf{U}_\nu(\boldsymbol{\phi})] = \prod_{i=1}^{N-K} |l_{ii}|^2 \tag{6-52}$$

where l_{ii} are the diagonal elements of $\mathbf{L}$. Since the diagonal elements of $\mathbf{L}$ can always be changed into positive real numbers by scaling them with appropriate complex constants, the objective function for optimization can be expressed as

$$\min_{\boldsymbol{\phi}\in\Omega} \log g(\boldsymbol{\phi}, \hat{\mathbf{R}}_M) = \min_{\boldsymbol{\phi}\in\Omega}\left(\sum_{i=1}^{N-K} \log l_{ii}\right) \tag{6-53}$$

Equations (6-48), (6-49), (6-50), and (6-53) represent an alternative method for evaluating the objective function and can be realized using a systolic array structure [33, 34], to be discussed shortly. This method avoids the explicit evaluation of the determinant of $\mathbf{U}_\nu^H(\boldsymbol{\phi})\hat{\mathbf{R}}_M\mathbf{U}_\nu(\boldsymbol{\phi})$ and is more efficient, since explicit computation of $\mathbf{W}$ may be avoided.

However, to locate $\hat{\boldsymbol{\phi}}$ by conventional numerical means, at which (6-53) is a minimum, we have to employ either a gradient method or a global search technique. The derivative $\partial l_{ii}/\partial \phi_k$ is not easy to evaluate and thus the gradient method may not be efficient computationally. On the other hand, if the range of the search is large, a search method can also be computationally intensive.

The following is a computationally efficient method to evaluate the approximate value of the cost function and to locate $\tilde{\hat{\boldsymbol{\phi}}}$ at which this approximate cost function is a minimum.

Let $\hat{\mathbf{R}}_M$ be resolved such that

$$\hat{\mathbf{R}}_M = [\mathbf{U}_s(\boldsymbol{\phi}) \quad \mathbf{U}_\nu(\boldsymbol{\phi})]\begin{bmatrix} \hat{\mathbf{Z}}_{ss} & \hat{\mathbf{Z}}_{s\nu} \\ \hat{\mathbf{Z}}_{\nu s} & \hat{\mathbf{Z}}_{\nu\nu} \end{bmatrix}\begin{bmatrix} \mathbf{U}_s^H(\boldsymbol{\phi}) \\ \mathbf{U}_\nu^H(\boldsymbol{\phi}) \end{bmatrix} \tag{6-54}$$

Then, the matrix $\mathbf{U}_\nu^H(\boldsymbol{\phi})\hat{\mathbf{R}}_M\mathbf{U}_\nu(\boldsymbol{\phi})$ in (6-46) becomes

$$\mathbf{U}_\nu^H(\boldsymbol{\phi})\hat{\mathbf{R}}_M\mathbf{U}_\nu(\boldsymbol{\phi}) = \hat{\mathbf{Z}}_{\nu\nu} \tag{6-55}$$

But

$$\begin{aligned}\log g(\boldsymbol{\phi}, \hat{\mathbf{R}}_M) &= \log \prod_{n=1}^{N-K} \hat{\lambda}_n(M) \\ &= \sum_{n=1}^{N-K} \log \hat{\lambda}_n(M) = \operatorname{tr} \log\{\operatorname{diag}[\hat{\lambda}_1(M), \ldots, \hat{\lambda}_{N-K}(M)]\}\end{aligned} \tag{6-56}$$

and diag $[\hat{\lambda}_1(M), \ldots, \hat{\lambda}_{N-K}(M)]$ is the eigenvalue matrix of $\mathbf{U}_\nu^H(\boldsymbol{\phi})\hat{\mathbf{R}}_M\mathbf{U}_\nu(\boldsymbol{\phi})$. Hence, (6-55) and (6-56) yield

$$\log g(\boldsymbol{\phi}, \hat{\mathbf{R}}_M) = \operatorname{tr}\{\log[\mathbf{U}_\nu^H(\boldsymbol{\phi})\hat{\mathbf{R}}_M\mathbf{U}_\nu(\boldsymbol{\phi})]\} = \operatorname{tr}\{\log \hat{\mathbf{Z}}_{\nu\nu}\} \tag{6-57}$$

On the other hand, if the vector $\boldsymbol{\phi}$ is close to the true direction $\boldsymbol{\phi}_0$, and if the signals are situated in the part of the noise spatial spectrum in which gradient of the spectrum is not too steep, then [35] $\hat{\mathbf{Z}}_{s\nu} = \hat{\mathbf{Z}}_{\nu s}^H$ is negligible compared to $\hat{\mathbf{Z}}_{ss}$ and $\hat{\mathbf{Z}}_{\nu\nu}$. Thus we can write

$$\begin{aligned}\operatorname{tr}\{\mathbf{U}_\nu^H \log \hat{\mathbf{R}}_M)\mathbf{U}_\nu\} &\approx \operatorname{tr}\left\{\mathbf{U}_\nu^H \log\left([\mathbf{U}_s \quad \mathbf{U}_\nu]\begin{bmatrix}\hat{\mathbf{Z}}_{ss} & \mathbf{0}\\ \mathbf{0} & \hat{\mathbf{Z}}_{\nu\nu}\end{bmatrix}\begin{bmatrix}\mathbf{U}_s^H\\ \mathbf{U}_\nu^H\end{bmatrix}\right)\mathbf{U}_\nu\right\} \\ &= \operatorname{tr}\left\{\mathbf{U}_\nu^H[\mathbf{U}_s \quad \mathbf{U}_\nu]\left(\log\begin{bmatrix}\hat{\mathbf{Z}}_{ss} & \mathbf{0}\\ \mathbf{0} & \hat{\mathbf{Z}}_{\nu\nu}\end{bmatrix}\right)\begin{bmatrix}\mathbf{U}_s^H\\ \mathbf{U}_\nu^H\end{bmatrix}\mathbf{U}_\nu\right\} \\ &= \operatorname{tr}(\log \hat{\mathbf{Z}}_{\nu\nu})\end{aligned} \tag{6-58}$$

Comparing (6-57) and (6-58) we see that the cost function can be approximated by

$$\log g(\boldsymbol{\phi}, \hat{\mathbf{R}}_M) \approx \operatorname{tr}\{\mathbf{U}_\nu^H(\log \hat{\mathbf{R}}_M)\mathbf{U}_\nu\} = \operatorname{tr}\{\mathbf{P}(\log \hat{\mathbf{R}}_M)\} \tag{6-59}$$

The last term in (6-59) is reminiscent of the maximum likelihood (ML) DOA estimation criterion [3, 9, 10]. The difference is that $\hat{\mathbf{R}}_M$ is replaced by $\log \hat{\mathbf{R}}_M$ in the approximated MAP criterion.

The minimization of the cost function in the form of (6-59) has been studied by various authors [4, 9, 10, 36, 37, 38]. A popular and effective algorithm for solving nonlinear least square problems is the Newton method [39] for which computation of both first and second order derivatives of the cost function are required. The first and second order derivatives of the cost function in the form of tr{**PA**} where **A** is Hermitian have been derived explicitly [40, 41], and thus the Newton method can be applied directly to the approximate cost function in (6-59). The initial estimates of $\boldsymbol{\phi}$ for the Newton method may be obtained in a manner similar to the alternating projection procedure as described in [10].

After obtaining $\tilde{\hat{\boldsymbol{\phi}}}$ which minimizes the approximate cost function of (6-59), we can then perform a search over a limited range for minimizing the exact cost function of (6-53). This local search can be carried out efficiently by a gradient-free type of algorithm such as the Nelder–Mead simplex method [43]. Since the search for $\hat{\boldsymbol{\phi}}$ which minimizes (6-53) is performed over a limited neighborhood of $\tilde{\hat{\boldsymbol{\phi}}}$ much saving in computation can be achieved over a global search for minimizing (6-53). Thus, the procedure for implementation of the MAP algorithm can be summarized as follows:

1. Obtain an initial estimate of $\boldsymbol{\phi}^{(0)}$ by the alternating projection method.
2. Use the Newton method to obtain $\tilde{\hat{\boldsymbol{\phi}}}$ which minimizes (6-59).
3. Perform a local search in the neighborhood of $\tilde{\hat{\boldsymbol{\phi}}}$ for $\hat{\boldsymbol{\phi}}$ which minimizes (6-53).

Although strictly speaking, (6-59) is a good approximation only when $\boldsymbol{\phi}$ is close to the true vector $\boldsymbol{\phi}_0$ and when the signals are coming from directions in the part of the noise spatial spectrum where the gradient is not too steep, in practice, this does not seem to be a harsh limitation. Extensive computer simulations have been carried out with signals situated at various parts of the colored noise spatial spectrum. It has been found that the above procedure invariably leads to an estimate $\hat{\boldsymbol{\phi}}$ close to the true value $\boldsymbol{\phi}_0$. The approximate estimate $\tilde{\hat{\boldsymbol{\phi}}}$ obtained by minimizing the approximate cost function is also in good agreement with the final estimate $\hat{\boldsymbol{\phi}}$ for a wide range of signal-to-noise ratios (SNRs).

Systolic Array Implementation. The main cost of solving the optimization problem associated with the MAP criterion is in the evaluation of the objective function. A parallel systolic array architecture for the calculation of the exact objective function (6-53) is now presented. This structure is comprised of three parts. In Part 1, the QR decomposition of the directional matrix $\mathbf{D}$ as indicated by (6-48) is performed. Part 2 applies the orthogonal transformations generated in Part 1 to the data matrix $\mathbf{X}$ to obtain $\bar{\mathbf{U}}_\nu^H \mathbf{X}$. In Part 3, an LQ decomposition of the matrix $\bar{\mathbf{U}}_\nu^H \mathbf{X}$ is performed and the outputs of Part 3 contain the elements l_{ii}, $i = 1, \ldots, N - K$, whose product forms the objective function.

The structure for Part 1 is shown in Fig. 6-2. This structure is a triangular array for QR decomposition proposed by Gentleman and Kung [33] and the functions of the cells are described in the figure. The boundary cells (represented by the circles) generate cosines and sines (denoted by c and s

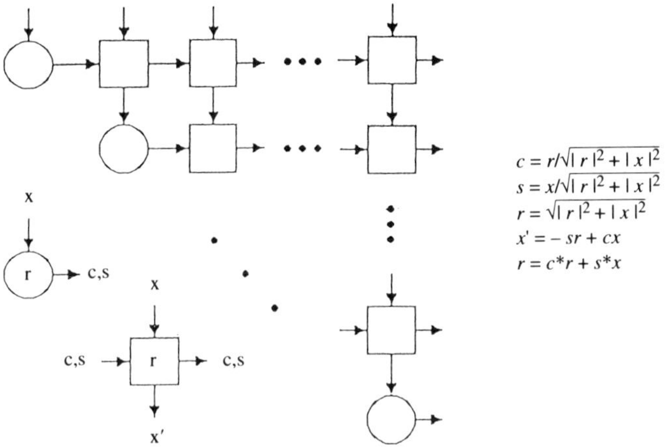

Figure 6-2 Structure of Part 1.

respectively) of the rotation angles, whereas the interior cells (represented by the squares) apply the rotations. The resulting triangular matrix $\mathbf{\Gamma}$ resides in the array and the information for the rotations are passed to Part 2 where they are applied to $\mathbf{X}$. Thus the function of cells in Part 2 is the same as that of the interior cells in Part 1, that is, Part 2 performs the following computation

$$\begin{bmatrix} \hat{\mathbf{U}}_s^H \\ \hat{\mathbf{U}}_\nu^H \end{bmatrix} \mathbf{X} = \begin{bmatrix} \hat{\mathbf{U}}_s^H \mathbf{X} \\ \hat{\mathbf{U}}_\nu^H \mathbf{X} \end{bmatrix} \tag{6-60}$$

The structure and cell function of Part 2 are shown in Fig. 6-3. Subsequently, $\hat{\mathbf{U}}_s^H \mathbf{X}$ stays in the rectangular array and the rows of $\hat{\mathbf{U}}_\nu^H \mathbf{X}$ come out from the bottom of Part 2 and are passed onto Part 3 where an LQ decomposition as indicated by (6-50) is performed. The structure of Part III and the functions of cells are given in Fig. 6-4. The boundary cells are simple reflectors. The interior cells (represented by the circles) have two states. Initially, they compute cosines and sines of the rotation angles (denoted by c and s respectively) and store them in the cells. After the initial state, the rotations are applied to input data from Part 2. To explain how it works, let us denote

$$\bar{\mathbf{U}}_\nu^H \mathbf{X} = \mathbf{L}^{(0)} = \begin{bmatrix} l_{1,1}^{(0)} & l_{1,2}^{(0)} & \cdots & l_{1,M}^{(0)} \\ l_{2,1}^{(0)} & l_{2,2}^{(0)} & \cdots & l_{2,M}^{(0)} \\ \vdots & \vdots & \cdots & \vdots \\ l_{N-K,1}^{(0)} & l_{N-K,2}^{(0)} & \cdots & l_{N-K,M}^{(0)} \end{bmatrix} \tag{6-61}$$

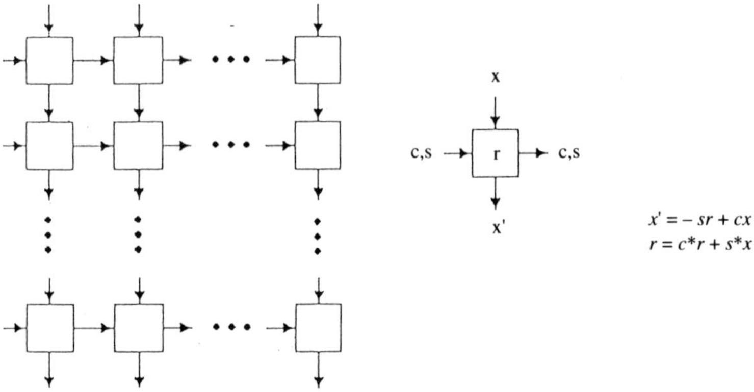

Figure 6-3 Structure of Part 2.

First, $l_{1,1}^{(0)}$ is passed to the right and rotations are computed using the pair of elements $(l_{1,1}^{(j-2)}, l_{1,j}^{(0)})$, $j = 2, \ldots, M$ to eliminate $l_{1,j}^{(0)}$. Then, $l_{1,1}^{(j-2)}$ is updated to $l_{1,1}^{(j-1)}$. The rotation parameters are stored in the cells. When the second row $(l_{2,1}^{(0)}, l_{2,2}^{(0)} \ \cdots \ l_{2,M}^{(0)})$ arrives, the cells in the first row have entered steady state and rotations are applied to the pair $(l_{2,1}^{(j-2)}, l_{2,j}^{(0)})$ which is then updated to $(l_{2,1}^{(j-1)}, l_{2,j}^{(1)})$ for $j = 2 \ \cdots \ M$. As the updated row $l_{2,2}^{(1)} \ \cdots \ l_{2,M}^{(1)}$ enters the cells in the second row, $l_{2,2}^{(1)}$ is passed to the right to generate rotations and for the elimination of $l_{2,2}^{(1)}$ for $j = 3, \ldots, M$. This procedure is repeated until all N rows are processed. The output of Part 3 is a lower triangular matrix:

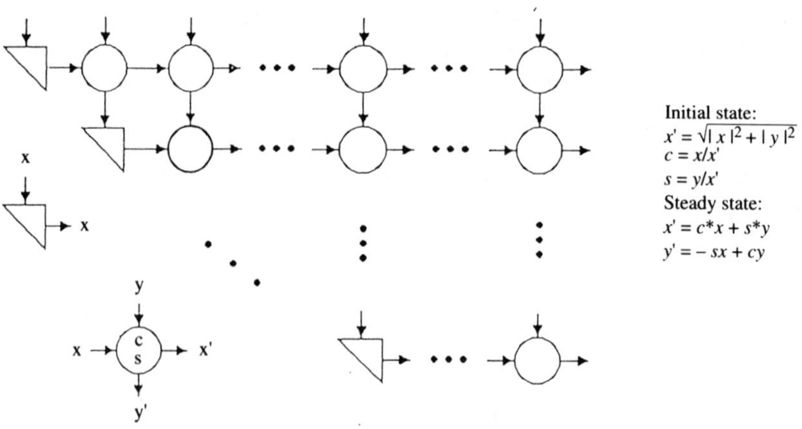

Figure 6-4 Structure of Part 3.

$$\mathbf{L} = \begin{bmatrix} l_{1,1} & 0 & \cdots & 0 & 0 & \cdots & 0 \\ l_{2,1} & l_{2,2} & \cdots & 0 & 0 & \cdots & 0 \\ \vdots & \vdots & \ddots & \vdots & \vdots & \cdots & \vdots \\ l_{N-K,1} & l_{N-K,2} & \cdots & l_{N-K,N-K} & 0 & \cdots & 0 \end{bmatrix} \tag{6-62}$$

The diagonal elements are then collected and the objective function can then be evaluated as indicated by (6-52). The overall structure combining Parts 1, 2, and 3 is shown in Fig. 6-5.

The calculation of the objective function may be updated to include new data as follows. Before the calculation begins, new snapshots of data in column format are fed into the **X**-data storage block from the right as shown in Fig. 6-5. Existing columns are shifted to the left, and the oldest data fall out the left side of the **X**-block.

It is of interest to note that the MUSIC algorithm, to be discussed further in Section 6-7, may also be implemented using systolic arrays. A systolic array for implementing the singular value decomposition, which renders numerically stable values of the eigenvalues and eigenvectors, is proposed by Brent and others [58, 59]. For uniformly spaced sensors, the denominator

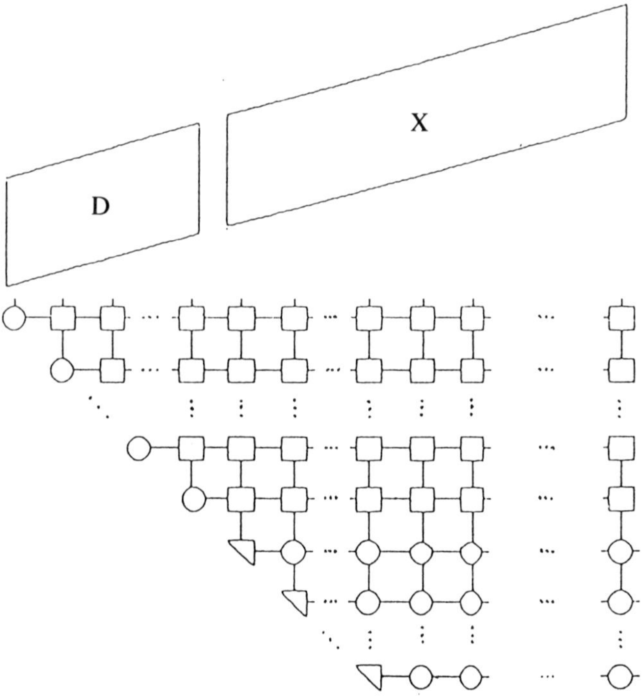

Figure 6-5 Overall structure.

of the MUSIC spatial spectrum given by (6-129) is a polynomial in $e^{j\phi}$. A systolic array for the evaluation of this polynomial is proposed in [60].

6-5 ASYMPTOTIC PROPERTIES OF THE MAP ESTIMATE

In this section we examine some asymptotic properties of the MAP estimate. Let the true covariance matrix of the received data $\mathbf{x}$ be $\mathbf{R}$, that is,

$$\mathbf{R} = E[\mathbf{x}\,\mathbf{x}^H] \tag{6-63}$$

With the sampled covariance matrix over M snapshots denoted by $\hat{\mathbf{R}}_M$ and defined as in (6-29), it is well known [48] that

$$\lim_{M\to\infty} \hat{\mathbf{R}}_M \underset{\text{a.s.}}{=} \mathbf{R} \tag{6-64}$$

where $\underset{\text{a.s.}}{=}$ denotes equality with probability 1. Since $g(\boldsymbol{\phi}, \hat{\mathbf{R}}_M)$ is measurable in $\hat{\mathbf{R}}_M$ and continuous in $\boldsymbol{\phi} \in \Omega$, then we conclude that

$$\lim_{M\to\infty} g(\boldsymbol{\phi}, \hat{\mathbf{R}}_M) \underset{\text{a.s.}}{=} g(\boldsymbol{\phi}, \mathbf{R}) \tag{6-65}$$

Furthermore, as $M \to \infty$, with probability 1, $g(\boldsymbol{\phi}, \hat{\mathbf{R}}_M)$ converges uniformily to $g(\boldsymbol{\phi}, \mathbf{R})$, that is,

$$\lim_{M\to\infty} \sup_{\boldsymbol{\phi}\in\Omega} |g(\boldsymbol{\phi}, \hat{\mathbf{R}}_M) - g(\boldsymbol{\phi}, \mathbf{R})| = 0 \tag{6-66}$$

This can be shown as follows: let $\{\lambda_n\}$ and $\{\hat{\lambda}_n(M)\}$, $n = 1, \ldots, N-K$, be the nonzero eigenvalues of the matrices $\mathbf{P}(\boldsymbol{\phi})\mathbf{R}\mathbf{P}(\boldsymbol{\phi})$ and $\mathbf{P}(\boldsymbol{\phi})\hat{\mathbf{R}}_M\mathbf{P}(\boldsymbol{\phi})$ respectively. Also, let

$$\hat{\mathbf{R}}_M = \mathbf{R} + \Delta\mathbf{R}_M \tag{6-67}$$

Then,

$$\mathbf{P}\hat{\mathbf{R}}_M\mathbf{P} = \mathbf{PRP} + \mathbf{P}\,\Delta\mathbf{R}_M\mathbf{P} \tag{6-68}$$

Now, if we let $\{\delta_n(M)\}$, $n = 1, \ldots, N-K$, be the eigenvalues of $\mathbf{P}\,\Delta\mathbf{R}_M\mathbf{P}$ where

$$\delta_1(M) \geq \delta_2(M) \geq \cdots \geq \delta_{N-K}(M) \tag{6-69}$$

then [49] we have

$$\lambda_n + \delta_{N-K}(M) \geq \hat{\lambda}_n(M) \geq \lambda_n - \delta_1(M) \tag{6-70}$$

But because of (6-64), we have

$$\lim_{M\to\infty} \Delta\mathbf{R}_M \underset{\text{a.s.}}{=} \mathbf{0} \Rightarrow \lim_{M\to\infty} \delta_n(M) \underset{\text{a.s.}}{=} 0 \tag{6-71}$$

Thus, using (6-70) and (6-71) in $g(\boldsymbol{\phi}, \hat{\mathbf{R}}_M)$, the uniform convergence in (6-66) follows. Because of the continuity and uniform convergence of $g(\boldsymbol{\phi}, \hat{\mathbf{R}}_M)$, $\forall\boldsymbol{\phi} \in \Omega$, then

$$\lim_{\boldsymbol{\phi}\to\boldsymbol{\phi}_a} \lim_{M\to\infty} g(\boldsymbol{\phi}, \hat{\mathbf{R}}_M) = \lim_{M\to\infty} \lim_{\boldsymbol{\phi}\to\boldsymbol{\phi}_a} g(\boldsymbol{\phi}, \hat{\mathbf{R}}_M) \tag{6-72}$$

where $\boldsymbol{\phi}_a$ is an arbitrary point in Ω. From this we can conclude that

$$\lim_{M\to\infty} \min_{\boldsymbol{\phi}\in\Omega} g(\boldsymbol{\phi}, \hat{\mathbf{R}}_M) \underset{\text{a.s.}}{=} \min_{\boldsymbol{\phi}\in\Omega} g(\boldsymbol{\phi}, \mathbf{R}) \tag{6-73}$$

Equation (6-73) tells us that the criterion used in the MAP method is consistent, that is, as $M\to\infty$, if we find a point $\boldsymbol{\phi}_a$ which minimizes the criterion evaluated with the sample covariance matrix, the same point will minimize the criterion evaluated with the true covariance matrix.

Let us now examine the process of projecting the data onto a subspace so that a noise subspace minimizing $g(\boldsymbol{\phi}, \hat{\mathbf{R}}_M)$ can be located. The projector $\mathbf{P}(\boldsymbol{\phi})$ can be expressed as

$$\mathbf{P}(\boldsymbol{\phi}) = \mathbf{U}_\nu(\boldsymbol{\phi})\mathbf{U}_\nu^H(\boldsymbol{\phi}) \tag{6-74}$$

where

$$\mathbf{U}_\nu(\boldsymbol{\phi}) = [\mathbf{u}_{K+1} \quad \mathbf{u}_{K+2} \quad \cdots \quad \mathbf{u}_N] \tag{6-75}$$

are the orthonormal basis vectors spanning the $N-K$ dimensional range subspace of $\mathbf{P}(\boldsymbol{\phi})$.

Consider the ideal situation in which we can find $\boldsymbol{\phi}_0$ that resolves the signal subspace exactly according to the true signal subspace. Applying the projector $\mathbf{P}(\boldsymbol{\phi}_0)$ to the received data vector $\mathbf{x}$, we obtain

$$\begin{aligned}\mathbf{P}(\boldsymbol{\phi}_0)\mathbf{R}\mathbf{P}(\boldsymbol{\phi}_0) &= \mathbf{U}_\nu(\boldsymbol{\phi}_0)\mathbf{U}_\nu^H(\boldsymbol{\phi}_0)\{\mathbf{D}(\boldsymbol{\phi}_0)E[\mathbf{a}\,\mathbf{a}^H]\mathbf{D}^H(\boldsymbol{\phi}_0) \\ &\quad + E[\boldsymbol{\nu}\,\boldsymbol{\nu}^H]\}\mathbf{U}_\nu](\boldsymbol{\phi}_0)\mathbf{U}_\nu^H(\boldsymbol{\phi}_0) \\ &= \mathbf{U}_\nu(\boldsymbol{\phi}_0)\mathbf{U}_\nu^H(\boldsymbol{\phi}_0)\boldsymbol{\Sigma}\mathbf{U}_\nu(\boldsymbol{\phi}_0)\mathbf{U}_\nu^H(\boldsymbol{\phi}_0)\end{aligned} \tag{6-76}$$

where $\boldsymbol{\Sigma}$ is the covariance matrix of the noise. The last step in (6-76) is possible because $\mathbf{P}(\boldsymbol{\phi}_0)\mathbf{D}(\boldsymbol{\phi}_0) = \mathbf{0}$. From (6-76), we conclude that the product of the nonzero eigenvalues of $\mathbf{P}(\boldsymbol{\phi}_0)\mathbf{R}\mathbf{P}(\boldsymbol{\phi}_0)$ is equal to the product of the eigenvalues of $\mathbf{U}_\nu^H(\boldsymbol{\phi}_0)\boldsymbol{\Sigma}\mathbf{U}_\nu(\boldsymbol{\phi}_0)$. In particular, if the noise is spatially white, then

$$\mathbf{\Sigma} = \sigma_v^2 \mathbf{I} \tag{6-77}$$

and the product of the nonzero eigenvalues of $\mathbf{P}(\boldsymbol{\phi}_0)\mathbf{R}\mathbf{P}(\boldsymbol{\phi}_0)$ becomes

$$\prod_{n=1}^{N-K} \lambda_n = (\sigma_v^2)^{N-K} . \tag{6-78}$$

Now, let us consider the situation that after M snapshots, the estimated direction vector is $\hat{\boldsymbol{\phi}}_M$ where $\hat{\boldsymbol{\phi}}_M \neq \boldsymbol{\phi}_0$. We let

$$\begin{aligned} \mathbf{U}(\hat{\boldsymbol{\phi}}_M) &= [\mathbf{U}_s(\hat{\boldsymbol{\phi}}_M) \;\vdots\; \mathbf{U}_\nu(\hat{\boldsymbol{\phi}}_M)] \\ &= [\mathbf{u}_1(\hat{\boldsymbol{\phi}}_M) \cdots \mathbf{u}_K(\hat{\boldsymbol{\phi}}_M) \;\vdots\; \mathbf{u}_{K+1}(\hat{\boldsymbol{\phi}}_M) \cdots \mathbf{u}_N(\hat{\boldsymbol{\phi}}_M)] \end{aligned} \tag{6-79}$$

where $\mathbf{u}_1(\hat{\boldsymbol{\phi}}_M), \ldots, \mathbf{u}_K(\hat{\boldsymbol{\phi}}_M)$ are the orthonormal basis vectors spanning the null subspace, and $\mathbf{u}_{K+1}(\hat{\boldsymbol{\phi}}_M), \ldots, \mathbf{u}_N(\hat{\boldsymbol{\phi}}_M)$ are the orthonormal basis vectors spanning the range subspace of $\mathbf{P}(\hat{\boldsymbol{\phi}}_M)$. At the mth snapshot, we can resolve the true signal vector $\mathbf{D}(\boldsymbol{\phi}_0)\mathbf{a}(m)$ and the true noise vector $\boldsymbol{\nu}(m)$ into components in the estimated signal subspace spanned by $\{\mathbf{u}_1(\hat{\boldsymbol{\phi}}_M), \ldots, \mathbf{u}_K(\hat{\boldsymbol{\phi}}_M)\}$ and components in the estimated noise subspace spanned by $\{\mathbf{u}_{K+1}(\hat{\boldsymbol{\phi}}_M), \ldots, \mathbf{u}_N(\hat{\boldsymbol{\phi}}_M)\}$ such that

$$\mathbf{D}(\boldsymbol{\phi}_0)\mathbf{a}(m) = \mathbf{U}_s(\hat{\boldsymbol{\phi}}_M) \cdot \mathbf{z}_s(m) + \mathbf{U}_\nu(\hat{\boldsymbol{\phi}}_M) \cdot \mathbf{z}_\nu(m) \tag{6-80a}$$

$$\boldsymbol{\nu}(m) = \mathbf{U}_s(\hat{\boldsymbol{\phi}}_M) \cdot \boldsymbol{\zeta}_s(m) + \mathbf{U}_\nu(\hat{\boldsymbol{\phi}}_M) \cdot \boldsymbol{\zeta}_\nu(m) \tag{6-80b}$$

where $\mathbf{z}_s(m) \in \mathscr{C}^{K\times 1}$ and $\mathbf{z}_\nu(m) \in \mathscr{C}^{(N-K)\times 1}$ are the component vectors of the true signal vector in the estimated signal and noise subspaces respectively, and $\boldsymbol{\zeta}_s(m) \in \mathscr{C}^{K\times 1}$ and $\boldsymbol{\zeta}_\nu(m) \in \mathscr{C}^{(N-K)\times 1}$ are the component vectors of the true noise vectors in the estimated signal and noise subspaces respectively. Then the estimated covariance matrix of the received data after M snapshots is given by

$$\hat{\mathbf{R}}_M = [\mathbf{U}_s(\hat{\boldsymbol{\phi}}_M) \quad \mathbf{U}_\nu(\hat{\boldsymbol{\phi}}_M)] \begin{bmatrix} \hat{\mathbf{Z}}_{ss}(M) & \hat{\mathbf{Z}}_{s\nu}(M) \\ \hat{\mathbf{Z}}_{\nu s}(M) & \hat{\mathbf{Z}}_{\nu\nu}(M) \end{bmatrix} \begin{bmatrix} \mathbf{U}_s^H(\hat{\boldsymbol{\phi}}_M) \\ \mathbf{U}_\nu^H(\hat{\boldsymbol{\phi}}_M) \end{bmatrix} \tag{6-81}$$

where

$$\begin{aligned} \hat{\mathbf{Z}}_{ss}(M) = \frac{1}{M} \sum_{m=1}^{M} \{&\mathbf{z}_s(M) \cdot \mathbf{z}_s^H(m) + \mathbf{z}_s(m) \cdot \boldsymbol{\zeta}_s^H(m) \\ &+ \boldsymbol{\zeta}_s(m) \cdot \mathbf{z}_s^H(m) + \boldsymbol{\zeta}_s(m) \cdot \boldsymbol{\zeta}_s^H(m)\} \end{aligned} \tag{6-82a}$$

$$\hat{\mathbf{Z}}_{s\nu}(M) = \hat{\mathbf{Z}}_{s\nu}^H(M) = \frac{1}{M} \sum_{m=1}^{M} \{\mathbf{z}_s(m)\mathbf{z}_\nu^H(m) + \mathbf{z}_s(m)\boldsymbol{\zeta}_\nu^H(m) + \boldsymbol{\zeta}_s(m)\mathbf{z}_\nu^H(m) + \boldsymbol{\zeta}_s(m)\boldsymbol{\zeta}_\nu^H(m)\} \tag{6-82b}$$

and

$$\hat{\mathbf{Z}}_{\nu\nu}(M) = \frac{1}{M} \sum_{m=1}^{M} \{\mathbf{z}_\nu(m)\mathbf{z}_\nu^H(m) + \mathbf{z}_\nu(m)\boldsymbol{\zeta}_\nu^H(m) + \boldsymbol{\zeta}_\nu(m)\mathbf{z}_\nu^H(m) + \boldsymbol{\zeta}_\nu(m)\boldsymbol{\zeta}_\nu^H(m)\} \tag{6-82c}$$

Then, applying (6-74) on (6-81), and using the orthonormality of the basis vectors in $\mathbf{U}_s(\hat{\boldsymbol{\phi}}_M)$ and $\mathbf{U}_\nu(\hat{\boldsymbol{\phi}}_M)$, we have

$$\mathbf{P}(\hat{\boldsymbol{\phi}}_M)\hat{\mathbf{R}}_M\mathbf{P}(\hat{\boldsymbol{\phi}}_M) = \mathbf{U}_\nu(\hat{\boldsymbol{\phi}}_M)\hat{\mathbf{Z}}_{\nu\nu}(M)\mathbf{U}_\nu^H(\hat{\boldsymbol{\phi}}_M) \tag{6-83}$$

Now, let us examine the behaviour of (6-83) as $M \rightarrow \infty$. We propose the following theorem

Theorem 6-1 Under the condition of spatially white noise, as $M \rightarrow \infty$, $\hat{\boldsymbol{\phi}}_M$ converges to the true point $\boldsymbol{\phi}_0$ with probability 1 such that

$$\lim_{M \rightarrow \infty} (\hat{\boldsymbol{\phi}}_M - \boldsymbol{\phi}_0) \underset{\text{a.s.}}{=} \mathbf{0} \tag{6-84}$$

Proof. As $\mathrm{M} \rightarrow \infty$, since the signal and noise components are uncorrelated, (6-82c) becomes

$$\lim_{M \rightarrow \infty} \hat{\mathbf{Z}}_{\nu\nu}(M) = \lim_{M \rightarrow \infty} \frac{1}{M} \sum_{m=1}^{M} \{\mathbf{z}_\nu(m)\mathbf{z}_\nu^H(m) + \boldsymbol{\zeta}_\nu(m)\boldsymbol{\zeta}_\nu^H(m)\} \tag{6-85}$$

and (6-83) becomes

$$\lim_{M \rightarrow \infty} \mathbf{P}(\hat{\boldsymbol{\phi}}_M)\hat{\mathbf{R}}_M\mathbf{P}(\hat{\boldsymbol{\phi}}_M) = \mathbf{U}_\nu(\hat{\boldsymbol{\phi}}_M)\,\Delta\mathbf{R}_s\mathbf{U}_\nu^H(\hat{\boldsymbol{\phi}}_M) + \mathbf{U}_\nu(\hat{\boldsymbol{\phi}}_M)\mathbf{R}_\nu\mathbf{U}_\nu(\hat{\boldsymbol{\phi}}_M) \tag{6-86}$$

where

$$\Delta\mathbf{R}_s = \lim_{M \rightarrow \infty} \frac{1}{M} \sum_{m=1}^{M} \mathbf{z}_\nu(m)\mathbf{z}_\nu^H(m) \tag{6-87a}$$

and

$$\hat{\mathbf{R}}_\nu = \lim_{M \rightarrow \infty} \frac{1}{M} \sum_{m=1}^{M} \boldsymbol{\zeta}_\nu(m)\boldsymbol{\zeta}_\nu^H(m) \tag{6-87b}$$

Also, premultiply the left hand side (6-80b) by $\mathbf{P}(\hat{\boldsymbol{\phi}}_M) = \mathbf{U}_\nu(\hat{\boldsymbol{\phi}}_M)\mathbf{U}_\nu^H(\hat{\boldsymbol{\phi}}_M)$, then postmultiply the resulting quantity with its Hermitian conjugate and average over M snapshots, we obtain

$$\mathbf{U}_\nu(\hat{\boldsymbol{\phi}}_M)\mathbf{U}_\nu^H(\hat{\boldsymbol{\phi}}_M)\{\lim_{M\to\infty} \frac{1}{M} \sum_{m=1}^{M} \boldsymbol{\nu}(m)\boldsymbol{\nu}^H(m)\}\mathbf{U}_\nu(\hat{\boldsymbol{\phi}}_M)\mathbf{U}_\nu^H(\hat{\boldsymbol{\phi}}_M)$$

$$= \mathbf{U}_\nu(\hat{\boldsymbol{\phi}}_M)\{\mathbf{U}_\nu^H(\hat{\boldsymbol{\phi}}_M)\boldsymbol{\Sigma}\mathbf{U}_\nu(\hat{\boldsymbol{\phi}}_M)\}\mathbf{U}_\nu^H(\hat{\boldsymbol{\phi}}_M) \tag{6-88}$$

Applying the same operation to the right hand side (6-80b), remembering the orthonormality of $\mathbf{U}_s(\hat{\boldsymbol{\phi}}_M)$ and $\mathbf{U}_\nu(\hat{\boldsymbol{\phi}}_M)$, and equating the result to (6-88), we have

$$\hat{\mathbf{R}}_\nu = \mathbf{U}_\nu^H(\hat{\boldsymbol{\phi}}_M)\boldsymbol{\Sigma}\mathbf{U}_\nu(\hat{\boldsymbol{\phi}}_M) \tag{6-89}$$

Substituting (6-89) into (6-86), we have

$$\mathbf{P}(\hat{\boldsymbol{\phi}}_M)\mathbf{R}\mathbf{P}(\hat{\boldsymbol{\phi}}_M) = \mathbf{U}_\nu(\hat{\boldsymbol{\phi}}_M)\,\Delta\mathbf{R}_s\mathbf{U}_\nu^H(\hat{\boldsymbol{\phi}}_M) + \mathbf{U}_\nu(\hat{\boldsymbol{\phi}}_M)\mathbf{U}_\nu^H(\hat{\boldsymbol{\phi}}_M)\boldsymbol{\Sigma}\mathbf{U}_\nu(\hat{\boldsymbol{\phi}}_M)\mathbf{U}_\nu^H(\hat{\boldsymbol{\phi}}_M) \tag{6-90}$$

Since $\Delta\mathbf{R}_s$ is Hermitian and nonnegative definite, the eigenvalues of $\{\mathbf{U}_\nu(\hat{\boldsymbol{\phi}}_M)\,\Delta\mathbf{R}_s\mathbf{U}_\nu^H(\hat{\boldsymbol{\phi}}_M)\}$ are nonnegative , and will all be equal to zero only if $\Delta\mathbf{R}_s = \mathbf{0}$. Also if the noise is spatially white, that is, $\boldsymbol{\Sigma} = \sigma_\nu^2\mathbf{I}$, then all the $N-K$ nonzero eigenvalues of the second term of the right hand side of (6-90) are equal to σ_ν^2 regardless of the value of $\hat{\boldsymbol{\phi}}_M$. Hence the product of the nonzero eigenvalues of the term on the left hand side (6-90) is always larger or equal to $(\sigma_\nu^2)^{N-K}$. Equality would hold only if $\hat{\boldsymbol{\phi}}_M = \boldsymbol{\phi}_0$ in which case $\Delta\mathbf{R}_s = \mathbf{0}$ and (6-90) is reduced to (6-76).

Hence we can conclude that

$$\lim_{M\to\infty} \hat{\boldsymbol{\phi}}_M \underset{\text{a.s.}}{=} \boldsymbol{\phi}_0 = \lim_{M\to\infty} \arg\min g(\boldsymbol{\phi}, \hat{\mathbf{R}}_M) \tag{6-91}$$

The above theorem tells us that the estimation of $\boldsymbol{\phi}$ using the minimization of $g(\boldsymbol{\phi}, \hat{\mathbf{R}}_M)$ is consistent under the condition of spatially white noise. Unfortunately, we cannot come to the same conclusion when the spatial noise is nonwhite because in that case the second term on the right hand side (6-90) may have very different eigenvalues from those of (6-76) and thereby the inequality relationship between the eigenvalues of both sides of (6-90) cannot be set up for $\hat{\boldsymbol{\phi}}_M = \boldsymbol{\phi}_0$. However, if as a result of the structure of $\boldsymbol{\Sigma}$, the second term in (6-90) has eigenvalues of similar magnitudes to those of (6-76), then the estimation of $\boldsymbol{\phi}$ may become consistent. Physically, this can be interpreted as follows. If the signal(s) is located in the part of the spatial spectrum where the noise spectrum has large variation, then high SNR is needed for the estimation of $\boldsymbol{\phi}$ to be consistent. On the other hand, if the signal(s) is located in the part where the noise spectrum is relatively flat, consistency of the estimation can be achieved under lower SNR.

6-6 PERFORMANCE ANALYSIS OF THE MAP ESTIMATE

The criterion of the MAP estimate is to minimize the function $g(\boldsymbol{\phi}, \mathbf{R}_M)$ which can also be written as

$$g(\boldsymbol{\phi}, \hat{\mathbf{R}}_M) = \prod_{n=1}^{N-K} \hat{\lambda}_n(M) = \text{tr}_{N-K}[\mathbf{P}(\boldsymbol{\phi})\hat{\mathbf{R}}_M\mathbf{P}(\boldsymbol{\phi})] \tag{6-92}$$

where $\{\hat{\lambda}_n(M)\}$, $n = 1, \ldots, N-K$, denote the non-zero eigenvalues of the matrix $\mathbf{P}(\boldsymbol{\phi})\hat{\mathbf{R}}_M\mathbf{P}(\boldsymbol{\phi})$, and $\text{tr}_{N-K}(\cdot)$ denotes the $(N-K)$th order trace of a matrix defined in Section 6-3. Since the criterion is to minimize $g(\boldsymbol{\phi}, \hat{\mathbf{R}}_M)$ with respect to $\boldsymbol{\phi}$, therefore, at $\boldsymbol{\phi} = \hat{\boldsymbol{\phi}}$, the gradient vector is zero. Equivalently, if we let

$$f(\boldsymbol{\phi}, \hat{\mathbf{R}}_M) = \ln g(\boldsymbol{\phi}, \hat{\mathbf{R}}_M) \tag{6-93}$$

then

$$\dot{\mathbf{f}}(\boldsymbol{\phi}, \hat{\mathbf{R}}_M)|_{\boldsymbol{\phi}=\hat{\boldsymbol{\phi}}} \stackrel{\Delta}{=} \nabla_{\boldsymbol{\phi}} f(\boldsymbol{\phi}, \hat{\mathbf{R}}_M)|_{\boldsymbol{\phi}=\hat{\boldsymbol{\phi}}} = 0 \tag{6-94}$$

where

$$\nabla_{\boldsymbol{\phi}} \stackrel{\Delta}{=} \left[\frac{\partial}{\partial \phi_1} \quad \cdots \quad \frac{\partial}{\partial \phi_k}\right]^T \tag{6-95}$$

and we have used the notation that the dot over a function denotes the partial derivative with respect to $\boldsymbol{\phi}$. Our aim is to express the performance of the MAP method, which is based on the criterion of (6-94), in terms of the true parameter $\boldsymbol{\phi}_0$ and the true data covariance matrix $\mathbf{R}$.

In Section 6-5, we have seen that under favorable conditions, the MAP estimation is asymptotically consistent. Therefore, under favorable conditions, we can expand (6-94) in a Taylor series, and can assume that $\Delta\boldsymbol{\phi}$ and $\Delta\mathbf{R}_M$ are small and ignore the terms involving their second and higher orders, where

$$\Delta\boldsymbol{\phi} \stackrel{\Delta}{=} \hat{\boldsymbol{\phi}} - \boldsymbol{\phi}_0 \tag{6-96}$$

$$\Delta\mathbf{R}_M \stackrel{\Delta}{=} \hat{\mathbf{R}}_M - \mathbf{R} \tag{6-97}$$

To express (6-94) as a Taylor series, we must take precautions in various aspects. The quantity $\dot{\mathbf{f}}$ is a vector function of the K-dimensional vector $\hat{\boldsymbol{\phi}}$ and of the $M \times M$ matrix $\hat{\mathbf{R}}_M$. The Taylor series expansion of $\dot{\mathbf{f}}$ must also be a K-dimensional vector, implying that the term involving the derivative with respect to $\mathbf{R}$ and $\Delta\mathbf{R}_M$ must yield a K-dimensional vector. This can be achieved by treating $\mathbf{R}$ and $\Delta\mathbf{R}_M$ as vectors. Using the standard notation in matrix calculus [24, 48] such that for an $N \times N$ matrix $\mathbf{A}$, the vec operator on $\mathbf{A}$ is the alignment of its column vectors in order to form an N^2-dimensional vector. Furthermore, we note that both $\mathbf{R}$ and $\Delta\mathbf{R}_M$ are complex and Hermitian, therefore, we can ignore the conjugate components and define,

$$\begin{aligned}\operatorname{vec}\mathbf{R} = [&\mathrm{Re}(r_{11})\,\mathrm{Im}(r_{11}) \quad \cdots \quad \mathrm{Re}(r_{N1})\,\mathrm{Im}(r_{N1}) \quad \mathrm{Re}(r_{22})\,\mathrm{Im}(r_{22}) \\ &\cdots \quad \mathrm{Re}(r_{NN})\,\mathrm{Im}(r_{NN})]^T \end{aligned} \tag{6-98}$$

and

$$\begin{aligned}\operatorname{vec}\Delta\mathbf{R}_M = [&\mathrm{Re}(\Delta r_{11})\,\mathrm{Im}(\Delta r_{11}) \quad \cdots \quad \mathrm{Re}(\Delta r_{N1})\,\mathrm{Im}(\Delta r_{N1}) \\ &\mathrm{Re}(\Delta r_{22})\,\mathrm{Im}(\Delta r_{22}) \quad \cdots \quad \mathrm{Re}(\Delta r_{NN})\,\mathrm{Im}(\Delta r_{NN})]^T\end{aligned} \tag{6-99}$$

where r_{nm} and Δr_{nm} are the nmth elements of $\mathbf{R}$ and $\Delta\mathbf{R}_M$ respectively. We note that both $\operatorname{vec}\mathbf{R}$ and $\operatorname{vec}\Delta\mathbf{R}_M$ are $(N(N+1))$-dimensional vectors. Therefore, the expansion of (6-94) in a Taylor series becomes

$$\begin{aligned}\dot{\mathbf{f}}(\hat{\boldsymbol{\phi}}, \hat{\mathbf{R}}_M) &= \dot{\mathbf{f}}(\boldsymbol{\phi}_0, \mathbf{R}) + \left\{\frac{\partial \dot{\mathbf{f}}(\boldsymbol{\phi}_0, \mathbf{R})}{\partial \boldsymbol{\phi}}\right\}^H \cdot \Delta\boldsymbol{\phi} + \left\{\frac{\partial \dot{\mathbf{f}}(\boldsymbol{\phi}_0, \mathbf{R})}{\partial \operatorname{vec}\mathbf{R}}\right\}^H \cdot \operatorname{vec}\Delta\mathbf{R}_M + \cdots \\ &= \mathbf{0}\end{aligned} \tag{6-100}$$

where we have ignored the terms involving the second and higher orders of $\Delta\boldsymbol{\phi}$ and $\Delta\mathbf{R}_M$. Note that $\partial\dot{\mathbf{f}}(\boldsymbol{\phi}_0, \mathbf{R})/\partial\boldsymbol{\phi}$ is a $K \times K$ matrix and $\partial\dot{\mathbf{f}}(\boldsymbol{\phi}_0, \mathbf{R})/\partial \operatorname{vec}\mathbf{R}$ is an $(N(N+1)) \times K$ matrix. From (6-100), we obtain

$$\Delta\boldsymbol{\phi} \simeq -\left[\left\{\frac{\partial \dot{\mathbf{f}}(\boldsymbol{\phi}_0, \mathbf{R})}{\partial \boldsymbol{\phi}}\right\}^H\right]^{-1}\left[\dot{\mathbf{f}}(\boldsymbol{\phi}_0, \mathbf{R}) + \left\{\frac{\partial \dot{\mathbf{f}}(\boldsymbol{\phi}_0, \mathbf{R})}{\partial \operatorname{vec}\mathbf{R}}\right\}^H \cdot \operatorname{vec}\Delta\mathbf{R}_M\right] \tag{6-101a}$$

$$= -(\ddot{\mathbf{f}}^H)^{-1}[\dot{\mathbf{f}} + \dot{\mathbf{f}}'^H \operatorname{vec}\Delta\mathbf{R}_M] \tag{6-101b}$$

where we have, for convenience, used the notation that a dot over a function denotes derivative with respect to $\boldsymbol{\phi}$ while a prime denotes the derivative with respect to $\operatorname{vec}\mathbf{R}$, and that all the derivatives of $\mathbf{f}$ are evaluated at $\boldsymbol{\phi}_0$ and $\mathbf{R}$.

We now endeavor to evaluate $\dot{\mathbf{f}}$ and its derivatives with respect to $\boldsymbol{\phi}$ and $\operatorname{vec}\mathbf{R}$ in (6-101). We propose the following theorem.

Theorem 6-2 For an $N \times N$ Hermitian matrix $\mathbf{A}$ of rank $(N-K)$, we have

$$\frac{\partial}{\partial \phi_k}\operatorname{tr}_{N-K}(\mathbf{A}) = \operatorname{tr}_{N-K}(\mathbf{A}) \cdot \operatorname{tr}\left(\mathbf{A}^{-I}\frac{\partial \mathbf{A}}{\partial \phi_k}\right) \tag{6-102}$$

where $\mathbf{A}^{-I}$ denotes the pseudo-inverse of $\mathbf{A}$.

Proof. The proof of the above theorem is given in Appendix B.

Employing Theorem 6-2, we can evaluate the kth element of $\dot{\mathbf{f}}$ such that

$$\dot{\mathbf{f}}_k = \mathrm{tr}\left\{(\mathbf{PRP})^{-I} \frac{\partial}{\partial \phi_k} (\mathbf{PRP})\right\} \tag{6-103}$$

where again, we omit the notational dependence of $\mathbf{P}$ on $\boldsymbol{\phi}_0$. We recall that

$$\mathbf{P} = \mathbf{I} - \mathbf{D}(\mathbf{D}^H\mathbf{D})^{-1}\mathbf{D}^H = \mathbf{I} - \mathbf{Q} \tag{6-104a}$$

where

$$\mathbf{Q} = \mathbf{D}(\mathbf{D}^H\mathbf{D})^{-1}\mathbf{D}^H \tag{6-104b}$$

Then, the second term in the braces of (6-103) becomes

$$\frac{\partial}{\partial \phi_k} (\mathbf{PRP}) = \dot{\mathbf{P}}_k\mathbf{RP} + \mathbf{PR}\dot{\mathbf{P}}_k \tag{6-105}$$

where

$$\begin{aligned}
\dot{\mathbf{P}}_k &\triangleq \frac{\partial \mathbf{P}}{\partial \phi_k} \\
&= -[\dot{\mathbf{D}}_k(\mathbf{D}^H\mathbf{D})^{-1}\mathbf{D}^H - \mathbf{D}(\mathbf{D}^H\mathbf{D})^{-1}(\dot{\mathbf{D}}_k^H\mathbf{D} + \mathbf{D}^H\dot{\mathbf{D}}_k)(\mathbf{D}^H\mathbf{D})^{-1}\mathbf{D}^H \\
&\quad + \mathbf{D}(\mathbf{D}^H\mathbf{D})^{-1}\dot{\mathbf{D}}_k^H] \\
&= -[\{\mathbf{I} - \mathbf{D}(\mathbf{D}^H\mathbf{D})^{-1}\mathbf{D}^H\}\dot{\mathbf{D}}_k(\mathbf{D}^H\mathbf{D})^{-1}\mathbf{D}^H + \mathbf{D}(\mathbf{D}^H\mathbf{D})^{-1}\dot{\mathbf{D}}_k^H \\
&\quad \times \{\mathbf{I} - \mathbf{D}(\mathbf{D}^H\mathbf{D})^{-1}\mathbf{D}^H\}] \\
&= -[\mathbf{P}\mathbf{Q}_k^{\cdot H} + \mathbf{Q}_k^{\cdot}\mathbf{P}]
\end{aligned} \tag{6-106}$$

where

$$\mathbf{Q}_k^{\cdot} \triangleq \mathbf{D}(\mathbf{D}^H\mathbf{D})^{-1}\mathbf{D}_k^H \tag{6-107}$$

and

$$\mathbf{D}_k \triangleq \partial \mathbf{D}/\partial \phi_k \tag{6-108}$$

Substituting (6-106) into (6-105) and employing the result in (6-103), we obtain

$$\begin{aligned}\dot{f}_k &= -\text{tr}[(\mathbf{PRP})^{-I}\{\mathbf{PRPQ}_k^{\cdot H} + \mathbf{Q}_k^{\cdot}\mathbf{PRP} + \mathbf{PRQ}_k^{\cdot}\mathbf{P} + \mathbf{PQ}_k^{\cdot H}\mathbf{RP}\}] \\ &= -\text{tr}[(\mathbf{PRP})^{-I}(\mathbf{PRP})\mathbf{Q}_k^{\cdot H} + (\mathbf{PRP})(\mathbf{PRP})^{-I}\mathbf{Q}_k^{\cdot} \\ &\quad + (\mathbf{PRP})^{-I}(\mathbf{PQ}_k^{\cdot H}\mathbf{RP} + \mathbf{PRQ}_k^{\cdot}\mathbf{P})]\end{aligned} \tag{6-109}$$

where the relationship $\text{tr}(\mathbf{AB}) = \text{tr}(\mathbf{BA})$ has been used.

Now, since $(\mathbf{PRP})^{-I}\mathbf{PRP}$ can be expanded in vectors which span only the true noise subspace $\mathcal{N}$ and which are orthogonal to $\mathbf{d}(\phi_k)$, $k = 1, \ldots, K$, then

$$\text{tr}[(\mathbf{PRP})^{-I}(\mathbf{PRP})\mathbf{Q}_k^{\cdot H}] = \text{tr}[\mathbf{Q}_k^{\cdot H}(\mathbf{PRP})^{-I}(\mathbf{PRP})] = 0 \tag{6-110}$$

Similarly,

$$\text{tr}[(\mathbf{PRP})(\mathbf{PRP})^{-I}\mathbf{Q}_k^{\cdot}] = 0 \tag{6-111}$$

Equation (6-109) is thus reduced to

$$\begin{aligned}\dot{f}_k &= -\text{tr}[(\mathbf{PRP})^{-I}(\mathbf{PRQ}_k^{\cdot}\mathbf{P} + \mathbf{PQ}_k^{\cdot H}\mathbf{RP})] \\ &= -2\,\text{Re}\,\text{tr}[(\mathbf{PRP})^{-I}(\mathbf{PQ}_k^{\cdot H}\mathbf{RP})]\end{aligned} \tag{6-112}$$

where the last step of (6-112) is facilitated by the fact that for two square matrices $\mathbf{A}$ and $\mathbf{B}$ with $\mathbf{A}$ being Hermitian

$$\text{tr}[\mathbf{A}(\mathbf{B} + \mathbf{B}^H)] = \text{tr}(\mathbf{AB}) + \text{tr}^*(\mathbf{BA}) = 2\,\text{Re}\,\text{tr}(\mathbf{AB})\,. \tag{6-113}$$

If the noise is spatially white, then the noise subspace $\mathcal{N}$ is $\mathbf{R}$-invariant meaning that for any vector $\mathbf{x} \in \mathcal{N}$, $\mathbf{Rx} \in \mathcal{N}$. Under such circumstances we have [44] $\mathbf{PRP} = \mathbf{PR} = \mathbf{RP}$, and the last term inside the square brackets of (6-112) vanishes since $\mathbf{Q}_k^{\cdot H}\mathbf{PRP} = 0$. This implies an unbiased estimate and confirms the results in Theorem 6-1.

We now turn our attention to the evaluation of the Hessian matrix in (6-101b)

$$\ddot{\mathbf{f}} \triangleq \frac{\partial \dot{\mathbf{f}}(\boldsymbol{\phi}_0, \mathbf{R})}{\partial \boldsymbol{\phi}} = \boldsymbol{\nabla}_{\boldsymbol{\phi}}\dot{\mathbf{f}}^H \tag{6-114}$$

Note that $\ddot{\mathbf{f}}$ is a $K \times K$ matrix the *ik*th element of which is given by

$$\begin{aligned}\ddot{f}_{ik} &= \frac{\partial}{\partial \phi_i}\dot{f}_k \\ &= -\text{tr}\left[\frac{\partial}{\partial \phi_i}\{(\mathbf{PRP})^{-I}(\mathbf{PQ}_k^{\cdot H}\mathbf{RP} + \mathbf{PRQ}_k^{\cdot}\mathbf{P})\}\right]\end{aligned} \tag{6-115}$$

The quantity in (6-115) is evaluated in [34, 35] and is given by

$$\ddot{f}_{ik} = 2\,\mathrm{Re}\,\mathrm{tr}[(\mathbf{PRP})^{-I}\{(\mathbf{P}\mathbf{Q}^{\cdot H}_{i} + \mathbf{Q}^{\cdot}_{i}\mathbf{P})\mathbf{R}\mathbf{Q}^{\cdot}_{k}\mathbf{P} - \mathbf{PR}(\mathbf{P}\,{}^{\cdot}\mathbf{Q}^{\cdot}_{ik} - \mathbf{Q}^{\cdot}_{i}\mathbf{Q}^{\cdot}_{k} + \mathbf{Q}^{\cdot\cdot}_{ik})\mathbf{P}$$
$$+ \mathbf{PR}\mathbf{Q}^{\cdot}_{k}(\mathbf{P}\mathbf{Q}^{\cdot H}_{k} + \mathbf{Q}^{\cdot}_{i}\mathbf{P})\}] - \mathrm{tr}[(\mathbf{PRP})^{-I}(\mathbf{P}\mathbf{Q}^{\cdot H}_{i}\mathbf{RP} + \mathbf{PRP}\mathbf{Q}^{\cdot H}_{i}$$
$$+ \mathbf{Q}^{\cdot}_{i}\mathbf{PRP} + \mathbf{PR}\mathbf{Q}^{\cdot}_{i}\mathbf{P})(\mathbf{PRP})^{-I}(\mathbf{P}\mathbf{Q}^{\cdot H}_{k}\mathbf{RP} + \mathbf{PR}\mathbf{Q}^{\cdot}_{k}\mathbf{P})] \tag{6-116}$$

where $\mathbf{Q}^{\cdot}_{k}$ is defined in (6-107) and where ${}^{\cdot}\mathbf{Q}^{\cdot}_{ik}$ and $\mathbf{Q}^{\cdot\cdot}_{ik}$ are given by

$${}^{\cdot}\mathbf{Q}^{\cdot}_{ik} = \dot{\mathbf{D}}_{i}(\mathbf{D}^{H}\mathbf{D})^{-1}\dot{\mathbf{D}}^{H}_{k} \tag{6-117a}$$

$$\mathbf{Q}^{\cdot\cdot}_{ik} = \begin{cases} \mathbf{0}\,, & \text{for } i \neq k \\ \mathbf{D}(\mathbf{D}^{H}\mathbf{D})^{-1}\ddot{\mathbf{D}}^{H}_{kk}\,, & \text{for } i = k \end{cases} \tag{6-117b}$$

respectively.

We now proceed to evaluate the derivative of $\dot{\mathbf{f}}$ with respect to vec $\mathbf{R}$, that is, the elements of the matrix $\dot{\mathbf{f}}'$ in (6-101b). The derivative of the kth element of $\dot{\mathbf{f}}$ with respect to $\mathrm{Re}(r_{il})$ is given by

$$\frac{\partial \dot{f}_{k}(\boldsymbol{\phi}_{0}, \mathbf{R})}{\partial\,\mathrm{Re}(r_{il})} = -\mathrm{tr}\left[\frac{\partial(\mathbf{PRP})^{-I}}{\partial\,\mathrm{Re}(r_{il})} \cdot (\mathbf{PR}\mathbf{Q}^{\cdot}_{k}\mathbf{P} + \mathbf{P}\mathbf{Q}^{\cdot H}_{k}\mathbf{RP})\right.$$
$$\left. + (\mathbf{PRP})^{-I}\,\frac{\partial}{\partial\,\mathrm{Re}(r_{il})}\,(\mathbf{PR}\mathbf{Q}^{\cdot}_{k}\mathbf{P} + \mathbf{P}\mathbf{Q}^{\cdot H}_{k}\mathbf{RP})\right]. \tag{6-118}$$

Applying a similar development as that leading to $\ddot{f}_{ik}$, we can write [34, 35]

$$\frac{\partial \dot{f}_{k}(\boldsymbol{\phi}_{0}, \mathbf{R})}{\partial\,\mathrm{Re}(r_{il})} = \mathrm{tr}\left[(\mathbf{PRP})^{-I}\mathbf{P}\,\frac{\partial \mathbf{R}}{\partial\,\mathrm{Re}(r_{il})}\,\mathbf{P}(\mathbf{PRP})^{-I}(\mathbf{PR}\mathbf{Q}^{\cdot}_{k}\mathbf{P} + \mathbf{P}\mathbf{Q}^{\cdot H}_{k}\mathbf{RP})\right.$$
$$\left. -(\mathbf{PRP})^{-I}\left(\mathbf{P}\,\frac{\partial \mathbf{R}}{\partial\,\mathrm{Re}(r_{il})}\,\mathbf{Q}^{\cdot}_{k}\mathbf{P} + \mathbf{P}\mathbf{Q}^{\cdot H}_{k}\,\frac{\partial \mathbf{R}}{\partial\,\mathrm{Re}(r_{il})}\,\mathbf{P}\right)\right] \tag{6-119a}$$

Similarly,

$$\frac{\partial \dot{f}_{k}(\boldsymbol{\phi}_{0}, \mathbf{R})}{\partial\,\mathrm{Im}(r_{il})} = \mathrm{tr}\left[(\mathbf{PRP})^{-I}\mathbf{P}\,\frac{\partial \mathbf{R}}{\partial\,\mathrm{Im}(r_{il})}\,\mathbf{P}(\mathbf{PRP})^{-I}(\mathbf{PR}\mathbf{Q}^{\cdot}_{k}\mathbf{P} + \mathbf{P}\mathbf{Q}^{\cdot H}_{k}\mathbf{RP})\right.$$
$$\left. -(\mathbf{PRP})^{-I}\left(\mathbf{P}\,\frac{\partial \mathbf{R}}{\partial\,\mathrm{Im}(r_{il})}\,\mathbf{Q}^{\cdot}_{k}\mathbf{P} + \mathbf{P}\mathbf{Q}^{\cdot H}_{k}\,\frac{\partial \mathbf{R}}{\partial\,\mathrm{Im}(r_{il})}\,\mathbf{P}\right)\right] \tag{6-119b}$$

The matrices $\partial\mathbf{R}/\partial\,\mathrm{Re}(r_{il})$ and $\partial\mathbf{R}/\partial\,\mathrm{Im}(r_{il})$ in (6-119) can be obtained easily such that

$$\frac{\partial \mathbf{R}}{\partial\, \mathrm{Re}(r_{il})} = [r'_{nm}] \quad \text{and} \quad \frac{\partial \mathbf{R}}{\partial\, \mathrm{Im}(r_{il})} = [\rho'_{nm}] \tag{6-120}$$

where

$$r'_{nm} = \begin{cases} 1\,, & \text{for } (n = i \text{ and } m = l) \text{ or } (n = l \text{ and } m = i) \\ 0\,, & \text{otherwise} \end{cases} \tag{6-121a}$$

and

$$\rho'_{nm} = \begin{cases} j\,, & \text{for } n = i \text{ and } m = l \\ -j\,, & \text{for } n = l \text{ and } m = i\,. \\ 0\,, & \text{otherwise} \end{cases} \tag{6-121b}$$

We are now in the position to examine the error of the estimated angles of arrival $\Delta\boldsymbol{\phi}$. From (6-101b), the covariance matrix of the estimates is given by

$$\begin{aligned} E[\Delta\boldsymbol{\phi}(\Delta\boldsymbol{\phi})^H] &= E[(\ddot{\mathbf{f}}^H)^{-1}\{\dot{\mathbf{f}} + \dot{\mathbf{f}}'^H \operatorname{vec} \Delta\mathbf{R}_M\}\{\dot{\mathbf{f}} + \dot{\mathbf{f}}'^H \operatorname{vec} \Delta\mathbf{R}_M\}^H(\ddot{\mathbf{f}})^{-1}] \\ &= (\ddot{\mathbf{f}}^H)^{-1}\{\dot{\mathbf{f}}\dot{\mathbf{f}}^H + \dot{\mathbf{f}}E[(\operatorname{vec} \Delta\mathbf{R}_M)^H]\dot{\mathbf{f}}' + \dot{\mathbf{f}}'^H E[(\operatorname{vec} \Delta\mathbf{R}_M)^H]\dot{f}^H \\ &\quad + \dot{\mathbf{f}}'^H E[(\operatorname{vec} \Delta\mathbf{R}_M)(\operatorname{vec} \Delta\mathbf{R}_M)^H]\dot{\mathbf{f}}'\}(\ddot{\mathbf{f}})^{-1} \end{aligned} \tag{6-122}$$

If we assume that the received data $\mathbf{x}(m)$ is zero-mean Gaussian, then $M\hat{\mathbf{R}}_M$ is of a complex Wishart distribution [27]. The moment generating function for the real Wishart distribution is well documented [48, 55]. By following similar procedures, the moment generating function for the complex case is derived in [34], and we obtain

$$E[\operatorname{vec}(\Delta\mathbf{R}_M)] = \mathbf{0} \tag{6-123}$$

and the covariance of the elements of $\operatorname{vec}(\Delta\mathbf{R}_M)$ are given by

$$E[\mathrm{Re}(\Delta r_{nm})\,\mathrm{Re}(\Delta r_{kl})] = \frac{1}{2M}\,\mathrm{Re}(r_{kn}r_{ml} + r_{km}r_{nl}) \tag{6-124a}$$

$$E[\mathrm{Im}(\Delta r_{nm})\,\mathrm{Im}(\Delta r_{kl})] = \begin{cases} \dfrac{1}{2M}\,\mathrm{Re}(r_{kn}r_{ml} - r_{km}r_{nl})\,, & k > l, n > m \\ 0\,, & k = l \text{ or } n = m \end{cases} \tag{6-124b}$$

$$E[\mathrm{Re}(\Delta r_{nm})\,\mathrm{Im}(\Delta r_{kl})] = \begin{cases} \dfrac{1}{2M}\,\mathrm{Im}(r_{kn}r_{ml} + r_{km}r_{nl})\,, & k > l, n \geqslant m \\ 0\,, & k = l \end{cases} \tag{6-124c}$$

$$E[\mathrm{Re}(\Delta r_{kl})\,\mathrm{Im}(\Delta r_{nm})] = \begin{cases} \frac{1}{2M}\,\mathrm{Im}(-r_{kn}r_{ml} + r_{km}r_{nl}), & n > m,\ k \geqslant l \\ 0, & n = m \end{cases} \tag{6-124d}$$

where Δr_{nm} and r_{nm} are the elements in $\Delta\mathbf{R}_M$ and $\mathbf{R}$ respectively. Thus we can write

$$E[\Delta\boldsymbol{\phi}(\Delta\boldsymbol{\phi})^H] = (\ddot{\mathbf{f}}^H)^{-1}\{\dot{\mathbf{f}}\dot{\mathbf{f}}^H + \dot{\mathbf{f}}'^H E[(\mathrm{vec}\,\Delta\mathbf{R}_M)(\mathrm{vec}\,\Delta\mathbf{R}_M)^H]\dot{\mathbf{f}}'\}(\ddot{\mathbf{f}})^{-1} \tag{6-125}$$

where the elements of $\dot{\mathbf{f}}$, $\ddot{\mathbf{f}}$, $\dot{\mathbf{f}}'$, and $E[(\mathrm{vec}\,\Delta\mathbf{R}_M)(\mathrm{vec}\,\Delta\mathbf{R}_M)^H]$ are given by (6-112), (6-116), (6-119), and (6-124) respectively. Equation (6-125) yields the theoretic mean square error of the MAP method of estimating the DOA of signals in spatially correlated noise.

6-7 COMPUTER SIMULATION RESULTS

In this section, we present some computer simulation results showing the performance of the MAP method under various correlated noise environments. We first compare the theoretical performance of the MAP method as developed in the last section to that obtained by computer simulations. Then, we present the comparison of the performance of the MAP method to other methods by computer simulations.

In all our cases of comparison, we employ the following signal and noise specifications

Signal

a. The array is linear with sensors uniformly spaced at $L = \pi c/\omega_0$ where c is the velocity of propagation and ω_0 is the angular frequency of the narrowband signals.
b. There are $N = 8$ sensors in the array, and $K = 2$ signals incident at angles of ± 0.25 standard beamwidths ($2\pi/N$ radians) to the array.
c. The signal vector $\mathbf{a}(m)$ at the mth snapshot is zero mean complex Gaussian such that

$$E[\mathbf{a}(m)\,\mathbf{a}^H(n)] = \delta_{mn}\sigma_s^2\mathbf{I} \tag{6-126}$$

Noise

a. The noise vector $\boldsymbol{\nu}(m)$ is generated as a spatially varying, first order, Gaussian autoregressive process with zero mean and $E[\boldsymbol{\nu}(m)\,\boldsymbol{\nu}^H(n)] = \delta_{mn}\sigma_\nu^2\boldsymbol{\Sigma}$. The signal-to-noise ratio (SNR) at the array is defined as

$$\text{SNR} = 10 \log \sigma_s^2/\sigma_\nu^2 \text{ dB} \tag{6-127}$$

b. The (ik)th element, σ_{ik}, of $\boldsymbol{\Sigma}$ is given by

$$\sigma_{ik} = \rho^{|i-k|} \exp[j\phi_p(i-k)] \tag{6-128}$$

where the parameter ρ controls the relative height of the peak of the noise spatial spectrum and the parameter ϕ_p controls the angle at which the peak occurs.

In all our examples presented in this section, we choose $\rho = 0.9$, and the noise spectrum is shifted along the spatial frequency axis to create various noise environments, by choosing ϕ_p to be $\pi/8$, $\pi/2$, and π respectively, while the incident angles of the signals are fixed at ± 0.25 standard beamwidth. The spatial noise power spectral densities having these characteristics are illustrated in Fig. 6-6 from which it can be observed that, for different values of ϕ_p the incident angles of the signals are situated in parts for which the gradients of the noise spatial spectrum are of different values, $\phi_p = \pi/8$ corresponding to the part with steep slope, $\phi_p = \pi/2$ the part with intermediate slope, and $\phi_p = \pi$ the part where the spectrum is almost flat.

The computer simulations for the evaluation of the performance of the MAP estimate are therefore carried out as follows. For a given SNR and a given noise spatial spectrum, M snapshots of the received data are taken and the MAP estimation procedure is carried out and a DOA estimate is obtained. The above is repeated 200 times using the same signal and noise environment. The mean square error (MSE), expressed in square of standard beamwidth, for each of the two signal DOA estimates is then calcu-

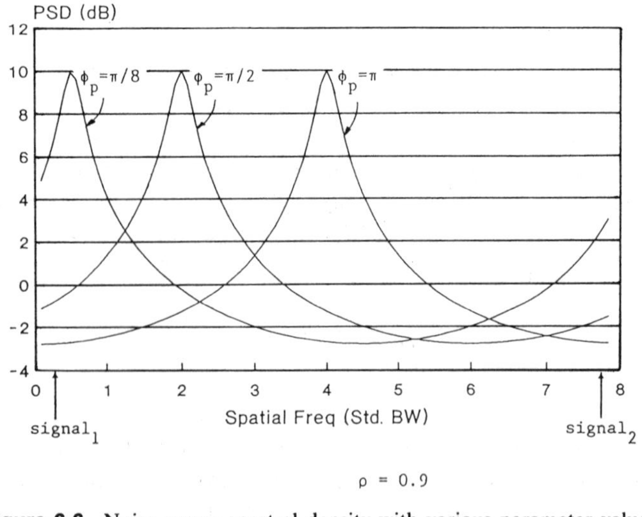

Figure 6-6 Noise power spectral density with various parameter values.

lated. The overall MSE, $\overline{e^2}$ which is the average of the MSE of the two estimates, is then obtained. The same procedure is repeated for a different signal and noise environment, and the results are plotted for comparison.

Figure 6-7 shows the variation of the overall MSE, $\overline{e^2}$, with the number of snapshots M. The comparison is between the theoretically evaluated overall MSE as obtained by employing (6-124) and that obtained by computer simulations, and is carried out at a SNR = 7 dB. There are three groups of performance curves corresponding to the three different locations of noise spectrum peak. As can be observed, the theoretical performance curves agree very well with those obtained by simulations when $M > 30$. This is because the theoretical performance is derived based on the assumption that M is sufficiently large. It is also observed that for $\phi_p = \pi/8$, which corresponds to the case when the signal DOA is situated in the part where the gradient of the noise spectrum is steep, there is a relatively high irreducible error even if M is large. This is, as discussed in Section 6-5, because of the fact $\phi_p = \pi/8$ corresponds to the case that the nonzero eigenvalues of (6-76) and those of the second term in (6-90) are dissimilar in magnitude, resulting in a biased estimate and hence the irreducible error. For $\phi_p = \pi/2$, the signal DOA is situated in the part where the gradient of the noise spectrum is less steep which corresponds to the case when the aforementioned eigenvalues are less dissimilar resulting in a smaller bias and hence a smaller irreducible error. For $\phi_p = \pi$, the signal DOA is situated in the part where the noise

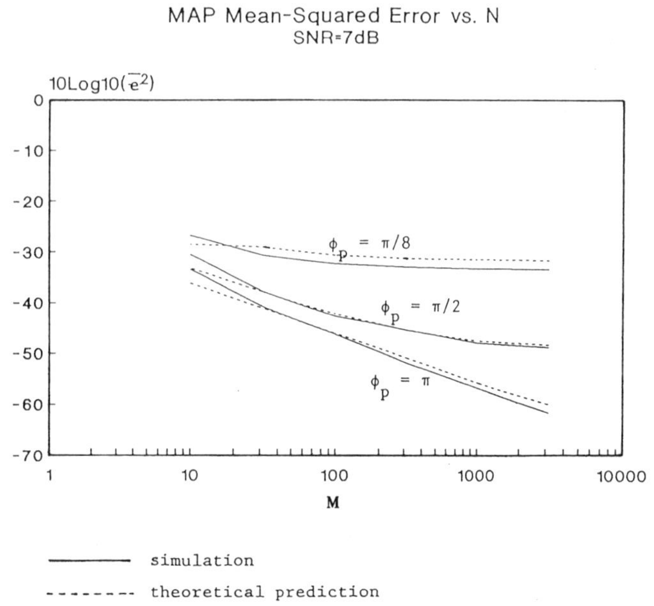

Figure 6-7 Variation of MAP performance with the number of snapshots. ———, simulation; – – – – –, theoretical prediction.

spectrum is almost flat which corresponds to the case when the eigenvalues are very similar resulting in an estimate that has almost no bias and is almost consistent.

Figure 6-8 shows the variation of the overall MSE with the signal-to-noise ratio, and M, the number of snapshots, is taken to be 100. The comparison is between the theoretical performance as given in (6-124) and that obtained by computer simulations. Again, the theoretical performance curves are in close agreement with those obtained by simulations. The case when $\phi_p = \pi/8$ has the highest MSE while the case when $\phi_p = \pi$ has the lowest. When the SNR decreases, the MSE increases steadily for all the cases. However, when the SNR decreases to below -5 dB (not shown), simulation results show a threshold effect in both cases when $\phi_p = \pi$ and $\phi_p = \pi/2$ which marks an abrupt deterioration of performance. This threshold effect has not been predicted by the theoretical evaluation since the theoretical performance was developed with the assumptions that both $\Delta\boldsymbol{\phi}$, the error of the DOA estimate, and $\Delta\mathbf{R}_M$, the error of covariance matrix estimate are sufficiently small such that the terms involving the second and higher orders of $\Delta\boldsymbol{\phi}$ and

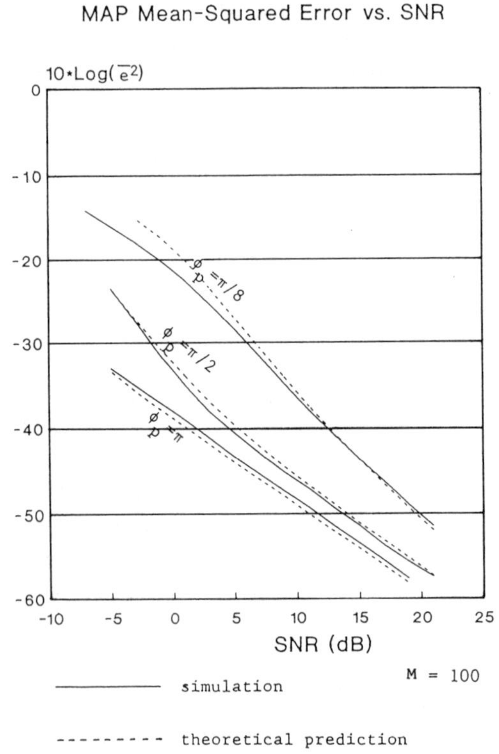

Figure 6-8 Variation of MAP performance with SNR. ———, simulation; – – – – –, theoretical prediction.

$\Delta\mathbf{R}_M$ can be neglected. This assumption is not valid at low SNR and therefore the simulated results depart from the theoretically evaluated performance.

We now compare the MAP method with the well-known MUSIC [18] and the maximum likelihood estimation (MLE) [3, 9, 10] methods. The MUSIC method (MUltiple SIgnal Characterization) is discussed in some detail in Chapter 3. Very briefly, the method consists of evaluating the signal covariance matrix $\hat{\mathbf{R}}_M$ from M snapshots of data. An $N \times (N-K)$ matrix $\mathbf{V}_N$ is then formed, the columns of which are the eigenvectors associated with the smallest $N-K$ eigenvalues of $\hat{\mathbf{R}}_M$. The values of ϕ, at which the peaks of the MUSIC spectrum $S_{\mathrm{MUS}}(\phi)$, defined as

$$S_{\mathrm{MUS}}(\phi) = \frac{1}{\mathbf{d}^H(\phi)\mathbf{V}_N\mathbf{V}_N^H\mathbf{d}(\phi)} \qquad (6\text{-}129)$$

occur, are the MUSIC DOA estimates.

The MLE method, on the other hand, is also a well-known technique, which is developed under the assumption that the background noise is spatially white. The values of $\boldsymbol{\phi}$ that satisfy the following objective function

$$\arg\min_{\boldsymbol{\phi}\in\Omega} \operatorname{tr} \mathbf{P}\hat{\mathbf{R}}_M \qquad (6\text{-}130)$$

are the MLE DOA estimates and the set Ω is defined by (6-16). In the above, $\mathbf{P}$ is defined by (6-7). The trace term (6-130) can be written

$$\begin{aligned} \operatorname{tr} \mathbf{P}\hat{\mathbf{R}}_M &= \operatorname{tr} \mathbf{P}\hat{\mathbf{R}}_M\mathbf{P} = \operatorname{tr} \mathbf{U}_\nu\mathbf{U}_\nu^H\hat{\mathbf{R}}_M\mathbf{U}_\nu\mathbf{U}_\nu^H \\ &= \operatorname{tr} \mathbf{U}_\nu^H\hat{\mathbf{R}}_M\mathbf{U}_\nu \end{aligned} \qquad (6\text{-}131)$$

Comparing (6-131) with the argument of the MAP objective function given by (6-30), we see that the MLE minimizes the *trace* of $\mathbf{U}_\nu^H\mathbf{R}_M\mathbf{U}_\nu$, whereas the MAP method minimizes the *determinant* of the same matrix.

We applied the MUSIC method firstly under the assumption that the noise environment is spatially white—the method is designated as MUSIC_1, and then under the assumption that the colored noise covariance matrix $\boldsymbol{\Sigma}$ is exactly known—the method is designated as MUSIC_2. In MUSIC_1, the eigen-decomposition of the data covariance matrix is performed and a matrix $\mathbf{V}_N$ based on the eigenvectors of the noise subspace is constructed. In MUSIC_2, the generalized eigen-decomposition of the matrix pencil $(\hat{\mathbf{R}}_M, \boldsymbol{\Sigma})$ is performed and a projector based on the generalized eigenvectors of the noise subspace is constructed. In both cases, the square magnitude of the inverse of the projected directional manifold $\mathbf{d}(\phi)$ as in (6-129), forms the respective spatial spectrum.

The four methods are tested with $M = 20$ for the various colored noise environments and each test is repeated 200 times so that the overall MSE for

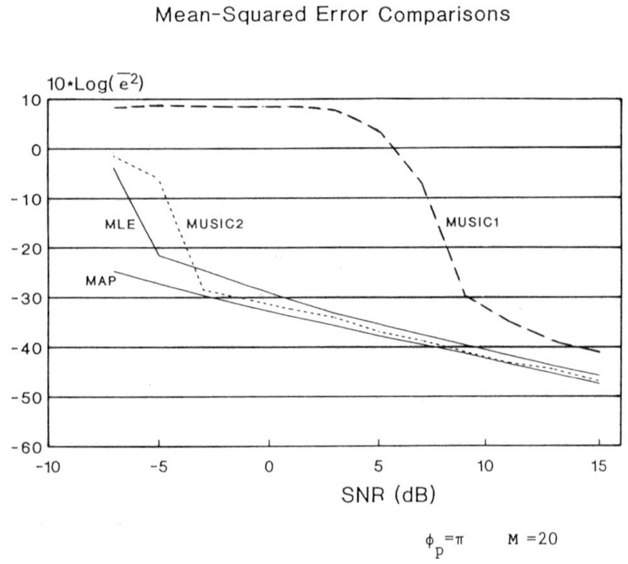

(a)

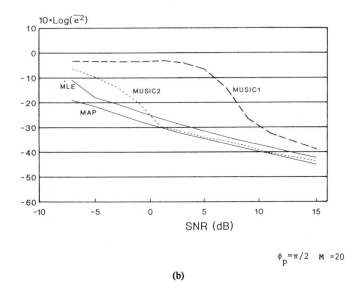

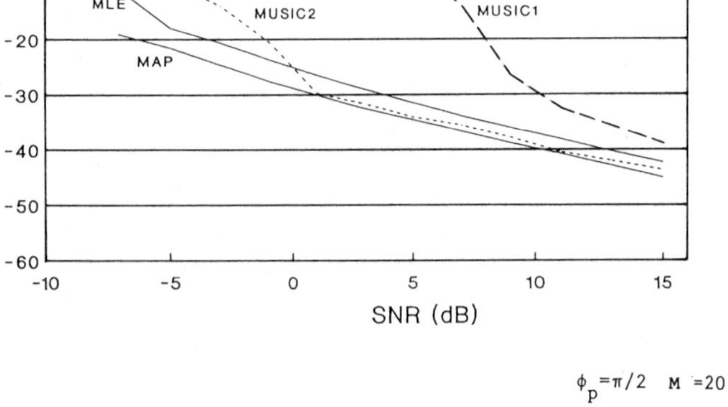

(b)

Figure 6-9 Comparison of performance of MAP, $MUSIC_1$, $MUSIC_2$, and MLE methods. (a) $\phi_p = \pi$; (b) $\phi_p = \pi/2$; (c) $\phi_p = \pi/8$.

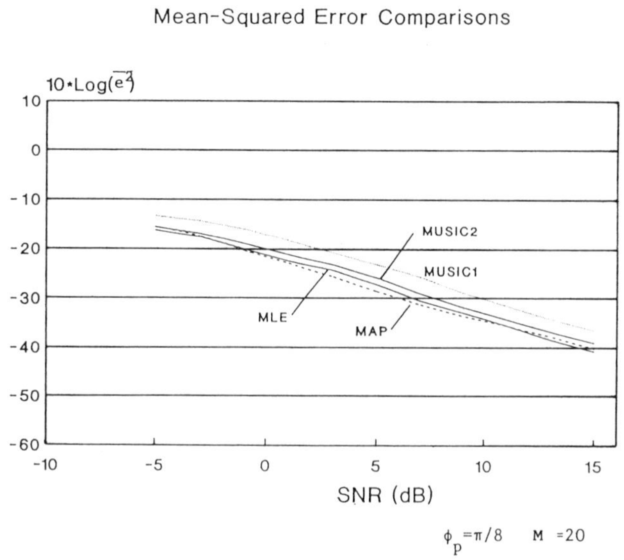

(c)

Figure 6-9 (*Continued*)

all the methods can be evaluated. Figures 6-9(a), (b), and (c) show the variation of the overall MSE, $\overline{e^2}$, with the SNR for the MAP, MUSIC_1, MUSIC_2, and MLE methods with $\phi_p = \pi$, $\phi_p = \pi/2$, and $\phi_p = \pi/8$ respectively. It is observed from Figures 6-9(a) and (b) that for the cases of $\phi_p = \pi$ and $\phi_p = \pi/2$, MAP consistently outperforms all the other methods. Furthermore, the MAP method has a lower threshold than all the other methods in both cases. In the case of $\phi_p = \pi/8$ (Fig. 6-9c)), the performance of all methods deteriorates. The performance of MAP, MLE, and MUSIC_2 does not differ very much throughout the whole range of SNR, but they are all superior to MUSIC_1.

Other values of ρ and ϕ_p have been used to obtain different noise conditions and the comparison repeated. Similar observations persist.

6-8 CONCLUSIONS

In this chapter, we have considered the problem of estimating the DOAs of narrowband signals in correlated noise. We approached the problem from a Bayesian point of view and examined the data in the noise subspace. The *a posteriori* PDF of the signal parameters is then obtained by averaging over the covariance matrix of the projected noise. To achieve that, we have selected a noninformative *a priori* PDF of the noise covariance matrix using Jeffrey's invariance principle. Then we chose the MAP criterion for the estimation of the signal parameters and arrived at a new and simple objective function for optimization in the noise subspace. We presented some properties of this objective function and showed, from a geometrical point of view, that the new criterion for estimating the DOA is geometrically equivalent to the location of a subspace in which the total volume of the parallelepipeds formed by the projected data row vectors is a minimum. Then, we presented a method of evaluating the approximate MAP estimate the result of which can be used as a starting point for a search for the exact MAP estimate in a much reduced region. Such a combination procedure grossly reduces the complexity of computation. The main attraction of this new method is that the objective function for optimization is relatively simple and that it performs well under unknown correlated noise.

We have also examined the asymptotic properties of the MAP approach for the estimation of the DOA in spatially correlated noise. By examining the leakage of signal and noise into the estimated signal and noise subspaces, we have found that the MAP estimate of the DOA is consistent when the noise is white, but may not be consistent under colored noise unless the SNR is high and/or the DOA of the signals is situated in a relatively flat part of the noise spectrum. An analysis of the performance of the MAP method of estimation has also been carried out presenting the MSE of the estimate with the assumptions that the errors of the DOA estimate and of the covariance matrix are sufficiently small so that terms containing the second and higher orders of these quantities are negligible. Under low SNR and/or small number of snapshots, these assumptions may not be valid. Computer simulation results show that the analysis is accurate. The comparison of the MAP method with MUSIC and MLE reveals the superiority of the MAP method. If, in the ignorance of the noise characteristics, the noise is assumed white and the MUSIC method is applied to estimate the DOA, the results are much inferior to those obtained by the MAP estimate which also assumes no knowledge of the noise characteristics. The MLE method performs much better than MUSIC under the assumption of white noise but is worse than the MAP method. Even with the exact knowledge of the noise covariance matrix, the results obtained by the MUSIC method are still slightly inferior to those obtained from the MAP estimate for most noise characteristics and for a large range of SNR. Therefore, it can be concluded that the MAP approach, especially in view of the possible parallel implementation, is a very attractive method of estimating the DOA of signals in unknown correlated noise.

As a footnote to this chapter, it is worthy of note that Wax [57] has, at time of writing, just recently proposed an extension to this MAP method. His extension involves a modified objective function which is the product of a determinant of a matrix involving a signal subspace projection, and the determinant of a noise subspace projection matrix. Our proposed MAP criterion involves only the latter determinant. His results show improved threshold performance over the method proposed here.

APPENDIX A A BRIEF NOTE ON THE CONCEPT OF NON-INFORMATIVE A PRIORI PDF [21]

An *a priori* PDF is supposed to represent knowledge about parameters before the outcome of an experiment is known. Since we assume that we know nothing about a particular parameter beforehand, the main issue is how to select an *a priori* PDF of the parameter which provides little information relative to what is expected to be provided by the intended experiment. Bayes suggested the following postulate: Where such knowledge was lacking concerning the nature of the prior distribution, it might be regarded as uniform.

In refutation of Bayes's postulate, it has been argued that if the PDF of a continuous parameter ψ were taken locally uniform, then the PDF of $\log \psi$, ψ^{-1}, or some other transformation of ψ (which might provide equally sensible bases for parametrization of the problem) would not be uniform. Thus, application of Bayes's postulate to different transformations of ψ would lead to *a posteriori* PDF from the same data which were inconsistent.

Data Translated Likelihood and Noninformative Prior. Our problem is to express the idea that little is known *a priori* to what the data has to tell us about a parameter ψ. What the data has to tell us about ψ is expressed by the likelihood function. Now, in general, suppose it is possible to express the unknown parameter ψ in terms of a transformation $\xi(\psi)$, so that the corresponding likelihood is *data translated*. This means that the likelihood curve for $\xi(\psi)$ is *completely determined a priori* (the shape and size of the likelihood curve in ξ is exactly known before the collection of the data) except for its *location* which depends on the data (in our particular case, the projected data) yet to be observed. Mathematically, a data translated likelihood must be expressible in the form

$$l(\psi \mid \mathbf{y}) = f_0[\xi(\psi) - f_1(\mathbf{y})] \tag{A-1}$$

where $f_0(\cdot)$ is a known function independent of the data $\mathbf{y}$ and $f_1(\mathbf{y})$ is a function of $\mathbf{y}$ only. Then to say that we know little *a priori* relative to what the data are going to tell us, may be expressed by saying that we are almost equally willing to accept one value of $\xi(\psi)$ as another. This state of indifference may be expressed by taking the *a priori* PDF of $\xi(\psi)$ to be locally uniform, and the resulting *a priori* PDF is called *noninformative* for $\xi(\psi)$ with respect to the data.

In general, if the noninformative *a priori* PDF is locally uniform in $\xi(\psi)$, then the corresponding noninformative *a priori* PDF for ψ is locally proportional to $|d\xi/d\psi|$, assuming the transformation is one-to-one, that is,

$$p(\psi) = p(\xi)\left|\frac{d\xi}{d\psi}\right| \propto \left|\frac{d\xi}{d\psi}\right| \tag{A-2}$$

The idea of a uniform *a priori* PDF needs clarification, since it is meaningless to consider a uniform *a priori* PDF on a parameter with unbounded range. However, in [21], this problem is addressed by suggesting that, for any physical experiment, any parameter is in fact bounded over a finite region of interest. The practical uniform *a priori* PDF is thus uniform over this region, but decays to zero outside.

The above argument dealing with noninformative priors can be easily extended to cases of multiple parameters. In the case of the projected noise covariance matrix, the distribution of the projected data $\mathbf{y}$ involves $(N-K)^2$ parameters $s^{11}, \ldots, s^{(N-K)(N-K)}$ which are the elements of $\mathbf{S}^{-1}$. A data translated likelihood must be of the form

$$l(\mathbf{S}^{-1} \mid \mathbf{y}) = f_0[\boldsymbol{\Xi}(\mathbf{S}^{-1}) - \mathbf{F}_1(\mathbf{y})] \tag{A-3}$$

where $\boldsymbol{\Xi}$ is a one-to-one transformation of $\mathbf{S}^{-1}$ having elements $\xi_{11}, \ldots, \xi_{(N-K)(N-K)}$ and $\mathbf{F}_1$ is an $(N-K) \times (N-K)$ matrix function of $\mathbf{y}$. A locally uniform *a priori* PDF of $\boldsymbol{\Xi}$ is noninformative if the likelihood function in (A-3) is data translated. The corresponding noninformative *a priori* PDF of the projected noise covariance matrix is then

$$p(\mathbf{S}^{-1} \mid \boldsymbol{\phi}) \propto \left| \frac{\nabla(\xi_{11}, \ldots, \xi_{(N-K)(N-K)})}{\partial(s^{11}, \ldots, s^{(N-K)(N-K)})} \right| \tag{A-4}$$

where $|\cdot|$ denotes the absolute value of a quantity and $\partial(\xi_{11}, \ldots, \xi_{(N-K)(N-K)}) / \partial(s^{11}, \ldots, s^{(N-K)(N-K)})$ is the Jacobian $((N-K)^2 \times (N-K)^2)$ determinant of the transformation involving the elements ξ_{ij} and s^{ij} of $\boldsymbol{\Xi}$ and $\mathbf{S}^{-1}$ respectively.

Jeffreys's Rule. By observing that the log-likelihood function for $\mathbf{S}^{-1}$ and $\boldsymbol{\Xi}$ are approximately quadratic asymptotically such that

$$\begin{aligned} L(\mathbf{S}^{-1} \mid \boldsymbol{\phi}, \mathbf{y}) = \log l(\mathbf{S}^{-1} \mid \boldsymbol{\phi}, \mathbf{y}) \approx L(\mathbf{S}^{-1} \mid \boldsymbol{\phi}, \mathbf{y}) \\ - \frac{M}{2} (\mathbf{S}^{-1} - \hat{\mathbf{S}}^{-1})^H C_{\hat{S}^{-1}} (\mathbf{S}^{-1} - \hat{\mathbf{S}}^{-1}) \end{aligned} \tag{A-5}$$

where

$$C_{\hat{S}^{-1}} \approx \frac{1}{M} \mathbf{J}_{\hat{S}^{-1}} = \frac{1}{M} E[-\nabla_{\hat{S}^{-1}} \cdot \nabla_{\hat{S}^{-1}} L] \tag{A-6}$$

Jeffreys [19] argues that ideally the transformation $\boldsymbol{\Xi}(\mathbf{S}^{-1})$ should be chosen such that $\mathbf{J}_{\hat{\Xi}}$ is a constant matrix independent of $\hat{\boldsymbol{\Xi}}$ so that the likelihood function would be data translated. Because this is not possible in general, we may seek a transformation $\boldsymbol{\Xi}$ which ensures

$$(\boldsymbol{\Xi} - \hat{\boldsymbol{\Xi}})^H \mathbf{J}_{\hat{\Xi}} (\boldsymbol{\Xi} - \hat{\boldsymbol{\Xi}}) < \text{constant} \tag{A-7}$$

to remain constant for different $\hat{\boldsymbol{\Xi}}$. Since the square root of the determinant, $\sqrt{\det(\mathbf{J}_{\hat{\Xi}})}$, measures the volume of the likelihood region, the above requirement is equivalent to seeking a transformation for which det $(\mathbf{J}_{\hat{\Xi}})$ is independent of $\hat{\boldsymbol{\Xi}}$. To find such a transformation, we note that

$$\left\{ \frac{\partial(\xi_{11}, \ldots, \xi_{(N-K)(N-K)})}{\partial(s^{11}, \ldots, s^{(N-K)(N-K)})} \right\}^2 \det(\mathbf{J}_{\Xi}) = \det(\mathbf{J}_{S^{-1}}) \tag{A-8}$$

whence, the above requirement is satisfied if

$$\frac{\partial(\xi_{11}, \ldots, \xi_{(N-K)(N-K)})}{\partial(s^{11}, \ldots, s^{(N-K)(N-K)})} \propto \sqrt{\det(\mathbf{J}_{S^{-1}})} \tag{A-9}$$

Using (A-9) in (A-4), the corresponding noninformative PDF of the projected noise covariance matrix is then

$$p(\mathbf{S}^{-1} \mid \boldsymbol{\phi}) \propto \sqrt{\det(\mathbf{J}_{S^{-1}})} \tag{A-10}$$

which is Jeffreys rule.

As in illustrative example in the evaluation of Jeffries' priors, and as a means of proving (6-25), we present the following theorem.

Theorem The determinant of the Fisher information matrix $\mathbf{J}_{S^{-1}}$ defined by (6-23) is given by $M(\det(\mathbf{S}^{-1}))^{-2(N-K)}$.

The result of this theorem is that the prior PDF of the projected noise covariance matrix is given by (6-26).

Proof. Using (6-22), we can write

$$\boldsymbol{\nabla}_{S^{-1}} L = M \boldsymbol{\nabla}_{S^{-1}} \{\log \det \mathbf{S}^{-1}\} - \boldsymbol{\nabla}_{S^{-1}} \left\{ \sum_{n=1}^{M} \mathbf{y}^H(m) \mathbf{S}^{-1} \mathbf{y}(m) \right\} \tag{A-11}$$

where $\boldsymbol{\nabla}_{\mathbf{S}^{-1}}$ is the complex differential operator defined in (6-24), $L(\cdot)$ is the log-likelihood function, $\mathbf{y}(m)$ is an N-vector of observed data, and $\mathbf{S}$ is the true covariance matrix of the projected data. The first term in (A-11) can be evaluated following standard matrix calculus procedures [24] extended to the complex operator, so that

$$\boldsymbol{\nabla}_{S^{-1}} \{\log \det \mathbf{S}^{-1}\} = \{(\mathbf{S}^{-1})^{-1}\}^T = \mathbf{S}^T \tag{A-12}$$

The second term in (A-11) can be evaluated as follows: let

$$\mathbf{y}(m) = [y_1(m) \quad y_2(m) \quad \cdots \quad y_{N-K}(m)]^T \tag{A-13}$$

Then,

$$\mathbf{y}^H(m)\mathbf{S}^{-1}\mathbf{y}(m) = \sum_{i=1}^{N-K}\sum_{k=1}^{N-K} y_i(m)y_k^*(m)s^{ik} \tag{A-14}$$

Employing the complex matrix differential operator defined in (6-24) on (A-14), we obtain the lnth element as

$$[\nabla_{S^{-1}}\{\mathbf{y}^H(m)\mathbf{S}^{-1}\mathbf{y}(m)\}]_{ln} = y_l(m)y_n^*(m) \tag{A-15}$$

Substituting (A-12) and (A-15) into (A-11), we have

$$\nabla_{S^{-1}}L = M\mathbf{S}^T - \sum_{m=1}^{M} \mathbf{y}(m)\mathbf{y}^H(m) \tag{A-16}$$

Using the definition of (6-23), the Fisher information matrix can be written as

$$\mathbf{J}_{S^{-1}} = -E[\nabla_{S^{-1}}\nabla_{S^{-1}}L] = -E[M\nabla_{S^{-1}}\mathbf{S}^T] \tag{A-17}$$

To evaluate $\nabla_{S^{-1}}\mathbf{S}^T$ in (A-17) we go back to the definition of the matrix differential operator operating on another matrix, that is, by definition, for any $(N-K)\times(N-K)$ matrix $\mathbf{A}$, we have

$$\nabla_{S^{-1}}\mathbf{A} \equiv \begin{bmatrix} \nabla_{s11}\mathbf{A} & \nabla_{s12}\mathbf{A} & \cdots & \nabla_{s1(N-K)}\mathbf{A} \\ \nabla_{s21}\mathbf{A} & \nabla_{s22}\mathbf{A} & \cdots & \nabla_{s2(N-K)}\mathbf{A} \\ \vdots & \vdots & \cdots & \vdots \\ \nabla_{s(N-K)1}\mathbf{A} & \nabla_{s(N-K)2}\mathbf{A} & \cdots & \nabla_{s(N-K)(N-K)}\mathbf{A} \end{bmatrix} = \sum_{l=1}^{N-K}\sum_{n=1}^{N-K} \mathbf{E}_{ln} \otimes \nabla_{sln}\mathbf{A} \tag{A-18}$$

where the differential ∇_{sln} is defined in (6-24), the matrix $\mathbf{E}_{ln}$ is an $(N-K)\times(N-K)$ matrix having unity as the lnth element and zero elsewhere, and the operation $\otimes$ is the Kronecker product such that for an $(N-K)\times(N-K)$ matrix $\mathbf{C}$ and another matrix $\mathbf{B}$,

$$\mathbf{C}\otimes\mathbf{B} \equiv \begin{bmatrix} c_{11}\mathbf{B} & c_{12}\mathbf{B} & \cdots & c_{1(N-K)}\mathbf{B} \\ c_{21}\mathbf{B} & c_{22}\mathbf{B} & \cdots & c_{2(N-K)}\mathbf{B} \\ \vdots & \vdots & \cdots & \vdots \\ c_{(N-K)1}\mathbf{B} & c_{(N-K)2}\mathbf{B} & \cdots & c_{(N-K)(N-K)}\mathbf{B} \end{bmatrix} \tag{A-19}$$

Now,

$$(\mathbf{S}\mathbf{S}^{-1})^T = (\mathbf{S}^{-1})^T\mathbf{S}^T = \mathbf{I} \tag{A-20}$$

where $\mathbf{I}$ is the $(N-K)\times(N-K)$ identity matrix.

Applying $\nabla_{S^{-1}}$ on (A-20) we have

$$\{\nabla_{S^{-1}}(\mathbf{S}^{-1})^T\}\{\mathbf{I}\otimes\mathbf{S}^T\}+\{\mathbf{I}\otimes(\mathbf{S}^{-1})^T\}\{\nabla_{S^{-1}}\mathbf{S}^T\}=\mathbf{0} \qquad \text{(A-21)}$$

Hence,

$$\begin{aligned}\nabla_{S^{-1}}\mathbf{S}^T &= -\{\mathbf{I}\otimes(\mathbf{S}^{-1})^T\}^{-1}\{\nabla_{S^{-1}}(\mathbf{S}^{-1})^T\}\{\mathbf{I}\otimes\mathbf{S}^T\}\\ &= -\{\mathbf{I}\otimes\mathbf{S}^T\}\{\nabla_{S^{-1}}(S^{-1})^T\}\{\mathbf{I}\otimes\mathbf{S}^{\mathrm{T}}\} \end{aligned} \qquad \text{(A-22)}$$

But from the definition of (A-18), substituting $(\mathbf{S}^{-1})^T$ for $\mathbf{A}$, we obtain

$$\begin{aligned}\nabla_{S^{-1}}(\mathbf{S}^{-1})^T &= \sum_{l=1}^{N-K}\sum_{n=1}^{N-K}\mathbf{E}_{ln}\otimes\nabla_{s^{ln}}(\mathbf{S}^{-1})^T\\ &= \sum_{l=1}^{N-K}\sum_{n=1}^{N-K}\mathbf{E}_{ln}\otimes\mathbf{E}_{ln}^T\equiv\mathbf{\Pi}\end{aligned} \qquad \text{(A-23)}$$

We note that $\mathbf{\Pi}$ in (A-23) is an $(N-K)^2\times(N-K)^2$ permutation matrix the determinant of which is -1. Thus substituting (A-23) into (A-22), we have

$$\nabla_{S^{-1}}\mathbf{S}^T=-\{\mathbf{I}\otimes\mathbf{S}\}\cdot\mathbf{\Pi}\cdot\{\mathbf{I}\otimes\mathbf{S}\} \qquad \text{(A-24)}$$

Using (A-24) in (A-17) and evaluating the determinant, we have

$$\det(\mathbf{J}_{S^{-1}})=-ME[\det\{(\mathbf{I}\otimes\mathbf{S})\mathbf{\Pi}(\mathbf{I}\otimes\mathbf{S})\}] \qquad \text{(A-25)}$$

But

$$\det(\mathbf{CB})=\det(\mathbf{C})\cdot\det(\mathbf{B}) \qquad \text{(A-26)}$$

and

$$\det(\mathbf{I}\otimes\mathbf{S})=(\det\mathbf{S})^{N-K} \qquad \text{(A-27)}$$

Therefore

$$\det(\mathbf{J}_{S^{-1}})=M(\det\mathbf{S})^{2(N-K)} \qquad \text{(A-28)}$$

APPENDIX B DERIVATION OF THE DERIVATIVE OF A GENERALIZED DETERMINANT

It is well known that the derivative of a determinant of a full-rank matrix $\mathbf{A}$ with respect to a parameter ϕ is given as [30]

$$\frac{\partial\det(\mathbf{A})}{\partial\phi}=\det(\mathbf{A})\cdot\mathrm{tr}\left[\mathbf{A}^{-1}\frac{d\mathbf{A}}{\partial\phi}\right]$$

In this appendix, we generalize the above formula to include singular matrices. We define the $(N-K)$th trace of an $N \times N$ matrix $\mathbf{A}$, as in Section 6-3, as the sum of the $(N-K)$th order principal minors of $\mathbf{A}$; that is,

$$\mathrm{tr}_{N-K}(\mathbf{A}) = \sum_{\vartheta} \det(\mathbf{A}_{i_1, i_2, \ldots, i_K})$$

where $\mathbf{A}_{i_1, i_2, \ldots, i_K}$ denotes a principal submatrix obtained by deleting the i_1th, i_2th, ..., i_Kth rows and columns of $\mathbf{A}$, and ϑ denotes the set of all possible combinations of K deletions from N rows. The quantity $\mathrm{tr}_{N-K}(\mathbf{A})$ of a matrix $\mathbf{A}$ of rank $N-K$ is nonzero and may be regarded as a generalization of a determinant of a rank deficient matrix.

In this appendix, we show that

$$\frac{\partial}{\partial \phi} \mathrm{tr}_{N-K}(\mathbf{A}) = \mathrm{tr}_{N-K}(\mathbf{A}) \, \mathrm{tr}\left[\mathbf{A}^{-I} \frac{\partial \mathbf{A}}{\partial \phi}\right]$$

where $\mathbf{A}^{-I}$ is the pseudo-inverse of $\mathbf{A}$.

Let the characteristic polynomial of $\mathbf{A}$ be given by

$$h(\varepsilon) = \det(\mathbf{A} - \varepsilon \mathbf{I}) = \sum_{n=0}^{N} \frac{1}{n!} \varepsilon^n h^{(n)}(0) \tag{B-1}$$

Since the $(N-K)$th order trace of $\mathbf{A}$ is equal to $(-1)^{N-K}$ times the coefficient associated with ε^K in the characteristic polynomial, then

$$\begin{aligned}
\frac{\partial}{\partial \phi} \mathrm{tr}_{N-K}(\mathbf{A}) &= \frac{(-1)^{N-K}}{K!} \left[\frac{\partial}{\partial \phi} h^{(K)}(0)\right] \\
&= \frac{(-1)^{N-K}}{K!} \frac{\partial}{\partial \phi} \left[\frac{d^K}{d\varepsilon^K} \det(\mathbf{A} - \varepsilon \mathbf{I})\right]_{\varepsilon=0} \\
&= \frac{(-1)^{N-K}}{K!} \frac{d^K}{d\varepsilon^K} \left[\frac{\partial}{\partial \phi} \det(\mathbf{A} - \varepsilon \mathbf{I})\right]_{\varepsilon=0}
\end{aligned} \tag{B-2}$$

But $(\mathbf{A} - \varepsilon \mathbf{I})$ is of full rank and [30]

$$\frac{\partial}{\partial \phi} \det(\mathbf{A} - \varepsilon \mathbf{I}) = \det(\mathbf{A} - \varepsilon \mathbf{I}) \, \mathrm{tr}\left\{(\mathbf{A} - \varepsilon \mathbf{I})^{-1} \frac{\partial \mathbf{A}}{\partial \phi}\right\} \tag{B-3}$$

Thus, (B-2) becomes

$$\frac{\partial}{\partial \phi} \mathrm{tr}_{N-K}(\mathbf{A}) = \frac{(-1)^{N-K}}{K!} \sum_{k=0}^{K} \left[\binom{K}{k} h^{(k)}(\varepsilon) \, \mathrm{tr}\left\{\frac{d^{K-k}}{d\varepsilon^{K-k}} (\mathbf{A} - \varepsilon \mathbf{I})^{-1} \frac{\partial \mathbf{A}}{\partial \phi}\right\}\right]_{\varepsilon=0} \tag{B-4}$$

Now, if we let $\{\lambda_n\}$, $n = 1, \ldots, N$ be the eigenvalues of $\mathbf{A}$ where

$$\lambda_{N-K+1} = \lambda_{N-K+2} = \cdots = \lambda_N = 0$$

then the eigenvalues of $(\mathbf{A}-\varepsilon\mathbf{I})$ are $(\lambda_n-\varepsilon)$, $n=1,\ldots,N$. From the property that the $(N-k)$th order trace is equal to the sum of the products of the eigenvalues taken $N-k$ at a time, then

$$h^{(k)}(\varepsilon)=(-1)^{N-k}k!\,\mathrm{tr}_{N-k}(\mathbf{A}-\varepsilon\mathbf{I})=(-1)^{N-k}k!\sum_{i\in\mathscr{C}}(\lambda_{i_1}-\varepsilon)\cdots(\lambda_{i_{n-k}}-\varepsilon) \tag{B-5}$$

where $\mathscr{C}$ is the combination set of N numbers taken $N-k$ at a time.

For $k<K$, because there are K out of N of the eigenvalues $\{\lambda_n\}$ are zero, the lowest order of ε in the product $(\lambda_{i_1}-\varepsilon)\cdots(\lambda_{i_{N-k}}-\varepsilon)$ is ε^{K-k} and there are $\binom{K}{k}$ of these terms so that (B-5) becomes

$$h^{(k)}(\varepsilon)=(-1)^{N-k}k!\binom{K}{k}\times\left[\varepsilon^{K-k}\prod_{n=1}^{N-K}(\lambda_n-\varepsilon)+(\text{terms involving higher orders of }\varepsilon)\right] \tag{B-6}$$

Also, if we let $\mathbf{v}_n$ be the eigenvector of $\mathbf{A}$ corresponding to the eigenvalue λ_n, then $\{\mathbf{v}_n\}$ will also be the eigenvectors of $(\mathbf{A}-\varepsilon\mathbf{I})$ and we have

$$(\mathbf{A}-\varepsilon\mathbf{I})^{-1}=\sum_{n=1}^{N-K}\frac{1}{\lambda_n-\varepsilon}\mathbf{v}_n\mathbf{v}_n^H+(-\varepsilon)^{-1}\sum_{n=N-K+1}^{N}\mathbf{v}_n\mathbf{v}_n^{\dagger} \tag{B-7}$$

and

$$\frac{d^{K-k}}{d\varepsilon^{K-k}}(\mathbf{A}-\varepsilon\mathbf{I})^{-1}=(K-k)!\left\{\sum_{n=1}^{N-K}(\lambda_n-\varepsilon)^{-(K-k+1)}\mathbf{v}_n\mathbf{v}_n^H+\varepsilon^{-(K-k+1)}\sum_{n=N-K+1}^{N}\mathbf{v}_n\mathbf{v}_n^H\right\} \tag{B-8}$$

Thus, for $k<K$, from (B-6) and (B-8), we have

$$\begin{aligned}
&\lim_{\varepsilon\to 0}h^{(k)}(\varepsilon)\,\mathrm{tr}\left\{\frac{d^{K-k}}{d\varepsilon^{K-k}}(\mathbf{A}-\varepsilon\mathbf{I})^{-1}\frac{\partial\mathbf{A}}{\partial\phi}\right\}\\
&=\lim_{\varepsilon\to 0}(-1)^{N-k}K!\,\varepsilon^{K-k}\sum_{n=1}^{N-k}(\lambda_n-\varepsilon)\left[\mathrm{tr}\left\{\sum_{n=1}^{N-K}(\lambda_n-\varepsilon)^{-(K-k+1)}\mathbf{v}_n\mathbf{v}_n^H\frac{\partial\mathbf{A}}{\partial\phi}\right\}\right.\\
&\quad\left.+\varepsilon^{-(K-k+1)}\,\mathrm{tr}\left\{\sum_{n=N-K+1}^{N}\mathbf{v}_n\mathbf{v}_n^H\frac{\partial\mathbf{A}}{\partial\phi}\right\}\right]\\
&\quad+(\text{terms involving higher orders of }\varepsilon)\\
&=\lim_{\varepsilon\to 0}(-1)^{N-k}K!\sum_{n=1}^{N-K}\lambda_n\left[\mathrm{tr}\left(\varepsilon^{-1}\sum_{n=N-K+1}^{N}\mathbf{v}_n\mathbf{v}_n^H\right)\frac{\partial\mathbf{A}}{\partial\phi}\right]\\
&=(-1)^{N-k}K!\prod_{n=1}^{N-K}\lambda_n\,\mathrm{tr}\left(\mathbf{0}^{-I}\frac{\partial\mathbf{A}}{\partial\phi}\right)=0, \qquad k=0,1,\ldots,K-1
\end{aligned} \tag{B-9}$$

where we have used the property of the generalized inverse of a null matrix [44] in the last step of (B-9). Thus, in (B-4) all terms vanish except when $k = K$.

Now for $k = K$, using (B-6) and (B-8) in (B-4), we have

$$\frac{\partial}{\partial \phi} \operatorname{tr}_{N-K}(\mathbf{A}) = \lim_{\varepsilon \to 0} \prod_{n=1}^{N-K} (\lambda_n - \varepsilon)$$

$$\times \operatorname{tr}\left[\left\{ \sum_{n=1}^{N-K} (\lambda_n - \varepsilon)^{-1} \mathbf{v}_n \mathbf{v}_n^H + \varepsilon^{-1} \sum_{n=N-K+1}^{N} \mathbf{v}_n \mathbf{v}_n^H \right\} \frac{\partial \mathbf{A}}{\partial \phi}\right]$$

$$= \operatorname{tr}_{N-K}(\mathbf{A}) \operatorname{tr}\left[\mathbf{A}^{-I} \frac{\partial \mathbf{A}}{\partial \phi}\right] \quad \text{(B-10)}$$

where the terms within the braces have been identified to be the pseudo-inverse of **A**.

REFERENCES

[1] J. Capon (1969) "High resolution frequency wavenumber spectrum analysis", *Proc. IEEE*, **57**, 1408–1418.

[2] V. F. Pisarenko (1973) "The retrieval of harmonics from a covariance function", *Geophys. J. R. Astron. Soc.*, **33**, 347–366.

[3] J. P. Reilly and S. Haykin (1982) "Maximum likelihood receiver for low-angle tracking radar", Part 1 and Part 2, *Proc. IEEE, Pt. F*, **129**, 261–272, 331–340.

[4] D. H. Johnson and S. R. Degraff (1982) "Improving the resolution of bearing in passive sonar arrays by eigenvalue analysis", *IEEE Trans. Acoust., Speech, Signal Process.*, **30**, 638–647.

[5] D. W. Tufts and R. Kumaresan (1982) "Estimation of frequencies of multiple sinusoids: Making linear predictions perform like maximum likelihood", *Proc. IEEE*, **70**, 975–989.

[6] G. Bienvenue and L. Kopp (1983) "Optimality of high resolution array processing using the eigensystem approach", *IEEE Trans. Acoust., Speech, Signal Process.*, **31**, 1235–1247.

[7] R. Kumaresan and D. W. Tufts (1983) "Estimating the angles of arrival of multiple plane waves", *IEEE Trans. Aerosp. Electron. Systems*, **19**, 134–139.

[8] R. O. Schmidt (1986) "Multiple emitter location and signal parameter estimation", *IEEE Trans. Antennas Propag.*, **34**, 276–280.

[9] Y. Bresler and A. Macovski (1986) "Exact maximum-likelihood parameter estimation of superimposed exponential signals in noise", *IEEE Trans. Acoust., Speech, Signal Process.*, **34**, 1081–1089.

[10] I. Ziskind and M. Wax (1988) "Maximum likelihood localization of multiple sources by alternating projection", *IEEE Trans. Acoust., Speech, Signal Process.*, **36**, 1553–1560.

[11] M. Wax and T. Kailath (1985) "Detection of signals by information theoretic criteria", *IEEE Trans. Acoust., Speech, Signal Process.*, **33**, 387–392.

[12] Y. Yin and P. Krishnaiah (1987) "On some nonparametric methods for detection of the number of signals", *IEEE Trans. Acoust., Speech, Signal Process.*, **35**, 1533–1538.

[13] L. C. Zhao, P. R. Krishnaiah, and Z. D. Bai (1986) "On detection of signals in presence of white noise", *J. Multivar. Anal.*, **20**, 1–25.

[14] W. G. Chen, K. M. Wong, and J. P. Reilly (1991) "Detection of the number of signals—a predictive eigen-threshold approach", *IEEE Trans. Signal Process.*, **39**, 1088–1098.

[15] Q. T. Zhang, K. M. Wong, P. C. Yip, and J. P. Reilly (1989) "Statistical analysis of the performance of information theoretic criteria in the detection of the number of signals in array processing", *IEEE Trans. Acoust., Speech, Signal Process.*, **37**, 1557–1567.

[16] K. M. Wong, Q. T. Zhang, J. P. Reilly, and P. Yip (1990) "On information theoretic criteria for the determination of the number of signals in high resolution array processing", *IEEE Trans. Acoust., Speech, Signal Process.*, **38**, 1959–1971.

[17] Q. Wu and D. R. Fuhrmann (1991) "A parametric method for determining the number of signals in narrowband direction-finding", *IEEE Trans. Acoust., Speech, Signal Process.* **39**, 1848–1857.

[18] A. Paulraj and T. Kailath (1986) "Eigenstructure methods for direction of arrival estimation in the presence of unknown noise fields", *IEEE Trans. Acoust., Speech, Signal Process.* **34**, 13–20.

[19] H. Jeffreys (1961) *Theory of Probability*, 3rd edition, Oxford University Press, London.

[20] H. L. Van Trees (1968) *Detection, Estimation, and Modulation Theory*, Pt. I, Wiley, New York.

[21] G. Box and G. Tiao (1973) *Bayesian Inference in Statistical Analysis*, Addison-Wesley, Reading, MA.

[22] L. E. Franks (1969) *Signal Theory*, Prentice Hall, Englewood Cliffs, NJ.

[23] P. Lancaster and M. Tismenetsky (1985) *The Theory of Matrices*, Academic Press, New York.

[24] A. Graham (1981) *Kronecker Products and Matrix Calculus with Applications*, Ellis Horwood, New York.

[25] F. A. Graybill (1969) *Matrices with Applications in Statistics*, Wadsworth.

[26] G. E. P. Box and N. R. Draper (1965) "The Bayesian estimation of common parameters from several responses", *Biometrika*, **52**, 355–365.

[27] N. R. Goodman (1963) "Statistical analysis based on a certain multivariate complex Gaussian distribution (an introduction)", *Ann. Math. Statist.*, **34**, 152–171.

[28] A. A. Giordano and F. M. Hsu (1985) *Least Square Estimation with Applications to Digital Signal Processing*, Wiley, New York.

[29] M. S. Srivastava and E. M. Carter (1983) *An Introduction to Applied Multivariate Statistics*, North-Holland, Amsterdam.

[30] R. Bellman (1970) *Introduction to Matrix Analysis*, McGraw-Hill, New York.

[31] D. Finkbeiner (1966) II, *Introduction to Matrices and Linear Transformations*, 2nd edition, W. H. Freeman, San Francisco, CA.

[32] G. H. Golub and C. F. Van Loan (1989) *Matrix Computations*, John Hopkins University Press, Baltimore, MD.

[33] W. M. Gentleman and H. T. Kung (1981) "Matrix triangularization by systolic arrays", in *Proc. SPIE*, Vol. 298, *Real-Time Signal Processing IV*, San Diego, CA, pp. 19–26.

[34] K. M. Wong, J. P. Reilly, Q. Wu, and S. Qiao (1990) "High resolution array processing in unknown correlated noise", Technical Report, CRL-226.

[35] K. M. Wong and J. P. Reilly (1992) "Estimation of the direction of arrival of signals in unknown correlated noise, Part II: Asymptotic properties and performance analysis of the MAP method", *IEEE Trans. Signal Process.* Vol. 40 Aug.

[36] M. Miller and D. Fuhrmann (1990) "Maximum likelihood narrow-band direction finding and the EM algorithm", *IEEE Trans. Acoust., Speech, Signal Process.* **38**, 1560–1577.

[37] K. C. Sharman and T. S. Durrani (1988) "Maximum likelihood parameter estimation by simulated annealing", in *Proc. ICASSP'88*, New York.

[38] A. Swindlehurst, B. Ottersten, R. Roy, and T. Kailath (1991) "Multiple invariance ESPRIT", *IEEE Trans. Acoust., Speech, Signal Process.* submitted.

[39] P. E. Gill, W. Murry, and M. H. Wright (1981) *Practical Optimization*, Academic Press, London.

[40] Q. Wu (1989) "Array signal processing at low signal to noise ratio", D Sc Dissertation, Department of Electronic Engineering, Washington University, St. Louis, MO.

[41] M. Viberg and B. Ottersten (1991) "Sensor array processing based on subspace fitting", *IEEE Trans. Signal Process.*, **39**, 1110–1121.

[42] M. Viberg, B. Ottersten, and T. Kailath (1991) "Performance analysis of the total least squares ESPRIT algorithm", *IEEE Trans. Signal Process.*, **39**, 1122–1135.

[43] J. E. Dennis, Jr., and D. J. Woods (1987) "New computing environments: microcomputer in large-scale computing", in *SIAM*, pp. 116–122.

[44] P. Lancaster and M. Tismenetsky (1985) *The Theory of Matrices with Applications*, 2nd edition, Academic Press, New York.

[45] L. C. Zhao, P. R. Krishnaiah, and B. D. Bai (1986) "On detection of number of signals in presence of white noise", *J. Multivariate Anal.*, **20**, 1–25.

[46] D. J. Jeffries and D. R. Farrier (1985) "Asymptotic results for eigenvector methods", *Proc. IEE, Pt. F*, **132**, 589–594.

[47] M. Kaveh and A. J. Barabell (1986) "The statistical performance of the MUSIC and the minimum-norm algorithms in resolving plane waves in noise", *IEEE Trans. Acoust., Speech, Signal Process.*, **34**, 331–341. Corrections: *IEEE Trans. Acoust., Speech, Signal Process.*, **34**, 633.

[48] R. J. Muirhead (1982) *Aspects of Multivariate Statistical Theory*, Wiley–Interscience, New York.

[49] J. H. Wilkinson (1965) *The Algebraic Eigenvalue Problem*, Oxford University Press.

[50] R. Bellman (1970) *Introduction to Matrix Analysis*, McGraw-Hill, New York.

[51] G. H. Golub and V. Pereyra (1973) "The differentiation of pseudo-inverses and nonlinear least squares problems whose variables separate", *SIAM J. Numer Anal.*, **10**, 413–432.

[52] A. E. Taylor and D. C. Lay (1980) *Introduction to Functional Analysis*, 2nd edition, Wiley, New York.

[53] A. Graham (1981) *Kronecker Products and Matrix Calculus with Applications*, Ellis Horwood, New York.

[54] N. R. Goodman (1963) "Statistical analysis based on a certain multivariate complex Gaussian distribution (an introduction)", *Ann. Math. Statist.*, **34**, 152–171.

[55] G. A. F. Seber (1984) *Multivariate Observations*, Wiley, New York.

[56] K. M. Wong, J. P. Reilly, Q. Wu, and S. Qiao (1990) "High resolution array signal processing in unknown correlated noise", Research Report CRL 226, McMaster University.

[57] M. Wax (1991) "Detection of coherent and noncoherent signals via the stochastic signals model", *ICASSP91*, Toronto, Canada.

[58] R. P. Brent and F. T. Luk (1985) "The solution of singular-value and symmetric eigenvalue problems on multiprocessor arrays", *SIAM J. Sci. Statist. Comput.*, **6**, 69–84.

[59] R. P. Brent, F. T. Luk, and C. Van Loan (1983) "Computation of the singular value decomposition using mesh-connected processors", *J. VLSI Comput. Systems*, **1**, 242–270.

[60] J. P. Reilly, "Systolic array evaluation of polynomials with application to nonlinear spectrum analysis", *Electron. Lett.*, **22** 1300–1302.

7

ADAPTIVE RADAR PARAMETER ESTIMATION WITH THOMSON'S MULTIPLE-WINDOW METHOD

A. DROSOPOULOS and S. HAYKIN
Communications Research Laboratory,
McMaster University,
Hamilton, Ontario, Canada

7-1 INTRODUCTION

Since its early inception in the 1930s and rapid development in the early 1940s, radar (RAdio Detection And Ranging) is designed to provide information on the detection and ranging of targets. A pulse of electromagnetic energy is emitted, with the radar antenna physically pointing in a certain direction, bounces off a target (if it is there) and returns to the radar. If the target is in close proximity to a scattering surface, the desired target signal is scattered, leading to a multipath component at the receiver, as illustrated in Fig. 7-1. Resolution limitations due to finite transmit power, finite receive antenna aperture, and frequency capabilities of the radar hardware exist as well.

Nowadays, with the more sophisticated electronics available and ingenious signal processing, radar capabilities have been pushed far beyond what they initially were. Demands have increased too. We would like the radar to

Adaptive Radar Detection and Estimation, Edited by Simon Haykin and Allan Steinhardt.
ISBN 0-471-54468-X

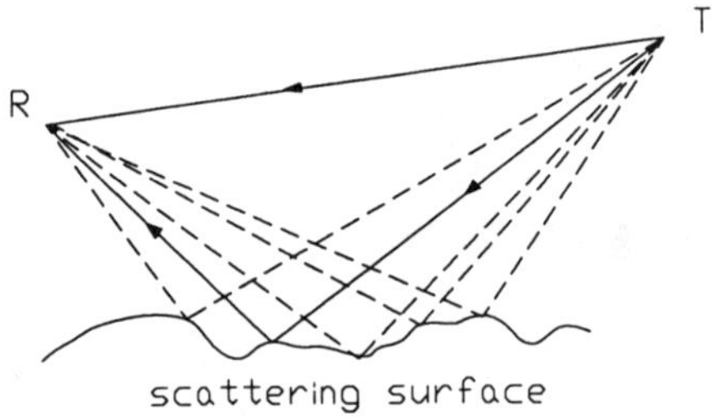

Figure 7-1 Illustrating the variety of paths (multipath) that a signal from a target (T) can reach the radar receiver (R) over a water surface.

be able to classify targets. In SAR (Synthetic Aperture Radar) the effective aperture is increased, providing very fine resolution radar images. Satellites employ radar too, probing weather patterns and doing remote sensing of the earth's surface. The spatial and temporal characteristics of all the possible received signals, are now more necessary than ever in the effective modeling and understanding of the physical phenomena involved in electromagnetic wave propagation. Mismatches from assumed models that describe what should be observed and the actual observations, can seriously affect overall system performance.

In this chapter we deal with the angle-of-arrival estimation problem in the presence of multipath. This is a difficult problem to solve, especially at low-grazing angles when it is difficult to separate the multipath from the desired signal. A simple model of specular reflection from an idealized smooth surface may account for the specular component of multipath. This model produces an *image* target beneath the real one, and the received signal from this image target has a magnitude related to the target signal by the reflection or Fresnel coefficient. A more accurate model takes into account surface roughness which modifies somewhat the specular component. Furthermore, nonspecular components (diffuse multipath) are then introduced as well. Since multipath is strongly correlated with the desired target signal, fading can occur (i.e., desired signal cancellation because of phase opposition with multipath). At low-grazing angles both components enter at the main lobe of the receiving antenna, which makes the resolution of the two components extremely hard.[1]

Vytas Kezys and Ed Vertatschitsch of the Communications Research Laboratory (CRL), McMaster University, designed and built a 32-element sampled aperture linear array for the purpose of collecting real multipath data at low-grazing angle conditions over the sea surface. To simplify the signal processing, a bistatic radar system was used, with a separate transmitter posing as a target and the sampled aperture antenna as the receiver. In this way the received signal is composed totally of the desired target

[1] This is due to practical limitations on the physical size of the antenna aperture. If the aperture could become arbitrarily large then there would be no resolution problem.

component and multipath. There is no other clutter component. Data collected with this array are used as a testbed later in this chapter.

Modern signal processing techniques that promise increased resolution are examined in this chapter. However, this increased resolution can only be achieved by having accurate models of the phenomena involved in order to be able to estimate the desired parameters (directions-of-arrival) from the data. Model simplicity usually assumes the existence of specular multipath only, ignoring diffuse multipath. Theoretical models of vector electromagnetic wave rough surface scattering can become quite complex at low-grazing angles when shadowing and diffraction effects are significant. In fact, a full solution has yet to be developed.

In this particular study, a relatively new, nonparametric technique is developed that estimates in an optimum and data-adaptive manner the spatial/temporal (wavenumber/frequency) characteristics of the received signal *without a priori model assumptions*. Our goal is first to describe the method in sufficient detail as to make it informative and useful to implement in any situation where accurate experimental spectra are desired. Second, we describe our results of applying this method to angle-of-arrival estimation at low-grazing angles with diffuse multipath taken into account. Finally, we compare some particular theoretical spectra with measured results.

7-2 LOW-ANGLE TRACKING RADAR PROBLEM

Data-adapative parameter estimation refers to the estimation of one or more of the radar parameters (e.g., range, angle-of-arrival) in a data-adaptive manner. In this chapter the focus will be on the angle-of-arrival (AOA) estimation. The instrumentation most suited for this purpose is a sampled-aperture antenna where concurrent data samples are taken at different points in space and subsequently combined in a suitable manner (beamforming) to estimate the AOA of an incoming signal. The data adaptivity comes in the signal processing involved with the beamforming process.

The simplest form of beamforming is the use of all the data samples to construct the wavenumber spectrum (analogous to the frequency spectrum of a time series), which can be one- or two-dimensional depending on whether the sampled aperture is one- or two-dimensional. In essence the sampled aperture can resolve incoming signals in a direction perpendicular to the aperture axis (a vertical aperture resolves elevation angles and a horizontal aperture resolves azimuth angles). The resolution limit is on the order of a beamwidth (which, for an M-element linear array, is defined as $2\pi/M$). The more sensors the aperture has, the better the resolution capability. Superresolution techniques can, in principle, achieve better performance than traditional techniques based on Fourier transforms at the cost of more intense signal processing.

At low grazing angles the situation becomes particularly difficult because, on the one hand, the direct and specular components are both at sub-beamwidth separations, and also the specular component, being coherent with the direct, can approach phase opposition with it; signal cancellation thereby occurs. This is particularly severe for the common monopulse radar which is normally incapable of performing well in such a case. The radar can then lose track of a low altitude flying target.

The most common way to deal with this problem is to perform beam-forming of the received signal in order to suppress signals coming from directions below the horizon, that is, to take care that the main lobe of the receiver antenna is always above the horizon, and to suppress sidelobe signals. This is where adaptivity is most useful if the radar is designed to adapt its beamforming technique depending on the data received.

MTI (Moving Target Indication) is another general technique, of great use in ordinary clutter suppression. With this technique, signals from nonmoving targets, which have a fixed Doppler spectrum, are canceled out and effectively the radar sees only targets in motion. However, MTI is useless in the case of multipath since both multipath and the desired target signal have the same Doppler shift. The MTI processed signal would still contain the multipath component.

Practical issues (such as cost and physical limitations) put constraints on the number of elements in a sampled aperture and the separation between them. Thomson's multiple-window method (MWM) described in this chapter, seems to perform in a robust enough manner for this case of correlated signals and is capable of extracting the desired signal information in an optimum way.

7-3 SPECTRUM ESTIMATION BACKGROUND

Consider the sequence $\{x(t_i)\}_{i=1}^{N}$ composed of sequential samples from a single realization of a complex-valued, weakly stationary, continuous-time, one-dimensional[2] random process $X(t)$. We further assume that the process is zero-mean with autocorrelation function $r_x(\tau)$; that is,

$$\mathsf{E}\{X(t)\} = 0$$
$$\mathsf{E}\{X^*(t)X(t+\tau)\} = r_x(\tau) \tag{7-1}$$

If $\bar{X}(t) = \mathsf{E}\{X(t)\} \neq 0$ then the above relations are satisfied by $X(t) - \bar{X}(t)$.

[2] The generalization to more dimensions is done by simply allowing the index parameter time t to become a d-dimensional vector $\mathbf{t}$ where d is the dimension of the process. See [5, 6] for a concise explanation. In the following we consider one-dimensional processes only, since they can adequately describe our experimental data.

The definition of the power spectrum in terms of the autocorrelation function is given by the Einstein–Wiener–Khintchine relations

$$r_x(\tau) = \int_{-\infty}^{+\infty} S(f)e^{j2\pi f\tau}\, df \quad \text{and} \quad S(f) = \int_{-\infty}^{+\infty} r_x(\tau)e^{-j2\pi f\tau}\, d\tau \qquad (7\text{-}2)$$

where $S(f)$ is the power spectral density (PSD) or simply spectrum of the process $X(t)$. $S(f)\, df$ represents the average (over all realizations) contribution to the total power (or process variance) from all possible components of $X(t)$ with frequencies between f and $f + df$.

The power interpretation of the spectrum is more evident from an alternate definition in terms of the random process itself [27]:

$$S(f) = \lim_{T\to\infty} \mathsf{E}\left\{ \frac{1}{T} \left| \int_{-T/2}^{T/2} X(t)e^{-j2\pi ft}\, dt \right|^2 \right\} \qquad (7\text{-}3)$$

Typical nonparametric spectrum estimators for finite data may be considered approximations to either (7-2) (Blackman and Tukey method [3]) or (7-3) (modified periodogram). The Blackman and Tukey spectrum estimate, for an N size data sample, is given by the formula

$$\hat{S}(f) = \sum_{m=1-N}^{N-1} \hat{r}(\Delta m)d(m)e^{-j2\pi f\,\Delta m}$$

where the autocorrelation sequence is estimated as

$$\hat{r}(\Delta m) = \frac{1}{N-m} \sum_{n=1}^{N-m} x[\Delta(n+m)]x^*(\Delta n)$$

with $0 \leqslant m \leqslant N-1$, $\hat{r}(\Delta m) = \hat{r}^*(-\Delta m)$ for $m<0$, and Δ is the sampling period. The weight sequence $\{d(m)\}$ has real positive elements satisfying $d(m) = d(-m)$ to ensure that the spectral estimate is real, and $d(0) = 1$ to ensure that it is unbiased when the true spectrum is flat across $\mathcal{B}$, where $\mathcal{B} = \{f: |f| < 1/2\Delta\}$.

The modified periodogram spectrum estimate is given by

$$\hat{S}(f) = \left| \sum_{n=1}^{N} x(\Delta n)c(n)e^{-j2\pi f\,\Delta n} \right|^2$$

where $f \in \mathcal{B}$ and the weight sequence $\{c(n)\}$ typically has real, positive elements satisfying $\Sigma_{n=1}^{N} c^2(n) = \Delta$ to ensure that the spectral estimator is unbiased when the true spectral density is flat across $\mathcal{B}$. To approximate the ensemble averaging operator E, the data record is usually segmented and individual results are averaged to reduce estimator variance.

Usually, the underlying distribution of $X(t)$ is assumed to be Gaussian, so that the second order statistics suffice for a complete process description. If not, a higher order statistics hypothesis should be made (see [37] for bispectrum estimation using multiple windows). The ergodicity property, which holds for a zero-mean Gaussian process *with no line components*, is also frequently invoked, so that ensemble averages can be replaced by time averages.

The spectrum of a stochastic process can also be estimated by assuming a process model (usually AR, MA or ARMA) and then estimating the model parameters from the data. This approach (called *modern spectrum estimation* in contradistinction to the nonparametric approach which is called classical) works well provided the process can be accurately described by such a model; the material presented in Chapter 2 on adaptive radar detection is a good example of this approach.

Nevertheless, Thomson and Chave [40] point out that they have "analyzed many different time series from scientific and engineering problems" and neither of them "has yet seen an example which could reasonably be described by an autoregressive model". This is a compelling argument not to make the a priori assumption that diffuse multipath can be accurately described by such a model.

However, it is only after Thomson's classic 1982 paper [36], where the power of MWM is demonstrated, that interest in classical spectrum estimation has been renewed [25]. Basically, Thomson has proved that a more fruitful approach to a spectrum estimator is through the spectral representation of $X(t)$ itself (Cramér representation). Ishimaru [12] gives a particularly lucid explanation of how this representation is defined. Following his arguments, consider a stationary complex random function $X(t)$ that satisfies (7-1). In attempting to develop a spectral representation for the random function $X(t)$, it is tempting to write down the Fourier transform

$$X(t) = \int_{-\infty}^{\infty} X(f) e^{j2\pi ft}\, df$$

However, the stationarity assumption is then violated, since Dirichlet's condition requires that $X(t)$ be absolutely integrable, that is, $\int_{-\infty}^{\infty} |X(t)|\, dt$ be finite. To avoid this difficulty, the random function is represented by a *stochastic Fourier–Stieltjes integral*

$$X(t) = \int_{-\infty}^{\infty} e^{j2\pi ft}\, dZ(f)$$

where $dZ(f)$ is called the random amplitude. Let us determine the properties of $dZ(f)$ by examining (7-1). First, we require that

$$\mathsf{E}\{dZ(f)\} = 0$$

and second that the covariance[3] function

[3] The terms autocorrelation and covariance are equivalent for zero-mean processes.

$$\mathsf{E}\{X(t_1)X^*(t_2)\} = \int_{-\infty}^{\infty}\int_{-\infty}^{\infty} e^{j2\pi f_1 t_1 - j2\pi f_2 t_2}\mathsf{E}\{dZ(f_1)\, dZ^*(f_2)\}$$

be a function of $t_1 - t_2$ only. This requires us to write

$$\mathsf{E}\{dZ(f_1)\, dZ^*(f_2)\} = S(f_1)\delta(f_1 - f_2)\, df_1\, df_2 \tag{7-4}$$

where $S(f)$ is the power spectral density of the process, representing the amount of power density at different frequencies. It is identical to the one given previously as can be seen by substituting the covariance (7-4) in the Einstein–Wiener–Khintchine relations above. Note also that at different frequencies $\mathsf{E}\{dZ(f_1)\, dZ^*(f_2)\}$ is zero, that is, the increments $dZ(f)$ are orthogonal (the energy at different frequencies is uncorrelated).

For the discrete-time case, the Einstein–Wiener–Khintchine relations become

$$r_x(n) = \int_{-1/2}^{1/2} S(f)e^{j2\pi fn}\, df \quad \text{and} \quad S(f) = \sum_{n=-\infty}^{\infty} r_x(n)e^{-j2\pi fn} \tag{7-5}$$

where $n = 0, \pm 1, \ldots$ and the time between samples is taken to be 1, so that the frequency f is confined in the principal domain $(-\frac{1}{2}, \frac{1}{2}]$. Similarly, the discrete-time spectral representation of the time series $\{x(n)\}$ is given by

$$x(n) = \int_{-1/2}^{1/2} e^{j2\pi fn}\, dZ(f) \tag{7-6}$$

The spectral representation concept is quite basic and in essence says that any stationary time series can be interpreted as the limit of a finite sum of sinusoids $A_i \cos(2\pi f_i t + \Phi_i)$ over all frequencies $f = f_i$. The amplitudes $A_i = A(f_i)$ and phases $\Phi_i = \Phi(f_i)$ are uncorrelated random variables with $S(f) \approx \mathsf{E}\{A^2(f_i)\}$ for $f_i \approx f \neq 0$ and can be related to $dZ(f)$ in a simple way (see [19]). This point of view leads us to write

$$\mathsf{E}\{dZ(f)\} = 0\,, \qquad S(f)\, df = \mathsf{E}\{|dZ(f)|^2\} \tag{7-7}$$

as the proper definition of the power spectrum.

When a number of line components is present, the above relations are easily generalized to include them. The first moment becomes

$$\mathsf{E}\{dZ(f)\} = \sum_i \mu_i \delta(f - f_i)\, df \tag{7-8}$$

where f_i are the frequencies of the periodic or line components, and μ_i are their amplitudes. The continuous part of the spectrum, or second moment, becomes

$$S(f)\, df = \mathsf{E}\{|dZ(f) - \mathsf{E}\{dZ(f)\}|^2\} \tag{7-9}$$

First moments are associated with the study of periodic phenomena (harmonic analysis). Typically, a few such lines will exist in a process, each described by their amplitude, frequency and phase.[4] Such parameters may be estimated using techniques based on maximum-likelihood. In classic methods of spectrum estimation—nonparametric, based on the periodogram— the resolution limit (termed also as the *Rayleigh resolution limit*) is $1/T$ where T is the total observation time. Super-resolution, that is, discrimination of frequencies spaced closer than the Rayleigh resolution, is possible, depending on the SNR, which is defined as the ratio of power in the first moment to power in the second moment as a function of frequency.

Second moments, on the other hand, are stochastic in character. In contrast to the line spectrum of the first moments, the second moment spectrum is typically continuous and often smooth. In this case, one is trying to estimate a function of frequency, not just a handful of parameters and therefore maximum-likelihood parameter estimation is not applicable here. Concerning resolution, it is impossible now to resolve details separated by less than twice the Rayleigh limit. Typically, resolution is between $2/T$ and $50/T$, much poorer than Rayleigh.

Thomson [39] states that "confusing the distinction between the two moment properties will result in absurdities like smoothing line spectra or applying super-resolution criteria to noise-like processes". Finally, a point that must be kept in mind is the fact that classically, spectra are defined only for stationary processes. For nonstationary processes the usual assumption that allows us to work with them is that of local stationarity.

7-3-1 The Fundamental Equation of Spectrum Estimation

We assume that we have a finite data set of N contiguous samples, $x(0), x(1), \ldots, x(N-1)$, which are observations from a stationary, complex, ergodic, zero-mean, Gaussian process. The problem of spectrum analysis is that of estimating the statistical properties of $dZ(f)$ from the finite time series $x(0), x(1), \ldots, x(N-1)$.

Taking the Fourier transform of the data, we obtain

$$\tilde{x}(f) = \sum_{n=0}^{N-1} x(n) e^{-j2\pi fn} \tag{7-10}$$

and substituting the Cramér representation for $x(n)$, we arrive at the *fundamental equation of spectrum estimation*

$$\tilde{x}(f) = \int_{-1/2}^{1/2} \mathsf{D}_N(f - \nu)\, dZ(\nu) \tag{7-11}$$

[4] For an excellent treatment of decaying sinusoids with the multiple-window method, see [29, 30].

where the kernel is given by

$$\mathsf{D}_N(f) = \sum_{n=0}^{N-1} e^{-j2\pi fn} = \exp[-j\pi f(N-1)] \frac{\sin N\pi f}{\sin \pi f} \tag{7-12}$$

We can interpret the fundamental equation as a convolution that describes the window leakage or smearing that is a consequence of the finite sample size. Clearly, there is no obvious reason to expect the statistics of $\tilde{x}(f)$ to resemble those of $dZ(f)$.

The fundamental equation can be regarded as a *linear Fredholm integral equation of the first kind* for $dZ(f)$. Since this is the frequency domain expression of the projection from an infinite stationary process generated by the random orthogonal measure $dZ(f)$ onto the finite observation sample, it does not have an inverse. This makes it impossible to find exact or unique solutions, and our goal becomes instead one of searching for approximate solutions whose statistical properties are in some sense close to those of $dZ(f)$.

The above observation is another way of saying that the problem of spectrum estimation from finite data is *ill-posed*. Mullis and Scharf [25] define both the time-limiting operation (windowing, finite data) and isolation in frequency (power in a finite spectral window) as projection operators on the data, P_T and P_F respectively. These operators do not commute, that is, $\mathsf{P}_T\mathsf{P}_F \neq \mathsf{P}_F\mathsf{P}_T$. If they did, then their product would also be a projection and it would then be possible to isolate a signal component in both time and frequency. However, under certain conditions, $\mathsf{P}_T\mathsf{P}_F \approx \mathsf{P}_F\mathsf{P}_T$ and the product operator is close to a projection having rank NW, the time–bandwidth product. It turns out that Thomson's MWM is equivalent to a projection of the data onto a subspace where the signal power in a narrow spectral band is maximized, that is, the conditions required for the above mentioned operators to approximately commute are found.

7-4 THOMSON'S METHOD OF MULTIPLE-WINDOWS

Generally, in a radar environment the received signal is composed of the desired direct signal(s) coming from the target(s), their multipath reflections and/or clutter and the receiver noise. In detection and tracking, the need exists to accurately estimate the angle(s)-of-arrival (AOA) of the desired signal(s). Data from a sampled aperture antenna lead to the estimation of the wavenumber spectrum as well. Multipath is commonly divided into specular and diffuse. The first, being essentially a plane wave, appears as an additional spectral line, while the latter, being stochastic, has a broader continuous shape. The need exists therefore to somehow estimate this mixed spectrum, in the best way possible, from a finite set of data samples. Ideally, the method used for this purpose should be non-parametric in nature (this refers to the continuous spectrum background) so as not to be influenced by any a priori assumptions about the signal structure, and in fact, it should be able to validate or disprove them.

Thomson's multiple-window method (MWM)[5] as expanded in [36, 38, 39] is our choice. Useful background information and implementation examples are also given in [8, 19, 20, 23, 28, 29, 35]. Some recent work [13, 26] should also be pointed out as it offers a point of view that is more familiar to the array signal processing community. MWM is nonparametric, provides a unified approach on spectrum estimation, is optimally suited for finite-data samples, can be generalized to irregularly sampled multidimensional processes [5, 6], is consistent and efficient, has good bias control and stability and finally, it also provides an analysis of variance test for line components.

The solution of the fundamental equation (7-11) is found in terms of the eigenfunction expansion of the kernel. This is recognized as the Dirichlet kernel with known eigenfunctions, the prolate spheroidal wavefunctions, fundamental to the study of time and frequency limited systems. Exploiting this finite bandwidth property of the process to be estimated, a search for solutions to (7-11) is carried out in some local interval about f, say $(f - W, f + W)$, using the prolate spheroidal wavefunctions as a basis.

7-4-1 Prolate Spheroidal Wavefunctions and Sequences

From Slepian,[6] the eigenfunction expansion of the Dirichlet kernel [33], is

$$\int_{-W}^{W} \frac{\sin N\pi(f-f')}{\sin \pi(f-f')} U_k(N, W; f')\, df' = \lambda_k(N, W) U_k(N, W; f) \qquad \text{(7-13)}$$

where $U_k(N, W; f)$, $k = 0, 1, \ldots, N-1$ are the discrete prolate spheroidal wavefunctions (DPSWF) and W, $0 < W < \frac{1}{2}$ is the local bandwidth, normally on the order of $1/N$.

Some of the properties of the discrete prolate spheroidal wavefunctions are:

- The functions are ordered by their eigenvalues

  $$1 > \lambda_0(N, W) > \lambda_1(N, W) > \cdots > \lambda_{N-1}(N, W) > 0$$

  the first $K = 2NW$ being very close to 1.

[5] Also called multitaper.

[6] In the most recent literature, the prolate spheroidal wavefunctions and sequences are also referred to as Slepian functions and Slepian sequences, respectively, to honor David Slepian who first described their properties for use in signal processing and statistical applications. The term prolate spheroidal first came about from the solution of the wave equation in a prolate spheroidal corrdinate system. The zeroth order solution of that differential equation is also the solution to the integral equation (7-13).

- They are doubly orthogonal

$$\int_{-W}^{W} U_j(N, W; f)U_k(N, W; f)\, df = \lambda_k(N, W)\delta_{j,k}$$

$$\int_{-1/2}^{1/2} U_j(N, W; f)U_k(N, W; f)\, df = \delta_{j,k}$$

- Their Fourier transforms are the discrete prolate spheroidal sequences (DPSS)

$$v_n^{(k)}(N, W) = \frac{1}{\varepsilon_k \lambda_k(N, W)} \int_{-W}^{W} U_k(N, W; f)e^{-j2\pi f[n-(N-1)/2]}\, df$$

or

$$v_n^{(k)}(N, W) = \frac{1}{\varepsilon_k} \int_{-1/2}^{1/2} U_k(N, W; f)e^{-j2\pi f[n-(N-1)/2]}\, df$$

valid for $n, k = 0, 1, \ldots, N-1$ and

$$\varepsilon_k = \begin{cases} 1, & \text{for } k \text{ even} \\ i, & \text{for } k \text{ odd} \end{cases}$$

The DPSWFs can be expressed as

$$U_k(N, W; f) = \varepsilon_k \sum_{n=0}^{N-1} v_n^{(k)}(N, W)e^{j2\pi f[n-(N-1)/2]}$$

and the DPSS's satisfy a *Toeplitz matrix eigenvalue equation*

$$\sum_{m=0}^{N-1} \frac{\sin 2\pi W(n-m)}{\pi(n-m)}\, v_m^{(k)}(N, W) = \lambda_k(N, W)v_n^{(k)}(N, W)$$

Both of these equations give a relatively straightforward way of computing the DPSSs and DPSWFs for moderate values of N. In matrix form, the above eigenvalue equation is written as

$$\mathbf{T}(N, W)\mathbf{v}_{(k)}(N, W) = \lambda_k(N, W)\mathbf{v}_{(k)}(N, W)$$

where

$$\mathbf{T}(N, W)_{mn} =$$

$$\begin{cases} \sin[2\pi W(m-n)]/[\pi(m-n)], & m, n = 0, 1, \ldots, N-1 \text{ and } m \neq n \\ 2W, & \text{for } m = n \end{cases}$$

Both Thomson [36] and Slepian [31–34] give asymptotic expressions for their computation; it is probably the complexity of these expressions that initially discourages people from using the prolate basis. If, however, only the eigenvectors are required, Slepian [33] notes that the DPSSs satisfy a *Sturm–Liouville differential equation* leading to

$$\mathbf{S}(N, W)\mathbf{v}_{(k)}(N, W) = \theta_k(N, W)\mathbf{v}_{(k)}(N, W) \tag{7-14}$$

The matrix **S** is tridiagonal in the sense that

$$\mathbf{S}(N, W)_{ij} = \begin{cases} \frac{1}{2}\, i(N-1)\,, & j = i - 1 \\ \left(\frac{N-1}{2} - i\right)^2 \cos 2\pi W\,, & j = i \\ \frac{1}{2}\,(i+1)(N-1-i)\,, & j = i + 1 \\ 0\,, & \text{otherwise} \end{cases}$$

Even though the eigenvalues θ_k are not equal to λ_k, they are ordered in the same way and the eigenvectors are the same. Tridiagonal systems are easier than Toeplitz to solve, and this offers a good practical way of numerically computing the eigenvectors. The single EISPACK routine IMTQL2 is sufficient if one desires all eigenvectors. Since, however, only a small number of eigenvalues and eigenvectors is actually needed, the combination RATQR and TINVIT is less wasteful of computer time. Given the eigenvectors, the eigenvalues can then be found[7] from

$$\lambda_k(N, W) = [\mathbf{v}_{(k)}(N, W)]^T \mathbf{T}(N, W)\mathbf{v}_{(k)}(N, W) \tag{7-15}$$

Note that in our case, Slepian's Dirichlet kernel is modulated by a complex exponential factor. This gives us the eigenfunction expansion

$$\int_{-W}^{W} \mathsf{D}_N(f-\nu)V_k(\nu)\, d\nu = \lambda_k V_k(f) \tag{7-16}$$

where, for notational simplicity, the dependence on N and W has been suppressed. The connection with Slepian's [39] is established by writing

$$V_k(f) = (1/\varepsilon_k)e^{-i\pi f(N-1)}U_k(-f)$$

so that the Fourier transform of Slepian's DPSSs gives us

[7] Thomson [38] uses BISECT and TINVIT to evaluate the Slepian sequences, and $\lambda_k(N, W) = \int_{-W}^{W} |V_k(f)|^2\, df > \int_{-1/2}^{1/2} |V_k(f)|^2\, df$ for the eigenvalues.

$$V_k(f) = \sum_{n=0}^{N-1} v_n^{(k)}(N, W)e^{-j2\pi fn} \tag{7-17}$$

A step-by-step procedure in computing the data, $v_n^{(k)}$'s and spectral, $V_k(f)$'s windows (Slepian sequences and functions) is given as:

1. Obtain first an N-point data sample; this specifies N.
2. Select a time-bandwidth product NW; this specifies the analysis window W.
3. Use (7-14) and (7-15) to compute the λ_k's and $v_n^{(k)}$'s; actually, only the first $K = 2NW$ terms with the largest eigenvalues are needed (Thomson [38] suggests the use of $K = 2NW - 1$ to $K = 2NW - 3$ to minimize higher order window leakage).
4. Finally, use (7-17) with an FFT (preferably zero padded) to compute the corresponding $V_k(f)$'s.

Figures 7.2 and 7.3 show an example of data and spectral windows (Slepian sequences and functions) that are used in the test dataset below. Note that the importance of using these windows is the fact that they are optimum in the sense of energy concentrated within the frequency band $(f - W, f + W)$. In essence, by using them we are maximizing the signal energy within the band $(f - W, f + W)$ and minimizing at the same time the energy leakage outside this band. They are therefore the ideal choice to use

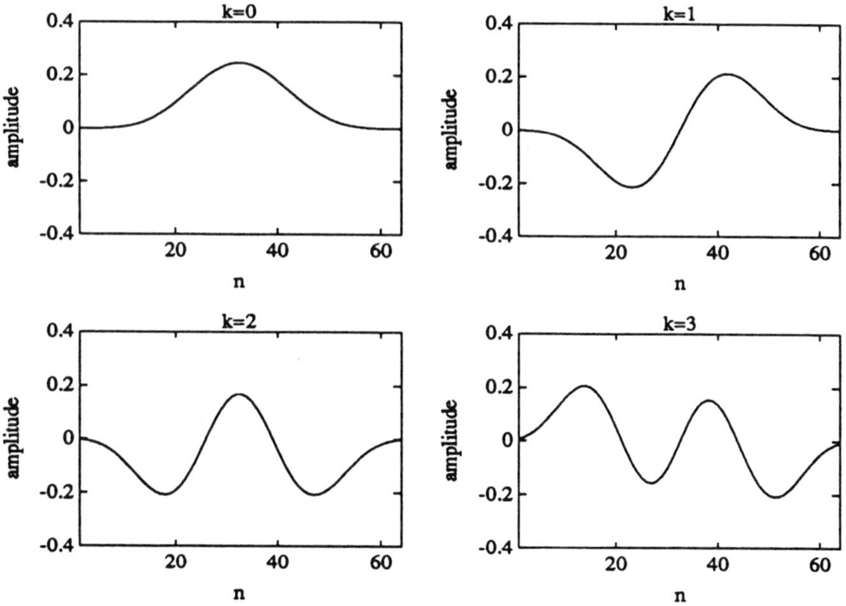

Figure 7-2 The first four data windows for the case $N = 64$ and $NW = 4$. They are simply displays of the first four eigenvectors $\mathbf{v}_{(k)}(N, W)$.

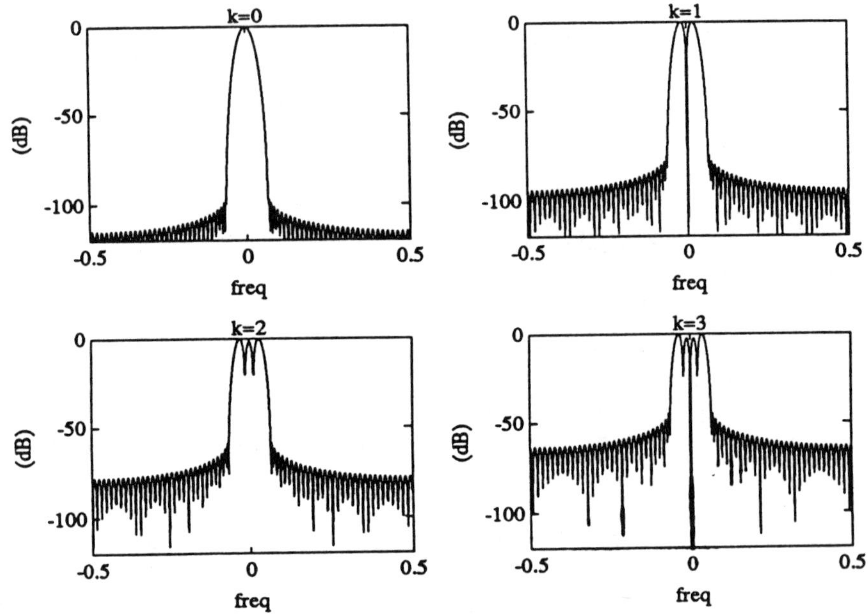

Figure 7-3 The first four spectral windows for the case $N=64$ and $NW=4$. These are the complex amplitudes squared (in dB) of the Fourier transforms of the above data windows.

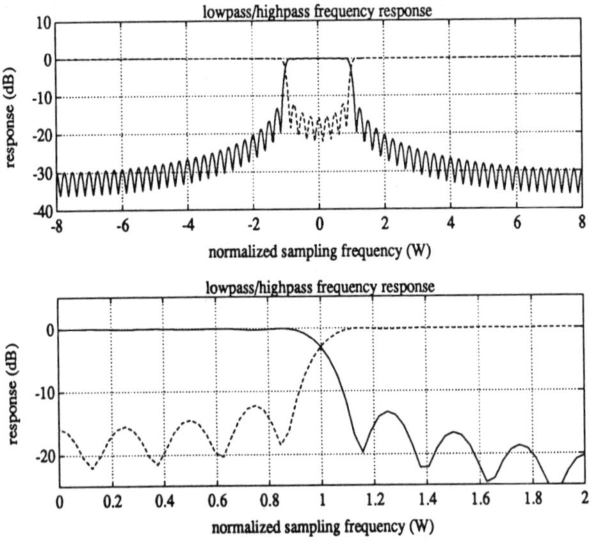

Figure 7-4 Low-pass/high-pass frequency response. The low-pass (baseband) response is the average of the first 8 spectral windows while the high-pass is the average of the other 56. The low-pass response approximates the ideal rectangular filter for spectrum estimation, while the high-pass one shows the leakage that occurs from outside the analysis window W. If sidelobe reduction is desired (e.g., in filter design), a smaller number of windows should be used. The top part of the figure shows the complete frequency response $-8W \leq f \leq 8W$, scaled in units of the window W, while the lower part expands on the range $0 \leq f \leq 2W$.

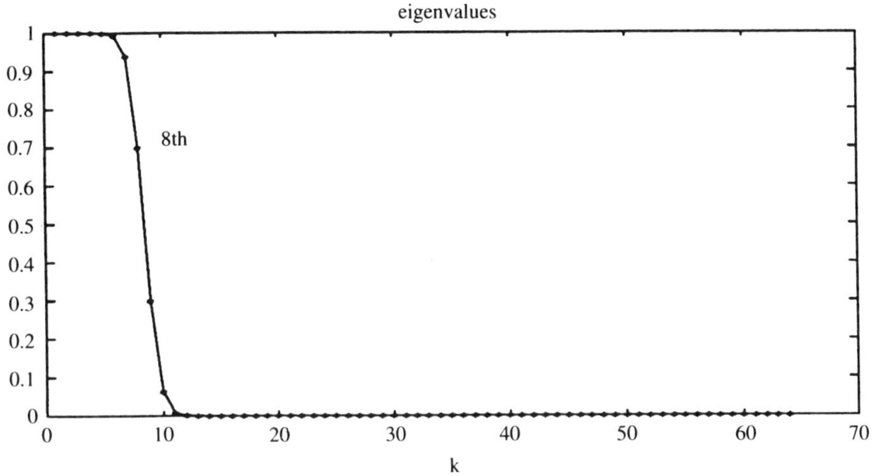

Figure 7-5 The eigenvalue spectrum for $NW = 4$ and $N = 64$. As we can see the first 8 eigenvalues are very close to 1 corresponding to the first $K = 2NW = 8$ windows that have a negligible effect on the bias of the spectrum estimator.

as a basis of expansion in the frequency domain for bandlimited processes. Actually, another way of viewing MWM, is by having the data pass through the baseband (low-pass) filter (Fig. 7-4) as it slides over all frequencies in $(-1/2, 1/2]$. Since spectrum estimation is essentially the estimation of signal power within a certain analysis window and this can ideally be done with a narrow rectangular filter, we see that the baseband filter above is the best possible approximation of such a window. The fact that more than one window is used makes for a smaller variance in the estimator. Also, since the signal power concentration within the analysis band is large (eigenvalues close to one), the bias introduced from the multiplicity of windows is kept small.

7-5 TEST DATASET AND A COMPARISON OF SOME POPULAR SPECTRUM ESTIMATION PROCEDURES

In order to check our understanding of the multiple-window method, at each stage of development, we try to implement it on a known test dataset. This set consists of a complex time series of $N = 64$ points as described in [22]. It is an extension to the complex domain of the famous real dataset given in [15] where 11 modern methods of spectrum estimation were tested. (As will be seen later on, none performed as well as MWM). The analytic spectrum of this synthetic dataset is composed of:

1. Two complex sinusoids of fractional frequencies 0.2 and 0.21 in order to test the resolution capability of a spectral estimator. Note that the

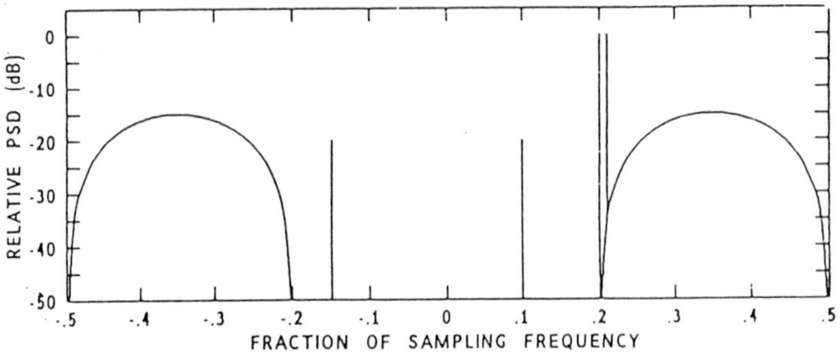

Figure 7-6 The exact known analytic spectrum of Marple's synthetic data set.

Rayleigh resolution limit is $1/N = 1/64 = 0.015625$, so that the difference of the two fractional frequencies of the doublet is just slightly below it.

2. Two weaker complex sinusoids of 20 dB less power at 0.1 and −0.15. These were selected to test a spectral estimator's capability to pick out weaker signal components among stronger ones.
3. A colored noise process, generated by passing two independently generated zero-mean real white noise processes through identical moving average filters to separately generate the real and imaginary components of the test data noise process. Each filter has the identical raised cosine response, seen in Fig. 7-6, between fractional frequencies 0.2–0.5 centered at 0.35 or between −0.2 and −0.5 centered at −0.35. The maximum power level of this process is 15 dB lower than the doublet and 5 dB higher than the singlets.

Note that even though the shape of the colored noise process is identical in the exact, analytic form of the spectrum for both positive and negative frequencies, this symmetry is not expected to be seen in the estimated spectrum because the real and imaginary components were generated independently.

7-5-1 Classical Spectrum Estimation

Following Marple [22], we first start with the classical spectrum estimation method of simply taking the discrete Fourier transform (implemented with a 4096-point FFT) and a rectangular window. To compare with a case where a window is implemented, the Hamming window (3 segments of 32 samples each, with 16 samples overlap between segments) is then used. Figure 7-7 displays the results and we can see that some of the spectrum features are picked out already. Of course, since the Fourier transform is essentially a cross correlation of the data sequence and a complex sinusoid, it tries to fit

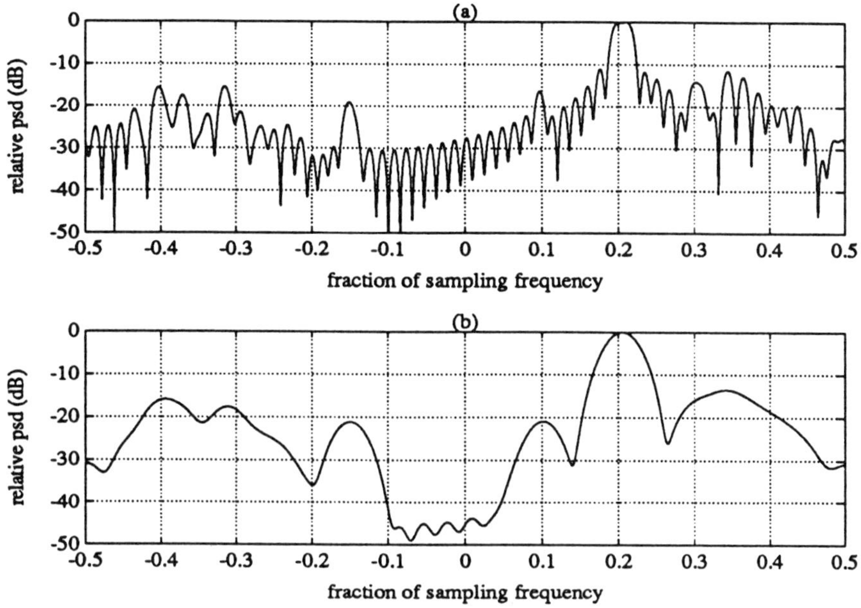

Figure 7-7 Classical spectrum estimation. (a) Periodogram with a 4096 point FFT. (b) Using a Hamming window and a 4096 point FFT.

sinusoids to the continuous part of the spectrum. The Hamming window alleviates the problem to some extent, with a smaller variance on the continuous part of the spectrum, but an increased bias on the line component estimates.

7-5-2 MUSIC and MFBLP

MUlitple SIgnal Classification (MUSIC) and *Modified Forward Backward Linear Prediction* (MFBLP) are two of the modern algorithms for estimating line components in a data sequence. They both use the concept of a signal and noise subspace. Projection operators are constructed that map the data onto one or the other subspace. The signal space component is thus optimized, effecting a higher SNR that leads to superresolution, provided the assumptions on which the construction of the projectors is based, are valid (i.e., background noise correlation matrix is known. SNR is above a certain threshold, and the data are properly calibrated).

Choosing the number of signals arbitrarily to be 10, we see (Fig. 7-8) that there is no problem in picking out the line components, but the methods, as expected, try to fit the continuous part of the spectrum with sinusoids as well. Without any a priori knowledge, it is easy to mistake a noise peak for a signal peak. We also observe a gradual decay of the eigenvalue spectrum, a clear indication of the existence of colored noise.

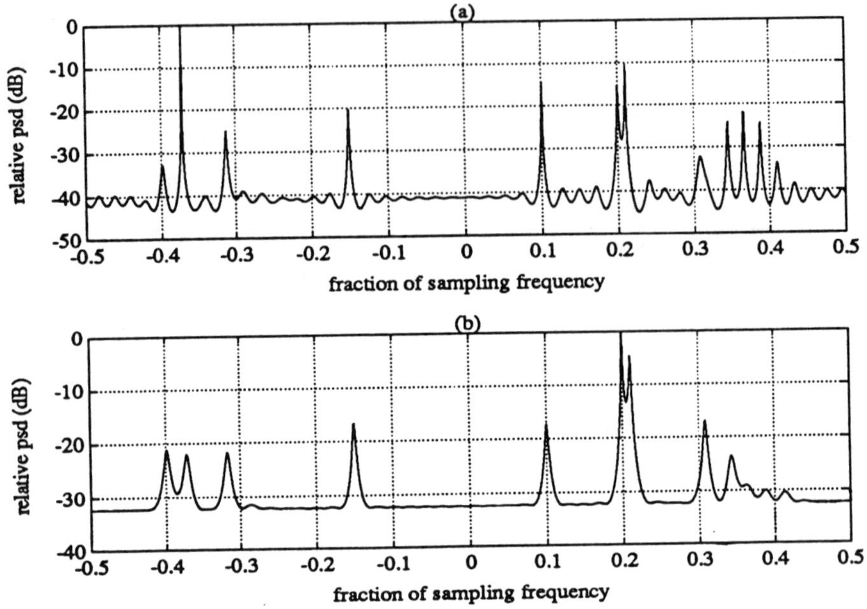

Figure 7-8 MFBLP (a) and MUSIC (b) spectra.

To summarize, both the classical and eigendecomposition methods mentioned above, fail in estimating fully and correctly both the line and the continuous part of the given spectrum.

7-6 MULTIPLE-WINDOW SPECTRUM ESTIMATION

We now turn our attention to MWM[8] and begin to solve the fundamental equation of spectrum estimation (7-11) by expanding its factors in $(f - W, f + W)$ using the Slepian basis. From *Mercer's theorem*, the kernel expansion is

$$\mathsf{D}_N(f-\nu) = \sum_{k=0}^{\infty} \lambda_k V_k(f) V_k^*(\nu) \tag{7-18}$$

and

$$dZ(f-\nu) = \sum_{k=0}^{\infty} x_k(f) V_k^*(\nu)\, d\nu \tag{7-19}$$

[8] The name multiple-window is the result of using a multiplicity of windows in the spectrum estimation process instead of just one, as is commonly the practice. This reduces the variance while increasing somewhat the bias in the estimation procedure. However, as long as we are using windows with corresponding eigenvalues $\lambda \approx 1$, the bias increase is negligible.

Using orthogonality properties carefully (with some help from Yaglom [41]) the coefficients of (7-19) are given by

$$x_k(f) = \sum_{n=0}^{N-1} x(n) v_n^{(k)}(N, W) e^{-j2\pi fn} \tag{7-20}$$

We call the $\{x_k(f)\}$ the *eigencoefficients* of the sample. Since they are computed by transforming the data multiplied by the kth data window $v_n^{(k)}(N, W)$, their absolute squares

$$\hat{S}_k(f) = |x_k(f)|^2 \tag{7-21}$$

are individually direct spectrum estimates and we call them *eigenspectra*. Using the first $K = 2NW$ terms, the ones with the largest eigenvalues, we can obtain a crude multiple-window spectrum estimate as

$$\bar{S}(f) = \frac{1}{K} \sum_{k=0}^{K-1} \frac{1}{\lambda_k(N, W)} |x_k(f)|^2 \tag{7-22}$$

7-6-1 The Adaptive Spectrum

While the lower order eigenspectra have excellent bias properties there is some degradation as k increases toward $2NW$. In his 1982 paper [36], Thomson introduces a set of weights $\{d_k(f)\}$ that downweight the higher order eigenspectra. He derives them by minimizing the mean square error between $Z_k(f)$, the exact coefficients[9] of the expansion of $dZ(f)$ and $d_k(f)x_k(f)$

$$\mathsf{E}\{|Z_k(f) - d_k(f)x_k(f)|^2\} \tag{7-23}$$

Given that

$$Z_k(f) = \frac{1}{\sqrt{\lambda_k}} \int_{-W}^{W} V_k(\nu)\, dZ(f - \nu)$$

and

$$x_k(f) = \int_{-1/2}^{1/2} V_k(\nu)\, dZ(f - \nu)$$

we can write

[9] Though unobservable, they are important in the sense that they are the expansion coefficients which would be obtained if the entire process were passed through an ideal bandpass filter from $(f - W)$ to $(f + W)$ before truncation to the finite sample size.

$$Z_k(f) - d_k(f)x_k(f) = \left[\frac{1}{\sqrt{\lambda_k}} - d_k(f)\right]\int_{-W}^{W} V_k(\nu)\, dZ(f-\nu)$$
$$- d_k(f) \fint V_k(\nu)\, dZ(f-\nu)$$

where the cut integral is defined as

$$\fint = \left(\int_{-1/2}^{1/2} - \int_{-W}^{W}\right)$$

and we name the intervals $(-W, W)$ and $(-1/2, -W) \cup (W, 1/2)$ the inner and outer bands, respectively.

It is helpful here to think of the energy in the band $(f-W, f+W)$ as a signal component and the energy outside it as a noise component. These two components are uncorrelated, making the expectation of their cross products zero and the minimization of (7-23) is then simply a process of finding the *optimum Wiener filter*.

Define the *broadband bias* of the kth eigenspectrum as

$$B_k(f) = \left|\fint V_k(\nu)\, dZ(f-\nu)\right|^2 \tag{7-24}$$

The expected value of the broadband bias is

$$\mathsf{E}\{B_k(f)\} = \fint |V_k(\nu)|^2 S(f-\nu)\, d\nu \tag{7-25}$$

and from the *Cauchy–Schwartz inequality* it can be bounded as

$$\mathsf{E}\{B_k(f)\} \leq \fint |V_k(\nu)|^2\, d\nu \fint \mathsf{E}\{|dZ(f-\nu)|^2\} \tag{7-26}$$

The first integral on the right side is just the energy of the Slepian function in the outer band with value $(1-\lambda_k)$. The second one, by filling the gap in the inner band, has as expected value the average power of the process, σ^2 (the process variance). We therefore have

$$\mathsf{E}\{B_k(f)\} \leq (1-\lambda_k)\sigma^2 \tag{7-27}$$

The weights that minimize (7-23) are

$$d_k(f) = \frac{\sqrt{\lambda_k} S(f)}{\lambda_k S(f) + \mathsf{E}\{B_k(f)\}} \tag{7-28}$$

The process variance is

$$\sigma^2 = \int_{-1/2}^{1/2} S(f)\, df = r_x(0) = \mathsf{E}\{|x(n)^2\} \approx \frac{1}{N}\sum_{n=0}^{N-1} |x(n)|^2 \tag{7-29}$$

A fair initial estimate of the expected value of the broadband bias $\mathsf{E}\{B_k(f)\}$, $\hat{B}_k(f)$, is given by

$$\hat{B}_k(f) = \mathsf{E}\{B_k(f)\} = (1 - \lambda_k)\sigma^2 \tag{7-30}$$

which is actually the upper bound for $\mathsf{E}\{B_k(f)\}$.

Note that in order to compute the adaptive weights $d_k(f)$ in (7-28) we need to know the true spectrum $S(f)$. Of course, if we did, there would be no need to do any spectrum estimation at all. Relation (7-28), however, is useful in setting up an iterative scheme to estimate $S(f)$ as $\hat{S}(f)$, by substituting (7-28) into

$$\hat{S}(f) = \frac{\sum_{k=0}^{K-1} |d_k(f)|^2 \hat{S}_k(f)}{\sum_{k=0}^{K-1} |d_k(f)|^2} \tag{7-31}$$

which leads to

$$\sum_{k=0}^{K-1} \frac{\lambda_k[\hat{S}(f) - \hat{S}_k(f)]}{[\lambda_k \hat{S}(f) + \hat{B}_k(f)]^2} = 0 \tag{7-32}$$

The solution can be found iteratively from

$$\hat{S}^{(i+1)}(f) = \left[\sum_{k=0}^{K-1} \frac{\lambda_k \hat{S}_k^{(i)}(f)}{[\lambda_k \hat{S}^{(i)}(f) + \hat{B}_k(f)]^2}\right]\left[\sum_{k=0}^{K-1} \frac{\lambda_k}{[\lambda_k \hat{S}^{(i)}(f) + \hat{B}_k(f)]^2}\right]^{-1}$$

taking as a starting value for $\hat{S}(f)$ the average of the two lowest order eigenspectra. Convergence is rapid with successive spectrum estimates differing by less than 5% in 5–20 iterations.

A simple implementation of the above scheme can be achieved by using (7-30) for $\hat{B}_k(f)$. In his original paper [36], Thomson goes to some length to find a tighter bound than this one. The basic idea is the following: note that the definition of $\mathsf{E}\{B_k(f)\}$ is a convolution in the frequency domain. Transforming into the time domain, we only need to multiply two time functions. After doing the multiplication, we can go back into the frequency domain. The whole procedure can be efficiently implemented with standard FFTs.

For this purpose, define the outer lag window

$$L_k^{(o)}(\tau) = \fint e^{j2\pi\tau\nu} |V_k(\nu)|^2 \, d\nu = \int_{-1/2}^{1/2} e^{j2\pi\tau\nu} |V_k'(\nu)|^2 \, d\nu$$

where

$$V'_k(\nu) = \begin{cases} V_k(\nu)\,, & \text{if } \nu \in (-1/2, -W) \cup (W, 1/2) \\ 0 & \text{if } \nu \in (-W, W) \end{cases}$$

The last integral can be approximated with an FFT.

Next, compute the autocovariance function $R^{(o)}(\tau)$ corresponding to the spectrum estimate of the current iteration

$$R^{(o)}(\tau) = \int_{-1/2}^{1/2} \hat{S}(\nu) e^{j2\pi\tau\nu}\, d\nu$$

This can also be approximated with an FFT.

Finally, transform back into the frequency domain

$$\hat{B}_k(f) = \sum_{\tau} e^{-j2\pi f\tau} L_k^{(o)}(\tau) R^{(o)}(\tau)$$

and this sum can also be computed with another FFT.

In implementing the above idea, a somewhat better resolution is achieved than by using (7-30) at the expense of more computer time. Note, however, that the condition (7-27) is not necessarily satisfied for the first few k. Note also that a good estimate of the autocovariance function can be computed by using the final spectrum estimate above.

A useful byproduct of this adaptive estimation procedure is an estimate of the stability of the estimates given by

$$v(f) = 2 \sum_{k=0}^{K-1} |d_k(f)|^2 \tag{7-33}$$

which is the approximate number of degrees of freedom for $\hat{S}(f)$ as a function of frequency. If the average $\bar{v}$, of $v(f)$ over frequency, is significantly less than $2K$, then either the window W is too small, or additional prewhitening should be used. This, together with a variance efficiency coefficient which is also developed in [36] can provide a useful stopping rule when W and K are varied. In more complicated cases, *jackknifed error estimates* for the spectrum can be computed as well [40]. (The jackknife, in the simplest form, refers to the following procedure. Given a set of N observations, each observation is deleted in turn, forming N subsets of $N-1$ observations. These subsets are used to form estimates of a given parameter, which are then combined to give estimates of bias and variance for this parameter, valid under a wide range of parent distributions. Thomson and Chave [40] discuss the extension of this concept to spectra, coherences and transfer functions.)

7-6-2 The Composite Spectrum

The use of adaptive weighting as developed above provides superior protection against leakage and bias. Thomson also offers a further refinement to achieve higher resolution by considering each specific frequency point f_0 as a free parameter in $(f - W \leqslant f_0 \leqslant f + W)$. A different choice of weights is the result, leading to a composite spectrum estimate

$$\hat{S}_C(f) = \frac{\int_{f-W}^{f+W} w(f_0)\hat{S}_h(f; f_0)\, df_0}{\int_{f-W}^{f+W} w(f_0)\, df_0} \tag{7-34}$$

where

$$\hat{S}_h(f; f_0) = \frac{2}{v(f)} \left| \sum_{k=0}^{K-1} V_k(f - f_0) d_k(f) x_k(f_0) \right|^2 \tag{7-35}$$

$$w(f_0) = \frac{v(f_0)}{\hat{S}(f_0)^2} \tag{7-36}$$

and $\hat{S}(f_0)$ is the adaptive spectrum developed earlier.

This choice of weights imposes the constraint that $\hat{S}(f_0)$ should have sufficient degrees of freedom for $w(f_0)$ to have a reasonable distribution. In practice, it is inadvisable to do a free parameter expansion over the full window, $|f - f_0| = W$, but rather step near $0.8W$ to $0.9W$ in order to minimize the outside leakage (see Fig. 7-4). Furthermore, in regions where the number of degrees of freedom is small, Thomson suggests rescaling $\hat{S}_h(f; f_0)$ by dividing by a factor proportional to

$$\sum_{k=0}^{K-1} |w_k(f_0)V_k(f - f_0)|^2$$

The implementation of this final version of Thomson's spectrum estimation method was done numerically. Function values of $w(f)$ at any desired frequency point from the already computed data table of $v(f)$ and $\hat{S}(f)$ were interpolated with splines. If, however, (7-30) is used for $\hat{B}_k(f)$ then it is easy to have an exact expression of $w(f)$ and no interpolation is necessary (the difference in the final composite spectrum estimate between this approach and the one where $w(f)$ is interpolated is almost negligible). The integration was performed numerically with *Romberg's method*[10] which, for a given numerical accuracy, requires the least number of function evaluations. The integration boundary was chosen to be $0.8W$ as this resulted in line components having a ratio closer to the known one. Finally, it is necessary to explicitly adjust the scaling[11] of $\hat{S}_C(f)$. Note that Thomson's

[10] Both the spline and Romberg integration subroutines were adapted from [30].

[11] That is, multiply by a proper scaling factor in order to get a variance estimate $\hat{\sigma}^2$ (the area under the spectrum computed by trapezoidal integration) close to the known one.

suggestion above on rescaling $\hat{S}_h(f; f_0)$ does not specify the proportionality constant so there is some justification for this ad hoc rescaling procedure.

A more recent discussion on this *high-resolution* spectrum estimate is given in [38]. Thomson points out that this is still an area being actively developed; for example, it is shown that although the estimate is unbiased for slowly varying spectra, it underestimates fine spectral structure. In the latter part of this chapter, where Thomson's method is implemented on real data, it is the adaptive spectrum estimator that is being used.

7-6-3 Computing the Crude, Adaptive, and Composite Spectra

The results of applying the crude, adaptive, and composite spectrum algorithms developed above are seen in Figs. 7-9, 7-10, and 7-11. The reason for varying the time–bandwidth product has to do with the bias-variance trade-off that is the result of the ill-posedness of the spectrum estimation problem. If the analysis window W is too small (to better resolve details) we have poor statistical stability (larger variance), but if W is too large, the estimate has poor frequency resolution. We see these effects, both here, in the spectra and in the F-tests considered in the following section. In practice therefore, we have to try a variety of time–bandwidth product values to pick out the features that are of interest. Thomson [39] recommends that W be between $1/N$ and $20/N$ with a time–bandwidth product of 4 or 5 being a common starting point.

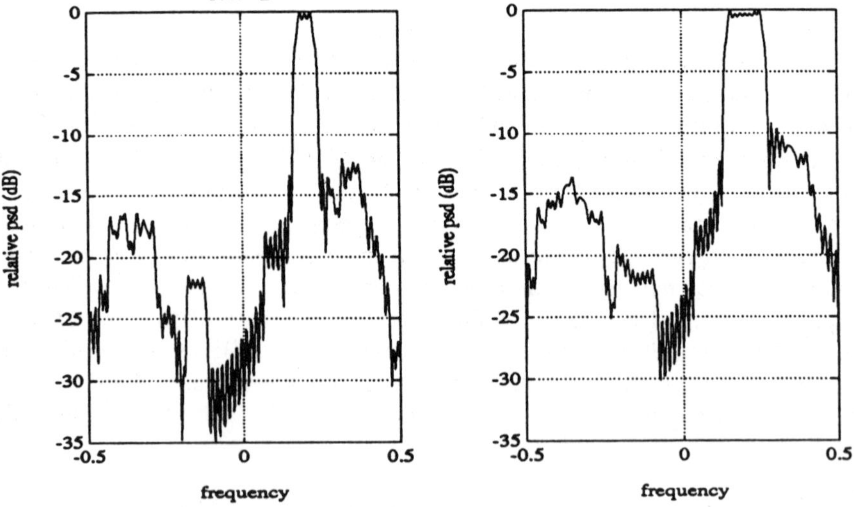

Figure 7-9 The crude spectra $\bar{S}(f)$ for $NW = 2$ and 4.

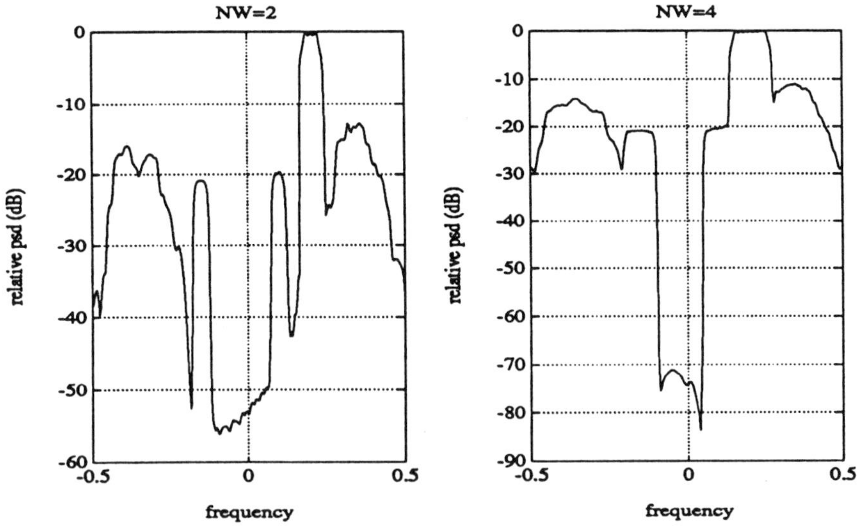

Figure 7-10 The adaptive spectra $\hat{S}(f)$ for $NW = 2$ and 4.

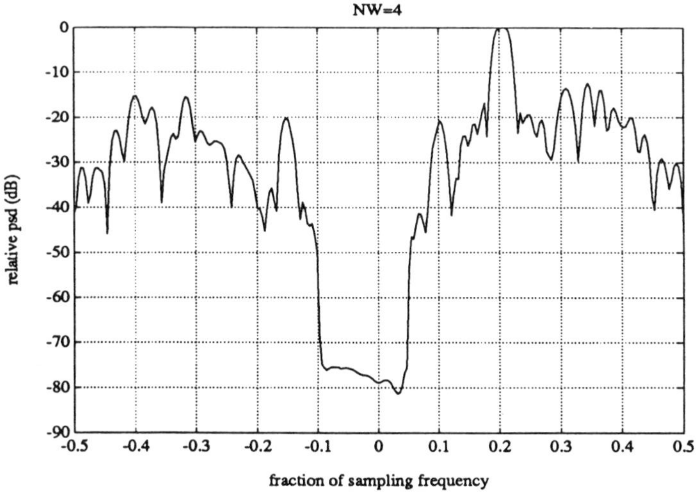

Figure 7-11 The composite spectrum for $NW = 4$, $K = 8$ and an integration boundary of $0.8W$. A larger boundary results in a larger mismatch of the power levels between the known signal components. Note the detail of the composite spectrum despite the larger spectral window.

7-7 F-TEST FOR THE LINE COMPONENTS

The spectra computed above are not expected to be good estimates, since we know that line components exist and the spectrum estimation techniques developed, implicitly assume none. With MWM we can apply the statistical F-test, to check for and estimate existing line components in the spectrum. The F-test is a statistical test that assigns a probability value on each of two hypotheses concerning samples taken from a parent population. These samples are assumed to follow a χ^2 distribution, which is the case for an unknown mean and variance sample, of sum of squares, taken from a normal population. A brief outline of this test as applied to linear regression models is given in the following subsection; for more details, see Draper and Smith [7].

7-7-1 A Brief Outline of the *F*-Test

Let us assume that we have a model described by

$$\mathbf{y} = \mathbf{A}\mathbf{x} + \mathbf{e}$$

that is linear with respect to the $p \times 1$ parameter vector $\mathbf{x}$, where the $n \times p$ coefficient matrix $\mathbf{A}$ and $n \times 1$ vector $\mathbf{y}$ are known or can be estimated from a given data set. We assume that the error vector $\mathbf{e}$ has independent components that come from $N(0, \sigma^2)$. Therefore, another way to express our assumed model is

$$\mathsf{E}\{\mathbf{y}\} = \mathbf{A}\mathbf{x}$$

In order to get the best possible estimate of our parameter vector $\mathbf{x}$ in the least squares sense, we have to find

$$\min_{\mathbf{x}} \|\mathbf{y} - \mathbf{A}\mathbf{x}\|^2$$

The squared error can be written as

$$e^2(\mathbf{x}) = \mathbf{e}^H\mathbf{e} = \|\mathbf{y} - \mathbf{A}\mathbf{x}\|^2 = \mathbf{y}^H\mathbf{y} - \mathbf{y}^H\mathbf{A}\mathbf{x} - \mathbf{x}^H\mathbf{A}^H\mathbf{y} + \mathbf{x}^H\mathbf{A}^H\mathbf{A}\mathbf{x}$$

and this becomes minimum at the well known *linear least squares solution*

$$\hat{\mathbf{x}} = (\mathbf{A}^H\mathbf{A})^{-1}\mathbf{A}^H\mathbf{y} = \mathbf{A}^+\mathbf{y}$$

where $\mathbf{A}^+$ is the pseudo-inverse of $\mathbf{A}$.

The F-test comes about from observing that we can break the observed total variance $\mathbf{y}^H\mathbf{y}$, of our model, into two components, one due to the regression itself, $\|\mathbf{A}\hat{\mathbf{x}}\|^2$ [7, p. 80], and the other, $\|\mathbf{y} - \mathbf{A}\hat{\mathbf{x}}\|^2$, due to the

TABLE 7-1 The Basic ANOVA Table

Variation Source	Degrees of Freedom	Sum of Squares (SS)	Mean SS
Regression	ν_1	$SS_1 = \|\|\mathbf{A}\hat{\mathbf{x}}\|\|^2$	$MS_{\text{reg}} = SS_1/\nu_1$
Residuals	ν_2	$SS_2 = \|\|\mathbf{y} - \mathbf{A}\hat{\mathbf{x}}\|\|^2$	$s^2 = SS_2/\nu_2$

residual errors. Each of these components has associated with it a number of degrees of freedom, ν_1 and ν_2, respectively. For K complex data points the total number of degrees of freedom is

$$\nu_1 + \nu_2 = 2K$$

It turns out now that, provided the errors are independent and zero-mean Gaussian random variables, each of the two variance components follows a χ^2 distribution with ν_1 or ν_2 degrees of freedom, respectively. Their ratio follows the $F(\nu_1, \nu_2)$ distribution and we can make hypothesis testing at a desired level of significance. Table 7-1 shows this simple analysis of variance (ANOVA) breakdown.

We test the hypothesis H_0: $\mathbf{x} = 0$ against H_1: $\mathbf{x} \neq 0$ at a significance level α as follows: If H_0 is true then, the ratio

$$F = \frac{MS_{\text{reg}}}{s^2} = \frac{\nu_2 SS_1}{\nu_1 SS_2}$$

should follow the $F(\nu_1, \nu_2)$ distribution, whose value for a significance level of α is found from statistical tables. If the computed ratio is larger than the table value[12] then our hypothesis is rejected with $100(1 - \alpha)\%$ confidence. This means that at least one of the components of $\mathbf{x}$ is different from zero.

It is also possible to test various linear hypothesis about the model under consideration. Focusing on the possible addition of extra parameters, consider the two models below.

1. $\mathsf{E}\{Y\} = A_1x_1 + A_2x_2 + \cdots + A_px_p$
2. $\mathsf{E}\{Y\} = A_1x_1 + A_2x_2 + \cdots + A_qx_q$

where $q < p$. The A's in both models are the same when the subscripts are the same. We simply have fewer parameters to fit in the second one. From the first model, we estimate $\mathbf{x}$ as

$$\hat{\mathbf{x}}_p = (\mathbf{A}_p^H \mathbf{A}_p)^{-1} \mathbf{A}_p^H \mathbf{y}$$

The corresponding residual sum of squares is

[12] In practice, of course, we require a computed F ratio much larger than the tabulated one.

$$S_1 = \|\mathbf{y} - \mathbf{A}_p \hat{\mathbf{x}}_p\|^2$$

This has $2(n-p)$ degrees of freedom (n is the total number of complex data points we have available for the regression analysis), and

$$s^2 = \frac{S_1}{2(n-p)}$$

is also an estimate for the squared variance for model 1. For the second model, we have respectively

$$\hat{\mathbf{x}}_q = (\mathbf{A}_q^H \mathbf{A}_q)^{-1} \mathbf{A}_q^H \mathbf{y}$$

and

$$S_2 = \|\mathbf{y} - \mathbf{A}_q \hat{\mathbf{x}}_q\|^2$$

where S_2 has $2(n-q)$ degrees of freedom. To test the hypothesis that the extra terms are unnecessary, we consider the ratio

$$\frac{\dfrac{1}{2(p-q)}(S_2 - S_1)}{\dfrac{1}{2(n-p)} S_1}$$

and refer it to the $F[2(p-q), 2(n-p)]$ distribution in the usual manner.

This provides us with enough information to set up partial F-tests in the following subsections, for multiple spectral lines.

7-7-2 The Point Regression Single-Line *F*-Test

Given a line component at frequency f_0, the expected value of the eigencoefficients is

$$\mathsf{E}\{x_k(f)\} = \mu V_k(f - f_0) \tag{7-37}$$

At a given frequency f, our $x_k(f)$'s correspond to $\mathbf{y}$ above, the parameter μ corresponds to $\mathbf{x}$, and the $V_k(f-f_0)$'s correspond to the matrix $\mathbf{A}$. The number of parameters p is equal to 1 here, and by setting it up as a least squares problem we have to minimize the residual error with respect to the 1×1 complex scalar parameter μ

$$\|\mathbf{x} - \mathbf{V}\boldsymbol{\mu}\|^2$$

where

$$\mathbf{x}=\begin{bmatrix} x_0(f) \\ x_1(f) \\ \vdots \\ x_{K-1}(f) \end{bmatrix}, \qquad \boldsymbol{\mu}=\mu\,, \qquad \mathbf{V}=\begin{bmatrix} V_0(f-f_0) \\ V_1(f-f_0) \\ \vdots \\ V_{K-1}(f-f_0) \end{bmatrix}$$

The term *point regression* comes about because we consider each frequency point f as a possible candidate for f_0. We set $f_0 = f$ and test the model (7-37) for significance.

The residual error at f can also be written as

$$e^2(\mu, f) = \sum_{k=0}^{K-1} |x_k(f) - \mu(f)V_k(0)|^2 \tag{7-38}$$

and the least squares solution is

$$\hat{\boldsymbol{\mu}} = (\mathbf{V}^H\mathbf{V})^{-1}\mathbf{V}^H\mathbf{x} = \hat{\mu}(f) = \frac{\sum_{k=0}^{K-1} V_k^*(0)x_k(f)}{\sum_{k=0}^{K-1} |V_k(0)|^2}$$

This can also be written as

$$\hat{\mu}(f) = \sum_{n=0}^{N-1} h_n(N, W)x(n)e^{-j2\pi fn} \tag{7-39}$$

where the harmonic data windows are

$$h_n(N, W) = \frac{\sum_{k=0}^{K-1} V_k^*(0)v_n^{(k)}(N, W)}{\sum_{k=0}^{K-1} |V_k(0)|^2} \tag{7-40}$$

We can now test the hypothesis H_0 that when $f_0 = f$ the model (7-37) is false, that is, the parameter μ is equal to zero, versus H_1, where it is not. In other words, the ratio of the energy explained by the assumption of a line component at the given frequency, to the residual energy, gives us the F-ratio as explained in the previous section, namely

$$F(f) = \frac{\frac{1}{\nu_1}\|\mathbf{V}\boldsymbol{\mu}\|^2}{\frac{1}{\nu_2}\|\boldsymbol{x} - \boldsymbol{V}\boldsymbol{\mu}\|^2}$$

or

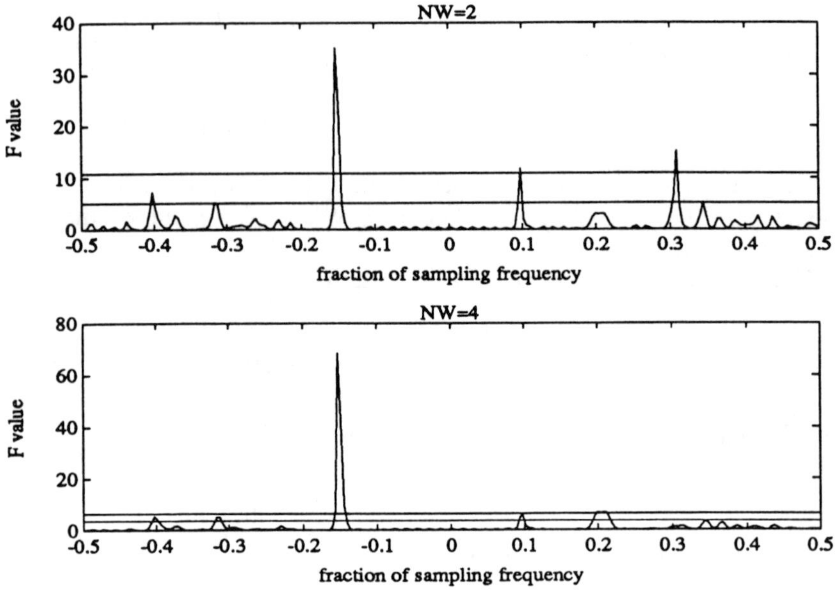

Figure 7-12 The single-line point regression F-test for $NW = 2$ and 4. The corresponding F values for 99% and 95% confidence levels are drawn as well.

$$F(f) = \frac{\dfrac{1}{\nu}\,|\hat{\mu}(f)|^2 \sum_{k=0}^{K-1} |V_k(0)|^2}{\dfrac{1}{2K-\nu}\, e^2(\hat{\mu}, f)} \tag{7-41}$$

where ν is equal to two degrees of freedom (real and imaginary parts of the complex line amplitude). The total number $2K$ of degrees of freedom come about from the K complex data points we have available to draw information from. If F is large at a certain frequency then the hypothesis is rejected, that is, a line component does exist there. The location of the maximum of F provides an estimate of the line frequency that has a resolution within 5–10% of the Cramér–Rao bound.

This test works well if the lines considered are isolated, that is, if there is only a single line in $(f - W, f + W)$. The total number of lines is not important as long as they occur singly. For lines spaced closer than W we may use a multiple-line test, a similar but algebraically more complicated regression of the eigencoefficients on a matrix of functions $V_k(f - f_i)$ with the simple F-test replaced by partial F-tests. Thomson notes that multiple-line tests should be used with caution, because the Cramér–Rao bounds for line parameter estimation degrade rapidly when the line spacing becomes less than $2/N$.

Note also that the F-test is a statistical test. This means that given a large number of different realizations of a data sequence, highly significant values can sometimes be seen, that in reality are only sampling fluctuations. A good rule-of-thumb, as Thomson points out, is not to get excited by significance levels below $1 - 1/N$. Experience also suggests to try the test for a variety of NW values. Line components that disappear from one case to the other are almost certainly sampling fluctuations.

In Fig. 7-12 we see the results of applying this F-test to the Marple dataset. Note the effect of the different values of NW and K. The 99% and 95% confidence levels are drawn as well.

7-7-3 The Integral Regression Single-Line *F*-test

Thomson [36] also suggests an integral regression test instead of the point regression at f_0 that was developed above. The criterion here is the minimization of the sum of integrals

$$e^2 = \sum_{k=0}^{K-1} \int_{f_0-W}^{f_0+W} |x_k(f) - \mu V_k(f - f_0)|^2 \, df \tag{7-42}$$

with respect to μ. The development is more complex now, but the underlying logic is the same. Again, we have an equivalent harmonic window, but here it is formed of convolutions of prolate functions that have exceedingly low sidelobes. The drawback is that as it uses information from a wider bandwidth it is also subject to noise from the same bandwidth. No further details are given for this approach or the multiple line F-test in [36], but the matrix formalism given earlier is a general enough framework that should be able to handle this situation.[13]

Differentiating e^2 with respect to μ and setting the result equal to zero, yields

$$\int_{f_0-W}^{f_0+W} (\mathbf{V}^H \mathbf{V} \boldsymbol{\mu} - \mathbf{V}^H \mathbf{x}) \, df = 0 \tag{7-43}$$

and the 1×1 complex parameter $\hat{\boldsymbol{\mu}}$ (scalar for this single-line case) is given by

$$\hat{\mu}(f_0) = \frac{\displaystyle\sum_{k=0}^{K-1} \int_{f_0-W}^{f_0+W} V_k^*(f - f_0) x_k(f) \, df}{\displaystyle\sum_{k=0}^{K-1} \int_{f_0-W}^{f_0+W} |V_k(f - f_0)|^2 \, df} \tag{7-44}$$

[13] In his more recent work [38], Thomson discusses the F-tests, both single and multiple, in some detail.

From known formulas, repeated here for convenience, we have

$$x_k(f) = \sum_{n=0}^{N-1} x(n) v_n^{(k)}(N, W) e^{-j2\pi fn} \quad \text{and} \quad V_k(f) = \sum_{n=0}^{N-1} v_n^{(k)}(N, W) e^{-j2\pi fn}$$

and the *Toeplitz matrix eigenvalue equation*,

$$\sum_{m=0}^{N-1} \frac{\sin 2\pi W(n-m)}{\pi(n-m)} v_m^{(k)} = \lambda_k v_n^{(k)}$$

We do the algebra required for each term in (7-44) and finally end up with the denominator having the form

$$\sum_{k=0}^{K-1} \int_{f_0-W}^{f_0+W} |V_k(f-f_0)|^2 \, df = \sum_{k=0}^{K-1} \lambda_k \tag{7-45}$$

while the numerator is

$$\sum_{k=0}^{K-1} \int_{f_0-W}^{f_0+W} V_k^*(f-f_0) x_k(f) \, df = \sum_{n=0}^{N-1} \sum_{k=0}^{K-1} \lambda_k [v_n^{(k)}]^2 x(n) e^{-j2\pi f_0 n} \tag{7-46}$$

These relations modify (7-44) into

$$\hat{\mu}(f_0) = \sum_{n=0}^{N-1} h_n x(n) e^{-j2\pi f_0 n}$$

where the harmonic windows now are

$$h_n = \frac{\sum_{k=0}^{K-1} \lambda_k [v_n^{(k)}]^2}{\sum_{k=0}^{K-1} \lambda_k}$$

Substituting $\hat{\mu}(f_0)$ in the squared error, we obtain

$$e^2(\hat{\mu}, f_0) = \int_{f_0-W}^{f_0+W} \left[\sum_{k=0}^{K-1} |x_k(f)|^2 - |\hat{\mu}|^2 \sum_{k=0}^{K-1} |V_k(f-f_0)|^2 \right] df$$

The first term gives us

$$Q = \sum_{k=0}^{K-1} \int_{f_0-W}^{f_0+W} |x_k(f)^2 \, df$$

$$= \sum_{k=0}^{K-1} \sum_{n=0}^{N-1} \sum_{m=0}^{N-1} x(n) x^*(m) v_n^{(k)} v_m^{(k)} \frac{\sin[2\pi(n-m)W]}{\pi(n-m)} e^{-j2\pi(n-m)f_0}$$

and a brute-force computation of this triple sum for each frequency point can take a few hours of CPU time, depending of course on the number of frequency points in question. Looking into the symmetries involved and after some algebra, we finally end up with the expression

$$Q = 2W \sum_{k=0}^{K-1} \sum_{n=0}^{N-1} |x(n)v_n^{(k)}|^2 + 2\,\mathrm{Re}\left\{ \sum_{k=0}^{K-1} \sum_{n=0}^{N-2} x(n)v_n^{(k)} \sum_{l=1}^{N-n-1} \times\, x^*(l+n)v_{l+n}^{(k)} \frac{\sin 2\pi l W}{\pi l} e^{j2\pi l f_0} \right\} \tag{7-47}$$

Some judicious precomputing, particularly of the exponentials, can drop the CPU time for this sum to only a few minutes and our estimate of the squared error becomes

$$e^2(\hat{\mu}, f_0) = Q - |\hat{\mu}|^2 \sum_{k=0}^{K-1} \lambda_k \tag{7-48}$$

Another, perhaps more straightforward way to compute Q faster, is to perform the summation with respect to k first, thus creating a data window matrix and reducing the triple sum to a double one. In complete analogy with the point regression test, the F ratio now is

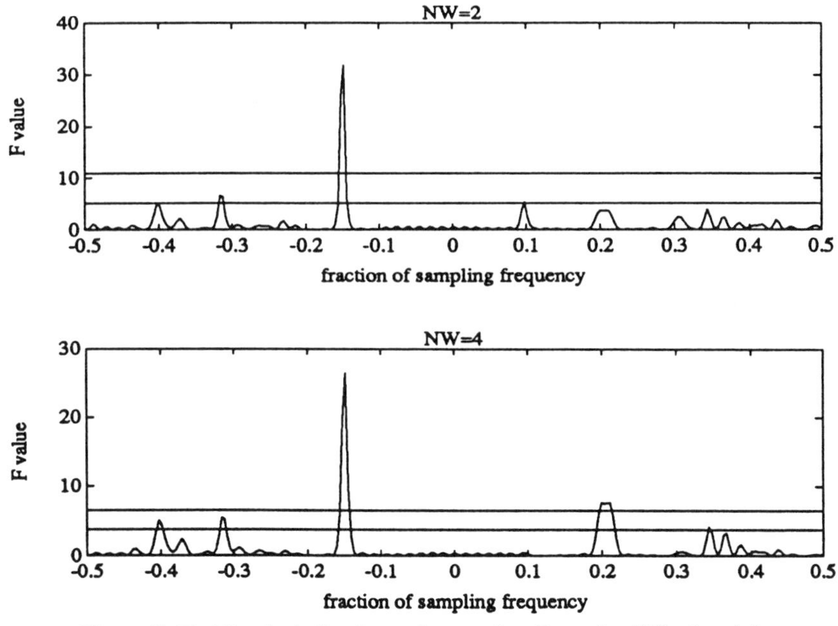

Figure 7-13 The single-line integral regression F-test for $NW = 2$ and 4.

$$F(f_0) = \frac{\frac{1}{\nu} |\hat{\mu}(f_0)|^2 \sum_{k=0}^{K-1} \lambda_k}{\frac{1}{2K-\nu} e^2(\hat{\mu}, f_0)} \tag{7-49}$$

The method should give similar results as compared to the point regression single-line F-test, and Fig. 7-13 shows that this is indeed the case.

7-7-4 The Point Regression Double-Line *F*-Test

The model we now have is described by

$$\mathsf{E}\{x_k(f)\} = \mu_1 V_k(f - f_1) + \mu_2 V_k(f - f_2) \tag{7-50}$$

We want to minimize the least squares residual error

$$e^2 = \|\mathbf{x} - \mathbf{V}\boldsymbol{\mu}\|^2$$

where now

$$\mathbf{x} = \begin{bmatrix} x_0(f) \\ x_1(f) \\ \vdots \\ x_{K-1}(f) \end{bmatrix}, \quad \boldsymbol{\mu} = \begin{bmatrix} \mu_1 \\ \mu_2 \end{bmatrix}, \quad \mathbf{V} = \begin{bmatrix} V_0(f-f_1) & V_0(f-f_2) \\ V_1(f-f_1) & V_1(f-f_2) \\ \vdots & \vdots \\ V_{K-1}(f-f_1) & V_{K-1}(f-f_2) \end{bmatrix} \tag{7-51}$$

If

$$\mathbf{V}^H\mathbf{V} = \begin{bmatrix} d_1 & c \\ c^* & d_2 \end{bmatrix} \tag{7-52}$$

the least squares solution is

$$\hat{\boldsymbol{\mu}} = (\mathbf{V}^H\mathbf{V})^{-1}\mathbf{V}^H\mathbf{x} = \frac{1}{d_1 d_2 - |c|^2} \begin{bmatrix} d_2 & -c \\ -c^* & d_1 \end{bmatrix} \begin{bmatrix} c_1 \\ c_2 \end{bmatrix} \tag{7-53}$$

where

$$d_1 = \sum_{k=0}^{K-1} |V_k(f - f_1)|^2, \qquad d_2 = \sum_{k=0}^{K-1} |V_k(f - f_2)|^2,$$

$$c = \sum_{k=0}^{K-1} V_k^*(f - f_1) V_k(f - f_2)$$

$$c_1 = \sum_{k=0}^{K-1} V_k^*(f - f_1) x_k(f), \qquad c_2 = \sum_{k=0}^{K-1} V_k^*(f - f_2) x_k(f)$$

and the F-test, for testing the hypothesis that at a given frequency we need at most two lines to explain the data, becomes

$$F(f;\, f_1,\, f_2) = \frac{1/\nu}{1/(2K-\nu)}\left[\frac{\|\mathbf{V}\boldsymbol{\mu}\|^2_{\text{double}}}{\|\mathbf{x}-\mathbf{V}\boldsymbol{\mu}\|^2_{\text{double}}}\right] \tag{7-54}$$

Here we have added two degrees of freedom for the extra μ parameter, making $\nu = 4$. This tests the hypothesis that at a certain frequency pair we have one or two line components. The modification to the hypothesis that we have two line components only is expressed as

$$F(f;\, f_1,\, f_2) = \frac{1/(4-2)}{1(2K-4)}\left[\frac{\|\mathbf{V}\boldsymbol{\mu}\|^2_{\text{double}} - \|\mathbf{V}\boldsymbol{\mu}\|^2_{\text{single}}}{\|\mathbf{x}-\mathbf{V}\boldsymbol{\mu}\|^2_{\text{double}}}\right] \tag{7-55}$$

The single-line model parameter μ is given by

$$\mu_{\text{single}} = c_1/d_1$$

and

$$\|\mathbf{V}\boldsymbol{\mu}\|^2_{\text{single}} = |c_1|^2/d_1$$

The introduction of the above single-line model term changes only the scale of the function F, if anything for the worst, without affecting the signal peak locations or μ estimates of the doublet. Note also that the double-line model assumes two lines, that is, $f_1 \neq f_2$. If this condition does not hold, then the denominator in the μ estimates could be zero. Care must be taken in implementing the computer program that this case is excluded.

If we now let $f_1 = f$ and $\Delta f = f_1 - f_2$ vary within $[-W, W]$, we have a two-dimensional surface for F that should be capable of resolving two lines that are spaced close together within the window bandwidth W. Figures 7-14–7-17 computed for a time-bandwidth product $NW = 2$ and 4 illustrate that this is indeed the case.

Zooming around the area of interest, we see that the doublet peaks line up at 0.2 and 0.21 (where we know they should). The fact that the surface plot displays two peaks is due to our mode of representation of using f and Δf as the independent variables. The two peaks are at $(f, f - \Delta f)$, and for $\Delta f = -0.01$ this corresponds to (f_1, f_2), while for $\Delta f = 0.01$ it corresponds to (f_2, f_1), where $f_1 = 0.2$ and $f_2 = 0.21$. The first pair is consistent with our model and has a larger F value, so it is the one we pick. The advantage with this particular representation is that we can project the maximum of $F(f, \Delta f)$ onto the f axis and resolve the doublet from a simpler, one-dimensional function.

Note, however, the appearance of spurious peaks particularly near the edges of the window boundary. These are explained by the fact that the

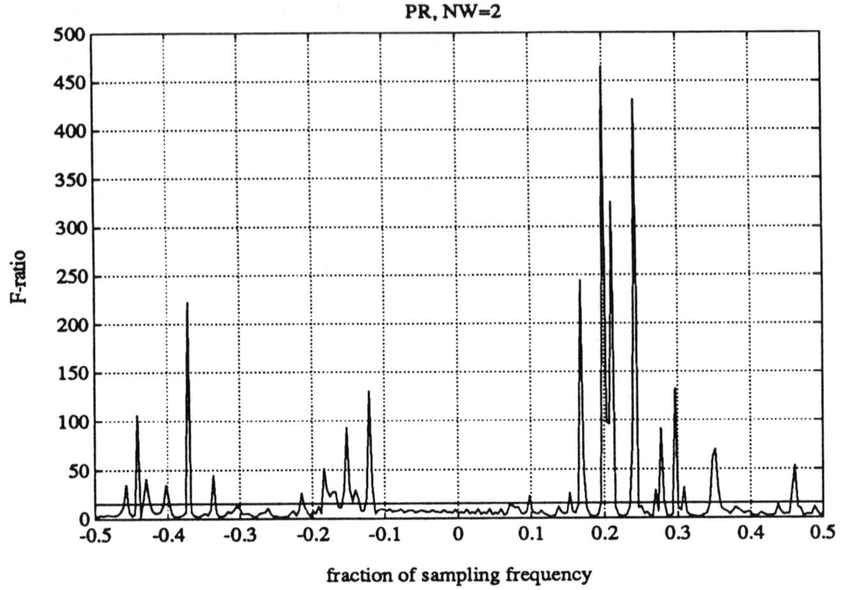

Figure 7-14 The projection of max $F(f, \Delta f)$, onto the f axis. The 99% confidence level is also drawn and we see some spurious (?) peaks above it. As the surface plot in Fig. 7-15 shows, the largest, are mostly due to leakage outside the window.

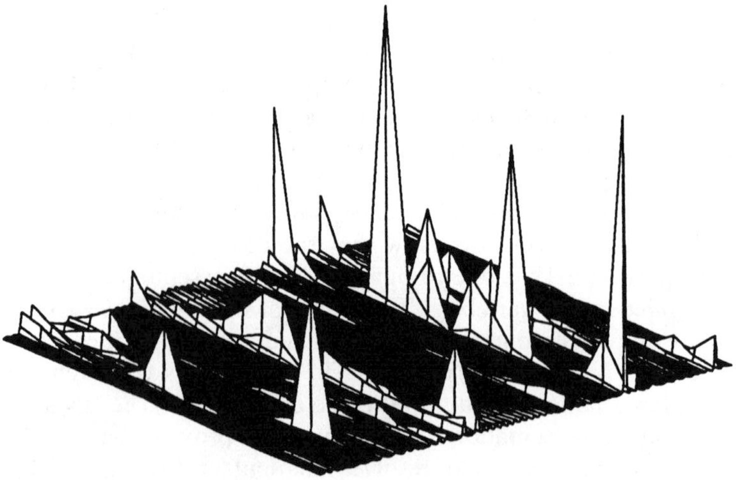

Figure 7-15 The double-line point regression F-test for $NW = 2$. The surface $F(f, \Delta f)$ is shown with a grid size 1/256 due to the 256-FFTs used. Note the large spurious peaks at the window boundary.

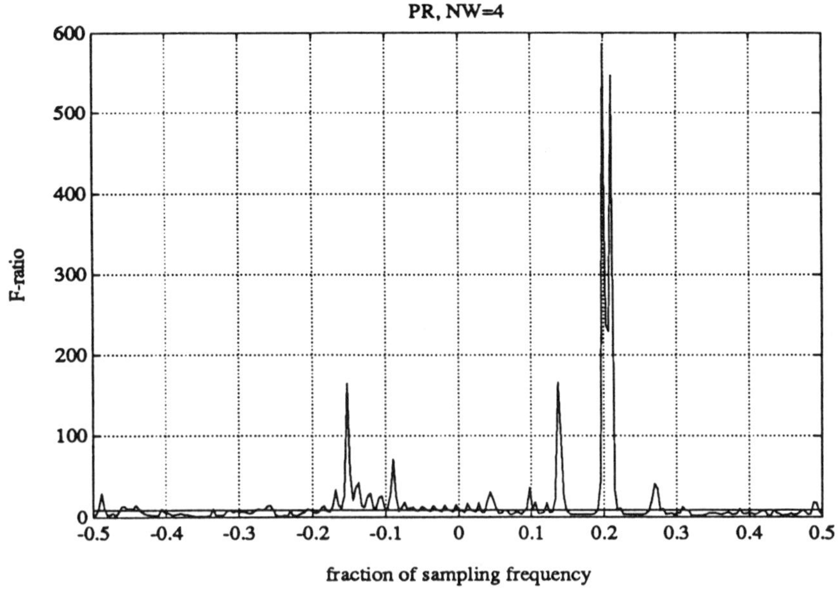

Figure 7-16 The projection of max $F(f, \Delta f)$, onto the f axis.

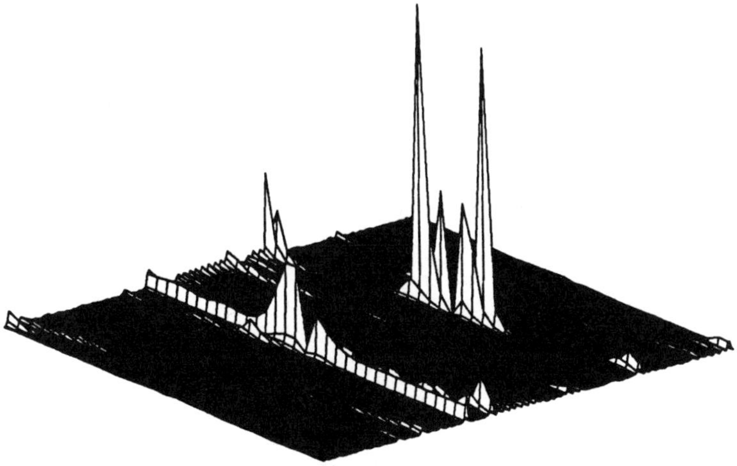

Figure 7-17 The double-line point regression F-test for $NW = 4$. The surface $F(f, \Delta f)$ is shown.

sliding window (Fig. 7-4) is not an ideal bandpass filter so that energy from outside the window, particularly near the window boundaries, can affect the estimation inside.

7-7-5 The Integral Regression Double-Line *F*-Test

The model we consider is again

$$\mathrm{E}\{x_k(f)\} = \mu_1 V_k(f - f_1) + \mu_2 V_k(f - f_2)$$

but the minimization criterion now becomes

$$\min_{\boldsymbol{\mu}} e^2(f_1, f_2) = \min_{\boldsymbol{\mu}} \int_{f_1 - W}^{f_1 + W} \|\mathbf{x} - \mathbf{V}\boldsymbol{\mu}\|^2 \, df \tag{7-56}$$

Differentiating e^2 with respect to $\boldsymbol{\mu}$ and setting the result equal to zero gives us

$$\int_{f_1 - W}^{f_1 + W} (\mathbf{V}^H \mathbf{V}\boldsymbol{\mu} - \mathbf{V}^H \mathbf{x}) \, df = 0$$

where the integration is performed directly on each term of the matrices or vectors involved. It is assumed that f_2 is within $(f_1 - W, f_1 + W)$. A detailed expression for each component is given in the previous section. The least squares solution now is

$$\hat{\boldsymbol{\mu}}(f_1, f_2) = \int_{f_1 - W}^{f_1 + W} (\mathbf{V}^H \mathbf{V})^{-1} \mathbf{V}^H \mathbf{x} \, df = \frac{1}{\delta_1 \delta_2 - |\gamma|^2} \begin{bmatrix} \delta_2 & -\gamma \\ -\gamma^* & \delta_1 \end{bmatrix} \begin{bmatrix} \gamma_1 \\ \gamma_2 \end{bmatrix} \tag{7-57}$$

where

$$\delta_1 = \int_{f_1 - W}^{f_1 + W} d_1 \, df = \cdots = \sum_{k=0}^{K-1} \lambda_k$$

$$\delta_2 = \int_{f_1 - W}^{f_1 + W} d_2 \, df = \cdots = \sum_{k=0}^{K-1} \sum_{n=0}^{N-1} \sum_{m=0}^{N-1} v_n^{(k)} v_m^{(k)} e^{j2\pi(n-m)(f_2 - f_1)} \frac{\sin 2\pi(n-m)W}{\pi(n-m)}$$

$$\gamma = \int_{f_1 - W}^{f_1 + W} c \, df = \cdots = \sum_{n=0}^{N-1} \sum_{k=0}^{K-1} \lambda_k [v_n^{(k)}]^2 e^{j2\pi n(f_2 - f_1)}$$

$$\gamma_1 = \int_{f_1 - W}^{f_1 + W} c_1 \, df = \cdots = \sum_{n=0}^{N-1} \sum_{k=0}^{K-1} \lambda_k [v_n^{(k)}]^2 x(n) e^{-j2\pi n f_1}$$

$$\gamma_2 = \int_{f_1 - W}^{f_1 + W} c_2 \, df = \cdots = \sum_{k=0}^{K-1} \sum_{n=0}^{N-1} \sum_{m=0}^{N-1}$$

$$\times\, v_n^{(k)} v_m^{(k)} x(m) e^{j2\pi[(n-m)f_1 - n f_2]} \frac{\sin 2\pi(n-m)W}{\pi(n-m)}$$

and (7-57) becomes

$$\hat{\boldsymbol{\mu}}(f_1, f_2) = \frac{1}{\delta_1\delta_2 - |\gamma|^2}\begin{bmatrix} \delta_2\gamma_1 - \gamma\gamma_2 \\ -\gamma^*\gamma_1 + \delta_1\gamma_2 \end{bmatrix} \tag{7-58}$$

We also have Q from (7.47). We just have to set $f_0 = f_1$. The regression error term is now

$$\zeta = \int_{f_1-W}^{f_1+W} \|\mathbf{V}\boldsymbol{\mu}\|^2_{\text{double}}\, df = \delta_1|\mu_1|^2 + \delta_2|\mu_2|^2 + 2\,\text{Re}\{\gamma\mu_1^*\mu_2\}$$

while

$$\int_{f_1-W}^{f_1+W} \|\mathbf{V}\boldsymbol{\mu}\|^2_{\text{single}}\, df = |\gamma_1|^2/\delta_1$$

We finally obtain for our F-test the expression

$$F(f_1, f_2) = \frac{1/4}{1/(2K-4)}\left[\frac{\zeta}{Q-\zeta}\right] \tag{7-59}$$

or

$$F(f_1, f_2) = \frac{1/(4-2)}{1/(2K-4)}\left[\frac{\zeta - |\gamma_1|^2/\delta_1}{Q-\zeta}\right] \tag{7-60}$$

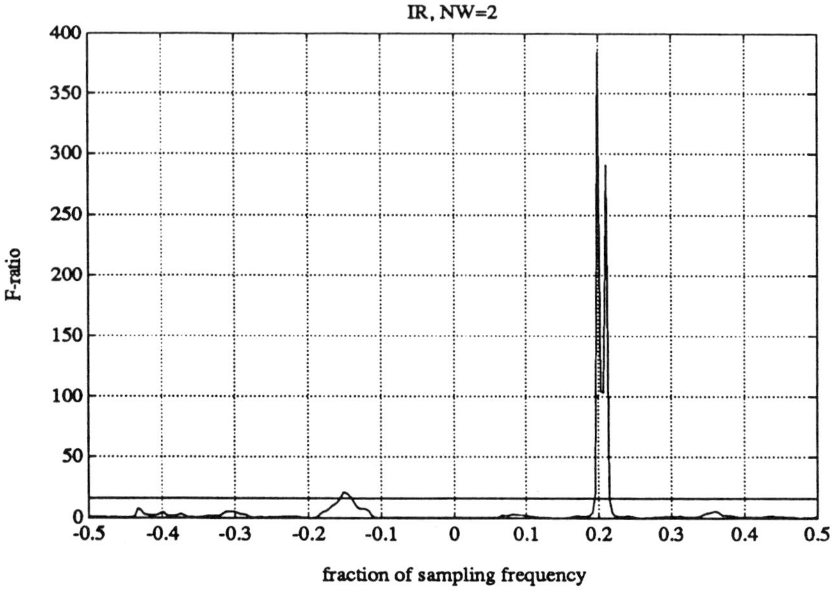

Figure 7-18 The projection of max $F(f, \Delta f)$, onto the f axis. Note the absence of spurious peaks.

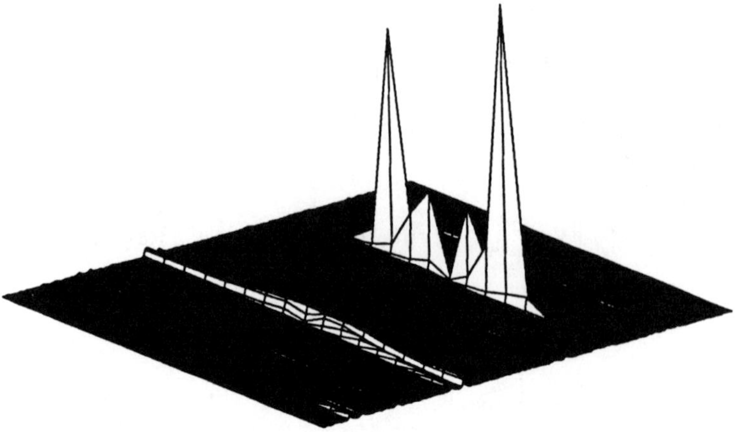

Figure 7-19 The double-line integral regression F-test for $NW = 2$. The surface $F(f, \Delta f)$ is shown and we see a small ridge indicating a singlet around -0.15.

Equation (7.59) tests the existence of one or two lines, whereas (7.60) tests the existence of two lines only.

If again we let f_1 vary over the entire frequency range, f_2 will vary in $(f_1 - W, f_1 + W)$. As we can see from Figs. 7.18 and 7.19, the doublet is resolved and the spurious peak problem disappears.

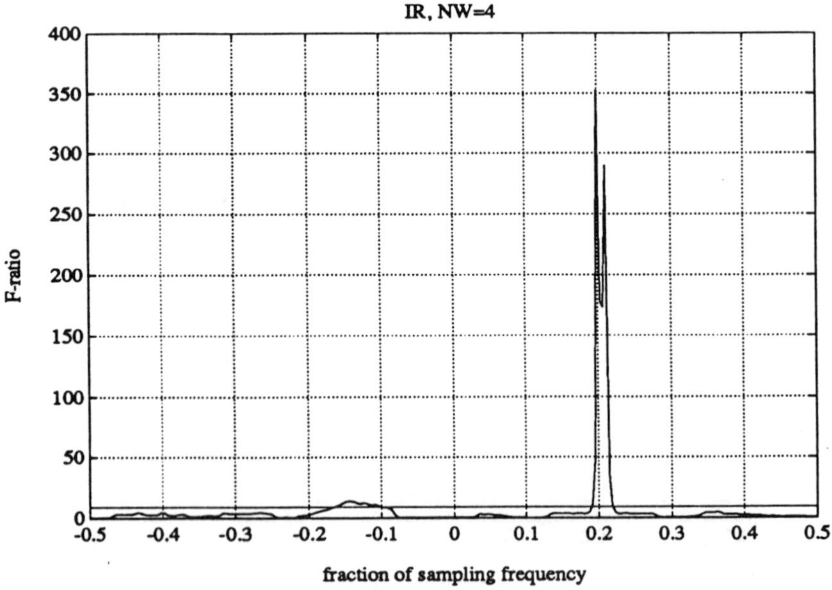

Figure 7-20 The projection of max $F(f, \Delta f)$, onto the f axis.

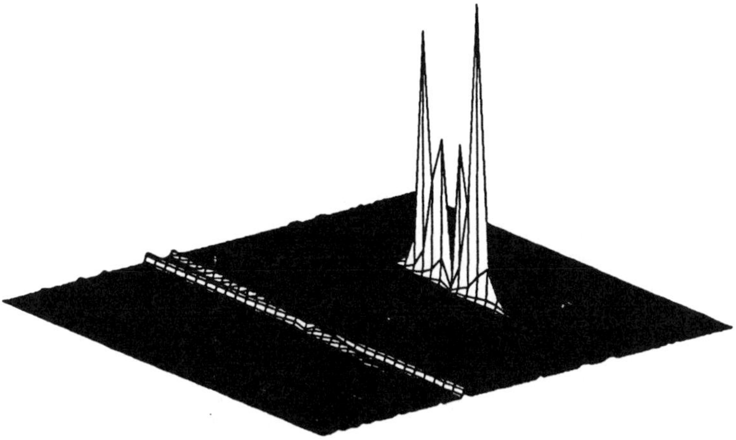

Figure 7.21 $F(f, \Delta f)$ for the double-line integral regression F-test for $NW = 4$.

7-7-6 Line Component Extraction

The next step in this spectrum estimation procedure is to extract the line components in order to be left with only the continuous part of the spectrum under investigation. For this residual spectrum, with no line components, the eigencoefficients $y_k(f)$ should obey

$$\mathrm{E}\{y_k(f)\} = 0$$

They should also be related to the original eigencoefficients $x_k(f)$ as

$$y_k(f) = x_k(f) - \sum_i \hat{\mu}(f_i)V_k(f - f_i) \tag{7-61}$$

This suggests that the data vector be modified as

$$y(n) = x(n) - \sum_i \mu_i e^{j2\pi n f_i} \tag{7-62}$$

for every n, where $\mu_i = \hat{\mu}(f_i)$.

Thomson [36] gives the formula

$$S(f) = |\hat{\mu}(f_0)|^2 \delta(f - f_0) + S_r(f) \tag{7-63}$$

for reshaping the spectrum around a line component at f_0, where $S_r(f)$ is the residual spectrum after extracting that particular line component. He also points out that care must be taken with this operation, so that power is conserved numerically. One cannot just stick $|\hat{\mu}(f_0)|^2$ at f_0 after estimating the residual spectrum. Since (7.63) is a power density relation, we obtain by integration

$$\hat{\sigma}^2 = \sum_i \hat{\sigma}_i^2 + \hat{\sigma}_r^2 \tag{7-64}$$

where $\hat{\sigma}^2$ is our estimate of the initial process variance, $\hat{\sigma}_r^2$ the corresponding estimate for the residual process generated by (7.62) and $\hat{\sigma}_i^2$ is the power that corresponds to the ith line component. Subtracting the residual power from the total in order to find a particular $\hat{\sigma}_i^2$ is very sensitive to roundoff errors. Hence, in implementation, it should only be used to find the total power that corresponds to all four line components (i.e., subtract initial variance from estimated variance when all lines are extracted). This power is then allocated into four components, according to the ratios of the $|\hat{\mu}(f_i)|^2$ which exhibit better robustness properties. Since the spectrum $S(f)$ is a power density, each of those power components is assumed to be equal to the area of a rectangle of width W, the spectral window used in the estimation of the μ's. The height of this rectangle is then the proper number to assign as the line power density at the estimated line frequency. Thomson also suggests that the line component be assigned a shape, similar to the F line shape. The width of the line would then be proportional to the frequency uncertainty of the estimate. (This is not done here as the FFT mesh used is only 256 points and the F-width is $\approx 1/256$.)

In order to perform the computations in (7.62), we have to have the best possible estimates of μ_i and f_i in order to minimize subtraction errors. For the univariate F function of the single-line tests the method indicated is the *golden section search* to find the maximum of F. This is similar in spirit to the bisection method for root finding. For the double-line tests, a multivariable optimization method is required. The one used in this implementation is the *Hooke and Jeeves direct search method* (see [11, 14]), but the specific choice of routine in both cases is a matter of personal taste.

The results of this parameter estimation of the μ_i's and f_i's are given in Tables 7-2 and 7-3. Fig. 7-22 shows the residual single-line F-test when the

TABLE 7-2 Initial estimates of μ_i and f_i. A time-bandwidth $NW = 2$ was used. The initial mean and variance are $-0.025 + j0.024$ and 1.780, respectively

	First singlet	Second singlet	First doublet	Second doublet
Known frequencies	−0.15	0.1	0.2	0.21
99% level	10.92	10.92	15.98	15.98
Point regression				
estimated frequency	−0.1508	0.0965	0.1995	0.2100
F-ratio	66.7	17.4	8978.0	8978.0
μ	$0.0734 - j0.0654$	$0.0042 + j0.1027$	$0.2158 + j0.9335$	$0.2977 + j0.9211$
Integral regression				
estimated frequency	−0.1501	0.0968	0.1997	0.2099
F-ratio	49.6	5.61	1185.0	1185.0
μ	$0.0662 - j0.0726$	$0.0027 + j0.1094$	$0.2491 + j0.9404$	$0.2632 + j0.9403$

TABLE 7-3 Final estimates of μ_i and f_i. The residual mean and variance are 0.0036 − j0.0007 and 0.1821 for point regression and 0.0038 − j0.0007 and 0.1817 for integral regression, respectively

	First singlet	Second singlet	First doublet	Second doublet
Point regression				
estimated frequency	−0.1499	0.1002	0.1996	0.2099
F-ratio	304.7	3040.0	10017.0	10017.0
μ	0.0596 − j0.0829	0.0839 + j0.0568	0.2357 + j0.9362	0.2774 + j0.9359
Integral regression				
estimated frequency	−0.1499	0.1001	0.1997	0.2098
F-ratio	201.7	915.9	1142.0	1142.0
μ	0.0606 − j0.0839	0.0843 + j0.0581	0.2712 + j0.9437	0.2380 + j0.9561

doublet is removed. Better estimates of the singlets are then obtained. These can then be extracted from the original dataset and the doublet can again be estimated (final table values). In fact, an iterative scheme of extracting line(s) with higher F values to improve the estimates of those with lower F values suggests itself. For more complex datasets, this may be a necessary approach.

Comparing point and integral regression parameter values we do not see a significant difference between the final estimates of the line components.

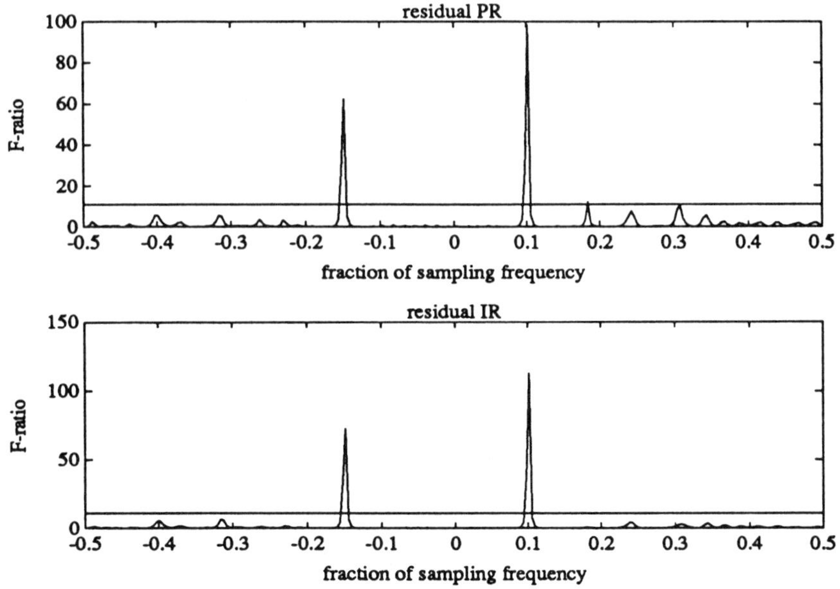

Figure 7-22 The residual single-line F-test when the initial estimate of the doublet is removed. The solid straight line indicates the 99% confidence level. Note that the spurious peak heights are reduced more in the integral regression version of the F-test.

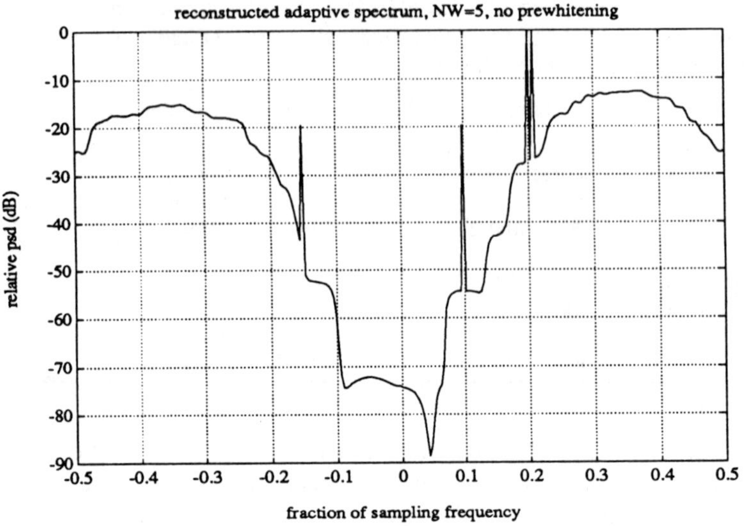

Figure 7-23 Adaptive spectrum reconstruction without prewhitening. The agreement with the theoretical analytic form is excellent except for some leakage of the colored noisy beyond the ±0.2 frequency boundaries.

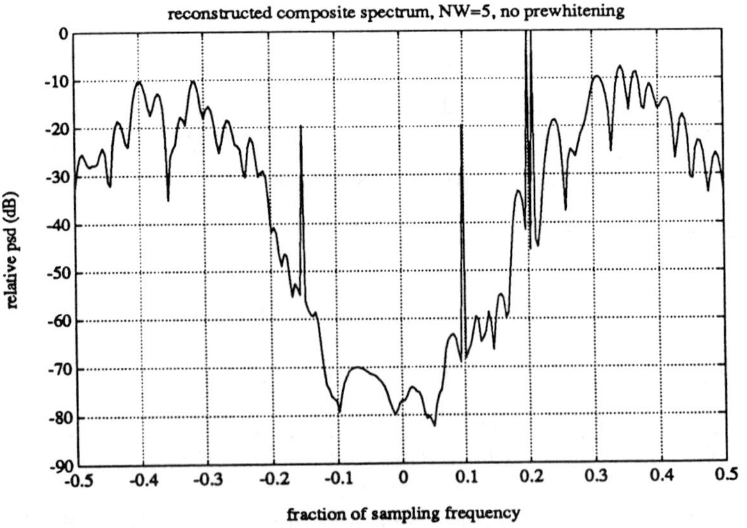

Figure 7-24 Composite spectrum reconstruction without prewhitening. The leakage problem is alleviated and more structure is evident in the colored noise component.

Point regression is, however, much faster in computer execution while integral regression is more stable regarding spurious peaks.[14] The reconstructed spectrum in Fig. 7-23 is in excellent agreement with the theoretical one in Fig. 7-6. The last one gives a range of values only down to −50 dB, and we see that the only problem is some leakage of the continuous colored noise component beyond the ±0.2 frequency boundaries. This is due to the low-pass properties of the prolate windows, and Thomson [36] offers the remedy of using his composite spectrum estimator to correct for this effect. Figure 7-24 shows the result. It does indeed alleviate the leakage problem showing a more structured colored noise as well. Note that this structure is actually more realistic to expect, since we are given only a single realization of a stochastic process.

7-7-7 Prewhitening

The importance of prewhitening cannot be stressed strongly enough for real-life cases. In essence, prewhitening reduces the dynamic range of the spectrum by filtering the data so that the residual spectrum is nearly flat or white. Leakage from strong components is then reduced so that the fine structure of weaker components is more likely to be resolved. Most of the theory behind spectrum estimation assumes smooth, almost white-like spectra to begin with, anyway.

In order to whiten the process $\{y(n)\}_{n=1}^{N}$, the method in vogue is to pass it through a linear prediction filter. This is equivalent to assuming an AR model for the process and if this is truly so and the filter order is the same as the AR process order, the output of the filter, $\{r(n)\}$, *the innovations process*, will be white noise. If not, the range of the spectrum will be decreased and any fine structure on the innovations process will be easier to pick out. In essence, a systems identification approach is taken. The reason that AR models are so popular is simply that they are linear and much easier to solve than MA or the full ARMA models.[15]

It should be stressed that the use of an AR filter for prewhitening does not imply an AR spectrum estimation approach. The AR filter simply performs some data preprocessing so that MWM (or any spectrum estimation technique for that matter) will operate on data having properties closer to the ones that the method or technique assumes. This makes for a better intermedium spectrum estimate. The effect of the AR filter is then removed by operating on this intermediate spectrum estimate with the inverse filter.

[14] It might be worthwhile to restructure integral regression so that the integral expressions are computed numerically using the known eigencoefficients. This should certainly improve execution time over the computation of the double and triple sums appearing in the test.

[15] In reference [19] the authors explain that they are currently investigating the generalization to the full ARMA case.

Note also, that the prewhitened data do not have to have a purely white spectrum. It need only be relatively flat.

Assuming therefore that

$$y(n) = -\sum_{k=1}^{p} a_k y(n-k) + r(n)$$

where the a_k's are the AR coefficients with p the model order, we compute the residual spectrum $\hat{S}_r(f)$ with Thomson's method and estimate $S_y(f)$ from

$$\hat{S}_y(f) = \frac{\hat{S}_r(f)}{\left|1 + \sum_{k=1}^{p} a_k e^{-j2\pi fk}\right|^2} \tag{7-65}$$

Estimation of the AR coefficients was done with the modified covariance method in which both the forward and backward prediction errors are minimized at the same time. Marple [22] claims that this method seems to be optimum although it does not guarantee minimum phase for the resultant filter. Marple also offers a routine that implements it (among other methods) and this is the one used here. Note that with this method the property that the backward error coefficients are complex conjugates of the forward error coefficients holds. The identification of the latter with the AR parameters helps us compute the residuals without assuming zero initial conditions, which introduce transient errors to the process as

$$r(n) = \begin{cases} y(n) + \sum_{k=1}^{p} a_k y(n-k)\,, & \text{for } n = p+1, \ldots, N \\ y(n) + \sum_{k=1}^{p} a_k^* y(n+k)\,, & \text{for } n = 1, \ldots, p \end{cases} \tag{7-66}$$

Observe also that in this way we do not reduce the sample size of the given time series.

A straightforward implementation of the above leads to Figs. 7-25–7-26 for $NW = 5$ and an AR(4) model. Prewhitening brings more structure to the colored noise, which is what we realistically expect. Note that in the case of the composite spectrum it is evident that the extraction of the singlets is not total since we see some increase in the residual spectrum at those frequencies. Prewhitening brings this detail out. Note that if prewhitening is performed before the line components are estimated and we then try to estimate them from the prewhitened data, some small improvement is indeed seen with the weaker, singlet components. However, the stronger doublet is removed by the AR filter. This indicates that this approach be used only for estimating weak in power line components.

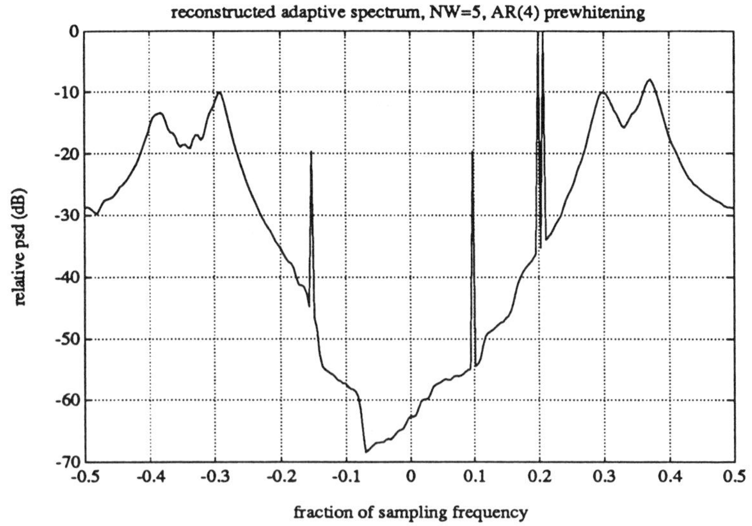

Figure 7-25 The adaptive reconstructed spectrum with prewhitening.

Comparing Figs. 7-23–7-26 with the ideal spectrum in Fig. 7-6 it appears that Fig. 7-23 is the best match, indicating that prewhitening does not seem to improve things. This was not the case with other datasets and spectra that were tried. Prewhitening always led to an improved spectrum estimate. The discrepancy observed with Marple's dataset is probably coincidental. It so happens that the smoothing that the prolate windows perform, give a final

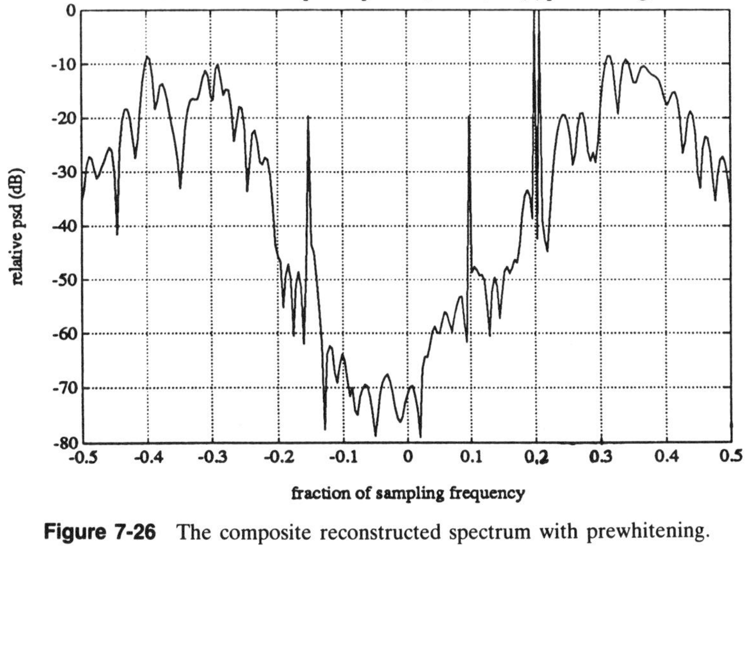

Figure 7-26 The composite reconstructed spectrum with prewhitening.

spectrum estimate that is very close to the ideal one. Normally, more than one snapshot is required for such an event to occur. Looking into the classical spectrum estimation results, in Fig. 7-7, we do see some similarities in the colored noise component, particularly with the Hamming spectrum. Clearly, the prewhitening gives a more accurate picture of the single snapshot spectrum estimate but, coincidentally in this case, the non-prewhitened spectrum estimate is closer to the ideal one.

In [35] and [19], the case of robust prewhitening is worked out for a simple additive outlier model, for real data. It is assumed that the observed data $\{y_n\}_{n=1}^N$, consist of the process of interest $\{x_n\}_{n=1}^N$, plus occasional outliers $\{e_n\}_{n=1}^N$

$$y_n = x_n + e_n$$

and an iterative scheme is developed. A further refinement is given in [23]. It would be nice to extend this technique to complex data, making prewhitening more robust (this is not done here).

7-7-8 Multiple Snapshots

In a practical situation we always have more than one snapshot of data available. Since the underlying process is stochastic in nature, it makes sense to combine all the available snapshots in estimating the spectrum into a single overall result. Note that simply averaging the individual adaptive or composite spectra for each snapshot is a nonoptimum way of doing so. The proper generalization is done as follows.

For a single data vector $\{x(n)\}_1^N$ the K eigencoefficients $x_k(f)$ are computed, and from them the individual eigenspectra $\hat{S}_k(f) = |x_k(f)|^2$ are estimated. Generalizing to N_s data vectors (snapshots) $\{x(n, n_s)\}$, $n = 1, \ldots, N$ and $n_s = 1, \ldots, N_s$, the average eigenspectra would be

$$\hat{S}_k(f) = \frac{1}{N_s} \sum_{n_s=1}^{N_s} |x_k(f|n_s)|^2$$

It is these values that are used in the adaptive and composite spectrum estimation procedures.

The F-tests are also slightly modified. In the following subsection the extension to multiple-snapshot point-regression F-tests is made. The integral regression case follows along similar lines.

7-7-9 Multiple-Snapshot, Single-Line, Point-Regression *F*-Tests

For the single line, point regression case, the criterion is the same. Given a line component at a frequency f_0 the expected value of the eigencoefficients is

$$\mathsf{E}\{x_k(f \mid n_s)\} = \mu V_k(f - f_0)$$

where now all N_s snapshots are included. The residual error that has to be minimized is

$$\|\mathbf{x} - \mathbf{U}\boldsymbol{\mu}\|^2$$

where

$$\mathbf{x} = \begin{bmatrix} x_0(f \mid 1) \\ \vdots \\ x_{K-1}(f \mid 1) \\ \vdots \\ x_0(f \mid N_s) \\ \vdots \\ x_{K-1}(f \mid N_s) \end{bmatrix}, \quad \boldsymbol{\mu} = \mu, \quad \mathbf{V} = \begin{bmatrix} V_0(f - f_0) \\ V_1(f - f_0) \\ \vdots \\ V_{K-1}(f - f_0) \end{bmatrix}, \quad \mathbf{U} = \begin{bmatrix} \mathbf{V} \\ \mathbf{V} \\ \vdots \\ \mathbf{V} \end{bmatrix}$$

The data vector $\mathbf{x}$ is now a $KN_s \times 1$ complex vector, $\boldsymbol{\mu}$ is again a 1×1 complex scalar, $\mathbf{V}$ is the same $K \times 1$ complex vector but $\mathbf{U}$ is now introduced, containing N_s identical $\mathbf{V}$ vectors.

This residual error at f can be written in component form as

$$e^2(\mu, f) = \sum_{n_s=1}^{N_s} \sum_{k=0}^{K-1} |x_k(f \mid n_s) - \mu(f)V_k(0)|^2$$

and the least squares solution is

$$\hat{\boldsymbol{\mu}} = (\mathbf{U}^H\mathbf{U})^{-1}\mathbf{U}^H\mathbf{x} = \hat{\mu}(f) = \frac{\displaystyle\sum_{n_s=1}^{N_s} \sum_{k=0}^{K-1} V_k^*(0)x_k(f \mid n_s)}{\displaystyle N_s \sum_{k=0}^{K-1} |V_k(0)|^2}$$

This can also be written as

$$\hat{\mu}(f) = \sum_{n=0}^{N-1} h_n(N, W)\bar{x}(n)e^{-j2\pi fn}$$

where the harmonic data windows are the same as in the single snapshot case

$$h_n(N, W) = \frac{\displaystyle\sum_{k=0}^{K-1} V_k^*(0)v_n^{(k)}(N, W)}{\displaystyle\sum_{k=0}^{K-1} |V_k(0)|^2}$$

but the data vector $\bar{x}(n)$ is the coherent average

$$\bar{x}(n) = \frac{1}{N_s} \sum_{n_s=1}^{N_s} x(n \mid n_s)$$

The F-ratio is now

$$F(f) = \frac{\frac{1}{\nu_1} \|\mathbf{U}\boldsymbol{\mu}\|^2}{\frac{1}{\nu_2} \|\mathbf{x} - \mathbf{U}\boldsymbol{\mu}\|^2}$$

or

$$F(f) = \frac{\frac{1}{\nu} |\hat{\mu}(t)|^2 \hat{\mu}(f)|^2 N_s \sum_{k=0}^{K-1} |V_k(0)|^2}{\frac{1}{2K - \nu} \sum_{n_s=1}^{N_s} \sum_{k=0}^{K-1} |x_k(f \mid n_s) - \mu(f) V_k(0)|^2}$$

where ν is again equal to two degrees of freedom. The total number of degrees of freedom, ν_{tot}, is set here to be equal to the old value $2K$. This is a worst case scenario. We do have KN_s complex data points now available to draw information from. Using however the value $\nu_{\text{tot}} = 2KN_s$ for multiple snapshot F-tests, with experimental data (see later sections), we obtain preposterously large F values, orders of magnitude larger than the 99% confidence level. There is no doubt that we do have more degrees of freedom but the data snapshots may not be independent enough, one from the other, in order to assign $\nu_{\text{tot}} = 2KN_s$. The appropriate value for ν_{tot} is probably somewhere between $2K < \nu_{\text{tot}} < 2KN_s$. Thomson [38, p. 565], points out that the effective number of degrees of freedom is given by a more complicated expression and is indeed less than $2KN_s$ (development of this formula is not done here).

Actually, the proper number of degrees of freedom is only needed to decide for the proper significance level. Using the test as is, we can check the relative heights of the various F peaks. Since we already know the number and approximate location we simply pick the largest peaks in that region. In the real data cases examined in the following sections, it is seen that excellent estimates are obtained with this approach, both for the line frequencies and their complex amplitudes. However, in order to test for the existence of other possible line components, this matter should be resolved (work is underway to do so).

7-7-10 Multiple-Snapshot, Double-Line Point-Regression *F*-Tests

Similar changes occur for the double-line tests. We have a similar criterion as before, namely,

$$\mathsf{E}\{x_k(f \mid n_s)\} = \mu_1 V_k(f - f_1) + \mu_2 V_k(f - f_2)$$

where now all N_s snapshots are included. The residual error to be minimized is

$$\|\mathbf{x} - \mathbf{U}\boldsymbol{\mu}\|^2$$

where

$$\mathbf{x} = \begin{bmatrix} x_0(f \mid 1) \\ \vdots \\ x_{K-1}(f \mid 1) \\ \vdots \\ x_0(f \mid N_s) \\ \vdots \\ x_{K-1}(f \mid N_s) \end{bmatrix}$$

$$\boldsymbol{\mu} = \begin{bmatrix} \mu_1 \\ \mu_2 \end{bmatrix}$$

$$\mathbf{V} = \begin{bmatrix} V_0(f - f_1) & V_0(f - f_2) \\ V_1(f - f_1) & V_1(f - f_2) \\ \vdots & \vdots \\ V_{K-1}(f - f_1) & V_{K-1}(f - f_2) \end{bmatrix}$$

$$\mathbf{U} = \begin{bmatrix} \mathbf{V} \\ \mathbf{V} \\ \vdots \\ \mathbf{V} \end{bmatrix}$$

If again, as in the single snapshot case

$$\mathbf{V}^H \mathbf{V} = \begin{bmatrix} d_1 & c \\ c^* & d_2 \end{bmatrix}$$

the least squares solution is

$$\hat{\boldsymbol{\mu}} = (\mathbf{U}^H \mathbf{U})^{-1} \mathbf{U}^H \mathbf{x} = \frac{1}{d_1 d_2 - |c|^2} \begin{bmatrix} d_2 & -c \\ -c^* & d_1 \end{bmatrix} \begin{bmatrix} c_1 \\ c_2 \end{bmatrix}$$

where now

$$d_1 = \sum_{k=0}^{K-1} |V_k(f - f_1)|^2, \qquad d_2 = \sum_{k=0}^{K-1} |V_k(f - f_2)|^2,$$

$$c = \sum_{k=0}^{K-1} V_k^*(f - f_1)V_k(f - f_2)$$

$$c_1 = \frac{1}{N_s} \sum_{n_s=1}^{N_s} \sum_{k=0}^{K-1} V_k^*(f - f_1)x_k(f \mid n_s),$$

$$c_2 = \frac{1}{N_s} \sum_{n_s=1}^{N_s} \sum_{k=0}^{K-1} V_k^*(f - f_2)x_k(f \mid n_s)$$

The F-test, for testing the hypothesis that at a given frequency we need two lines to explain the data, becomes

$$F(f; f_1, f_2) = \frac{1/(\nu_1 - \nu_2)}{1/(2K - \nu_1)} \left[\frac{\|\mathbf{U}\boldsymbol{\mu}\|^2_{\text{double}} - \|\mathbf{U}\boldsymbol{\mu}\|^2_{\text{single}}}{\|\mathbf{x} - \mathbf{U}\boldsymbol{\mu}\|^2_{\text{double}}} \right]$$

The single-line model parameter μ is now given by

$$\mu_{\text{single}} = c_1/d_1$$

and

$$\|\mathbf{U}\boldsymbol{\mu}\|^2_{\text{single}} = N_s|c_1|^2/d_1$$

while

$$\|\mathbf{U}\boldsymbol{\mu}\|^2_{\text{double}} = N_s \sum_{k=0}^{K-1} |V_k(0)\mu_1 + V_k(f_1 - f_2)\mu_2|^2$$

Similar observations, as given previously, hold for ν_{tot}, the effective number of degrees of freedom.

7-8 EXPERIMENTAL DATA DESCRIPTION FOR A LOW-ANGLE TRACKING RADAR STUDY

In this section we apply the method of multiple-windows to real-life sampled-aperture data corresponding to a low-angle tracking radar environment. The experimental data collection was performed at a site on the mouth of Dorcas Bay, which opens onto the eastern end of Lake Huron on the west coast of the Bruce Peninsula, close to Tobermory, Ontario. This particular location was chosen because of the high sea states normally encountered there, which are caused by a combination of westerly winds, the shallow water offshore[16] and the long fetch across Lake Huron. The

[16] The greatest depth along the transmission path was 12 m.

transmitter was situated at a distance $L = 4.61$ km from the receiver, both being within 10 m of the water's edge. Figure 7-27 shows the path profile. The receiver was secured at the top of a tower and its center was at a height h_R above the water's edge while that of the transmitter was at h_T (adjustable).

The transmitter consists of two antenna horns, one above the other, the top one being set for horizontal (H) and the bottom one for vertical (V) polarizations, respectively. Each horn was also capable of transmitting at different frequencies in order to implement frequency agility. For the latter, two different frequency signals were transmitted simultaneously. One of them was always at the fixed frequency of 10.2 GHz, while the other's frequency was varied from 8.02 to 12.34 GHz with a 30 MHz step size according to the formula: $8.02 + p0.03$ GHz. The picket number p controlled the frequency step size in the agile channel.

The available options for the transmitted signal were therefore:

- When a horizontally polarized signal was desired, the top H horn would transmit the mixture of the fixed and agile frequency signals while the bottom V horn was idle.
- For an H and V dual-polarized signal, the H horn would transmit the agile and the V horn the fixed frequency signals.
- Finally, for a vertically polarized signal, the top H horn remained ideal while the bottom V horn transmitted the mixture of fixed and agile frequency signals.

The microwave source was produced by phase locking to a highly stable 5 MHz crystal oscillator operating at a constant temperature. A travelling wave tube amplifier (TWTA) with a maximum gain of 40 dB was then used to amplify the source signal before feeding it to the transmitter horns. System control at the transmitter was achieved with a PC-XT microcomputer that adjusted transmission frequency, TWTA gain and carriage height.

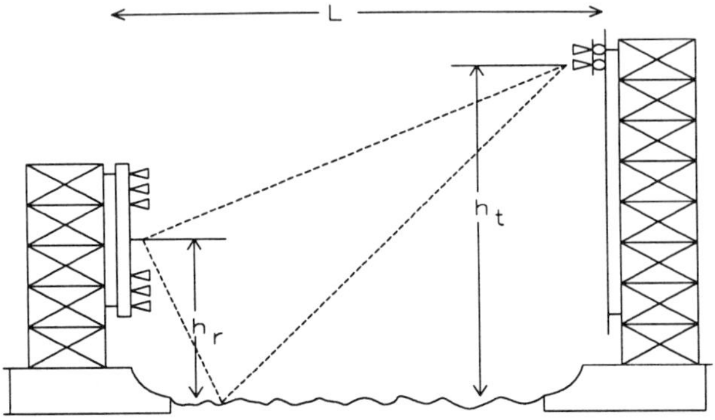

Figure 7-27 The path profile. Transmitter at the right and receiver array at the left.

The receiver is a 32-element linear array designed and built by V. Kezys and E. Vertatschitsch at the Communications Research Laboratory (CRL) of McMaster University and forms the main part of MARS, a Multi-parameter Adaptive Radar System. It is a coherent, sampled aperture, linear receiver array, vertically oriented, composed of 32 pairs of horizontally polarized standard gain horn antennas. Each pair of horns corresponds to a single array element and offers the opportunity to always implement frequency agility and when desired, by choosing one of the options for the polarization of the transmitted signal, polarization diversity, HH or VH.

The RF and IF local oscillators and the RF testing signal generator were shared by all the receiver elements. Since we have coherent reception, each receiver horn is followed by a quadrature demodulator block so that phase information can be extracted.

System control at the receiver is provided with a PC-AT microcomputer that offers a completely automated data collection environment. For each trial, only the parameters defining the desired physical conditions need to be fed into the computer. The computer accordingly adjusted the frequency of the RF oscillator for the agile channels, the IF amplifier gains, the band-

TABLE 7-41 Specifications of MARS, Used in the Lake Huron Experiments

Transmitter

- 100 mW CW into one or two 22 dB horns
- Simultaneous dual frequencies
- Option of dual polarization (H) or (V)
- Adjustable transmitter height h_T

Antenna Array

- 32 element linear array vertically oriented
- Array aperture 1.82 m, interelement spacing 5.715 cm
- Array machined to ± 0.1 mm tolerance
- Multifrequency capability 8.02–12.34 GHz with 30 MHz steps

Receiver Elements

- 2 receiving channels per array element
- 10 dB horizontally polarized horns as array elements
- $\sim$25 dB nominal cross polarization rejection ratio
- Coherent demodulation with nominal frequency stability to 10^{-12} over a few seconds
- 0.1 Hz Doppler resolution
- 1 Hz–2 kHz sampling rate capability

width of the low-pass filters in the I and Q channels and the sampling rate of the samplers. It also sent commands via a radio link to the transmitter computer that controlled the transmission operation. Most of the data were sampled with a 62.5 Hz sampling rate, corresponding to 8 sample points per cycle at the baseband frequency. They were then digitized by A/D converters with 12-bit precision and stored onto hard disk for off-line processing.

The data were divided into three groups with respect to polarization: like-polarized, dual-polarized, and cross-polarized. The first four characters in their name refers to the collection date, for example, *nov3*. The next two characters define the group they belong to; *dh* for the like-polarized group where both transmission and reception are H-polarized; *dd* for the dual-polarized group where both polarizations are transmitted; *dv* for the cross-polarized group where only a cross polarized signal is received; and *cff* for the far-field, aperture calibration datasets.

The numeral following the polarization designation is simply a dataset index providing order among similar datasets collected within the same day. It is followed by the data subset number. As an example, *nov3:dd1.dat;2* stands for the second subset of a dual polarized dataset, collected on November 3, being the first dataset in this group collected on that particular day. We normally have 127 snapshots per data subset (offering us about 2 s worth of data) and 16 subsets per dataset.

The data are carefully calibrated, both in phase and quadrature (IQ calibration) using known injected tone signals, and in aperture (aperture calibration) using a far-field technique (details are given in [8]). In the following section, we focus on some HH polarized data at different frequencies and grazing angles. The particular datasets used are shown in Table 7.5. Note that a spherical earth model [4, 16] where normal propagation conditions are assumed—earth radius is 4/3 the actual value—is used to calculate the expected angular separation between the direct and specular components. This quantity provides a better comparison measure between expected and estimated results, since it is not affected by the tilt of the array antenna occurring from wind action on the array tower.

TABLE 7-5 A sample of individual experimental datasets used in the MWM implementation. The distance between transmitter and receiver is 4.61 km in all cases

	Dataset	Frequency (GHz)	Separation (BW)	h_r (m)	h_t (m)
1	*nov4:dh9.dat;1*	8.05	0.320	8.64	15.53
2	*nov3:dh4.dat;3*	8.62	0.218	8.67	9.59
3	*nov3:dh6.dat;7*	9.76	0.247	8.67	9.59
4	*nov3:dd1.dat;2*	10.12	0.468	8.67	18.06
5	*nov3:dh4.dat;16*	12.34	0.312	8.67	9.59

7-9 ANGLE-OF-ARRIVAL ESTIMATION

Angle-of-arrival estimation on the MARS database is performed using both MWM and maximum likelihood (ML) on a snapshot-by-snapshot basis. ML was chosen because it provides the yardstick against which all modern parametric methods are compared. To single out the effect, if any, of including diffuse multipath in the AOA estimation process, which MWM implicitly does, the only a priori knowledge that we assume for ML is that we have two incoming plane waves in a white noise background. The noise covariance matrix is then diagonal and the maximization of the logarithm of the likelihood function leads to

$$\min_{\mathbf{\Phi}} \|\mathbf{x} - \mathbf{s}\|^2 \tag{7-67}$$

This is a *nonlinear least squares* problem, where $\mathbf{\Phi}$ is the parameter vector, $\mathbf{x}$ is the data vector and $\mathbf{s}$ is the signal model. Let us decompose the signal vector into

$$\mathbf{s} = \mathbf{A}\mathbf{a} \tag{7-68}$$

where $\mathbf{A}$ is a function of the parameters that enter nonlinearly into the model, while $\mathbf{a}$ is a function of the ones that do so linearly. Then, (7-67) becomes

$$\min_{\mathbf{\Phi}} \|\mathbf{x} - \mathbf{A}\mathbf{a}\|^2 \tag{7-69}$$

The separation of $\mathbf{\Phi}$ into a linear and a nonlinear subset helps us break the nonlinear least squares estimation problem of (7-67) into a linear least squares problem and a smaller nonlinear one that is hopefully easier to solve. This is done as follows.

At the minimum, when $\mathbf{a} = \hat{\mathbf{a}}$, the residual vector $\mathbf{x} - \mathbf{A}\hat{\mathbf{a}}$ is orthogonal to the surface $\mathbf{A}\mathbf{a}$. Therefore,

$$(\mathbf{A}\mathbf{a})^H(\mathbf{x} - \mathbf{A}\hat{\mathbf{a}}) = 0 \tag{7-70}$$

and

$$\hat{\mathbf{a}} = \mathbf{A}^+\mathbf{x} = (\mathbf{A}^H\mathbf{A})^{-1}\mathbf{A}^H\mathbf{x} \tag{7-71}$$

Also, from (7-69) we have

$$\begin{aligned}\|\mathbf{x} - \mathbf{A}\mathbf{a}\|^2 &= (\mathbf{x} - \mathbf{A}\mathbf{a})^H(\mathbf{x} - \mathbf{A}\mathbf{a}) = [\mathbf{x}^H - (\mathbf{A}\mathbf{a})^H](\mathbf{x} - \mathbf{A}\mathbf{a}) \\ &= \mathbf{x}^H\mathbf{x} - \mathbf{x}^H\mathbf{A}\mathbf{a} - (\mathbf{A}\mathbf{a})^H(\mathbf{x} - \mathbf{A}\mathbf{a})\end{aligned}$$

At the minimum, $\mathbf{a} = \hat{\mathbf{a}}$ and from (7-70) the last part is zero. Therefore, (7-69) becomes

$$\max \mathbf{x}^H \mathbf{A}\mathbf{A}^+ \mathbf{x} \tag{7-72}$$

where the maximization is done over the nonlinear parameter subset. In problems of this type, we first solve (7-72) and then (7-71), which is now a simple linear problem.

For the case of two plane waves impinging on an M-element array, the signal model at the mth element, referred to the array center, is

$$s(m) = a_1 e^{j[m-(M+1)/2]\phi_1} + a_2 e^{j[m-(M+1)/2]\phi_2}, \qquad m = 1, \ldots, M \tag{7-73}$$

where ϕ, the electrical phase angle [10], or spatial wavenumber, is defined as

$$\phi = \frac{2\pi d}{\lambda} \sin \theta$$

with d the array interelement spacing, λ the wavelength of the signal, and θ the physical direction of arrival of the incoming plane wave.

It can be seen that the signal is linear in the complex amplitudes a_1 and a_2 and nonlinear in the phase angles ϕ_1 and ϕ_2. To transform to vector notation, define the *direction of arrival* (DOA) *vector*

$$\mathbf{d}(\phi) = [e^{j[1-(M+1)/2]\phi} \quad e^{j[2-(M+1)/2]\phi} \quad \cdots \quad e^{j[M-(M+1)/2]\phi}]^T \tag{7-74}$$

The *direction matrix* for this case, where the number of plane waves is $K = 2$, becomes

$$\mathbf{A} = [\mathbf{d}(\phi_1) \quad \mathbf{d}(\phi_2)]$$

and the *signal amplitude vector* is

$$\mathbf{a} = [a_1 \quad a_2]^T$$

The signal vector is then expressed as

$$\mathbf{s} = \mathbf{A}\mathbf{a}$$

and we end up having to solve (7-72) and (7-71). Some further simplification can be done by defining the complex scalars

$$D_k = \mathbf{d}^H(\phi_k)\mathbf{x} \quad \text{and} \quad p = \mathbf{d}^H(\phi_1)\mathbf{d}(\phi_2)$$

where $k = 1, 2$. We have then

$$\mathbf{A}^H\mathbf{x} = [D_1 \quad D_2]^T$$

and

$$(\mathbf{A}^H\mathbf{A})^{-1} = \begin{pmatrix} M & p \\ p^* & M \end{pmatrix}^{-1} = (M^2 - pp^*)^{-1}\begin{pmatrix} M & -p \\ -p^* & M \end{pmatrix}$$

The nonlinear problem is transformed into a maximization problem where we have to maximize the real scalar function

$$f(\phi_1, \phi_2) = \frac{|D_1|^2 + |D_2|^2 - 2\,\mathrm{Re}\{pD_1^*D_2/M\}}{M^2 - |p|^2} \tag{7-75}$$

The desired AOAs are the ϕ_1, ϕ_2 values that maximize $f(\phi_1, \phi_2)$. The function $f(\phi_1, \phi_2)$ is a particularly tough one to handle numerically, since in the neighborhood of $\Delta\phi = \phi_2 - \phi_1 \sim 0$, the denominator becomes zero. Recently, it was pointed out to us [17] that as $\Delta\phi \rightarrow 0$ the function takes the indeterminate form $0/0$. A double application of l' Hospitals's rule gives us then the finite limit

$$\lim_{\phi_2 \rightarrow \phi_1} f(\phi_1, \phi_2) = \frac{12|D_1'|^2 + (M^2 - 1)|D_1|^2}{M^2(M^2 - 1)}$$

where the prime indicates differentiation with respect to ϕ_1.

Looking at the behavior of the objective function for a given ϕ_1 and a particular data snapshot, despite double precision arithmetic, we see the onset of oscillations (Fig. 7-28) as ϕ_2 approaches ϕ_1 from both left and right. An optimization package, given this behavior, could converge to a nonoptimum pair of values. An asymptotic expansion in the troublesome region suggests itself. The one used here, is the Taylor's expansion of $f(\phi_1, \phi_2)$ up to the linear order term:

$$f(\phi_1, \phi_2) \sim f(\phi_1, \phi_2)|_{\phi_2 = \phi_1} + (\phi_2 - \phi_1)\left.\frac{\partial f}{\phi_2}\right|_{\phi_2 = \phi_1}$$

The linear term is another $0/0$ indeterminate form requiring the application of l' Hospital's rule five times to obtain

$$\left.\frac{\partial f}{\phi_2}\right|_{\phi_2 \rightarrow \phi_1} = \frac{2\sum_k \sum_l x_k^* x_l e^{j(k-l)\phi_1} j\left\{k - l\left[3\left(k - \frac{M+1}{2}\right)\left(l - \frac{M+1}{2}\right) + \frac{M^2 - 1}{4}\right]\right\}}{M^2(M^2 - 1)}$$

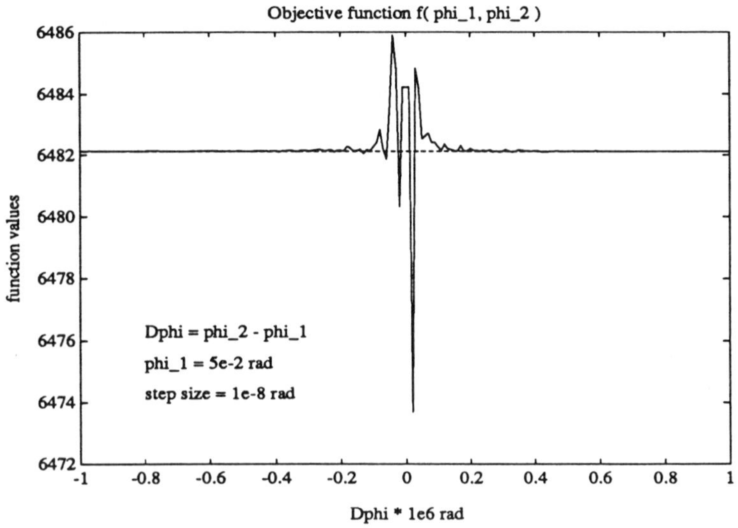

Figure 7-28 Behavior of the objective function (solid line) and its linear asymptotic expansion (dashed line) as $\Delta\phi \rightarrow 0$. Note that the function value at $\Delta\phi = 0$ is removed in the unmodified form of the function.

an antisymmetric expression giving a real value. As we can see in Fig. 7-28, the numerical problem is resolved.

Given the pair (ϕ_1, ϕ_2) that maximize the objective function f, we can estimate the signal amplitude vector as

$$\hat{\mathbf{a}} = \frac{\begin{pmatrix} M & -p \\ -p^* & M \end{pmatrix}\begin{pmatrix} D_1 \\ D_2 \end{pmatrix}}{M^2 - |p|^2}$$

Unfortunately, the method becomes more complex when the number of assumed plane waves is higher than two. The function $f(\phi_1, \phi_2)$ again, is almost flat in the neighborhood of the maximum when ϕ_1 is close to ϕ_2. In many cases also, where the model mismatch is particularly strong (strongly colored noise) the optimization algorithm converges to a solution where $|\phi_1 - \phi_2|$ is less than 0.001 of a beamwidth (such solutions are rejected).

Applying both MWM and ML to the datasets in Table 7-5 we obtain the results shown in Figs. 7-29–7-38. The signal ratio $\rho = a_2/a_1$ (magnitude and phase) is also included. This is related to the reflection coefficient R of the scattering surface. Ideally, for a plane water surface, at 10 GHz, for HH polarization and low grazing angles, the reflection coefficient R should have a magnitude approximately equal to 0.9 and a phase of approximately 180°. For a rough surface, the mean magnitude is modified by [2]

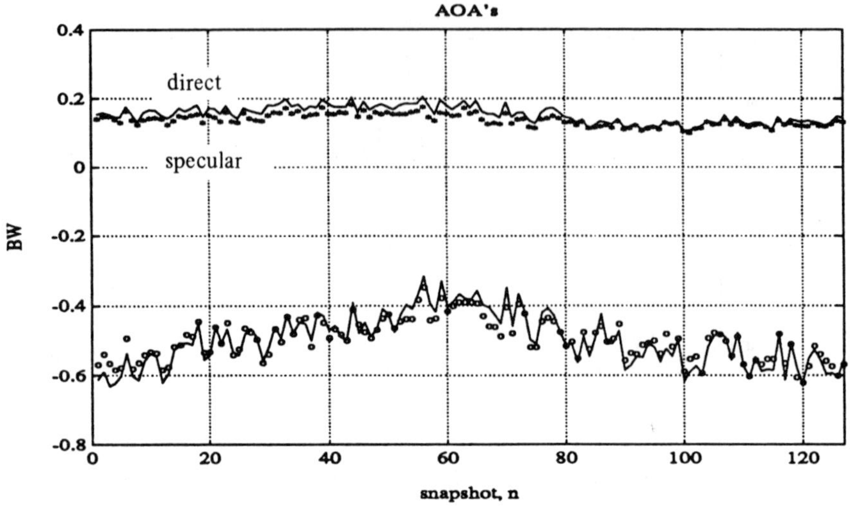

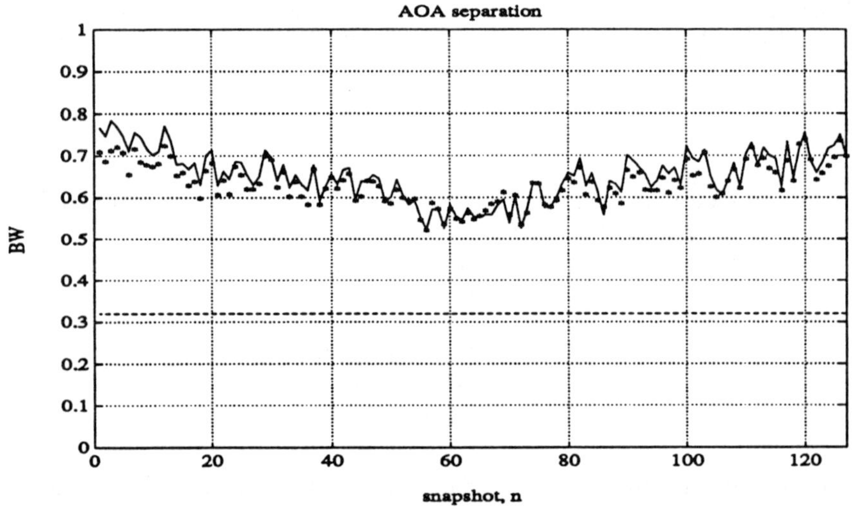

Figure 7-29 Dataset 1, *nov4:dh9.dat;1*, $f = 8.05$ GHz, far-field calibrated by *nov2:cff1.dat;1*. Solid lines always correspond to MWM estimates. In the top graph the direct component is above the specular, and $*$ and $\circ$ correspond to the ML direct and specular components, respectively. The second graph shows the estimated angular separation between direct and specular components compared with the expected one (dashed line).

$$\exp\left[-\left(\frac{4\pi\sigma_h \sin\psi}{\lambda}\right)^2\right]$$

where ψ is the grazing angle and σ_h is the rough surface height variance. The magnitude of ρ can be modeled by the modified R above, and possibly, the effects of the earth's curvature can be included by using the divergence factor.

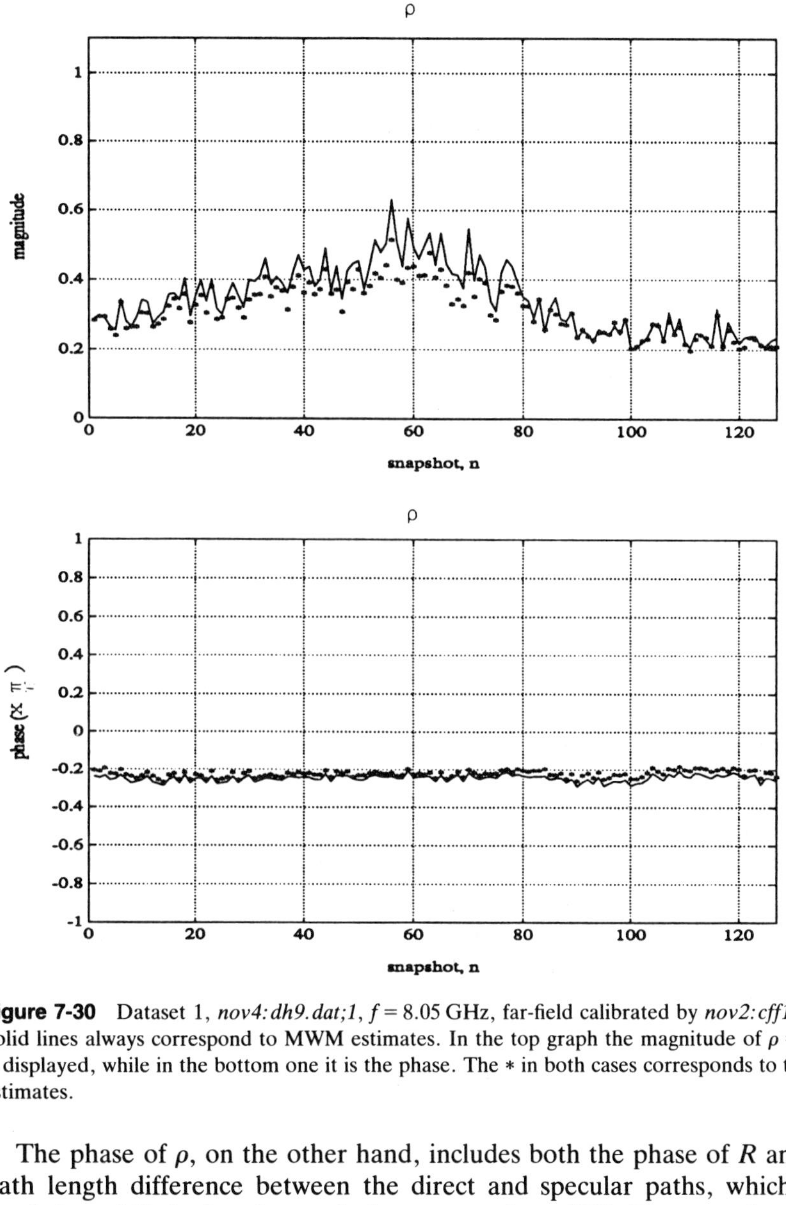

Figure 7-30 Dataset 1, *nov4:dh9.dat;1*, $f = 8.05$ GHz, far-field calibrated by *nov2:cff1.dat;1*. Solid lines always correspond to MWM estimates. In the top graph the magnitude of $\rho = a_2/a_1$ is displayed, while in the bottom one it is the phase. The $*$ in both cases corresponds to the ML estimates.

The phase of ρ, on the other hand, includes both the phase of R and the path length difference between the direct and specular paths, which may result in a shift in the observed phase, away from 180°. The quantity ρ, the specular model reflection coefficient, provides a way to check the data calibration. It happens, occasionally, that some far-field calibration data are so erroneous, as to give $|\rho| > 1$. As this would imply that the specular component is more powerful than the direct signal, data calibrated with these tables are rejected.

Examining the estimated AOAs we can make the following observations for each dataset individually:

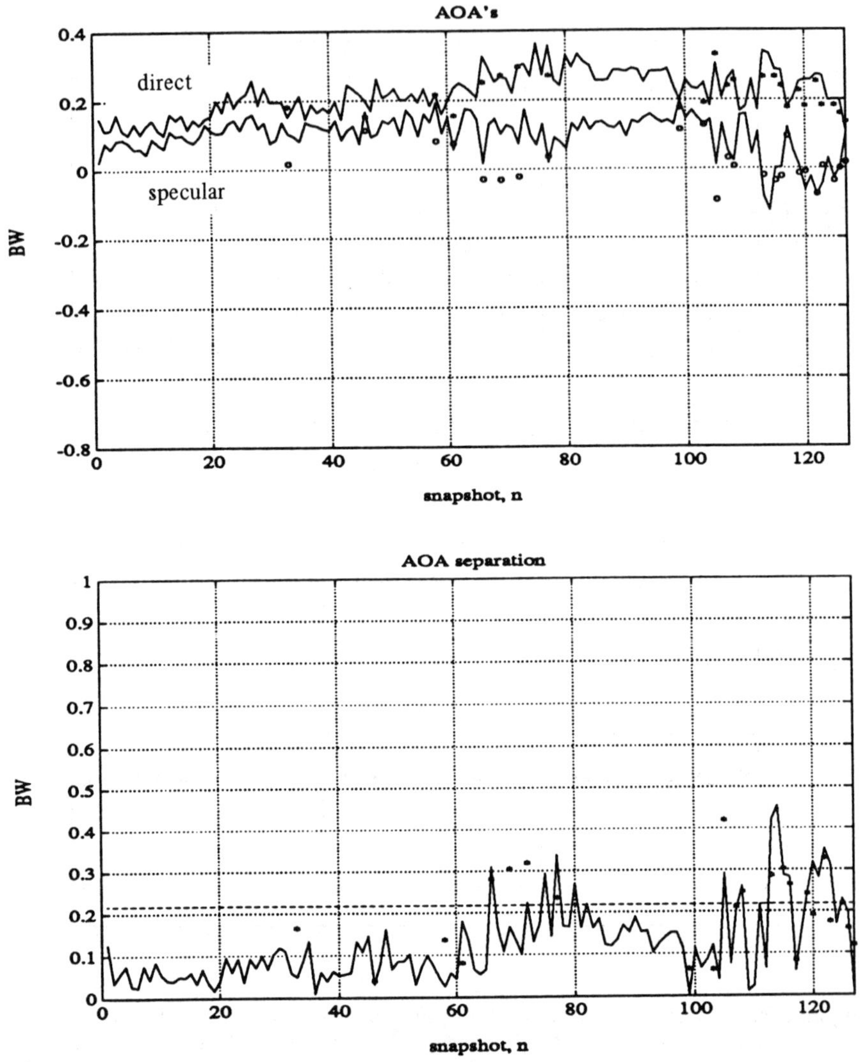

Figure 7-31 Dataset 2, *nov3:dh4.dat;3*, $f = 8.62$ GHz, far-field calibrated by *nov1:cff6.dat;3*. Solid lines always correspond to MWM estimates. In the top graph the direct component is above the specular, and $*$ and $\circ$ correspond to the ML direct and specular components, respectively. The second graph shows the estimated angular separation between direct and specular components compared with the expected one (dashed line).

Dataset 1 No significant difference is observed between MWM and ML estimates. The direct component, as more powerful, displays a smaller variance than the specular. A rather large bias is observed between the expected and estimated AOA separation. Comparing with [21] where

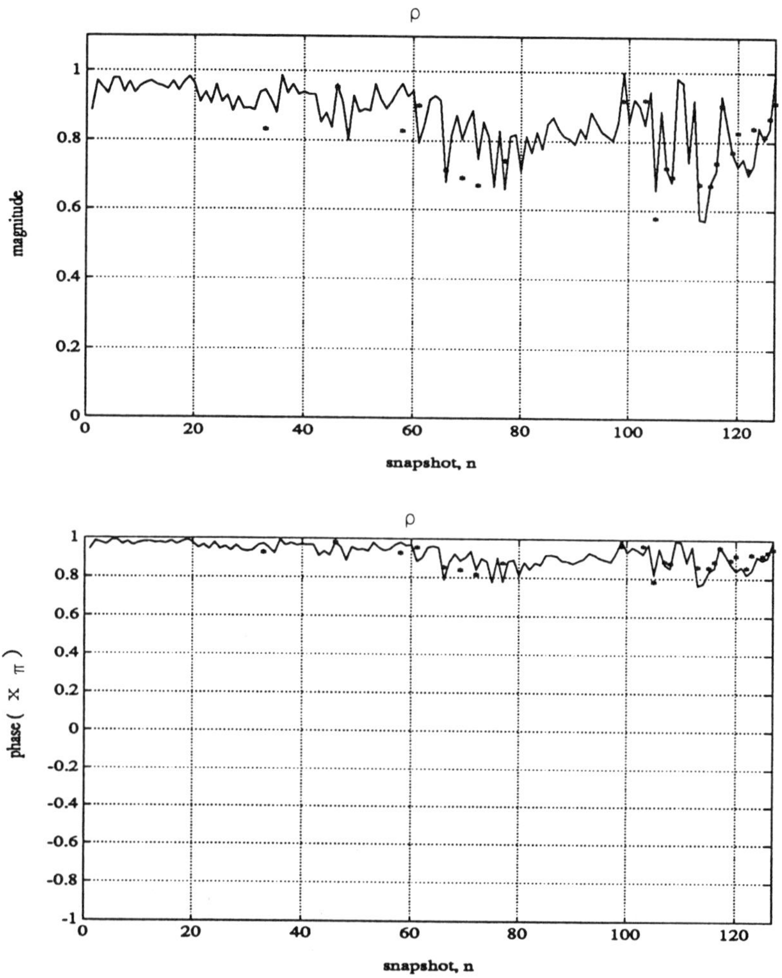

Figure 7-32 Dataset 2, *nov3:dh4.dat;3*, $f = 8.62$ GHz, far-field calibrated by *nov1:cff6.dat;3*. Solid lines always correspond to MWM estimates. In the top graph the magnitude of $\rho = a_2/a_1$ is displayed, while in the bottom one it is the phase. The $*$ in both cases corresponds to the ML estimates.

this same dataset is used with a more refined ML approach, we see that the bias appears in the estimation of the specular component (at 8.05 GHz 1 BW corresponds to 1.17°). Note that the estimated ρ is quite small. This means the presence of a weaker in power specular component, which explains the estimation difficulty.

Dataset 2 The situation here appears to the tougher for ML. Convergence occurred at much fewer points than with MWM. We see again in the first half of this dataset a bias in the AOA separation.

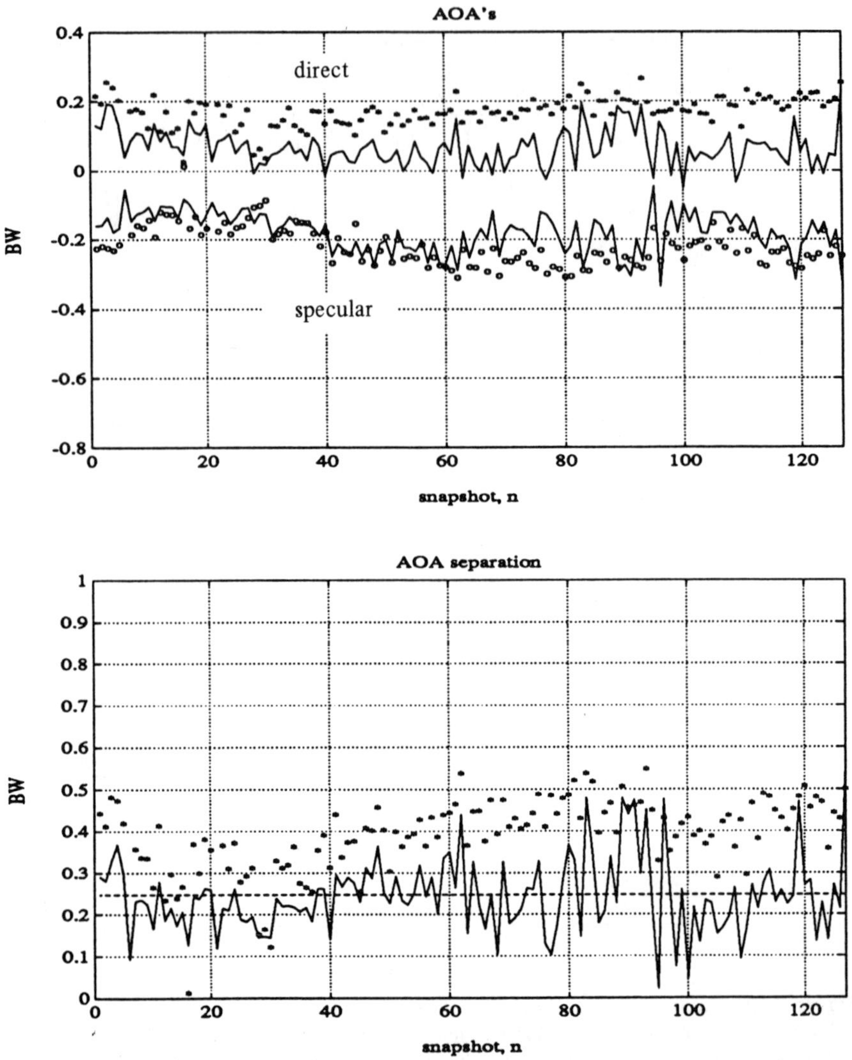

Figure 7-33 Dataset 3, *nov3:dh6.dat;7*, $f = 9.76$ GHz, far-field calibrated by *nov1:cff7.dat;7*. Solid lines always correspond to MWM estimates. In the top graph the direct component is above the specular, and $*$ and $\circ$ correspond to the ML direct and specular components, respectively. The second graph shows the estimated angular separation between direct and specular components compared with the expected one (dashed line).

Dataset 3 The behavior of both methods is quite good here, with MWM having a slight edge in the observed AOA separation.

Dataset 4 The bias observed in this case seems to have a periodic behavior. In this situation the antenna tower was being tilted by the force of the wind during all observation periods, and the periodic

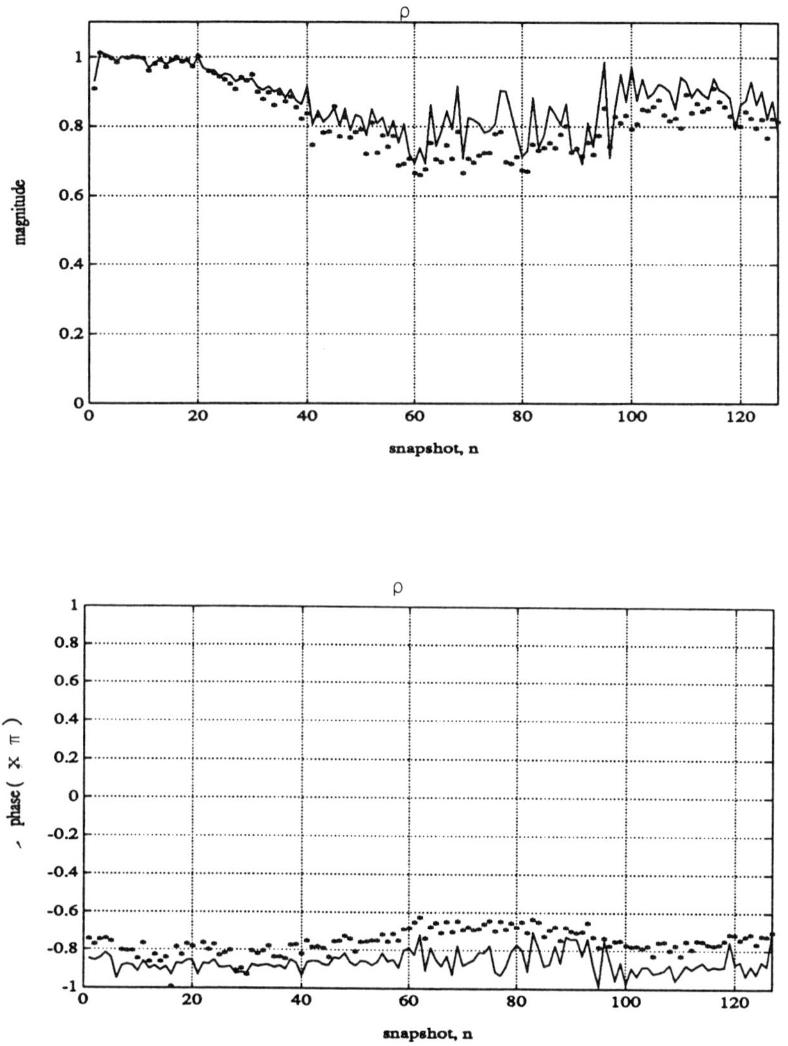

Figure 7-34 Dataset 3, *nov3:dh6.dat;7*, $f = 9.76$ GHz, far-field calibrated by *nov1:cff7.dat;7*. Solid lines always correspond to MWM estimates. In the top graph the magnitude of $\rho = a_2/a_1$ is displayed, while in the bottom one it is the phase. The $*$ in both cases corresponds to the ML estimates.

variation observed could well be caused by the tower swaying away from the vertical, making the actual separation angles larger than expected.

Dataset 5 The estimates here seem to be quite noisy but they agree quite well in all cases, and the periodic disturbance in the AOA separation is smaller.

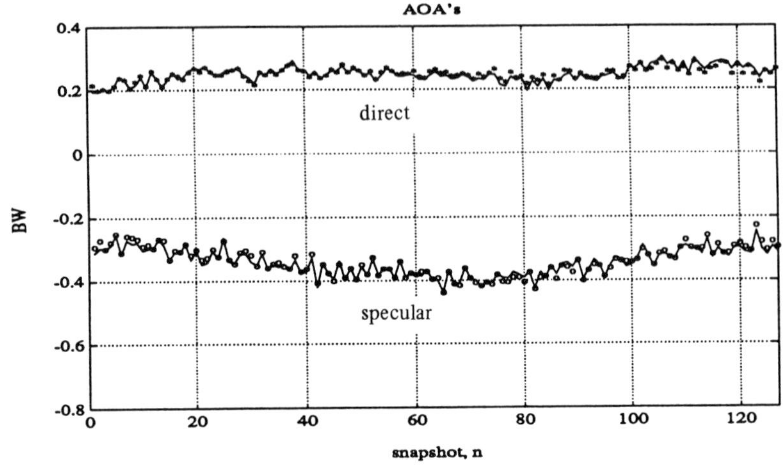

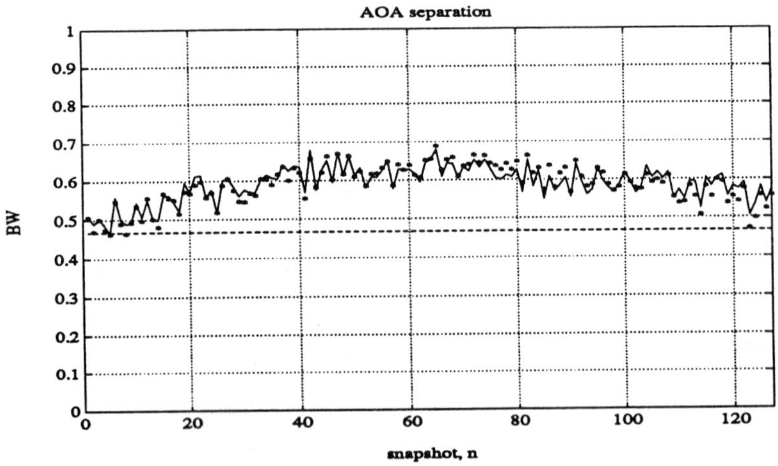

Figure 7-35 Dataset 4, *nov3:dd1.dat;2*, $f = 10.12$ GHz, far-field calibrated by *nov1:cff6.dat;8*. Solid lines always correspond to MWM estimates. In the top graph the direct component is above the specular, and $*$ and $\circ$ correspond to the ML direct and specular components, respectively. The second graph shows the estimated angular separation between direct and specular components compared with the expected one (dashed line).

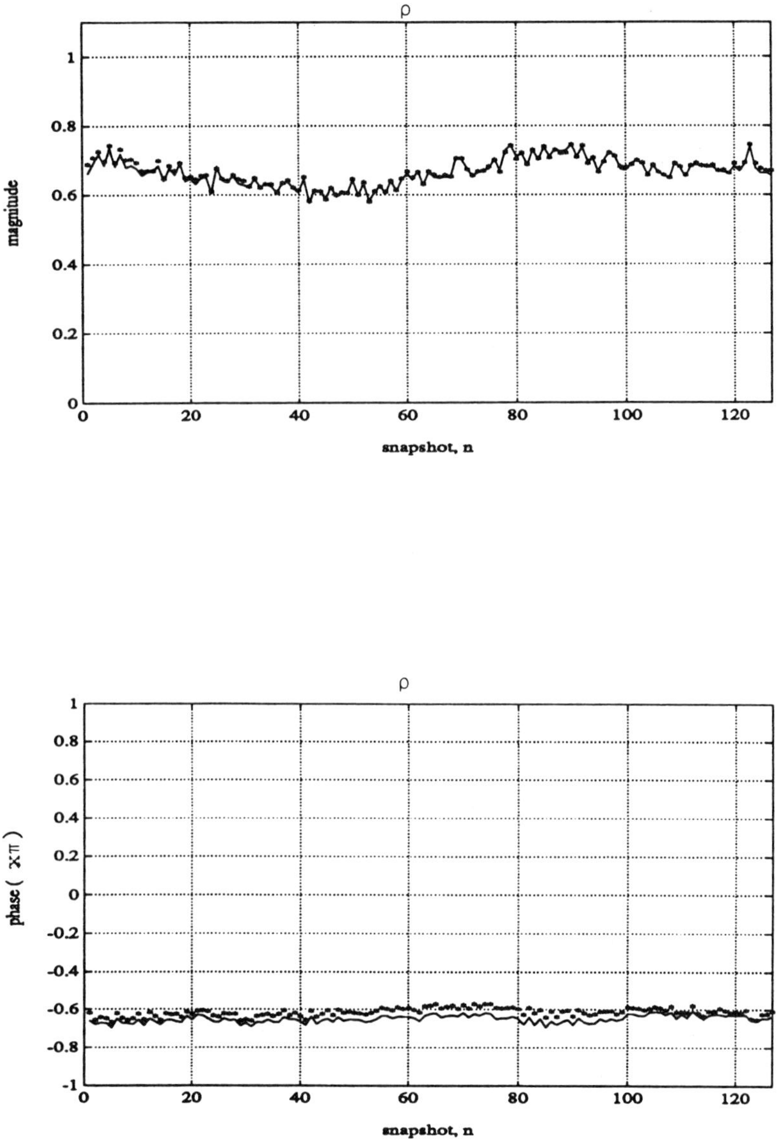

Figure 7-36 Dataset 4, *nov3:dd1.dat;2*, $f = 10.12$ GHz, far-field calibrated by *nov1:cff6.dat;8*. Solid lines always correspond to MWM estimates. In the top graph the magnitude of $\rho = a_2/a_1$ is displayed, while in the bottom one it is the phase. The $*$ in both cases corresponds to the ML estimates.

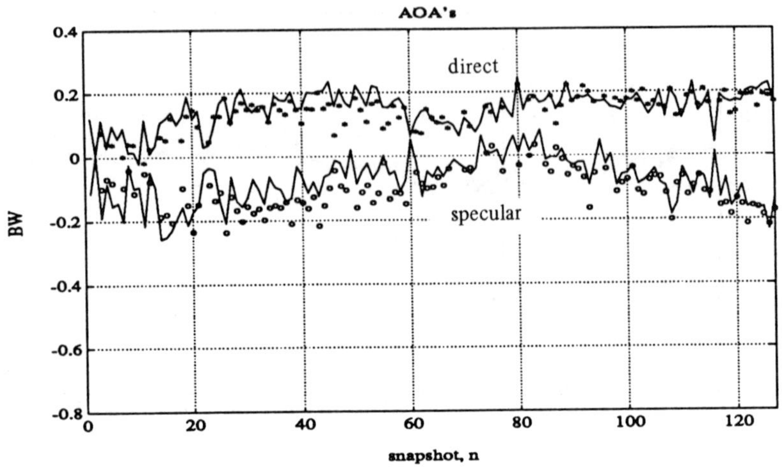

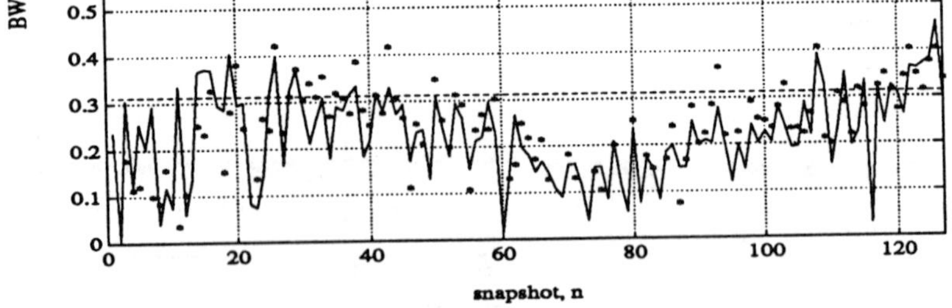

Figure 7-37 Dataset 5, *nov3:dh4.dat;16*, $f = 12.34\,\text{GHz}$, far-field calibrated by *nov1:cff6.dat;16*. Solid lines always correspond to MWM estimates. In the top graph the direct component is above the specular, and $*$ and $\circ$ correspond to the ML direct and specular components, respectively. The second graph shows the estimated angular separation between direct and specular components compared with the expected one (dashed line).

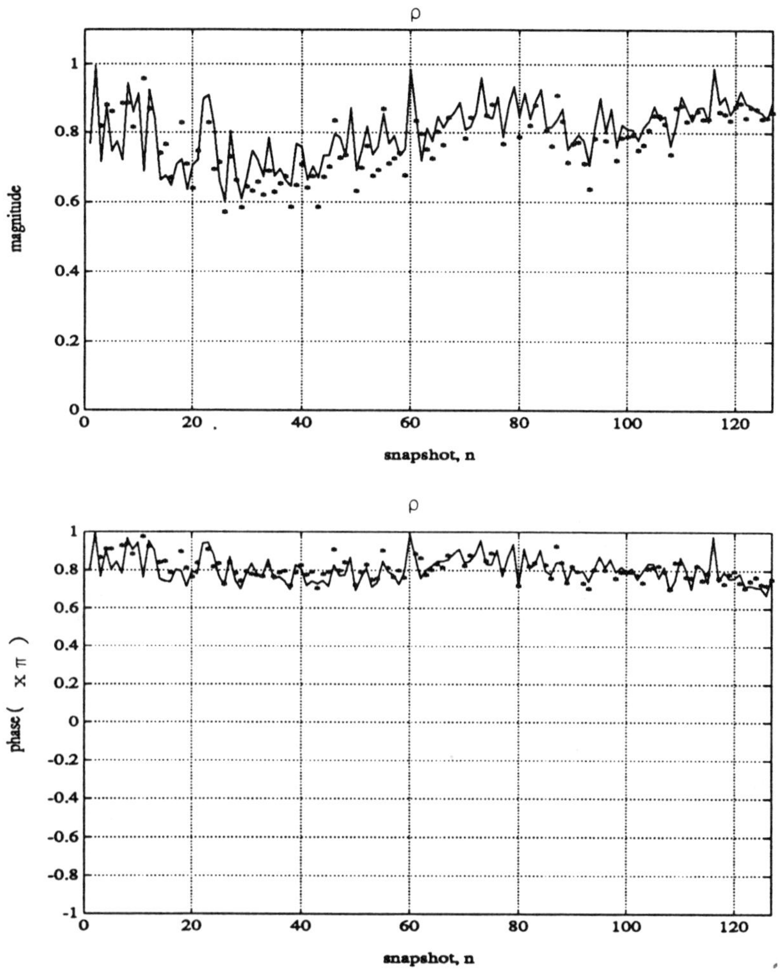

Figure 7-38 Dataset 5, *nov3:dh4.dat;16*, $f = 12.34$ GHz, far-field calibrated by *nov1:cff6.dat;16*. Solid lines always correspond to MWM estimates. In the top graph the magnitude of $\rho = a_2/a_1$ is displayed, while in the bottom one it is the phase. The $*$ in both cases corresponds to the ML estimates.

7-10 DIFFUSE MULTIPATH SPECTRUM ESTIMATION

If we remove the estimated plane waves (direct + specular components) from the measured spectrum, the residual spectrum should correspond to the diffuse component. In this way we can estimate the diffuse component in both the frequency (array element samples in time) and wavenumber (array element samples in space; snapshots) domains. An example of a fully reconstructed wavenumber spectrum is given in Fig. 7-39. This is the first

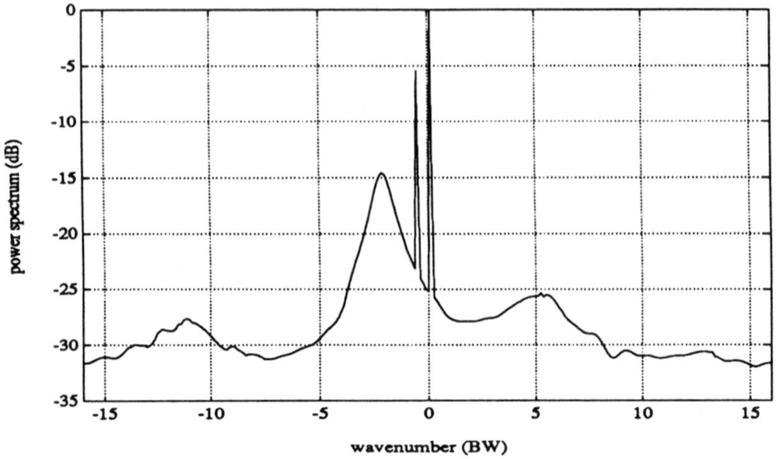

Figure 7-39 The fully reconstructed wavenumber spectrum of the first snapshot of *nov3:dh3.dat;2* far-field calibrated with *nov2:cff2.dat;1*.

snapshot of *nov3:dh3.dat;2*. Note, that as far as we are aware, this is the first time that a full picture of an experimentally measured wavenumber spectrum has been reported for a low-angle tracking radar environment. Both multipath components are displayed, together with the direct signal, and it is clearly seen that assuming a white noise background is wrong.

A variety of different HH subsets of data from the fixed 10.2 GHz channel of *nov3:dh3.dat*, *nov3:dh4.dat* and *nov4:dh9.dat* were also investigated. Their residual frequency and wavenumber spectra are given in Figs.

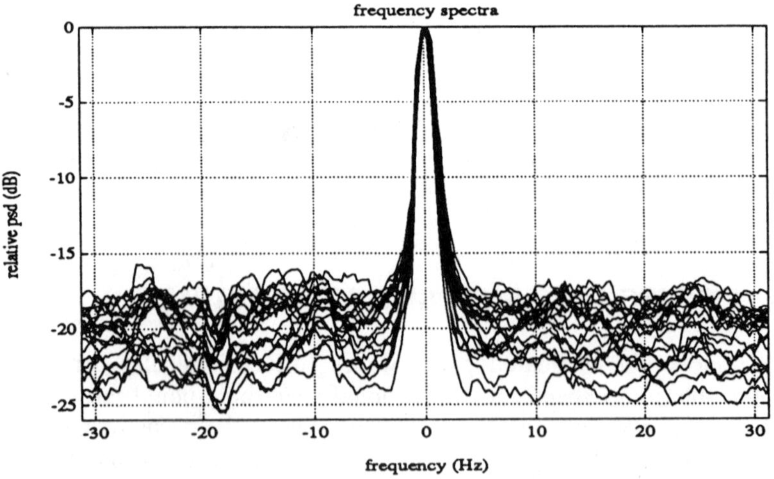

Figure 7-40 The frequency spectra of a variety of similar datasets from *nov3:dh3.dat;** and *nov3:dh4.dat;** all at 10.2 GHz.

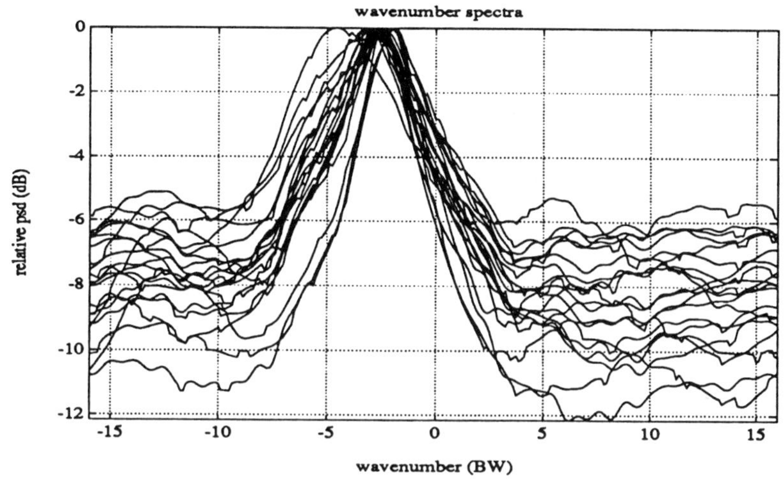

Figure 7-41 The wavenumber spectra of a variety of similar datasets from *nov3:dh3.dat;** and *nov3:dh4.dat;**.

7-40–7-43. Finally, similar residual spectra are computed from the datasets examined in the previous section (Figs. 7-44–7-49).

In both cases the frequency spectra display a straightforward behavior. The broad peak observed at baseband can be modeled as a Gaussian spectrum. However, for the wavenumber spectra, the situation appears to be more complex. The first set of spectra (Figs. 7-41 and 7-43), show a pronounced peak in the negative part of the spectrum (corresponding to directions of arrival from below the horizon) while such is not the case for

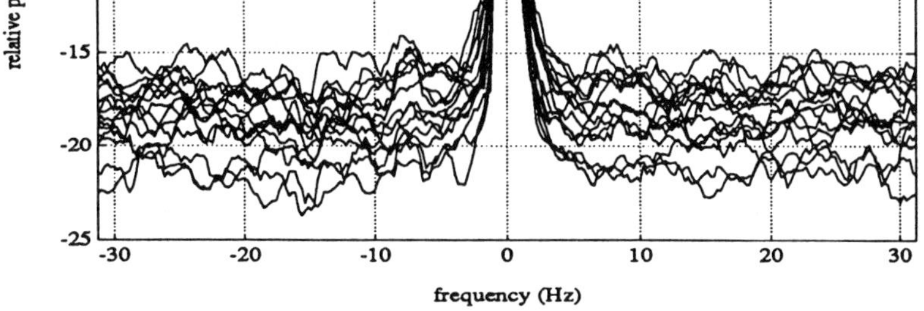

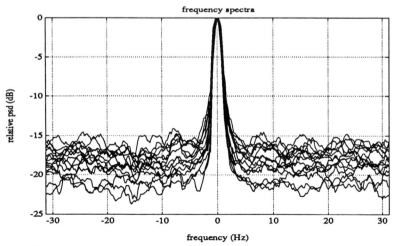

Figure 7-42 The frequency spectra of a variety of datasets from *nov4:dh9.dat;** at 10.2 GHz.

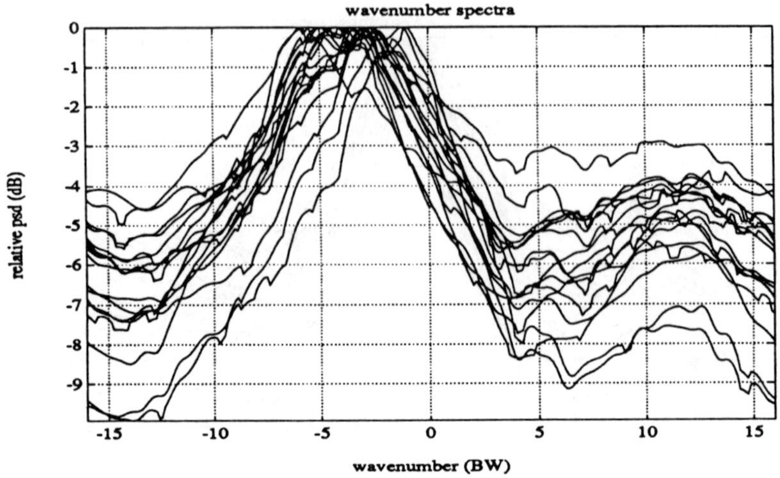

Figure 7-43 The wavenumber spectra corresponding to the frequency spectra of Figure 7-42. Note that the diffuse peak center is now less well defined, due to the estimated AOA separation being smaller than that for the previous case.

the second set. Several strong peaks are observed in both sides of the spectrum. Assuming for the moment that the behavior depicted is true (see next section) we can now explain some of the observations in the previous section. Specifically, we may make the following observations:

Dataset 1 We have two, strong, diffuse peaks appearing in the wavenumber spectrum, which could influence the behavior of the

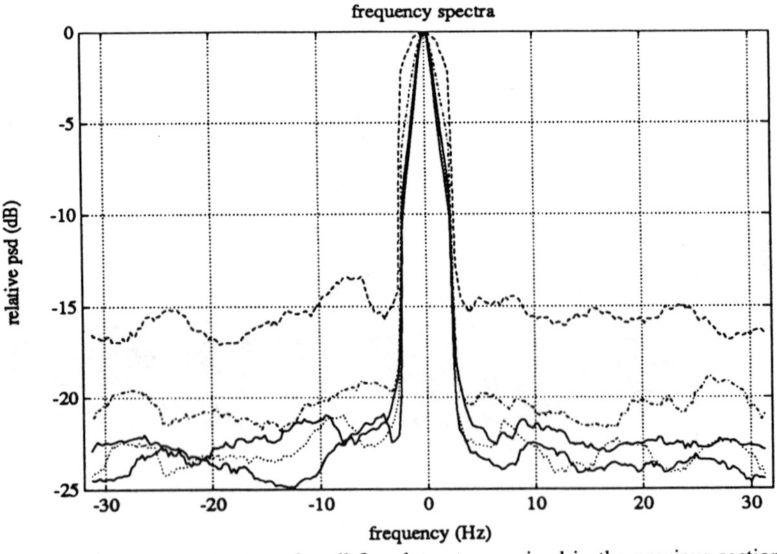

Figure 7-44 Frequency spectra for all five datasets examined in the previous section.

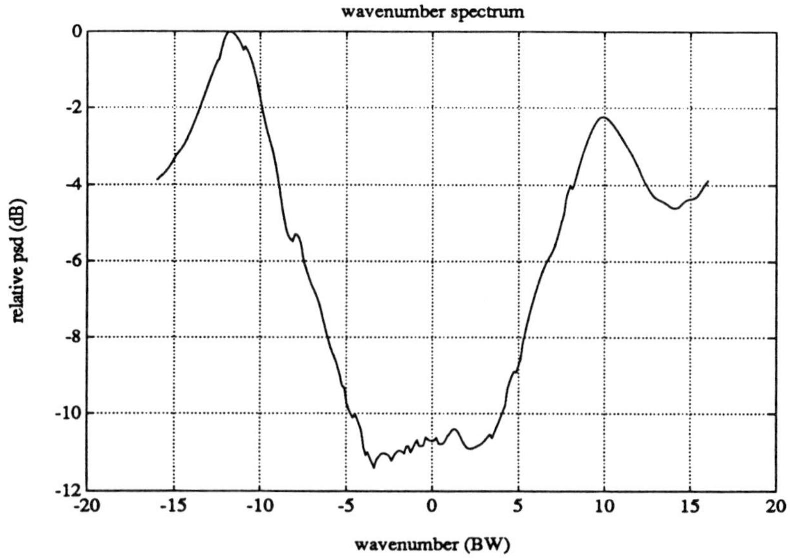

Figure 7-45 Wavenumber spectrum of dataset 1, *nov4:dh9.dat;1.*

estimated AOAs compared to that actually observed. Note also, that the expected separation assumes a specular model. This should be a better approximation the smaller the assumed model mismatch (i.e., weak diffuse component near the direct and specular AOAs).

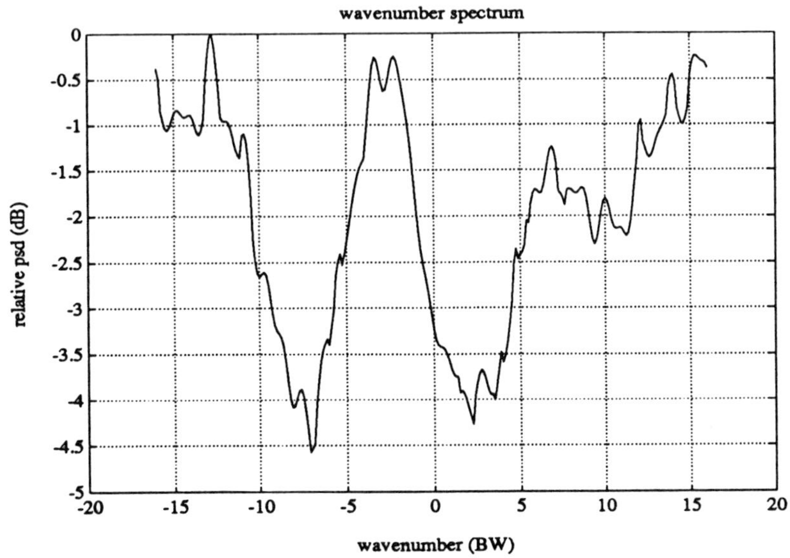

Figure 7-46 Wavenumber spectrum of dataset 2, *nov3:dh4.dat;3.*

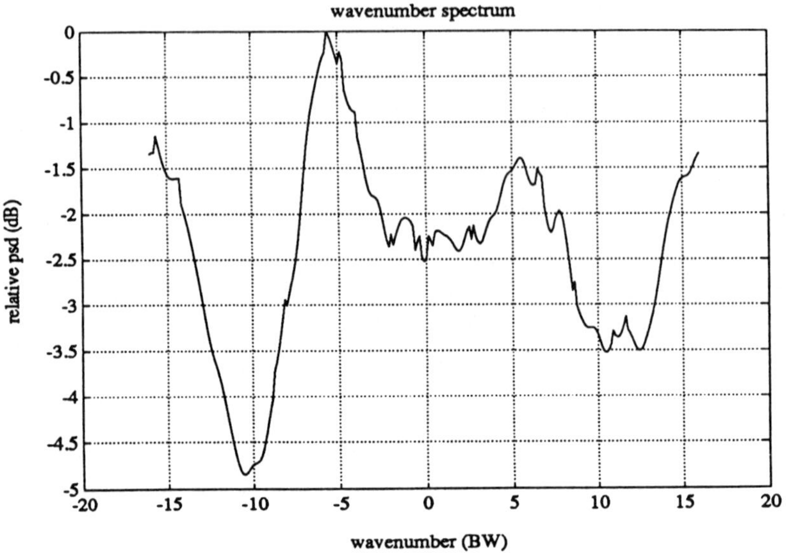

Figure 7-47 Wavenumber spectrum of dataset 3, *nov3:dh6.dat;7.*

Dataset 2 More than one region in the wavenumber domain shows a strong diffuse component. In addition, the frequency-domain spectrum for this dataset displays a higher noise floor by about 5 dB from all other cases. No wonder that ML often breaks down here.

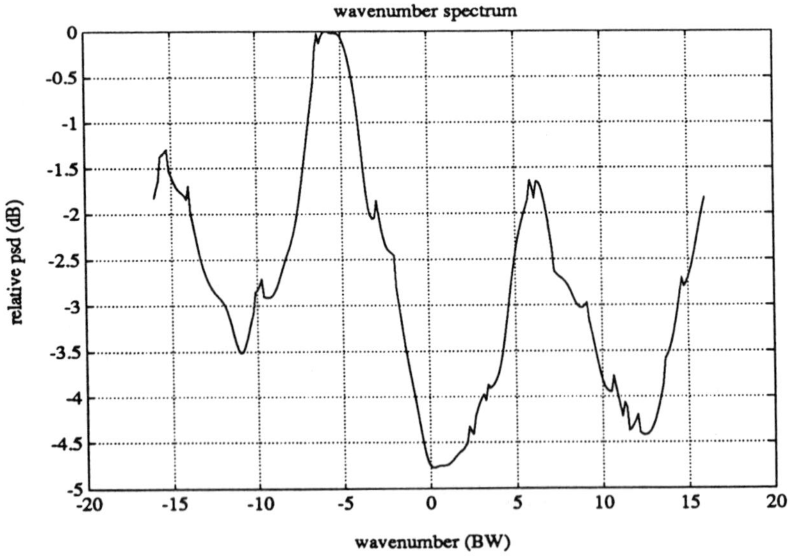

Figure 7-48 Wavenumber spectrum of dataset 4, *nov3:dd1.dat;2.*

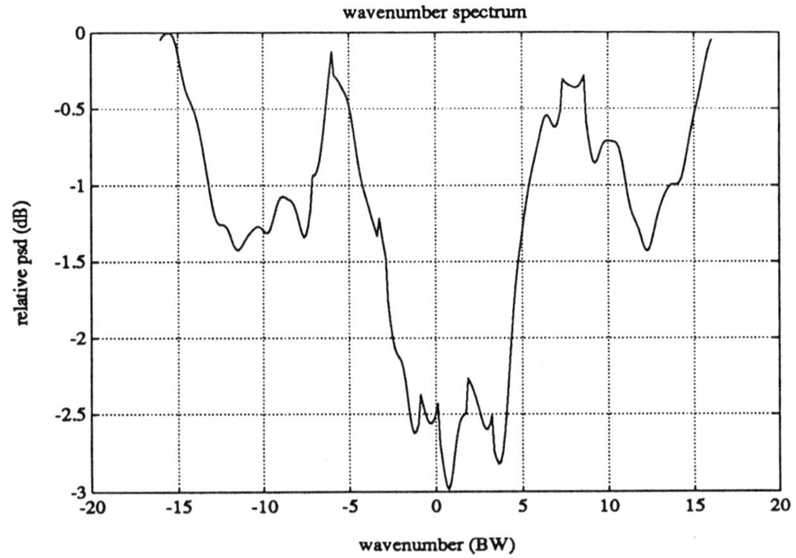

Figure 7-49 Wavenumber spectrum of dataset 5, *nov3:dh4.dat;16.*

Dataset 3 There appears to be a plateau around the origin which, locally at least, is closer to the specular model assumption, so that both the expected and the observed separations agree well enough. Note that MWM, which does not assume a white noise background, performs slightly better.

Dataset 4 Similar behavior to dataset 1 is observed for this dataset, except that the specular component power is larger here.

Dataset 5 Similar behavior to dataset 2 is observed here. However, since the observed AOA separation is slightly larger, the overall behavior is better.

7-11 DISCUSSION

Examining both the estimated AOAs and the diffuse component spectra reported in Sections 7-9 and 7-10, we may make the following observations:

- The AOA estimate often shows a strong variability during the observation interval, which justifies the snapshot-by-snapshot set-up implemented.
- MWM, which does not assume a white noise background, always gives AOA estimates with separation larger than 0.001 BW, while ML occasionally does not. This breakdown is more likely to occur when the separation between the direct and specular components is very small ($\sim$0.1 BW). Model mismatch in neglecting the diffuse component is

then significant. Note that ML can be improved by incorporating additional information in the form of a priori knowledge or frequency agility [18, 21]. Similar modifications can also be done for MWM [9].

- In the context of AOA estimation, there does not seem to be overwhelming evidence in favor of MWM being superior to ML (when ML does not break down). It is encouraging, however, to see that MWM does perform so well compared with ML. In addition, MWM is the only nonparametric way to get a good estimate of the background diffuse component. Note also, that multipath is correlated with itself both spatially and temporally. MWM can be extended in a natural way to obtain coherence estimates between the individual components; here again, work is underway to investigate this issue.

Concerning the shape of the spectra, the frequency domain one can be modeled as a Gaussian spectrum, the reason being that the contribution of a large number of moving scatterers (law of large numbers) leads to the peak broadening observed.

In the wavenumber domain, positive wavenumber values correspond to the volume space above the physical horizon, while negative values correspond to the space below the horizon, scattering surface included. Two rough surface specular point models were investigated, Barton's [1] and McGarty's [24]. However, only the latter model (when shadowing is included) yields qualitative results similar to some of the experimental cases observed.

Before a comparison can be made between theoretical models and experimentally observed spectra, the theoretically obtained spectra must be modified to take into account the wrapping and grating lobe effects that occur since $d/\lambda > 0.5$ in our experimental set-up. The choice of $d/\lambda > 0.5$ was made due to the practical limitation of how closely together the horn antenna elements could be placed. For a frequency of 10.2 GHz, we have $d/\lambda \sim 1.94$. This leads to the correspondence between physical angle θ and wavenumber ϕ seen in Fig. 7-50. The five domains observed should be convolved with the antenna field pattern (array factor $\times$ array element field pattern) in order to obtain the power spectrum that should be observed at the receiver. Figure 7-51 shows a sample case from McGarty's model [24] with a roughness scale $s = 0.1$ (the ratio of the surface height variance to the surface correlation length) and a geometry similar to that described in Section 7-8.

It is only when shadowing is included that we see a broad, diffuse peak between -10 and 0 BW. This agrees with some of the experimentally observed spectra. Most spectra, however, display additional peaks indicating (unless there is a strong calibration problem) other scattering centers, beyond the unambiguous field of view of the antenna receiver array. Returns from these centers would be wrapped into the positive wavenumber part of the spectrum, leading to the observed additional peaks. Since the

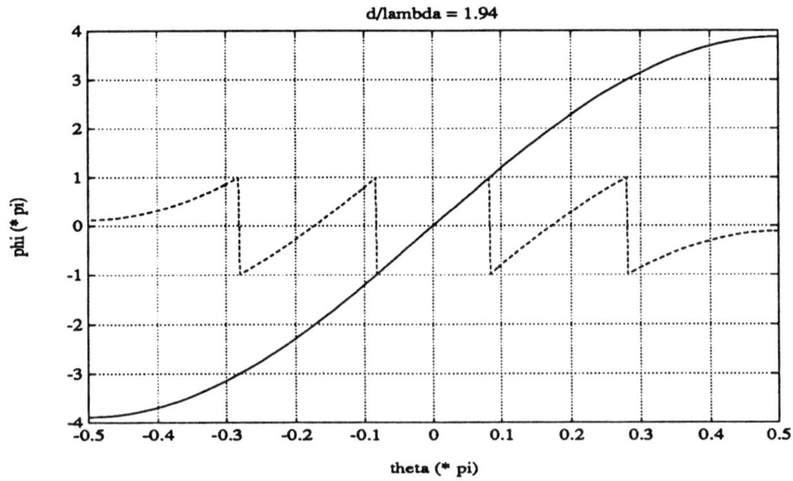

Figure 7-50 The wavenumber ϕ as a function of θ for the case $d/\lambda = 1.94$. The analytic form of the mapping is $\phi = (2\pi d/\lambda)\sin\theta$.

model employs only a simplified version of shadowing and only a single scale rough surface, we should not expect perfect agreement. Current ideas on sea surface modeling indicate that a coherent, composite model of two roughness scales is necessary for good predictions. Work is currently underway investigating such models.

In conclusion we would like to highlight the following points:

- MWM is seen to offer an exceptional approach to the problem of high resolution full-spectrum estimation. The nonparametric nature in estimating the continuous background spectra (colored noise) is particularly useful when no a priori knowledge, or accurate model of the underlying process(es) is available. In fact, this is the only method that performs so well in this area.
- Regarding AOA estimation, it is seen that MWM and ML (maximum likelihood with white noise background) give consistent results for the low grazing angle case examined here. MWM is seen to be slightly superior although the difference is not overwhelming. However, it should be pointed out that ML is the best parametric method currently available for AOA estimation. All modern parameter estimation techniques use ML as the yardstick against which their performance is evaluated. The results seen so far with MWM applied to real data are particularly promising.
- The F-tests used for AOA or tone estimation of harmonic components come in two varieties, point and integral regression. Our experience so far suggests that the latter is more stable than the former, although the final estimates in the cases examined here do not differ appreciably.

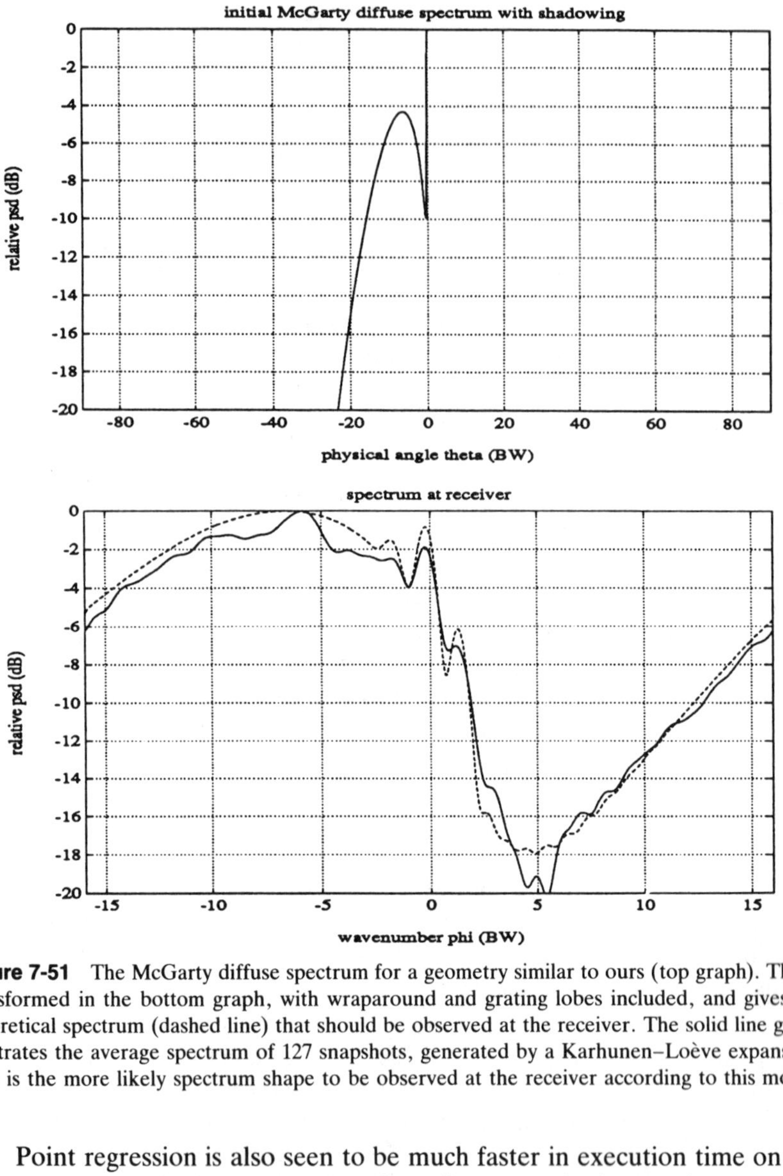

Figure 7-51 The McGarty diffuse spectrum for a geometry similar to ours (top graph). This is transformed in the bottom graph, with wraparound and grating lobes included, and gives the theoretical spectrum (dashed line) that should be observed at the receiver. The solid line graph illustrates the average spectrum of 127 snapshots, generated by a Karhunen–Loève expansion. This is the more likely spectrum shape to be observed at the receiver according to this model.

Point regression is also seen to be much faster in execution time on the computer, offering an excellent first glimpse on the harmonic content of a data sequence.

A final point to make in this discussion: the promising results seen so far with MWM lead us to suggest that this is an exceptional tool for full-spectrum estimation. It is our belief, based on our experience with it, that one cannot obtain better results with any other nonparametric approach. In fact, MWM is also a better general purpose spectrum estimator than many parametric ones.

ACKNOWLEDGMENTS

The authors would like to thank Vytas Kezys, Ed Vertatschitsch and the other members of the MARS group for the raw data from the Lake Huron experiments that appear in this chapter. Special thanks are also extended to Mr. Kezys for his many helpful comments and suggestions for improving the material presented here. Similarly, we thank Dr. Steinhardt and his students for some stimulating discussions and comments about the general nature of MWM. We also give special thanks to Dr. Thomson, for looking over this material on the use of his method, his comments and suggestions, and for bringing to our attention some of his most recent publications [37, 38]. Finally, the financial support of The Natural Sciences and Engineering Research Council (NSERC) of Canada, and the Telecommunications Research Institute of Ontario (TRIO) during this work, is gratefully acknowledged.

REFERENCES

[1] D. K. Barton (1974) "Low-angle radar tracking", *Proc. IEEE*, **62**, 687–704.

[2] P. Beckmann and A. Spizzichino (1987) *The Scattering of Electromagnetic Waves from Rough Surfaces*, Artech House (a reprint of the classic 1963 monograph).

[3] R. B. Blackman and J. W. Tukey (1959) *The Measurement of Power Spectra*, Dover (a reprint of papers appearing in the January and March issues of the B.S.T.J. in 1958).

[4] L. V. Blake (1986) *Radar Range-Performance Analysis*, Artech House.

[5] T. P. Bronez (1988) "Nonparametric spectral estimation of irregularly sampled multidimensional random processes", Ph.D. dissertation, Arizona State University, Tempe, AZ (also Univ. Microfilms Publ. No. 8602835).

[6] T. P. Bronez (1988) "Spectral estimation of irregularly sampled multidimensional processes by generalized prolate spheroidal sequences", *IEEE Trans. Acoust., Speech, Signal Process.*, **36**, 1862–1873.

[7] N. R. Draper and H. Smith (1981) *Applied Regression Analysis*, Wiley, New York.

[8] A. Drosopoulos (1991) "Investigation of diffuse multipath at low-grazing angles", Ph.D. dissertation, McMaster University, Department of Electrical Engineering, Hamilton, ONT (in preparation).

[9] A. Drosopoulos and S. Haykin (1991) "Angle-of-Arrival Estimation in the Presence of Multipath, Electronics Letters.

[10] S. Haykin (1985) "Radar array processing for angle of arrival estimation", in *Array Signal Processing*, S. Haykin (Ed.), Prentice Hall, Englewood Cliffs, NJ, Chap. 4.

[11] R. Hooke and T. A. Jeeves (1961) "Direct search solution of numerical and statistical problems", *J. ACM*, **8**, 212–229.

[12] A. Ishimaru (1978) *Wave Propagation and Scattering in Random Media*, Vol. II, Academic Press, New York.

[13] J. O. Jonsson and A. O. Steinhardt (1991) "The total probability of false alarm of the multi-window harmonic detector and its application to real data", submitted to IEEE Trans. on Signal Processing.

[14] A. F. Kaupe (19) "Algorithm 178: direct search", Collected algorithms from *Commun. ACM*, Vol. 178-P1-R1.

[15] S. M. Kay and S. L. Marple, Jr. (1981) "Spectrum analysis—a modern perspective", *Proc. IEEE*, **69**, 1380–1419.

[16] D. E. Kerr (Ed.) (1951) *Propagation of Short Radio Waves. M.I.T. Radiation Laboratory Series*, Vol. 13, McGraw Hill, New York.

[17] V. Kezys, personal communication.

[18] V. Kezys and S. Haykin (1988) "Multi-frequency angle-of-arrival estimation: an experimental evaluation", *Proc. SPIE, Advanced Algorithms and Architectures for Signal Processing III*, Vol. 975, pp. 93–100.

[19] B. Kleiner, R. D. Martin and D. J. Thomson (1979) "Robust estimation of power spectra", *J. R. Statist. Soc. B*, **41**, 313–351.

[20] C. R. Lindberg and J. Park (1987) "Multiple-taper spectral analysis of terrestrial free oscillations: Part II", *Geophys. J. R. Astron. Soc.*, **91**, 795–836.

[21] T. Lo and J. Litva (1991) "Use of a highly deterministic multipath signal model in low-angle tracking", *IEE Proc., Part F*, **138**, 163–171.

[22] S. L. Marple, Jr. (1987) *Digital Spectral Analysis with Applications*, Prentice Hall, Englewood Cliffs, NJ.

[23] R. D. Martin and D. J. Thomson (1982) "Robust-resistant spectrum estimation", *Proc. IEEE*, **70**, 1097–1115.

[24] T. P. McGarty (1976) "Antenna performance in the presence of diffuse multipath", *IEEE Trans. Aerosp. Electron. Systems*, **12**, 42–54.

[25] C. T. Mullis and L. L. Scharf (1991) "Quadratic estimators of the power spectrum", in *Advances in Spectrum Analysis and Array Processing*, Vol. I, S. Haykin (Ed.), Prentice Hall, Englewood Cliffs, NJ, Chap. 1.

[26] R. Onn and A. O. Steinhardt (1991) "A multi-window method for spectrum estimation and sinusoid detection in an array environment", *Proc. SPIE, Advanced Algorithms and Architectures for Signal Processing*, San Diego.

[27] A. Papoulis (1984) *Probability, Random Variables and Stochastic Processes*, 2nd edition, McGraw-Hill, New York.

[28] J. Park (1987) "Multitaper spectral analysis of high-frequency seismographs", *J. Geophys. Res.*, **92** (B12), 12,675–12,684.

[29] J. Park, C. R. Lindberg, and D. J. Thomson (1987) "Multiple-taper spectral analysis of terrestrial free oscillations: Part I", *Geophys. J. R. Astron. Soc.*, **91**, 755–794.

[30] W. H. Press, B. P. Flannery, S. A. Teukolsky, and W. T. Vetterling (1986) *Numerical Recipes*, Cambridge University Press.

[31] D. Slepian (1965) "Some asymptotic expansions for prolate spheroidal wave functions", *J. Math. Phys.*, **44**, 99–140.

[32] D. Slepian (1968) "A numerical method for determining the eigenvalues and eigenfunctions of analytic kernels", *SIAM J. Numer. Anal.*, **5**, 586–600.

[33] D. Slepian (1978) "Prolate spheroidal wave functions, Fourier analysis, and uncertainty—V: The discrete case", *Bell System Tech. J.*, **57**, 1371–1430.

[34] D. Slepian and E. Sonnenblick (1965) "Eigenvalues associated with prolate spheroidal functions of zero order", *Bell System Tech. J.*, **44**, 1745–1760.

[35] D. J. Thomson (1977) "Spectrum estimation techniques for characterization and development of WT4 waveguide", *Bell System Tech. J.*, **56**, 1769–1815, 1983–2005.

[36] D. J. Thomson (1982) "Spectrum estimation and harmonic analysis", *Proc. IEEE*, **70**, 1055–1096.

[37] D. J. Thomson (1989) "Multiple-window bispectrum estimation", in *Proc. Workshop on Higher-Order Spectral Analysis*, Vail, Colorado, pp. 19–23.

[38] D. J. Thomson (1990) "Quadratic-inverse spectrum estimation: applications to paleoclimatology", *Philos. Trans. R. Soc. London, Ser. A*, **332**, 539–597.

[39] D. J. Thomson (1990) "Time series analysis of Holocene climate data", *Philos. Trans. R. Soc. London, Ser. A*, **330**, 601–616.

[40] D. J. Thomson and A. D. Chave (1991) "Jackknifed error estimates for spectra, coherences and transfer functions", in *Advances in Spectrum Analysis and Array Processing*, Vol. I, S. Haykin (Ed.), Prentice Hall, Englewood Cliffs, NJ, Chap. 2.

[41] A. M. Yaglom (1987) *Correlation Theory of Stationary and Related Random Functions I: Basic Results*, Springer-Verlag, New York.

INDEX